PRODUCT AND PROCESS MODELLING
IN BUILDING AND CONSTRUCTION

PROCEEDINGS OF THE THIRD EUROPEAN CONFERENCE ON PRODUCT AND
PROCESS MODELLING IN THE BUILDING AND RELATED INDUSTRIES
LISBON/PORTUGAL/25 – 27 SEPTEMBER 2000

Product and Process Modelling in Building and Construction

Edited by

Ricardo Gonçalves & Adolfo Steiger-Garção
Universidade Nova de Lisboa-Uninova, Portugal

Raimar Scherer
University of Technology, Dresden, Germany

Catalogue No:
10,347

A.A.BALKEMA/ROTTERDAM/BROOKFIELD/2000

The texts of the various papers in this volume were set individually by typists under the supervision of each of the authors concerned.

Authorization to photocopy items for internal or personal use, or the internal or personal use of specific clients, is granted by A.A. Balkema, Rotterdam, provided that the base fee of US$ 1.50 per copy, plus US$ 0.10 per page is paid directly to Copyright Clearance Center, 222 Rosewood Drive, Danvers, MA 01923, USA. For those organizations that have been granted a photocopy license by CCC, a separate system of payment has been arranged. The fee code for users of the Transactional Reporting Service is: 90 5809 179 1/00 US$ 1.50 + US$ 0.10.

Published by
A.A. Balkema, P.O. Box 1675, 3000 BR Rotterdam, Netherlands
Fax: +31.10.413.5947; E-mail: balkema@balkema.nl; Internet site: www.balkema.nl
A.A. Balkema Publishers, 2252 Ridge Road, Brookfield, VT 05036-9704, USA
Fax: 802.276.3837; E-mail: info@ashgate.com

ISBN 90 5809 179 1
© 2000 A.A. Balkema, Rotterdam
Printed in the Netherlands

Table of contents

Human centred concurrent engineering

Virtual enterprise and e-Business

e-Business and e-Commerce

IT in the early phase of construction projects

Application of artificial intelligence methods

Modelling and visualisation in design

Modelling and visualisation in construction

IT in construction

Preface

The construction industry has been undergoing a remarkable change during the last two years, since the 2nd ECPPM at Watford, UK in 1998. Large construction companies are re-engineered in order to cope with the modern information society, the globalisation demands and the new ways of working, including virtual team work, teleworking and e-business to business. SMEs, and in particular design offices are carefully watching this process with great concerns with respect to their timely migration to these new ways of working because their reaction, if wrongly timed, can easily lead to loss of competitiveness. Internet portals are built up, mainly by new and agile venture capital companies and also by traditional publishers who have meanwhile migrated to multi-media companies. Especially e-commerce platforms are very promising to become beneficial to all participants, those who invest in buiding up the platform, and those who use it. The whole construction industry, being a traditional industry branch, undergoes a paradigm shift to an information society industry, and integration reaches beyond the traditional borders of AEC, involving multi-media companies and service providers as well.

Today, information is available in such a quantity that the question about its quality is obvious for everybody.

The basic requisite technology for all these processes is doubtlessly product and process modelling, i.e. establishing the right data structures, agreed by all players and standardized world-wide. The efforts of this sophisticated and time consuming process, namely agreed and standardized product and process data models have to be very much appreciated, and the progress made under STEP (ISO 10303), IFC (IAI) and CALS in the last years is remarkable, even though the results are lacking behind the state of the art in research. But this is a natural time gap.

Information overflow, information waste and information noise are the new terms, which characterize the new problems the solutions of which are yet to be found.

However, when looking for such solutions, we immediately come back to the old question that already stood at the first ECPPM in 1994, namely:

What are we dealing with?

Information?

Knowledge?

Intelligence?

This question is still unanswered although we meanwhile know more about it. Information quality is a subtopic of this more general question, and if we want to build information filters in order to filter out information noise, we have to distinguish between:

Data – information – knowledge, and

Value on an objective level – value on a subjective level.

Both need appropriate data, i.e. product and process data models which contain more than a simple mapping of the material product, and the material design and construction processes. They demand the mapping – or the representation – of the single individual concerning his scale of measure, too.

This is the challenge of the future. Are we prepared to cope with it?

Raimar J. Scherer
Dresden, September 2000

Product and Process Modelling in Building and Construction, Gonçalves, Steiger-Garção & Scherer (eds)
© 2000 Balkema, Rotterdam, ISBN 90 5809 179 1

The 3rd edition of ECPPM

The Architecture, Engineering and Construction (A/E/C) industry is on the edge of the information society.

The third edition of the European Conference on Product and Process Modelling in the building and related industries aims to help the A/E/C and related industries, e.g. architecture and furniture, to become a fully-fledged member of the information society by showcasing the benefits of appropriate IT-based product and process modelling.

This conference draws on the experience of researchers, the industry and international initiatives to highlight the steps required for successful re-engineering and integration of this industry.

The industry, research, and EC project coming to this conference provide a forum for these views to be analysed, presented and discussed. These aims are complemented with special sessions devoted to IFC and STEP, Industrial Applications and Case studies, where real industry cases and needs are presented and discussed.

This conference brings together researchers, developers, implementers and end-users actively involved in the development of practical and theoretical approaches to product and process modelling in building and related industries.

The approaches discussed are expected to form the basis of future design, integration and management decision-making tools, including CAD, CADD and CIM, and to be a bridge between Research & Development and Industry.

ECPPM 2000 also explores research issues and requirements related to the methodologies for all life cycle phases of buildings – from initial briefing, architectural and decoration matters, to demolition – as well as the information workflow inside and between companies organizations.

Contributions from research and development in product and process modelling applied in other industrial sectors were very welcome to this conference, though its results can stimulate technology transfer and enable news ideas and opportunities to Building and Construction related sectors.

Ricardo Jardim Gonçalves
Lisboa, September 2000

The 2nd edition of DOTBM

Product and Process Modelling in Building and Construction, Gonçalves, Steiger-Garção & Scherer (eds)
© 2000 Balkema, Rotterdam, ISBN 90 5809 179 1

ECPPM 2000 Committees

STEERING COMMITTEE

Raimar Scherer, Germany
Ricardo Gonçalves, Portugal
Robert Amor, United Kingdom

ORGANISING COMMITTEE

Adolfo Steiger-Garção
Anikó Costa
André Mora
João Paulo Pimentão
João Bento

José Fonseca
Luís Gomes
Ricardo Gonçalves
Rui Tavares

PROGRAM COMMITTEE

Adolfo Steiger-Garção, Portugal
João Paulo Pimentão, Portugal
João Bento, Portugal
José Fonseca, Portugal

Luís Gomes, Portugal
Raimar Scherer, Germany
Ricardo Gonçalves, Portugal
Robert Amor, United Kingdom

SCIENTIFIC COMMITTEE

Amor, R., United Kingdom, R
Andersen, T., Denmark, Univ
Anumba, C., United Kingdom, Univ
Augenbroe, G., USA, Univ
Bacon, M., United Kingdom, Ind
Björk, B., Sweden, Univ
Borras, M., Spain, Ind
Buckley, E., Ireland, Ind
Cooper, G., United Kingtom, Univ
da Silva, G., Portugal, Univ
de Martino, T., Belgium, EC
Drogemuller, R., Australia, Gov
Filos, E., Belgium, EC
Garas, F., United Kingdom, Ind
Garrett, Jr J., USA, Univ
Ghodous, P., France, Univ
Gonçalves, R., Portugal, R
Gudnason, G., Iceland, Gov
Haas, W., Germany, Ind
Hannus, M., Finland, R.

Howard, R., Denmark, Univ
Isobe, T., Japan, Ind
Junge, R., Germany, Univ
Los, R., Netherlands, R
Mangini, M., Italy, Ind
Muigg, P., Austria, Ind
Poyet, P., France, R
Protopsaltis, B., Germany, Ind
S.K.Ong, P., Singapore, Univ
Smith, I., Switzerland, Univ
Steiger-Garção, A., Portugal, R
Steinmann, R., Germany, Ind
Storer, G., United Kingdom, Ind
Terai, T., Japan, R
Tolman, F., Netherlands, Univ
Trapey, A., Taiwan, R.O.C., Univ
Turk, Z., Slovenia, Univ
Tzanev, D., Bulgaria, Univ
Wassermann, K., Germany, Ind

SPONSORED BY

EAPPM
European Association on
Product and Process Modelling

TECHNICAL SPONSORSHIP

IEEE
Networking the World™
Robotics and Automation Society

ORGANISED BY

UNINOVA
*Institute for the Development of
New Technologies*

*New University of Lisbon
Faculty of Sciences and Technology
Electrical Engineering Department*

Keynote lectures

Product and Process Modelling in Building and Construction, Gonçalves, Steiger-Garção & Scherer (eds)
© 2000 Balkema, Rotterdam, ISBN 90 5809 179 1

Moving construction towards the digital economy

E. Filos
European Commission, Information Society Directorate-General, Brussels, Belgium

ABSTRACT: Virtual organisations, or "organisations without walls", the collection of geographically distributed, functionally and culturally diverse organisational entities linked through information and communication technologies (ICTs) are key signposts of the shift from the industrial to the digital era. The construction industry, with its long tradition in forming virtual business alliances with suppliers and sub-contractors, can benefit most from the offers of the digital paradigm. The article considers technologies and trends towards interoperable tools for organisational agility and flexibility based on distributed business operations. The European Commission has been supporting these developments through its various research programmes since the early 1980s. Besides the notable research and development (R&D) achievements, important progress has also been made in the area of standardisation and through consensus building between the research and business communities in Europe.

Keywords: construction, virtual organisations, business-to-business electronic commerce, standards, R&D

1 INTRODUCTION

Since the dawn of history, construction has shaped economies, and economies have shaped construction. This applies to today's digital economy, as it did to any master mason creating a cathedral in the medieval world. Building methods and materials have changed down the ages, but human ingenuity, adaptability and skill are still vital to the construction industry's competitiveness and its interaction with the wider economy.

With and output of 780 million Euros, construction is the largest industrial sector, ahead of foodstuffs and chemical industries. Construction represents around 6 % of the EU value added and 58 % of the gross fixed capital formation. The EU is the principal world exporter, capturing 52 % of international markets. Construction is also Europe's largest employer, providing jobs for nearly 11 million workers. That is 7 % of the working population. Every new construction job generates another two in related sectors. 33 million workers in Europe depend, directly or indirectly, on construction. The key players in this sector are small and medium-sized enterprises (SMEs), which are the most job-creative: 93 % of the 2 million construction companies have less than 10 employees, each, whilst only 100 companies have more than 2000 employees.

But while many other European industries have made huge gains in competitiveness, those made in the construction sector have lagged behind them for the past 15 years. This is a crucial issue, not just for the construction industry itself, but for the competitiveness of the economy as a whole. Every economy depends on high-quality infrastructure, roads, plants, and office spaces. So there is enormous potential for enhancing competitiveness, ICTs and business-to-business electronic commerce in particular, will help revolutionise the construction business.

The new "digital" economy, arising as a result of advances in information and communication technologies (ICTs), necessitates a new set of rules and values, which determine the behaviour of its actors. The digital marketplace offers new and unlimited opportunities for those operating through it. Players in the digital market realise that to leverage the benefits to be derived, traditional attitudes and perspectives to doing business need to be redefined.

During the industrial revolution, enterprises changed dramatically from close-knit rural communities to a core of structured and independent urban organisations. In the 1980's and 90's these became more global and collaborative, a transition encouraged by fiercer competition, the introduction of information and communication technologies (ICTs) and the rapid emergence of the electronic business paradigm. The trend towards virtual collaborative scenarios gives rise to a blurring of organisational boundaries. For many business organisations, strategic partnerships have become central to competitive success in fast changing global environments. Since many of the skills and resources essential to an organisation's capabilities lie outside its boundaries, and as such, outside management's direct control, collaborations are no longer considered an option, but a necessity. Business will rely on an increased ability to conceive, shape and sustain a wide variety of virtual collaborations.

2 TECHNOLOGIES FOR DIGITAL ECONOMY ORGANISATIONS

The technologies that helped shape the Information Society, namely the information, communication and content technologies, have created a single "digital space" characterised by the following distinct features: connectivity and the emergence of networked environments, organisational "boundarylessness", and higher speed and quality of communication and information flow. Technologies and standards for interoperability are key to electronic business. Business requirements in the digital economy are diverse and complex and the technological capabilities are still emerging,

- TCP/IP will be established as the ubiquitous network protocol for open Internet transactions, with HTTP as the protocol for web-related transactions. For mobile terminals, the Wireless Access Protocol (WAP) will play a similarly significant role;

- Distributed object-oriented platforms and middleware services will become more stable, offering better performance, scalability and distribution, the most dominant platforms being the Java and CORBA frameworks,

- The rapid acceptance of Enterprise Java Beans in the business world raises expectations that this technology will play a decisive role in providing the basis for distributed business applications under a single administrative domain;

- In the area of workflow systems, the concepts, framework, architecture and interfaces proposed by the Workflow Management Coalition are the most stable, well-accepted and concrete;

- Mobile intelligent agents have significantly changed the way distributed systems are working. All mature agent platforms are based on Java and take advantage of the Java framework, which seems to be the most appropriate one for mobile agent applications

and systems. The intelligence of agents will be part of their internal architecture. The role of XML in defining ontologies, i.e. the beliefs, knowledge and expectations for agents will be significant, as well as standard ACLs.

- Due to the advent of XML, messaging middleware systems have gained much attention recently. The Java Messaging System (JMS) seems to be the most favourable among those in existence because it is integrated into the Java framework, and harmonisation activities with CORBA 3 will raise their deployment probability. XML also plays a significant role in specifying open Internet value-added protocols for e-business applications;

- Although in the past, standardisation efforts under STEP tried to solve inter-operability problems on an applications level, key developments like standardised product models and application protocols, as well as the modelling language EXPRESS, have triggered virtual product development applications for product life cycle management.

2.1 R&D Aspects: The Contribution of EU Research Programmes

The European R&D Programmes have played a substantial role in supporting the development of key technologies and applications relevant to the Information Society. Most notable are, R&D efforts in electronic commerce and electronic business on the basis of virtual organisation concepts, concurrent engineering [1], computer-supported collaborative work (CSCW) and product and process data modelling. The ACTS [2] and Telematics [3] Programmes were successful in setting up strong CSCW pilots. These and other activities provided the first steps towards remote working (including telework) and distributed collaborative engineering. Virtual enterprise concepts were, until 1998, strongly coupled to the Esprit Programme [4].
Facilitating electronic commerce, and in particular the emergence of new electronic business paradigms based on distributed enterprise concepts, have been a priority within the European Commission's R&D Programmes over the past five years [5]. Until 1999 R&D support was concentrated mainly in the Esprit programme and its international co-operation branches INCO and IMS - Intelligent Manufacturing Systems [6]. More than fifty industry-led projects, receiving more than eighty million Euro of support (shared cost funding, 50% industrial contribution) were set up under the Fourth Framework Programme, 1994 - 1998 [7]. Within Esprit, and in addition to regular consultations with industry, a number of user group reference projects were established

which brought together major industrial users of IT and the vendor community. The common aim of these projects was to set long-term targets for, and give direction to the research efforts of the IT industry, in order to meet well formulated industrial needs. The user group reference projects dealt with the automotive and aerospace industries [8], the process industries (PRIMA), and the construction or large-scale engineering industries, (ELSEWISE) [9]. Of similar impact for the furniture industry was project FUNSTEP [10]. The AIT initiative was successful in setting up twenty-two collaborative R&D projects funded under the Esprit and Brite-Euram Programmes and had major impact on the development of standards, (e.g. CORBA, STEP, PDM Enablers, WfMC). These projects were, (some of them are still ongoing) operating concurrently within a harmonisation framework. Of particular relevance to the distributed enterprise is the work performed under project AIT-IP [11], which set up an open integration platform, based on standards like STEP, CORBA 2 and MMS.

In the IST Programme [12], which emerged as an integrated programme from the previous Esprit, ACTS and Telematics programmes, the perspective widened from "virtual enterprises" to include all types of virtual organisations, profit and non-profit alike. In the programme's Key Action II ("New Methods of Work and Electronic Commerce") the focus is now on "virtual organisations". Twelve projects, directly related to the VO concept, were launched as a result of a first call for proposals in 1999. The R&D Programme on "Competitive and Sustainable Growth", in its workprogramme for 2000 [13] dedicates Targeted Research Action 1.7 to the extended manufacturing enterprise.

2.2 Socio-economic Aspects

The European Commission has, through its R&D Programmes, actively supported the technological developments and standardisation efforts mentioned above, and accompanied these with relevant policies, legislative measures and socio-economic analyses.

The emergence of electronic business, virtual organisational forms, and the digital economy at large, as key constituents of the Information Society, need to be explored in their entirety by involving organisation science, economics, law, the social sciences as well as by developing an appropriate policy framework. However, due to their unique characteristics, a range of issues other than technological, are beginning to emerge that threaten to inhibit their implementation. The transition towards a global information society requires a regulatory framework involving co-ordinated legislative initiatives, self-regulation, and international agreements.

Socio-economic research relevant to the above issues is currently ongoing in various contexts [14], [15], [16]. The 1999 workprogramme of IST contained an action line, II.1.1 - "New Perspectives for Work and Business", which focused on socio-economic research related to networked organisational structures, such as virtual enterprises, based on benchmarking, econometric models, new statistical indicators, technology foresight, legal issues such as liability and intellectual property rights (IPR) protection. Nine projects were retained for contract negotiation as a result of this call for proposals. The IST work programme for the year 2000 bundles major socio-economic research activities under cross-programme action 7 - "Socio-economic Analysis for the Information Society" and aims at studying the interplay between a broad range of technological, human, social, economic, environmental and policy issues that critically impact effective use and adoption of new IST solutions as well as developing novel approaches aimed at identifying and quantifying the many new facets and trends of the Information Society and the emerging digital economy [12].

To illustrate, Esprit project ELSEWISE focused on the use of virtual enterprise technologies in the construction sector and investigated relevant socio-economic effects on society [33].

2.3 Legal and Policy Issues

Legal and policy-related activities so far have focused on issues such as security, privacy, IPR protection and electronic trade [17], [18]. However, issues related to new and flexible organisational forms, e.g. virtual organisations, are now gaining more attention. The point of departure in identifying the legal challenges involved, is in considering the fundamental characteristics of virtual organisations (VOs). To begin with, the VO is a co-operative alliance, often between entities with individually distinct legal identities, which come together to exploit a particular business opportunity. The VO provides participants with a framework whereby they can share the risks and returns of bringing new products and services to market. The key actor in the development of a VO is the "business integrator" whose role is to identify new market opportunities, draw together skills and expertise necessary to exploit them, identify appropriate partners and establish a common communication infrastructure to facilitate VO operations.

Issues regarding the internal operation of a VO, the relationships between the different partners, and those regarding the way in which the VO deals with the external business environment are not well distinguished and may even be inter-related. The legal challenges stem from the fact the VO appears to be a single entity but may not have a legal identity. Therefore, questions arise as to the definition of liability with respect to third parties. This underlines the necessity for a consumer protection framework

5

as well as appropriate contractual arrangements between the VO and potential suppliers. Conditions regarding termination of contract by the customer and liability for goods produced are needed. Given the fact that the virtual organisation is both dynamic and temporary, questions arise as to the extent and duration of liability, and where conflict on IPR issues may arise, what recourse partners may have in law and what procedures would have to be followed, given the diversity amongst national legal frameworks. Furthermore, with VO partners dispersed geographically and operating under diverse national legal frameworks, the question of differing taxation regimes and accounting practices needs to be considered, as does compliance with employment and health and safety requirements.

At a workshop on the legal issues concerning virtual organisations held in Brussels in October 1999 [19], it was recognised that an essential difficulty with regard to the current legal framework is that the development of the digital marketplace is moving much faster than the abilities of regulatory bodies to respond. To facilitate convergence between them, it was suggested to encourage inter-disciplinary approaches like techno-legal research towards technical solutions and practices, like on-line dispute resolution (ODR), or appropriate digital ways of contract preparation and set-up in order to encourage collaboration.

A number of EU directives are in force or will soon come into force, which seek to address some of the issues, in particular, those pertaining to protection and privacy of data, electronic signatures and electronic commerce. A range of other issues is currently under consideration [20].

2.4 Smart Organisations

The technological developments described above have turned business upside down. The new economy is "digital" and organisations that participate in it have replaced "bytes for bricks".

As more and more organisations prepare to "go digital", only few of them really seem to recognise what implications this transition will have on their business processes and on the organisation as a whole. The transformation involves much more than setting up a digital infrastructure and requires even more than the ability to enter into a virtual collaboration with other partners. Virtual organisational forms are thus only one element of what is required from organisations in the digital economy.

The implications of the above trends for organisations have led to a proliferation in terminology applied primarily to enterprises, terms such as, agile enterprise, networked organisation, virtual company, extended enterprise, ascendant organisation, knowledge enterprise, learning organisation and smart or-

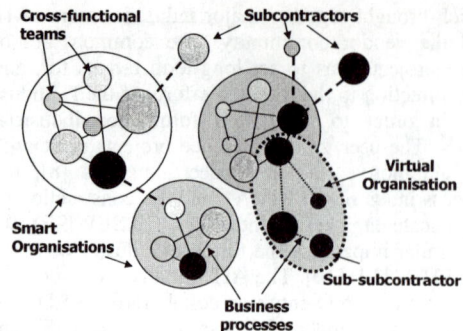

Figure 1. Collaborating smart organisations (includes virtual organisations) in an economic web.

ganisation. Each definition has its nuance, depending on what particular characteristic, or combination of traits, is given emphasis. The term "smart organisation" for instance, is used for organisations that are knowledge-driven, *internetworked*, dynamically adaptive to new organisational forms and practices, learning as well as agile in their ability to create and exploit the opportunities offered by the new economy. Ultimately, however, they all point to the need to respond to the changing landscape of the digital economy in dynamic and innovative ways (figure 1). However, the constantly changing customer and market opportunities in the digital economy ensure that there can be no single universal formula for describing the smart organisation. Nevertheless, Steven Goldman et al. [21], have described four strategic dimensions of agile behaviour that are crucial to smart organisations. These are: customer focus, commitment to intra- and inter-organisational collaboration, organising to master change and uncertainty, and leveraging the impact of people (entrepreneurial culture) and knowledge (intellectual capital).

3 VALUE CREATION IN THE DIGITAL ECONOMY

The emergence of the Internet and related technologies has led to new ways of interacting and interlinking between individuals as well as between organisations. Some characteristics [22], include,

A networked structure and self-reinforcing mechanisms. Nodes and connections are the basic ingredients of networks, and they increase in complexity exponentially with the growing number of nodes. Adding intelligence to the nodes enables a network to reach levels of unforeseeable "smartness". The number of possible interactions and options grows exponentially. In a networked environment, small efforts can lead to large effects. Mathematically, the sum value of a network increases as the square of the number of members, which means that adding a few members can dramatically increase the value of the network for all members, which in turn acts as a generator for growth.

The law of abundance. Value is created through the opportunities inherent in the potential for relationships in a network. The network economy rewards the plenitude open systems offer, more than the scarcity of closed systems, and therefore, the larger a user community becomes, the more value a relationships-enabling product has. The incentive for some commercial vendors to give technology away happens only because they realise that there is more to gain from adhering to an open standard than from maintaining a unique competitive advantage. If goods and services become more valuable as they become more plentiful, and if they become cheaper as they become valuable, consequently, *the most valuable things of all become those that are ubiquitous and free.*

Relationships. In a network environment "spaces" aren't bound by proximity. The advantage of spaces is rooted less in (non-geographical) virtuality and more in their unlimited ability to accommodate connections and relationships. Networks do not eliminate intermediaries, they are a "cradle" for intermediaries, since by definition, each node in a network is a node between others. The more connections there are between members in a net, the more intermediary nodes there can be. Everything in a network is intermediating something else. Since a relationship involves at least two members investing in it, its value increases exponentially as fast as one's investment. For this reason, such relationships, once established, are costly to dissolve. The network is the structure that thrives on relationships. Therefore with the development of the networked economy, the key source of value creation has shifted emphasis from productivity to relationships. The new economy relies on the synergies inherent in the multi-faceted interactive relationships of its players. Time changes in a network from sequential to random, enabling a skimming between nodes. Web time isn't just seven times faster than normal time. It's also a thousand times more random [23].

In a digital economy, the foundation for value creation is no longer primarily dependent on tangible assets. Whereas industrial age organisations derived value from investment in tangibles such as plant and machinery, smart organisations leverage the power of "smart" resources to identify and develop new opportunities. They can be leveraged to generate innovation. Some of the value creating, core competencies of smart organisations are given below.

3.1 Collaborative and Networking Competencies

For most business organisations, large and small, collaborative partnerships have become central to competitive success in fast changing global markets. As many of the skills and resources essential to an organisation's competence lie outside its boundaries, and outside management's direct control, partnerships are not an option any more but a necessity. Smart organisations today have the ability to conceive, shape and sustain a wide variety of collaborative partnerships. Hence the challenge: the "capacity to collaborate" needs to become a core competence of the organisation [24].

Organisations involved in partnerships are held together because of the added value that such partnerships offer them. There are a variety of strategic goals that organisations may pursue by entering into co-operation with others, goals such as,

Resource optimisation. Sharing investment with regard to infrastructure, R&D, market knowledge, etc. and the sharing of risks, while maintaining the focus on one's own core competencies;

Synergy creation by linking complementary competencies (i.e. to offer customers a solution rather than a mere product or service);

Achieve critical mass in terms of capital investment, shared markets and customers;

Increased benefits. Achieve shorter time-to-market, higher quality, with less investment.

Networking on the ICT level enables organisations to move into extended or virtual organisational forms. This may not be enough, though, if the organisational structure and the management culture need to move beyond steep hierarchies and related business processes towards flat hierarchies paired with networked cross-functional teams. The next step would be to involve the knowledge dimension into the networking, to empower the individuals in those teams to dynamically link up with each other and to share information and knowledge [25]. Figure 2 illustrates how smart organisations have a networked structure involving all three dimensions, thus leveraging the offers of ICTs, demonstrating their collaborative capabilities while focusing on their business core, and also exploiting their knowledge potential.

Most organisations are not designed, they evolve. Biological analogies provide an appropriate means to describe organisation phenomena. But not all organisations adapt equally well to the environment within which they evolve. Many, like dinosaurs of great size but with little brain, remain unchanged in a changing world. In the new economy the law of survival of the fittest will evidence its relevance to organisations as it does in the biological domain

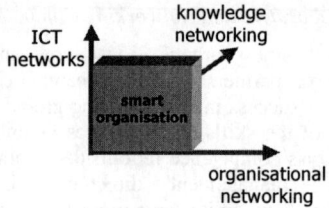

Figure 2. Smart organisations are networked in three dimensions

Charles Handy [26], remarks that the old understanding of alliances with suppliers, consultants, retailers and agents is changing into new types, i.e. stakeholder alliances with suppliers, customers and employees, as well as alliances with competitors. As no organisation today can afford to remain an "island entire unto itself", *every organisation is a network of other organisations*. No discussion of structure can therefore rest content with the inside of the organisation.

3.2 Managing Knowledge

Managing knowledge is a core competence of smart organisations. Thomas Stewart uses the term "intellectual capital" to denote the collective brainpower that comes in the form of knowledge, information, intellectual property, experience. In the digital economy knowledge becomes the primary raw material and result of economic activity [27].

The initial challenge in moving towards the smart organisation is that, to leverage the power of *knowledge*, one must know where to find it and once found, know what to do with it. Knowledge can be either explicit or tacit [28]. In the case of the former, knowledge is formal and systematic and thus easy to capture, store and communicate. Tacit knowledge on the other hand is personal, a combination of experience and intuition, and as such, the organisation's ability to capture and communicate it is heavily dependent on the individual owner's commitment to the organisation and to its need to generate value from it. In this sense, a great deal of trust and loyalty between the individual and the organisation is necessary to leverage organisational knowledge, including its tacit dimension.

Nonaka and Takeuchi see as a basic precondition for organisational knowledge the creation of a *hyper-linked organisation,* which is made up by three interconnected layers or contexts, such as the business system, the project teams and the (corporate) knowledge base. Its key characteristic is the capability to shift contexts. The bureaucratic structure efficiently implements, exploits, and accumulates new knowledge through internalisation and combination. The project teams generate (via externalisation) conceptual and (via socialisation) sympathised knowledge. The efficiency and stability of the bureaucracy is combined in this model with the effectiveness and dynamism of the task force or project team. Moreover, they add another context, the knowledge base, which serves as a "clearinghouse" for the new knowledge generated in the business system and the project team contexts. In addition to that, the hyper linked organisation has the organisational capability to convert knowledge from outside the organisation by being an open system that features also continuous and dynamic knowledge interaction with partners outside the organisation [29].

3.3 Managing Relationships

In the digital age, a key factor for success is the ability to innovate and innovation results from the clash of ideas. Networks, provide a natural environment to encourage this clash of ideas. In the networked economy, where everything and everyone is, or can be, connected, relationships are undoubtedly a very powerful source of value creation. In the industrial economy, relationships were limited in scope and important only in so far as they contributed to increased productivity. With ubiquitous network technologies there is little limitation to the nature and scope of relationships.

Connectivity breeds relationships. In a networked economy, the connections are "hyperlinks". It's not just documents that are hyperlinked over the Internet. People are. Organisations are. Real business is the set of connections among people [23]. Increased productivity was the "raison d'être" of organisations in the industrial age, in the digital age however, where information is in abundance, building relationships, and networks to facilitate and maintain them, is a key factor for success. One important element of relationships is emotion. The information and emotional engagement make up a growing proportion of the value that is being exchanged in a network.

When relationships are fostered via networks, roles become blurred: The seller also becomes a buyer of valuable feedback on his product. As Alvin Toffler illustrates [30], the distinction between producer and consumer diminishes as consumers begin to play an important role, e.g. in the development or further improvement of a product. Mass customisation enables smart organisations to see customers, suppliers, regulators, and even competitors as stakeholders who make meaningful and positive contribution.

Brands are a source of value, not unlike capital and knowledge. For the owner organisation they represent accumulated surplus value turned into client loyalty, which translates into lower marketing costs, higher prices, or larger market share [31]. In digital markets brands are an invaluable source of trust and orientation to consumers who are looking for quality and security. Many organisations invest heavily in building a reputation that is conveyed by an associated brand image. Some of them even have outsourced almost all other activities and keep a focus on managing the brand as their core competence.

4 CONSTRUCTION AND BUSINESS-TO-BUSINESS ELECTRONIC COMMERCE

What are the implications of the Information Society technologies for construction? Looking at the potential for global sourcing or the trend towards distributed engineering and construction and prefabrication, the digital economy can offer more transparency in the value chain, higher performance through optimised information flows and electronic data exchange, higher scheduling accuracy, reduced transaction costs, and optimisation of resources.

Figure 3 comes from a recent study on "Collaborative Commerce" by Morgan Stanley Dean Witter [32]. It shows how electronic business, which was traditionally confined to proprietary networks and point-to-point communications enabling Electronic Data Interchange (EDI), has moved to the (open) Internet. Business-to-business (B2B) makes business transactions more transparent, easier, and substantially cheaper, because buyers and sellers are brought together in a single medium.

Buyers discover new suppliers and suppliers discover new customers. B2B commerce enables buyers to determine the market price and product/service alternatives. It also enables them to determine the availability of the product across the market. Online B2B collaboration can bring suppliers and buyers closer. Although collaboration may not generate purchase transactions immediately, it is still critical to B2B relationships, e.g. planning, scheduling, product life cycle management, and support. All events before, during and after the order can also move online.

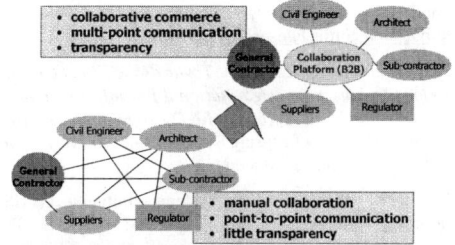

Figure 3. Business-to-business platforms for construction [32]

5 CONCLUSIONS

As more and more businesses prepare to "go digital", it will not be enough to set up a digital infrastructure. They need to become "smart", i.e. knowledge driven, *internet*worked, dynamically adaptive to new organisational forms and practices, learning as well as agile in their ability to create and exploit the opportunities offered by the new economy.

Construction companies will benefit from the new teaming and collaborative possibilities of the digital economy. They will be able to arrange their capabilities in a flexible way by building upon their ability to define and re-define multiple cross-functional and inter-organisational (virtual) teams as needed and thus respond to changing market requirements rapidly and flexibly.

6 ACKNOWLEDGEMENTS

The views expressed in this article are the author's only.

REFERENCES

1   Fan, I.-S., Filos, E. (1999) *Concurrent Engineering: Esprit-supported R&D Projects in a World-wide Context*, in: Wognum, N., Thoben, K.-D., Pawar, K. S. (1999) *Proceedings of ICE'99, International Conference on Concurrent Enterprising*, The Hague, The Netherlands, 15-17 March 1999, Nottingham: University of Nottingham, ISBN 0 9519759 86, 177-189.

2   Advanced Communications Technologies and Services Programme website, http://www.uk.infowin.org/ACTS/

3   Telematics Applications Programme website, http://www2.echo.lu/telematics/

4   Esprit (1997) *European Strategic Programme for Research in Information Technologies. Building the Information Society*, Workprogramme 1997, European Com-

mission, DG III - Industry, Brussels, http://www.cordis.lu/esprit/

5   Filos, E., Banahan, E. (2000), *Towards the Smart Organisation. An Emerging Organisational Paradigm and the Contribution of the European R&D Programmes*, Journal of Intelligent Manufacturing Systems. Special Issue: "Virtual Organisations", forthcoming.

6   Intelligent Manufacturing Systems Initiative (IMS), http://www.ims.org/

7   Filos, E., Ouzounis, V.K. (2000), *Virtual Organisations. Technologies, Trends, Standards and the Contribution of the European R&D Programmes*. International Journal of Computer Applications in Technology, Special Issue: "Applications in Industry of Product and Process Modelling Using Standards", 2000, forthcoming

8   Advanced Information Technology (AIT), http://www.ait.org.uk/

9   McCaffer, R., Garas, F. (eds) (1999) *eLSEwise: European Large Scale Engineering Wide Integration Support Effort*, Engineering Construction and Architectural Management, Special Issue, 6 (1), ISSN 0969 9988.

10   Jardim-Gonçalves, R., Sousa, P.C., Pimentão, J.P., Steiger-Garcão, A. (1999) *Furniture Commerce Electronically Assisted by Way of a Standard-based Integrated Environment*, in: [11], 129-136.

11   Ducroux, F. (1999) *The IT Integration Supporting the Extended Enterprise*, in [11], 137-145.

12   Information Society Technologies (IST) Programme, Workprogramme 2000, European Commission, DG Information Society, Brussels, http://www.cordis.lu/ist/

13   Competitive and Sustainable Growth (GROWTH) Programme, Workprogramme 2000, European Commission, DG Research, Brussels, http://www.cordis.lu/growth/

14   Journal of Computer-mediated Communication, http://jcmc.huji.ac.il/, in particular volume 3, issue 4, http://jcmc.huji.ac.il/vol3/issue4/

15   An overview of finished and ongoing IST projects is available at http://www.ispo.cec.be/serist/

16   VO-Net is focused on theoretical and empirical research related to Virtual Organisations, Virtual Teams, Network Organisation and Electronic Commerce; http://www.virtual-organization.net/

17   Julia-Barcelo, R. (1998) *Proposal for a Directive Establishing a Common Framework for Electronic Signatures: An Overview*, in: [38], 63-70.

18   Meinköhn, F. (1998) *Electronic Trade of Intangible Commodities: A Technological and Legal Challenge*, in: [38], 82-86.

19   Banahan, E., Banti, M. (1999) *Report on Workshop on Legal Aspects of Virtual Organisations*, Brussels, 30 November 1999, http://www.ispo.cec.be/serist/

20   Electronic Commerce Website of the European Commission, http://www.ispo.cec.be/ecommerce/legal/legal.html

21   Goldman, S.L., Nagel, R.N., Preiss, K. (1995), *Agile Competitors and Virtual Organizations. Strategies for Enriching the Customer*, New York: Van Nostrand Reinhold.

22   Kelly, K. (1998) *New Rules for the New Economy*, London: Fourth Estate.

23   Levine, R., Locke, Chr., Searls, D., Weinberger, D. (2000), *The Cluetrain Manifesto. The End of Business as Usual*, Cambridge, Mass.: Perseus.

24   Doz, Y.L., Hamel, G. (1998) *Alliance Advantage. The Art of Creating Value through Partnering*, Boston: Harvard Business School Press.

25   Savage, C. M. (1996), *5th Generation Management. Co-creating through Virtual Enterprising, Dynamic Teaming, and Knowledge Networking*. Newton, Mass.: Butterworth-Heinemann.

26   Handy, C. (1999), *Understanding Organizations*, 4th edition. London: Penguin.

27   Stewart, T.A. (1998) *Intellectual Capital. The Wealth of Organisations*, London: Nicholas Brealey.

28   Polanyi, M. (1966), *The Tacit Dimension*, London: Routledge & Kegan Paul.

29   Nonaka, I., Takeuchi, H. (1995), *The Knowledge-Creating Company. How Japanese Companies Create the Dynamics of Innovation*. New York: Oxford University Press.

30   Toffler, A. (1981), *The Third Wave*. London: Pan Books.

31   Davis, S., Meyer, Chr. (1998), *Blurr - The Speed of Change in the Connected Economy*. Oxford: Capstone.

32   Morgan Stanley Dean Witter (2000), Collaborative Commerce, April 2000, http://www.msdw.com

33   Damoradan, L., Hansen, J.R., Hassan, T.M., Olphert, C.W. (1999) *Impact of Large Scale Engineering Products and Processes on Society – The ELSEWISE View*, in: [9], 63-70.

Product and Process Modelling in Building and Construction, Gonçalves, Steiger-Garção & Scherer (eds)
© 2000 Balkema, Rotterdam, ISBN 90 5809 179 1

SCADEC, a Japanese practical approach to CAD data exchange in construction field

T. Terai
Chiba Institute of Technology, Japan

ABSTRACT: This paper deals with a Japanese national R&D project called SCADEC whose primary objective is to develop a neutral CAD data exchange format based on STEP AP202 applied to MOC, the Ministry of Construction, construction projects in conjunction with construction CALS/EC commission. The first half of the paper describes MOC's strategy for CALS/EC establishment in the Japanese construction industry and the outline of the project. The second half shows major intermediate activity results in relation to the structure of the data exchange system to be required. Those explained are a basic standard of CAD drawing practice, a common API library for efficient data processing with a SCADEC specific featuring concept, a series of features devised to skip STEP expertise at the development of an individual data exchange translator from a CAD system to STEP system and vice versa, and conformance test system. Validity of the SCADEC data exchange system and STEP/AP202 are also discussed.

1 INTRODUCTION

From the standardization point of view, the Japanese construction industry shall be regarded still be in an immature stage even in the highly industrialized home country caused by its gigantic but chaotic practical activities on too many platforms. There used to be almost no construction-related company or design office to take a leadership for the establishment of the common technical standards specific to the industry, which have the possibility to enhance the productivity and interoperability of the industry-wide practices on the whole.

Most of the influential organizations in a private sector have primarily pursued their profit mostly by using their own closed systems without utilizing populated core technology. So many systems with low interoperability have been developed and used in the whole industry, and brought various kinds of significant problems into the industry in an implicit way.

This situation is, however, getting changed gradually but steadily through popularization of information-related practices. The change is augmented by number of national R&D or implementation projects with a governmental IT-oriented strategy to introduce an e-commerce way of business and practice into the major industrial works and governmental administration tasks. Japanese government has re-

peatedly announced that IT is a key technology to reconstruct the society and the industries.

2 CONSTRUCTION CALS/EC AND SCADEC

2.1 *Mission of Construction CALS/EC*

The construction CALS/EC action program of MOC, the Ministry of Construction, set up in 1997 is one of those projects and publicly proposed a schedule plan how and when individual subprograms should be undergone. Here, CALS/EC stands for Continuous Acquisition and Life-cycle Support / Electronic Commerce.

The SCADEC, Standard Development for CAD Data Exchange in Japanese Construction Field, project is formulated to meet one subprogram of the MOC's action program to establish a neutral CAD data exchange format in place of current DXF de facto standard until the end of the fiscal year 2000.

One of the basic concepts of the action program is to utilize ISO standards and the like as much as possible to meet the international trade rules. Additionally the program has a successive subprogram to establish an automatic cost estimation system by analyzing CAD drawings in a digital way. The CAD data exchange subprogram has therefore a strong linkage with this subprogram and is obliged to de-

R&D Stage	Phase 1	Phase 2	Phase 3
R&D term	1996 - 1998	1999 - 2001	2002 - 2004
Overall target	Establishment of a framework for electronic data transmission in all the organizations in the Ministry of Construction	Implementation of an electronic procurement system in the projects of a relatively large scale	Establishment of electronic data utilization system which enables effective data interchange, sharing and correlation in all the processes involved at the research, planning, design, construction and management stages of the projects under ministerial justification
CAD related implementation items	Set up of a criteria for CAD drawing practice	-Drafting of the intermediate CAD data format for electronic delivery -Production of the standard design drawing by CAD -Practical test in regard to the recycling of CAD drawings	-Standardization of the three-dimensional data model -Construction of the system for automatic quantity pick-up from CAD data

Figure 1. Brief outline of the construction CALS/EC action program by MOC

velop a highly neutral and effective data exchange format. The long-term maintenance of the CAD drawings is also a target of the subprogram.

Fig.1 shows the brief outline of the action program scheduled by MOC. The plan consists of three phases and has a primary objective to introduce e-construction practices into MOC's construction projects by the end of the fiscal year 2004. The current stage is in the second phase and several subprograms have been undergone. The major ones are the SCADEC project and an electronic tendering system R&D project. The first phase as a feasibility study stage ended in the fiscal year 1998 and brought significant constitutional changes partly into the MOC's construction projects as follows;

- utilization of the Internet facilities for construction management tasks in the selected sites for case study called field experiment
- permission of digital memory camera for official report
- introduction of one-stop application system with the Internet for qualification for construction projects
- formulation of a prototype of CAD drawing standard for some civil engineering works.

2.2 SCADEC organization and standpoint

SCADEC was set up as an ad-hoc consortium in May 1999 to cope with the CAD data exchange subprogram in a public foundation called JACIC, Japan Construction Information Center, under the auspices of the engineering affairs management division of MOC with the IT R&D fund from MITI, the Ministry of International Trade and Industries.

The fund amounts to three hundred million yen and is to be primarily used for the development of related common software products and systems, those of which should be handed over to MITI by the end of August 2000. The annual fee of the participating company is decided to be one hundred fifty thousand yen, and is used for the management expense of the organization and information gathering.

MOC intends to introduce an electronic delivery system for the results of public works projects utilizing CAD and/or XML data with higher interoperability in stages from April 2001. Therefore, it is considered that SCADEC related works should be divided into two parts, the one planed for MITI and the other for MOC. At the present, the life span of the SCADEC project is from spring 1999 to August 2000, but some post-SCADEC works should be required to meet the requirements of MOC.

The number of the participating members of the consortium is 193 companies and 37 public organizations in January 2000. Nearly 40 CAD vendors are included. Japanese major construction or general-purpose CAD vendors mostly participate in the project, presumably because they regard it necessary for them to do so in order to put their products mainly into the public construction projects.

Also participate four IT vendors, NEC, Fujitsu, Toshiba and CTI Engineering who should develop

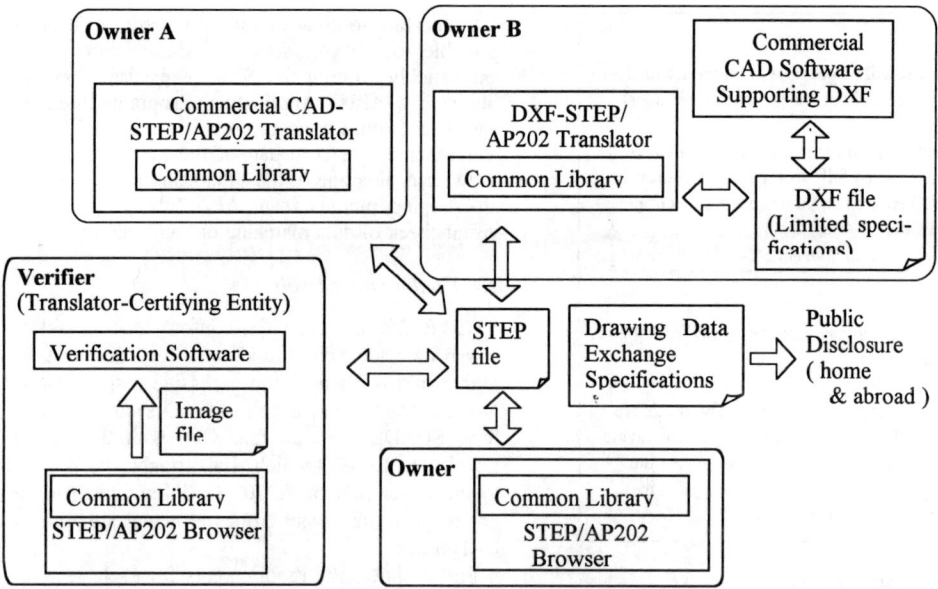

Figure 2. General outline of development

the related tools for CAD data exchange respectively. MITI's fund is used for their development works, the resulting systems of which will be delivered in Autumn 2000.

The goals of SCADEC are both to standardize CAD data exchange formats applicable to construction projects and to develop and establish a practical tools for the data exchange and cost estimation analysis based on a standard format, which is tentatively called SXF to stand for SCADEC eXchange Format. The format and the tools are to be developed and brushed up through the experimental exchanges of 2D-CAD drawing data between 30 and more CAD systems.

2.3 SCADEC project scheme

Fig. 2 shows a basic structure of SCADEC system. There are two standard formats to have affected the formulation of the R&D scheme, that is, AP202 and DXF.

MOC has an intention to introduce an automatic query system for BQ, Bill of quantity, derived directly from CAD drawing data. But three-dimensional CAD systems cannot yet get the majority in practical use in the Japanese construction industry. MOC has additionally enormous amount of conventional two-dimensional drawings for FM, facility management, certification and so on.

Accordingly, SCADEC decided to put the basis of the system to be developed onto ISO/STEP AP202,

Application Protocol: Associative Draughting, in consideration of its specific features to fit the requirements of the project. Major reasons of the decision are listed below.

- AP202 has both two-dimensional and three-dimensional aspects for drawing.
- AP202 has been investigated in a German STEP CDS project.
- STEP is acceptable by MOC as an international standard not to be affected by technical innovation in relatively longer period.
- SCADEC has a possibility to develop highly effective standards in a short time.

The scope of AP202 is wider than that of the SCADEC project, so that its standards may be regarded as a sort of subset of AP202. How to formulate the subset is also an important issue of the project.

DXF, Data eXchange Format, by Autodesk is another factor to be considered, which is a de facto standard for CAD data exchange especially in the domestic construction industry. Putting aside the incompleteness of exchange. DXF data can be imported from and/or exported to between most of the practical CAD systems. SXF, the SCADEC format to be developed can be a driving force to encourage many active CAD venders to develop their own direct translator to and from the neutral format authorized by MOC.

These works shall be done with their own expense. But one special translator had to be developed with

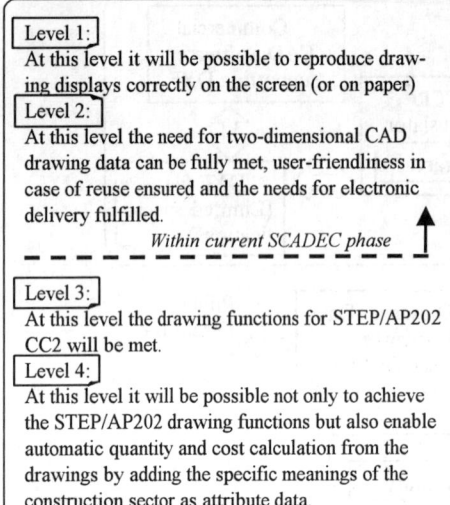

Level 1:
At this level it will be possible to reproduce drawing displays correctly on the screen (or on paper)

Level 2:
At this level the need for two-dimensional CAD drawing data can be fully met, user-friendliness in case of reuse ensured and the needs for electronic delivery fulfilled.

Within current SCADEC phase

Level 3:
At this level the drawing functions for STEP/AP202 CC2 will be met.

Level 4:
At this level it will be possible not only to achieve the STEP/AP202 drawing functions but also enable automatic quantity and cost calculation from the drawings by adding the specific meanings of the construction sector as attribute data.

Figure 3. SCADEC development levels

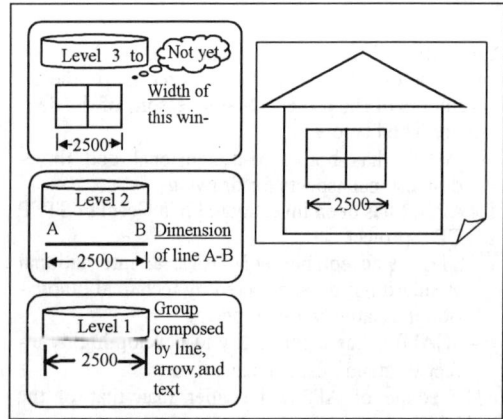

Figure 4. Graphic presentation of levels

the SCADEC fund, which is a DXF translator to and from SXF data, for not a few vendors cannot afford to cope with the speedy circumference change, and also, MOC and the construction industry have stocked enormous amount of DXF drawing data.

The other main tools to be developed are those for certification or verification by the authorities and for supporting of the development by vendors. A certification system of CAD software is essential to the R&D works of the project and also to the introduction of the SCADEC scheme into the practical construction projects.

Supporting tools are required in order to make it possible to develop direct translators more easily and quickly without full STEP expertise. Common library of APIs, Application Program Interface, based on feature concept specific to SCADEC project scheme is a core part of the tools. An AP202 browser is also one of the supporting tools to generate graphic images from AP202/Part 21 data for visual check of data matching or confirmation.

2.4 *Conformance levels*

STEP/AP202 has ten CCs, conformance classes, to cover its wide scope, and it is regarded that the higher part of them is too excessive a specification in the current practical use of CAD systems. Therefore, SCADEC set up four CCs with the title of Level 1 to Level 4,which shall roughly correspond to the lower part of AP202 CC hierarchy in order not to go so far away from the actual situation of CAD usage.

Fig. 3 shows the requirements for each development level. Level 2 is a target one of the SCADEC project within its R&D period until August 2000. It is considered that level 3 or 4 should be suitable and required for automatic quantity and cost calculation, but is regarded as a target in the next R&D term. Only their technological frameworks shall be formulated in the current project. Fig. 4 illustrates an data structure in each level.

2.5 *SCADEC R&D progress*

Eight Working groups have been set up for effective R&D works according to the schematic approach. The scopes and objectives of WG are listed below:
- AP202 WG
    to clarify a two-dimensional geometry data model as a subset of AP202 and to exchange information with the German STEP-CDS project
- Platform Software Developing WG
    to clarify the specification of the software package composed of STEP/AP202 browser, certification tool and common library, and to evaluate the related products to be developed
- Common Translator Developing WG
    to clarify the specification of the DXF-STEP/AP202 translator, and to evaluate the product to be developed
- CAD system WG
    to coordinate the issues of development between CAD venders and IT venders, and to coordinate the benefits among CAD venders
- Data Exchange Specification Standardization WG
    for Civil Engineering

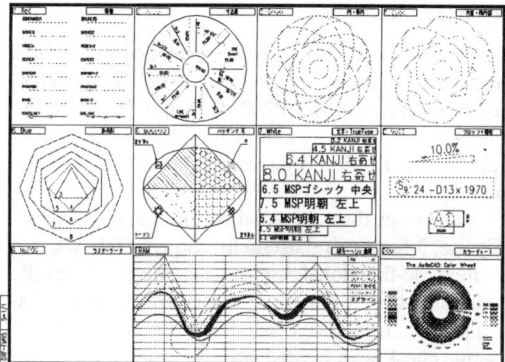

Figure 5 Sample drawing for evaluation test:
test data

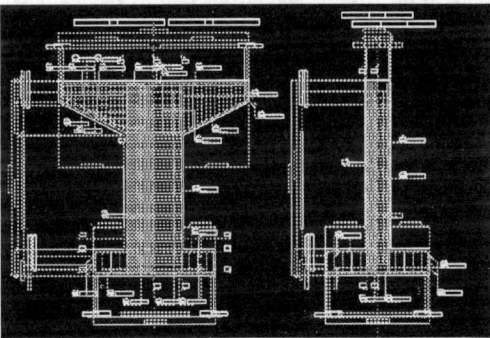

Figure 6 Sample drawing for evaluation test:
civil engineering data

to analyze the characteristic features of two-dimensional CAD data in civil engineering field, and to specify the standardized way of CAD data exchange
- Practical Testing WG for Civil Engineering
to verify the usability and interoperability of the translators developed by civil engineering CAD venders, and to specify the quality of SCADEC data exchange
- Data Exchange Specification Standardization WG for Building Construction
to analyze the characteristic features of two-dimensional CAD data in building construction field, and to specify the standardized way of CAD data exchange
- Practical Testing WG for Building Construction
to verify the usability and interoperability of the translators developed by building construction CAD venders, and to specify the quality of SCADEC data exchange.

As for the time schedule so far, the SXF specification was formulated and the platform software package was developed in their prerelease version by around the end of 1999. Thereafter, those outputs have been successively brushed up their usability.

CAD vendors who declared to join a series of evaluation test in plan have developed their direct translators step by step to meet the specifications to be issued, and presented issue logs against problems to be found to IT vendors in responsible for the development of specifications and tools for correction and reconstruction.

As most of the potential direct translators have reached to the level of practical use, a series of evaluation test is about to start in May. Below shows the time schedule of the evaluation tests in plan for implementation:
-Evaluation test to use test data: June-July 2000
-Evaluation test to assess translator level: July 2000
-Evaluation test to use actual data: July-Aug 2000
-Finalization of development: Aug 2000-Mar 2001
-Marketing of CAD system: April 2001

Final debugging of the whole products is in progress. 29 CAD systems of 24 vendors are of implementation of SCADEC scheme using common library. Only one vendor has been developing a STEP-DXF cross translator, for the company used to be well familiar with STEP architecture.

Fig. 5 shows an evaluation sample data for precheck test and Fig. 6 shows an actual civil engineering CAD drawing for actual data exchange test stage.

3 MAJOR R&D RESULTS

At the present, in the end of June 2000, SCADEC R&D works are just at the second start point to begin a series of evaluation test against the prototype products. Major development results gained so far are relevantly shown in this section very briefly. Note that R&D works are progressing rapidly and deeply in practice and the results reported here should be changeable.

3.1 Standard for CAD drawing

A draft standard of CAD drawing practice for civil engineering centered on road and bridge construction has been analyzed and finally approved as a basis for the experiment. The draft was developed in a MOC's R&D project managed by PWRI, the Public Works Research Institute.

3.2 Classified conformance levels

How to approach to the advanced Level 3 and 4 is clarified to some extent. MOC and the related public organization to have the possibility to adopt the SCADEC scheme have strong demand for these levels to make it possible to introduce effective CAD usage including automatic quantity calculation.

According to the specification to be formulated, most of the drawing functional elements provided by general-purpose CAD systems including compound dimensioning entity should be exchanged in Level 3 as they are, and all other elements to be required for introduction of automatic quantity calculation and electronic certification should be covered in Level 4.

Clothoid curve and drawing annotation specific to civil engineering works are the key in these levels. Management of luster data and drawing handling data are also the case.

3.3 SCADEC specific Features

According to an examination of data exchange, it is realized that STEP/AP202 file needs almost thousand times slower than DXF file with the same contents and may not be acceptable for practical use at the present, which have led SCADEC to devise a sort of newly-grouped entities called "feature".

Translators shall import and/or export CAD data via a set of features without any consideration of STEP architecture. Features have been defined and checked for Level 1 and Level 2 respectively.

3.4 Common library

The common library is used for the conversion between the feature information and Part21 data, and consists of data management portion and the common library API, the latter is a library of functions which provide access to the STEP exchange file according to the rule of STEP/AP202. The functions fall roughly into two categories, basic and additional functions. Feature comment portion described in STEP file can transform relevant data quite speedily.

3.5 DXF-STEP/AP202 Translator

A cross translator for SXF and DXF/R13J has been successfully developed by NEC and brushed up by WGs. The version of DXF is determined to be R13J, which is regarded populating and steady one in the Japanese construction industry. It is composed of three functional parts as listed below:
- drawing structure exchange functional part
- elementary annotation exchange functional part
- structural element exchange functional part.

The common library is used to bridge DXF with comments and STEP/AP202

3.6 STEP/AP202 browser

An AP202 browser has been developed by Toshiba and brushed up by the members of WGs, which is used widely to see AP202 drawings by MOC, to check AP202 data by contractors and to test translators by certification organization, as well as vendors and users.

As to meet their potential requirements for versatile use, it has several useful functions as follows:
- multiple-functional display including zoom/scroll, highlighting, printing of displayed elements, display of specified layers and drawing structure
- rule check with EXPRESS
- image file export

3.7 Verification system

A verification or certification system has been formulated by Toshiba and brushed up by the members of WGs. Its two main functions are comparable image file presentation function and STEP/Part 21file difference check function.

A data set for testing has been developed to be composed of seven STEP physical files as follows:
- Conformance with SCADEC Level 1 scheme
  a small normal file and a file that contains maximum number of instances and variations
- Conformance with SCADEC Level 2 scheme
  a small normal file and a file that contains maximum number of instances and variations
- Conformance with SCADEC Level 2 to 1
  down conversion
- Conformance with Japanese character set
  a file for Level 1 and a file for Level 2

Japanese character set is not dealt directly within AP202, so that this issue is considered quite important to input the test results into STEP R&D tasks.

4 CONCLUSIONS

Although the project is still in an intermediate development stage, SCADEC has brought various and valuable outputs for the implementation of its scheme. It has kept a close and dense contact with ISO/TC184/SC4(STEP) and the related overseas projects like STEP-CDS, ICIS, and presumably IAI.

A practical conformance testing and an analysis of the so-called Japanese flavor like two byte character set are one of the original contribution work to be beneficial in STEP development. The project is considered to have a possibility to change the peculiar nature of Japanese construction practices.

ACKNOWLEDGMENT

Most of the information on SCADEC and all figures included in this paper are come from JACIC and the related organizations. The author would like to give great thanks to them. Especially to IT vendors described above.

REFERENCES

Documents of SCADEC steering committee and CAD system WG(in Japanese only)
PPT Presentations from SCADEC at STEP meetings

Product and Process Modelling in Building and Construction, Gonçalves, Steiger-Garção & Scherer (eds)
© 2000 Balkema, Rotterdam, ISBN 90 5809 179 1

Development of the methodology on data modeling for land management

H. Mitsuhashi, N. Aoyama & T. Oshita
Public Works Research Institute, Ministry of Construction, Japan

T. Isobe
CTI Engineering Company Limited, Tokyo, Japan

ABSTRACT: In this paper, we examined to establish the methodology for digitalizing information which is not depend on application software, in order to define common information which is applicable to every section such as road, river, disaster prevention and so forth as a basis information for national land management. On a process of developing the methodology for data modeling, we would like to harmonize the international activities especially relating the harmonization between STEP and GIS.

1 INTRODUCTION

In order to properly cope with major issues confronting Japan such as disaster prevention and relief and environmental conservation, it is necessary to develop a technology for quickly identifying the present situations of, for example, topographical features, vegetation and social capital prevalence and for utilizing the obtained information for projects, disaster prevention plans and the like, to ensure more efficient and effective land management.

The information concerning land management is used by many people. So, to avoid double investment of costs required for collecting and preparing information, it is important to create an environment where the information concerning land management can be shared, that is, to establish a methodology for managing and operating the land management information through an information network, and to standardize the information.

Mainly for the disaster prevention and environmental conservation confronting as major issues of land management, the Ministry of Construction is developing, from 1999 to 2002, 1) information collection technology including GPS and remote sensing, 2) information processing technology including GIS for managing the information obtained there, and 3) land management technology contributing to the advanced management of the social capital by utilizing the results of the foregoing, under an organization consisting of industrial, governmental and academic experts.

Under this project, efforts are made to establish a methodology for compiling data in a form irrelevant to applications, based on the definition that the information commonly used in respective fields of roads, rivers, disaster prevention, etc. is the land management foundation information. It is intended to enable different organizations to share information and to link data to each other on a GIS base in the future, by promoting the digitization of information in the respective areas based on a standard methodology of data modeling.

The basic concept of the mechanism and system for widely using the land management information and the policy for information standardization are examined in this paper, and the results are reported hereunder.

2 POLICY FOR STANDARDIZING LAND MANAGEMENT INFORMATION FOUNDATION

2.1 Concept of land management

The purpose of land management is to efficiently construct abundant, safe and comfortable land under a limited investment. For this purpose, it is important to comprehensively redevelop, use and preserve land while effectively utilizing existing housing and social capital. To achieve this, it is necessary to enrich the information environment by upgrading the land observation network, information network, information bases and information infrastructure and by promoting information exchange, and also to particularize the land management scheme, keeping it transparent, while promoting the scheme under an agreement between the parties concerned.

It can be considered that land management includes information service (security of accountability) and real service (facilities maintenance and operation service, and aid and rescue service) aspects.

In this study, the hitherto unheeded "information service" as a land management (service) will be defined, and the framework as to what information is to be collected for it and how to process, manage and offer the information will be clearly designed, if possible, including the division of roles between the public and private sectors. In this case, especially the "technical rules" concerning the sharing and distribution of information will be clarified and authorized as a guideline for enrichment of information environment in future.

Similar efforts will also be made for the collection, processing and management of the information for maintenance and operation of facilities, and a framework will be made (to ensure that the activities for information service are also useful for enhancing the efficiency and grade of maintenance and operation of facilities). The framework will also be authorized as a guideline for enrichment of the information environment.

2.2 Land management information foundation

The land management information foundation includes 1)the information consisting of position data and basic attribute data and used as the foundation of various other data (foundation map data), 2)the information especially important for land management, 3)the information frequently used among respective fields, and 4)the information standardized in quality and capable of being reliably maintained.

If the CALS' idea is introduced, the information foundation can be considered to be a system in which the land management information existing in dispersed environments as shown in Fig.1 is connected by an information network so that the information can be retrieved and acquired freely. Furthermore, since the land management information

mostly relates to the positions on the earth, it can be considered that the information foundation based on GIS is effective as the foundation for expressing and storing the information.

2.3 Land management information foundation standard

The land management information foundation standard is standard rules for upgrading the land management information foundation. The authors consider the standard consists of the following five elements.

a) Foundation space data

These include foundation map data, basic ledgers and basic statistical information for managing land, basic image data (aerial photographs and satellite images), land use and natural environment conditions, basic information of respective fields such as roads and rivers, and so on respectively clarified in information item, definition, quality, etc.
They will be compiled according to the catalog standard of geographic information standard.

b) Exchange model

The modules, data constitutions, description methods, etc. for information exchange concerning the exchange methods of the information specified in the data model will be compiled according to the geographic information standard, etc.

c) Meta-data

Reference data concerning the upgrading organizations, upgraded items, upgraded regions, accuracy acquisition methods, etc. of the land management information foundation upgraded by respective organizations will be compiled according to the geographic information standard.

d) Operation guide

Standard means and methods for step-wise upgrading, utilization, maintenance, etc. of information will be compiled.

e) Data development model

The information items to be added to the land information foundation standard, procedure for setting the quality, etc. thereof, and method for registering standards will be compiled. The basic procedure is in the order of service definition, preparation of logic model and selection of information items and quality, but actually they will be decided through case studies. The service definition will be made particularly for a typical case of land management, i.e., comprehensive, wide-area or strategic redevelopment, utilization or preservation of land. The extracted information items will be classified for the common foundation and field-wise foundations, and will be added to the data model, exchange model, etc.

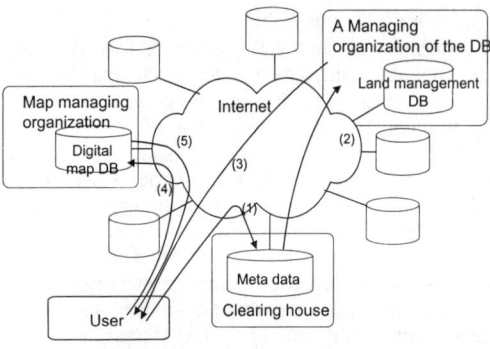

Note: The user can display the land management @@@@data required on a personal computer through @@@@(1)-(2)-(3)-(4)-(5)

Figure 1. Image of land management information foundation.

2.4 Relation between TC211 and land management information foundation standard

The data necessary for land management contains the space data handled in GIS. For the space data, a standard for efficiently exchanging GIS data between dissimilar systems is being examined at ISO /TC211, and the GIS data in the land management information are also exchanged according to the standard.

3 EXAMINATION OF STANDARD DATA MODEL DEVELOPMENT METHOD

3.1 Method for setting the to-be-modeled range

The standard data model for land management covers a wide range. So it is reasonable to develop an expandable model in which the data with higher necessity of meeting needs can be incorporated into the foundation information in the order of necessity, without limiting the to-be-modeled range from the beginning. Therefore, like the AP (Application Protocol) 202 of STEP (ISO 10303), it is desirable to set some of the anticipated foundation information at the to-be-modeled range, without clearly defining the land management tasks for limiting the range.

Furthermore, for the data items defined as the to-be-modeled range, the level at which necessary specifications can be defined must be assumed beforehand. There are three basic concepts for defining the level in the to-be-modeled range: (1) common matters only are defined as basic specifications, (2) the data used only in specific fields are defined as detailed specifications, (3) the whole of the range is defined in detail. In land management, the information in the to-be-modeled range has presently never been defined for modeling. So, it has been decided to adopt the method of defining the whole in detail.

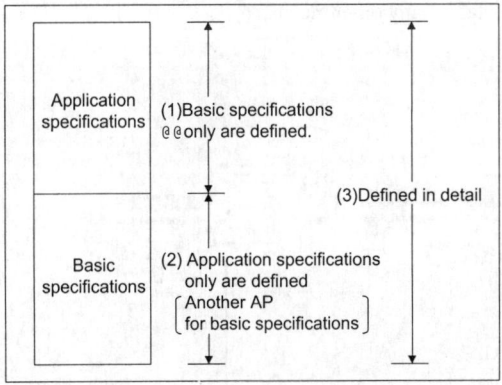

Figure 2. Setting method in the to-be-modeled range.

3.2 Verification of universality in reference to boring data

The above method is applied to boring data to verify the universality of the method.

The boring data as results of geological survey is different in submitted contents and format depending on the owner of the data (Ministry of Construction, Ministry of Transport, Ministry of Agriculture, Forestry and Fisheries, private enterprise). The data item levels observed in major patterns in Japan were compared, and as a result, the boring data items were classified as follows.

A: Data items common among all patterns
B: Data items common among plural patterns
C: Data items existing in specific patterns only

In the above, items A refer to the items adopted by 7 or more organizations of the 12 organizations, and items C are adopted by 3 or fewer of the 12 organizations.

Boring data are a constituent of the foundation information in land management, and the following must be clarified.

1) Is it necessary to authorize all the boring data already prepared as digital information by respective organizations, as foundation information?

2) If it is decided in the above 1) that all the information should not be authorized as foundation information, how should the unauthorized information be handled on the data model?

3) Are there any missing data items in the existing digital information when the future use in land management is considered? Or are there any unnecessary items?

The present boring data are different in data item level from organization to organization. However, actually, since the "Geological Survey Data Compilation Procedure (draft)" used by the Ministry of Construction is most popularly used, it can be considered that there is no problem if the data items covered by this form are handled as foundation information. Particularly, in addition to the 55 items in conformity with the JACIC form, there are 3 items classified as A.

Furthermore, for the data items that are not adopted in the JACIC form but are necessary for specific applications (in this case, classes B and C), it is necessary to construct the model in such a manner that such data items can be added and used for specific organizations or applications as an external (extension) function of the data model.

It is expected that the land management foundation information is integrally used on GIS, and the information has contents compatible with the concept of the separately established geographical information standard. Considering these, it can be considered that the following data items become necessary.

Examples of items necessary for position reference

Boring No.	Class A
Mesh No.	Class A
Position (longitude, latitude)	Class A
Bore altitude	Class A
Boring method	Class B

On the contrary, if the integral use on GIS is expected, the following data items will be doubled, and they can be treated as unnecessary data items for unification of data and simplification of data preparation.

Examples of items not necessary for position information (it can also be considered that they are allowed exist on the model)

Administrative district	Class A
Administrative classification	Class C
Coordinates	Class B

3.3 *Examination on the method for preparing a task model*

When constructing a data model, it is absolutely necessary to prepare a task model. In this report, the method for preparing the task model is particularized according to the following procedure. It is expected that the description form of the task model will conform to IDEF0.

a) Method for preparing the present task model
As an approach for preparing the present task analysis model, when the task flow of the to-be-modeled range is approximately compiled, the to-be-modeled information is extracted, and the task flow, the to-be-modeled information and the relation between both are compiled.

b) Examination on the method for preparing a future task model
If a future model is prepared according to the "as-is" method, tasks can be performed more rationally if the tasks and the future land management foundation information are upgraded as expected. However, the model is likely to depend on the present tasks.

In this report, for preparing a "to-be" model assuming that there exists an "as-is" model, two major methods different in the approach for analysis (in reference to data or in reference to tasks) are discussed as methods for analyzing the land management foundation information.

A: Anticipating situations where the compiled land management information is used, the portions to be changed in the "as-is" task flow are identified (concept in reference to data).

B: A task model as expected is prepared, and the land management foundation information is applied to the model. Lacking information is added to the land management information standard (concept in reference to tasks).

Method A is simple in approach if the "as-is" model is perfectly compiled since the data do not change.

On the other hand, method B requires, first of all, the identification of the expected tasks, making the approach difficult, and since the cases where the national management foundation data are utilized are not limited to the tasks compiled in the "as-is" model, it is difficult to identify the tasks. Therefore, it has been decided to use method A in this study.

c) Verification of methodology
The above methodology is verified in reference to the data compilation task of geological survey results.

At first, let's consider an "as-is" model as shown in Fig. 3. At present, the order is 1) boring work, 2) carrying out various tests, 3) writing various data into the format, 4) data input.

To detail the subsequent process, since generated data cannot be identified only by the detailing of the process by hearing, it is difficult to distinguish the tasks to be detailed from those not required to be further examined. On the other hand, if attention is simply paid to what operations generate the information aimed at, it is difficult to identify the relation among respective detailed tasks. So, it can be seen that it is necessary to perform the work of extracting tasks and information by using two approaches together.

Now, let's consider what is necessary to prepare a "to-be" model based on an "as-is" model. Since it is intended to verify the validity of procedure in this case, it is assumed that "to rationalize boring work by sharing boring data by constructing a shared database and by specifying the position information using GIS" is a future task image.
1) Sharing boring data by constructing a shared database
2) Specifying position information using GIS
It is difficult to select the above two improvement ideas with attention paid to the process. That is, it can be seen that it is reasonable to construct an expected task image by collecting "presently collected data" by any other means.

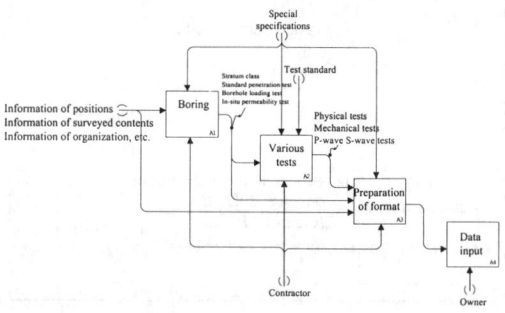

Figure 3. Present model of boring work.

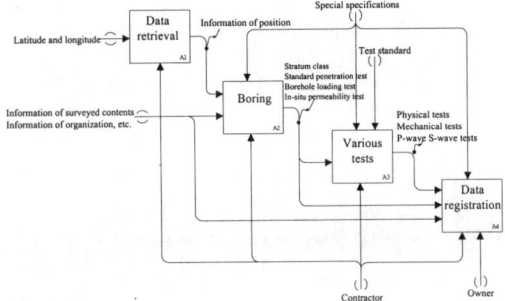

Figure 4. Future model of boring work.

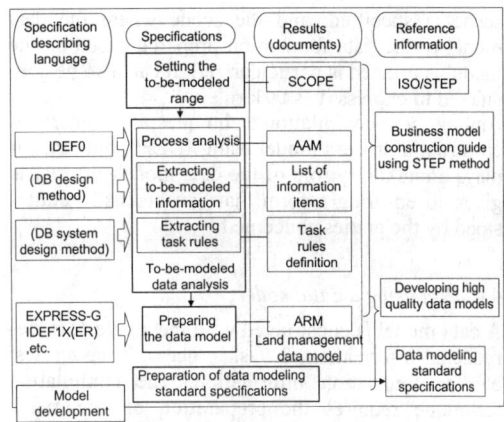

Figure 5. Data modeling flow (whole)

4 EXAMINATION ON THE METHODOLOGY FOR DATA MODELING

4.1 *Particularizing the methodology for data modeling*

Based on the above results, the methodology for data modeling is particularized. For data modeling, the documents to be preserved are important. So, data are compiled according to a format in conformity with the international standard.

For the data modeling in this case, it is intended to cover the development process up to ARM (Application Reference Model) in the AP development of STEP, and data are particularly compiled based on the above results as a general methodology in the area of land management. The particular methodology of model development is described below in an orderly manner for respective work (specification) units in the general flow (Fig. 5) of general model development.

4.2 *Setting the to-be-modeled range*

To specify the to-be-modeled range for land management foundation information, "some of the anticipated foundation information is set as the to-be-modeled range, without clearly defining the land management tasks for limiting the range". That is, to clarify the to-be-modeled range, it is necessary to show the range of foundation information without clarifying the land management tasks and the to-be-modeled products. So, the following procedure is proposed as a method for setting the to-be-modeled range".

a) Extracting the information (needs) considered necessary at this stage from the existing systems, existing tasks, etc.

b) Compiling the information required for respective needs, by positioning the information in the land foundation information conceptual view as illustrated below.

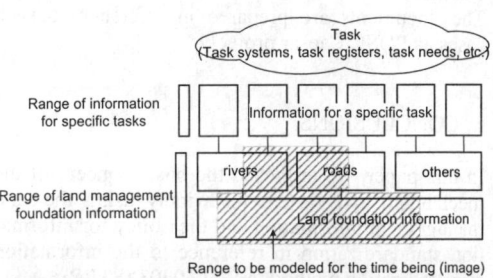

Figure 6. To-be-modeled range setting image

c) Examining the priority order in information compilation and whether to adopt for the data

d) To clarify the to-be-modeled range.

4.3 *To-be-modeled area*

To construct a task model (process model), at first, the tasks that use the land management information to be modeled are specified (what is to be noted here is that tasks are specified in one utilization scene of land management foundation information, and it does not mean that the use for other tasks is excluded.)

Then, "when the task flow of the to-be-modeled range is approximately compiled, the to-be-modeled information is extracted, and the task flow, the to-be-modeled information and the relation between both are compiled." The information extracted at this stage is used as the basic information for the subsequent data modeling stage.

In succession, "based on the extracted information, a future task process is estimated." Thus, a future task scene is specified, and the scene where the data model or the data foundation prepared based on the model is utilized is particularized. It can also be considered to express it as task rules.

Among these compilation techniques, it is general to construct a process model using IDEF0, but since it plays an auxiliary role of the data model, priority is given to ensuring a form that can be easily understood by the parties concerned.

4.4 Preparing a data model

A data model is constructed with the above 1. to-be-modeled information, 2. task process, 3. task rules, etc. as foundation information. The compilation technique requires the preparation of documents equivalent to those of STEP, 1. to ensure that no misunderstanding occurs among the parties concerned, and 2. to be prepared for international disclosure.

The documents are prepared in reference to such cases as POSC/Caesar project.

5 CONCLUSIONS

In this paper, we proposed the basic concept of the mechanism and system for widely using the land management information and the policy for information standardization in reference to the information standardization method in ISO 10303 (STEP).

From 2000 to 2002, the experiment in which Ministry of Construction, local government and private sectors share the land management information on the assumption of the scene of the public works, will be carried out. We will verify the modeling of this study through this experiment.

REFERENCES

Y.Tsukada, T.Isobe 1997. *A Basic Study of Application Protocol (AP) Development for Civil Engineering Fields,* CALS Expo International 1997, Tokyo

T.Ohita, N.Aoyama & H.Mitsuhashi 1999. *The Second STEP International Workshop Report,* Technical Memorandum of PWRI No.3607, Tsukuba: PWRI.

T.Ohita, N.Aoyama & H.Mitsuhashi 2000. *Business Model Methodology for Public Works based on STEP,* Technical Memorandum of PWRI No.3723, Tsukuba: PWRI.

Product and Process Modelling in Building and Construction, Gonçalves, Steiger-Garção & Scherer (eds)
© 2000 Balkema, Rotterdam, ISBN 90 5809 179 1

STEP and its application in the construction industry

Wolfgang R. Haas
Haas + Partner Engineering, Stuttgart, Germany

ABSTRACT: ISO/STEP (ISO 10303 family) standards are available for use in the construction industry. For 2D-CAD data exchange/sharing Conformance Class 2 of ISO 10303-202 and the corresponding subset of Conformance Class 4 of ISO 10303-214 have been implemented in Japan and Germany. In Germany they are already used in industrial practice. For 3D-CAD data exchange ISO10303-225 implementations, based on the committee draft version have been used in industrial practice for the exchange of major buildings. This paper gives an overview of STEP, its backbone data architecture and the state of art of its application in the construction industry.

1 INTRODUCTION

Approx. 80-90% of building design is still done using 2D-CAD technology, even when the systems would allow 3D-CAD design. Consequently CAD-Data exchange is still mainly done using 2D-CAD data exchange formats.

Clients usually require, that data exchange is done using the native format of the system, which they use. Typical formats of this kind are DXF, DGN and DWG.

Native formats have the advantage that the client can directly import the exchange file without conversions and loss of data. However they also have severe disadvantages. They may change from release to release of the corresponding CAD system. Moreover they limit the range of design partners which may be involved in a design/constructi project.

To accommodate these requirements design offices had to purchase the systems of their clients as so called satellite CAD systems. The only purpose of these satellite CAD systems is to convert the data of the CAD system, which the design office mainly uses for design projects, into the format, required by their clients. This approach does not solve the problem of CAD data exchange between different systems – it only shifts it into the design offices. So overall cost for design basically do not change.

In this situation the working group for computer aided facility management of VDA, the association of the German car manufacturers, started an initiative, to enable high quality 2D-CAD data exchange. They choose STEP, i. e. an ISO 10303 standard

(ISO 1995) as the neutral 2D-CAD data exchange format. This decision is based on the strategic goal of the German automotive industry to use STEP technology for the exchange and sharing of industrial product data.

3D-CAD systems building design systems are increasingly used by architects and engineers. Main applications areas are early design stages including visualization/animation and computer aided facility management. Especially facilities with complex building services equipment are candidates for 3D-CAD design. It enables advanced computations such as automatic clash detection and supports the creation of consistent drawings, derived from a spatial building model.

1999 the STEP standard ISO 10303-225 with the title "Building elements using explicit shape representation" has been approved. Its development was initiated and sponsored by the German construction ministry. One of its goals was to enable neutral 3D-CAD data exchange between the owners of facilities and design offices.

2 WHAT IS STEP?

2.1 Scope and architecture

STEP stands for Standard for the Exchange of Product Model Data. It is the unofficial abbreviation for the ISO 10303 family of standards for the representation of product model data.

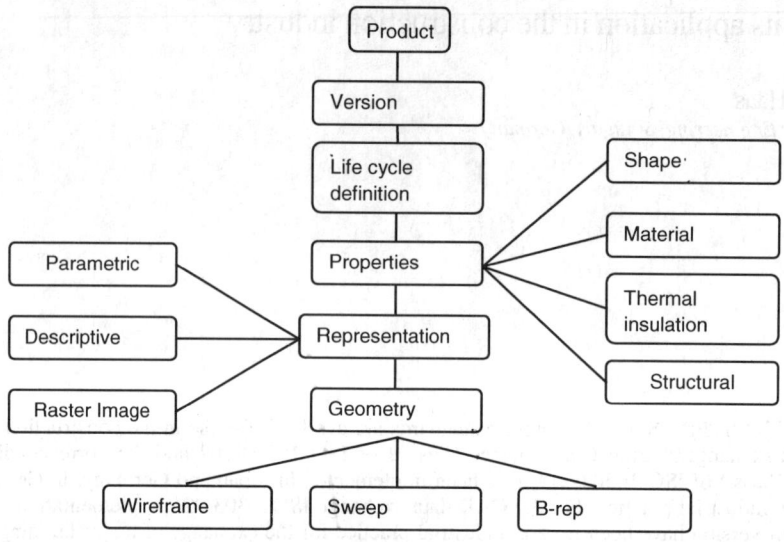

Figure. 1: STEP's data structure

The STEP standardization is a worldwide effort and consists of a series of standards for different fields of application like shipbuilding, mechanical design, plant design, automotive design and building/construction.

One of the objectives of the STEP family of standards is to enable concurrent engineering and design through high quality exchange and sharing of product model data. Concurrent engineering and design allows to reduce the time span for the development of new products. So the new product can be delivered to the market earlier and consequently, there is an earlier and better return of investment.

All STEP standards for these different application domains share a common architecture. The backbone of this common data architecture is shown in Figure 1 (PDIT 1996). This backbone data architecture enables interoperability between different applications. This is particularly important for building/construction industry with neighboring areas such as process plant and electrotechnical installations.

Besides this common data architecture all STEP standards use the same constructs to represent similar items such as geometry, topology, measure, units, tolerances, material properties, etc. This common data architecture and these common constructs are documented in the so called STEP integrated resources. They are ISO 10303 standards too and form the 40 and 100 series.

Another important aspect of STEP standardization is the separation of the data definition from implementation forms. Data definition is done using the EXPRESS language (ISO 1994).

This data models can be implemented as exchange formats and as shared data bases. Other implementation forms such as language bindings for C++ and Java are also available. So as part of the STEP standard there exists a family of 10303 standards which define standard forms of data instances and the mappings between these formats and EXPRESS. They are documented in the 20 series of STEP standards.

2.2 Specifications for application domains

As already mentioned, the STEP family of ISO standards consists of many standards for specific application domains. These so called application protocols are at least for the users the most interesting parts of the ISO 10303 standards. More than 30 application protocols have been standardized or under development.

The development of an application protocol starts with the collection of information requirements for a specific application domain such as building/construction or process plant. These information requirements are documented in a so called application reference model (ARM).

In a next step, the ARM is mapped to the STEP integrated resources. This process is called interpretation and results in the application interpreted model (AIM). So the AIM represents the interpretation of the integrated resources to satisfy the information requirements for a specific application domain.

The benefits of such an approach are reduced implementation efforts when several standards for neighboring applications are implemented and better interoperability between neighboring applications.

The development process of STEP standards is described in a series of documents (ISO96b, ISO96c, ISO96d).

A good and easy to understand overview of STEP is given in (FOWL95).

3 THE STEP-CDS INITIATIVE

3.1 State of the art of 2D-CAD data exchange

Today formats such as IGES or DXF are widely used for 2D-CAD data exchange. At a first glance the exchanged data seem to of a good quality. However if one inspects the data more closely, one can detect substantial loss of quality such as:

- circles and circular arcs are transformed into polygons
- cross hatchings are transformed into sets of lines, association to areas gets lost
- dimensions is transformed into text and lines
- groups and blocks are dissolved
- symbols get changed
- line styles, line thicknesses and colours get changed
- text fonts get changed
- layer information gets lost
- file sizes increase.

This loss of quality leads in many cases to the fact, that the exchanged data are manually re-entered in the receiving system. The imported data are used as a reference to ease the manual re-creation of the data.

The following two sources quantify the cost of the poor quality of CAD data exchange. In a survey of 1996 (PROS96) ProSTEP quantifies the cost of poor quality of CAD data exchange between car manufacturers and suppliers to 190 million DM per year in Germany.

Kazmierczak estimates in his DXF news (DXF98) the cost of poor quality of DXF based data exchange to 2.16 Billion DM per year in Germany.

Both estimates vary widely. The reality might be between both figures.

The amount of exchanged CAD data has increased dramatically during the last years. Figure 2 shows the increase in exchanged CAD data of a supplier of manufacturing equipment for the automotive industry. Within 10 years, between 1990 and 1999, the volume of exchanged CAD data increased by a factor of more than 100.

3.2 Launching companies

To accomplish high quality implementations and broad acceptance, influential interest groups have proven to achieve this goal.

In the case of STEP-CDS, such interest groups are
- End users and owner of buildings as customers, construction companies and design offices,
- CAD-system vendors, and
- Consulting companies with special STEP-knowledge.

Users provide knowledge concerning the appropriate functionality to be covered by the specification and the broad acceptance by industry. System vendors ensure high quality implementations and STEP consultants support implementation and acceptance by

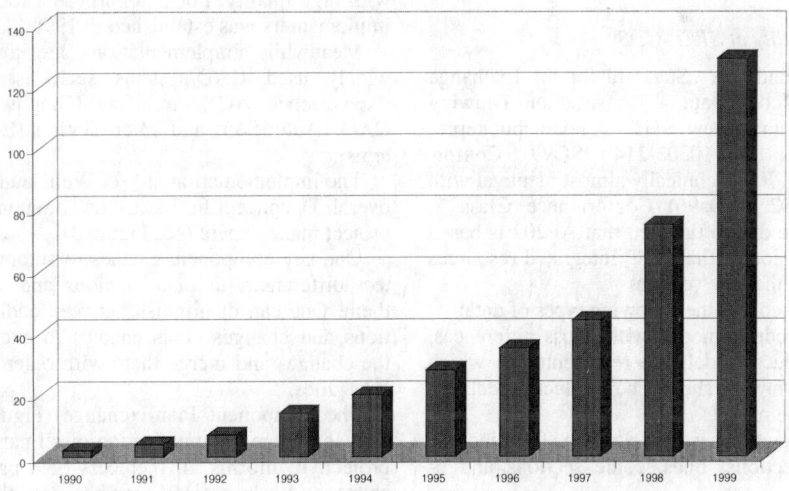

Figure 2 Increase of CAD Data exchange volume at a supplier of manufacturing equipment

industry by their know-how, tools facilitating the implementation, quality control of the implementation.

The STEP-CDS initiative was launched by the following companies:
– ABB
– Audi
– BETEK
– BMW
– Bosch
– DaimlerChrysler
– Ford
– IVECO
– KUKA
– MAN
– Porsche
– Volkswagen

It will be further supported by the working group „Computer Aided Facility Management" of VDA, and by VDMA.

3.3 Goals of STEP-CDS

By the broad acceptance and implementation of STEP-CDS, the following goals can be achieved:
– Cost savings by high-quality 2D-CAD data exchange. The cumbersome go over imported CAD-data, its subsequent debugging, and the variety of exchange formats can be reduced.
– There is no need to use the same CAD system for different design tasks. Extra cost for the purchase and maintenance of additional "satellite" CAD systems can be avoided.
– More design offices can participate in tendering for design projects. This leads to increased competition and to reduced costs.
– The problem of long-term archiving is solved.

3.4 Functionality of STEP-CDS

STEP-CDS stands for „Standard for the Exchange of Product Model Data – Construction Drawing Subset". It is not a new STEP standard but represents a subset ISO 10303-214 (ISO99), Conformance Class 4. It is technically almost identical with ISO 10303-202 (ISOa96) Conformance Class 2. Differences are due to the fact, that AP202 is based on the old version of the STEP integrated resources and AP214 on the new version.

STEP-CDS covers the following types of data:
– Generic product model with cross references, which product models are represented in which view of a drawing sheet. The product models includes shape models.
– 2D-CAD geometry in model space. They include points, directions, lines, conic sections and b-splines.

– Model structures such as groups and layers. They can be used to represent structures of factories, plants and buildings.
– A viewing pipeline which displays selected parts of the shape model in a drawing view.
– Annotation including text, symbols, various kinds of fill area such as cross hatching and tiling, different types of dimensions such as linear dimensions and angular dimensions. Dimensions may have tolerances.
– One can distinguish between view dependent and sheet dependent annotation.
– Administrative data such as identification, version and release.
– Non geometric attributes such as material, surface and performance.
– References to external data and documents. These external data can be for example bill of quantities, external symbol libraries, classification documents or shape models.
– The structure of technical drawings into drawing sheets and drawing views.

The functionality of STEP-CDS is well tailored to the needs of construction industry.

The application of STEP-CDS in conjunction with other STEP standards such as ISO 10303-212 for electrotechnical design and installation, ISO 10303-225 for spatial building models and ISO 10303-227 for plant spatial configuration is provided by the STEP architecture. References to the above mentioned ISO STEP standards are provided in (HAAS98).

3.5 What was achieved

Architect and engineers will only use the exchange format if it is implemented into his CAD-System with high quality. For this purpose a round table of implementers was established in 1998.

Meanwhile implementations are available for widely used CAD-systems such as ALLPLAN (Nemetschek AG), ArchiCad (Graphisoft), Auto-CAD (Autodesk), and MicroStation (Bentley Systems).

The implementation at WeltWeitBau is part of an overall IT concept for electronic communication and project management (see Figure 3).

One key component enables to automatically detect differences in plan versions and to highlight them. One can distinguish between additions, deletions and changes. This enables to exchange only the changes and merge them with older versions of CAD files.

The component PlanExchange (Figure 3) manages the entire communication of all partners of the project. It informs all members of a project about changes. Each project member can then decide, when to integrate changes of the original drawing into his copy. One can configure it to enforce drawing consistency at any time. The system keeps track of all activities and exchanges (send and receive) and so enables the retrieval of versions at any time.

The usage of STEP-CDS translators for factory design in the automotive industry will start in 2000.

3.6 Similar initiatives in other countries

In Japan the SCADEC consortium is implementing STEP for draughting in Building and Construction. Implementations are based on Conformance Class 2 of ISO 10303-202. The SCADEC consortium consists of 230 member institutions (193 industrial companies, 37 public institutions). 40 system vendors are participating in the SCADEC initiative. 17 are currently implementing the STEP specification.

Similar to the STEP-CDS initiative, the objective of the SCADEC initiative is to use STEP not only for the exchange of 2D-CAD files but also as a format for system-independent data management and archiving.

A good working cooperation exists between both initiatives, STEP-CDS and SCADEC. It includes the exchange of software tools and test files.

Similar co-operations exist with other initiatives to implement STEP for draughting related data exchange for other industries.

4 STEP-BASED 3D-CAD DATA EXCHANGE – THE AP225 SOLUTION

4.1 Scope and functionality

STEP AP225 or ISO 10303-225 (IOS99a) has the title „Building elements using explicit shape representation". 1999 it was approved as an ISO standard.

As the title „Building elements using explicit shape representation" already indicates, 3D building models are exchanged as aggregations of building elements. Emphasis is put on the exact representation of the geometry of the building elements and thus of the geometric 3D building models.

ISO 10303-225 enables the exchange of the shape, property, and spatial arrangement for the following types of building data:
- Structure and enclosing elements such as walls, beams, columns, slabs, roofs and stairways;
- Building services elements such as HVAC elements, plumbing, and piping elements;
- Fixture and equipment elements such as doors, frames, windows, shelving, and furniture;
- Spaces such as rooms, halls, and corridors;
- Shape of the construction site.

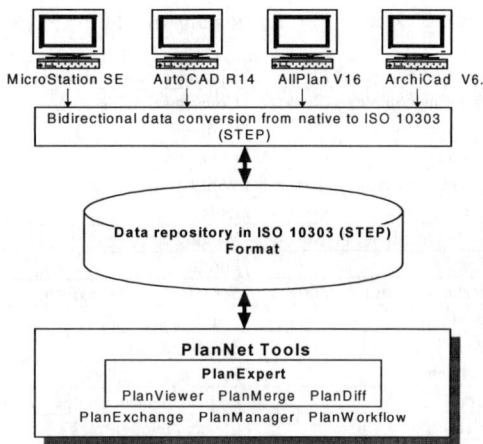

Figure 3. IT System provided by WeltWeitBau

In addition to the shape of building elements, properties can be exchanged. These Property data is not specified semantically in ISO 10303-225. It provides a mechanism for the exchange of property data which can be used in combination for example with existing national standards.

A similar approach was chosen to exchange the classification information of the building elements. ISO 10303-225 provides a mechanism for the exchange of classification data which can be used in combination for example with existing national standard.

Similar to building elements, spaces can be exchanged together with properties and classification data.

Building models produced with 3D CAD systems are highly structured. A building element „wall" may for example have an internal structure consisting of one or several layers of material, openings for doors, windows, and building service, and additional parts such as lintels and shutter boxes.

Building elements can be aggregated into element assemblies such as roofs and stairways.

The whole building can be structured into building sections and building levels. All these structures can be preserved when the building model is exchanged with ISO 10303-225.

The exchange of administrative data is a necessary prerequisite for project coordination. With each building element, request for change, change and approval information can be exchanged. The background is that the construction process is associated with a considerable amount of changes. In practical use, this causes numerous CAD data exchange operations. It would be disadvantageous to exchange the whole building model each time. It is appropriate to exchange only those elements which should be changed.

Table 1 Overview of functionality of STEP AP225

High level concepts	Fixture/equipm. elements
Building complex	Covering element
Construction site	Ceiling
building	Floor covering
Structure/enclosure element	Wall covering
Service element	Door
Fixture/equipment element	Window
Space	Furniture
Struct./enclosure elements	**Structuring Mechanisms**
Wall	Building section
Beam	Building level
Column	Element assemblies
Brace	Logical element groups
Slab	Element relationships
Foundation	**Administrative data**
Structural wire	Change request
General	Approval
Service elements	Change
Electrical	**Attribute data**
HVAC	Properties
Plumbing	Classification
Transport	Ref. to external documents

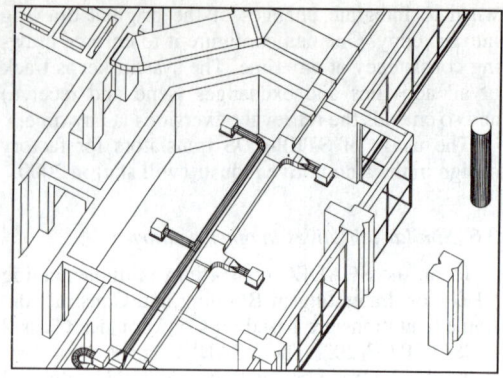

Figure 4 Part of a building model

Table 1 provides an incomplete overview of the functionality of ISO 10303-225. This table reflects the results of a market survey. In this survey questionnaires were sent out to 72 CAD System vendors acting in Germany. They include all major international system vendors. 17 Questionnaires were filled out and sent back. They include all major system vendors acting in Germany.

The results of this market survey confirmed the functionality as shown in Fig. 2. For example the types structure and enclosing elements of Fig. 2 are the high level classes available in today's CAD-systems.

The results of the market survey were also used to structure the functionality of STEP AP225 in conformance classes, representing implementation levels. These conformance classes define, which elements of STEP AP225 are supported by an implementation and thus characterize its functionality. Conformance classes are structured along the following aspects:
- Types of elements i. e.
- „physical" elements such as walls, beams, columns, service elements etc.
- Spaces,
- Geometric capabilities (elements with planar faces, faces which can be represented by analytic surfaces such as circular cylinder, cone, sphere, torus, and free form surfaces),
- CAD-geometric modeling methods (B-rep, CSG, sweeps)
- Construction site

- Structuring capabilities with or without nested logical groups and assemblies

4.2 Pilot implementations and practical applications

As part of the validation of ISO 10303-225, pilot implementations and extensive pilot testing took place. Three leading suppliers of CAD systems for building construction from Germany and Belgium participated. Pilotimplementation were based on the fully attributed ARM of the committee draft version. Fig. 4 shows a part of a typical building model which was exchanged between the three systems. The building model was created with the Belgian system ICAADS and then exchanged to the two German systems ALLPLAN and RIBCON.

In this example all building elements are properly joined together as they were in the sending CAD-system. Moreover all structure enclosing elements, i. e. walls, beams and columns etc. became similar types of elements in their parametric form in the receiving system. This is by no means trivial since the exchange is not done parametrically but using explicit shape representation.

The translators developed as part of the pilot implementations performed so satisfactory that they are also used in operational data exchange, for example to exchange building models of governmental buildings to be erected in Berlin. They have also been used to export 3D CAD models created by CAD systems for architectural design to a high performance animation system. Figure 5 shows a typical example of such a visualization.

Such visualizations were done for a wide range of projects.

June 1998, 13 system vendors had implemented STEP AP225, 6 from Germany 2 from Belgium 1 from USA and 4 from Japan. They are all based on the fully attributed ARM of the committee draft version.

Figure 5 Visualization of the new Post office area in the City of Esslingen, Design: Architectural office Mueller Benzing und Partner, Visualization: Wiedenhoefer&Kuch

Since STEP AP225 has been approved as an ISO Standard, NASA has announced, that it will adopt it for the exchange/representation of their building facilities. At least 2 system vendors have started to implement the AIM of the IS version.

5 CONCLUSIONS

1 STEP-CDS is a basis for high quality 2D-CAD data exchange
2 It can also be used for archiving.
3 Based on STEP-CDS, a WEB-based IT system for Project management and drawing administration is available
4 Similar initiatives have been established in other countries.
5 The standard can be extended to cover 3D-CAD data including properties such as material data or data, provided by the manufacturer e. g. performance and inspection data.
6 It is part of the STEP standards family, such as STEP AP225 for spatial building models, STEP AP227 for plant spatial configuration and STEP AP212 for electrotechnical equipment – to mention only a few.
7 STEP-AP225 is available as an ISO standard for 3D-building models.
8 Implementations of the fully attributed ARM of committee draft have been extensively used in pilot implementations and practical projects.
9 NASA has adopted STEP AP225 for their building facilities.
10 ISO standards are stable and as such a safeguard for the future.

REFERENCES

DXF98 Kazmierczak, A., *DXF news* Kazmierczak GmbH, Ostfildern, Germany

FOWL95 Fowler, J. 1995, *STEP for Data Management, Exchange and Sharing,* Technology Appraisals, Twickenham, UK
HAAS98 Haas, W. 1998, *The exchange of 3D CAD Building Models Using STEP AP225 – Scope, Functionality and Industrial Practice,* Proceedings of the second European Conference on Product and Process Modelling in the Building Industry, UK
ISO 1994 *ISO 10303-11, EXPRESS Language Reference Manual,* International Standardization Organization, Geneva, Switzerland
ISO 1995 *ISO 10303-1, Overview and fundamental principles,* International Standardization Organization, Geneva, Switzerland
ISOa 1996 ISO, 1996, ISO 10303-202, *Application protocol: Associative draughting,* International Standardization Organization, Geneva, Switzerland
ISO96b ISO, 1996, *Guidelines for the Development and Approval of STEP Application Protocols, Version 1.2,* Document ISO TC184/SC4/ N512, Geneva, Switzerland
ISO96c ISO, 1996, *Guidelines for AIM Development,* Document ISO TC184/SC4/ N435, Geneva, Switzerland
ISO96d ISO, 1996, *Guidelines for the Development of Mapping Tables,* Document ISO TC184/SC4/WG4 N507, Geneva, Switzerland
ISO 1999, ISO/FDIS 10303-214, *Application protocol: Core data of automotive design,* Geneva, Switzerland
ISO99a. *ISO 10303-225, Application protocol: Building elements using explicit shape representation,* International Standardization Organization, Geneva, Switzerland
PDIT96 *STEP data-file system, concept of operation,* Product Data Integration Technologies, Long Beach, 1996
PROS96 *The cost avalanche of CAD-data exchange – current situation and required actions* (in German), ProSTEP 1996

Product and Process Modelling in Building and Construction, Gonçalves, Steiger-Garção & Scherer (eds)
© 2000 Balkema, Rotterdam, ISBN 90 5809 179 1

Exchanging IFC content information using the XML protocol

Thomas Liebich
TLCon Munich, Germany

Yoshinobu Adachi
SECOM Company, Tokyo, Japan

ABSTRACT: The latest work of the International Alliance of Interoperability, IAI, on promoting interoperable solutions for the building industry, is the version 2*x* of the Industry Foundation Classes, IFC. For the first time, the IAI included a mechanism to exchange IFC based information using the XML protocol within the IFC 2*x* specification. The paper explores the XML definitions of IFC 2*x* and the methods used to correlate the XML definitions with the IFC model definitions specified using the EXPRESS language.

1 INTRODUCTION

The IAI is a worldwide federation of over 600 companies and software houses from over 25 countries involved in the AEC/FM industry. The IAI is represented by chapters in the following regions: North America, Great Britain, Germany, Scandinavia, France, Japan, Singapore, Korea and Australia.

To enable software interoperability in the AEC/FM industry, the mission of the IAI is to define, promote and publish a specification for sharing data throughout the project lifecycle globally, across disciplines and across technical applications. For that reason the IAI has developed a formal specification of requirements to represent data being exchanged between different systems of the AEC/FM industry: the IFC model [1].

1.1 Background

The International Alliance for Interoperability, IAI, has been issuing the IFC model since several years, and this year the leading software companies on the Building/Construction market, Autodesk, Graphisoft and Nemetschek, have released their flagship products with an IFC R1.5.1 exchange capability.

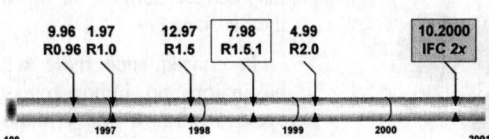

Figure 1: Timeline of the IFC releases

In meantime, IAI continued to enhance and improve the IFC exchange standard and is going to release the next major release, the IFC 2*x*, in October 2000. See figure 1 for a timeline of the release dates.

The IFC specifications have been developed following the STEP based implementation methods, in particular the EXPRESS definition language (ISO10303-11:1994) [2] and the STEP physical file format (ISO10303-21:1994) [3]. The adaptation of STEP technology has been proved being a success to define a rigid data exchange protocol, but on the other hand the use of STEP based technology remained to be restricted on special application areas.

The breath-taking development around the commercial deployment of the world-wide net let to an urgent need to transport structured information to exchange content over the net, which exceeded the presentation oriented capabilities of HTML. As a subset of the established, but not widely used SGML standard, the XML [4] language has caused a "hype" amongst the www community and the major business players. In conjunction with style definitions, as XSL, it proposes to be a universal language for content provision on the web.

Obviously both languages, EXPRESS and XML, can be used to specify data structures of the same information requirements. Even the same structure (with some limitations) can be encoded by both languages. Therefore there is no need to reinvent data structure, when switching to a different specification language.

Inside the STEP initiative, work is carried out to bridge between both worlds. A working draft on XML representation of EXPRESS-driven data has been submitted [5]. This part 28 is still in the specification phase.

1.2 IFC 2x – the newest developments

When the IAI was founded in 1996 the goal had been declared to release one major IFC release within each calendar year. This resulted into the following release cycle:
- 1996 : Pre-release 0.98 (proof-of-concept)
- 1997 : Release 1.0
- 1998 : Release 1.5 (1.5.1)
- 1999 : Release 2.0

Meanwhile some problems of such a release cycle became obvious. The IFC model specifications of the IFC releases are not backward compatible, which creates an additional burden to the implementation efforts. It also makes the marketing of IFC more difficult, as several releases with different tool support may co-exist at the same time.

Consequently the IAI decided to adjust the release strategy during the April 1999 summit. The new strategy is to release an IFC platform, which is stable and remain unchanged for a period of at least 2 years. The IFC 2x is the first platform release of the IAI

Incentive for IFC 2x is to provide a frozen core as the central part of IFC, which is referred to as the platform. The advantages are that it

- preserves investments in code development
- preserves data compatibility
- allows extensions by special domain model

In addition IFC 2x offers an extended quality assurance cycle, which includes:
- business cases to control scope and assure sufficient support of exchange scenarios
- extended feedback from major software implementers

1.2.1 IFC 2x architecture

The specification development of the IFC Model has been governed by the IFC architecture, which lays out a layered approach to schema development.

The IFC Model architecture provides a modular structure for the development of model components, the 'model schemas'. There are four conceptual layers within the architecture, which use a strict referencing hierarchy. Within each conceptual layer a set of model schemas is defined. The first conceptual layer provides Resource Classes. These classes are used by classes in the higher levels. The second conceptual layer provides a Core project model, containing the Kernel and several Core Extensions. The third conceptual layer is the Interoperability Layer and provides a set of modules defining concepts or objects common across multiple application types or AEC industry domains. Finally, the fourth and highest layer is the Domain/Applications Layer. It provides a set of modules tailored for specific AEC industry domain or application types.

The figure 2 beside shows the IFC 2x architecture.

The IFC architecture provides the backbone of the integration process that leads into a new release. The schemas of the resource, core and interoperability layer define the platform layer, which is required to remain unchanged (at least on the level of STEP physical file exchange). The schemas on the domain layer are extensible and new schemas can be defined on top of the platform.

In consequence there will be major and minor release cycles from now one. A major release incorporates a platform change, a minor release only changes on the domain layer.

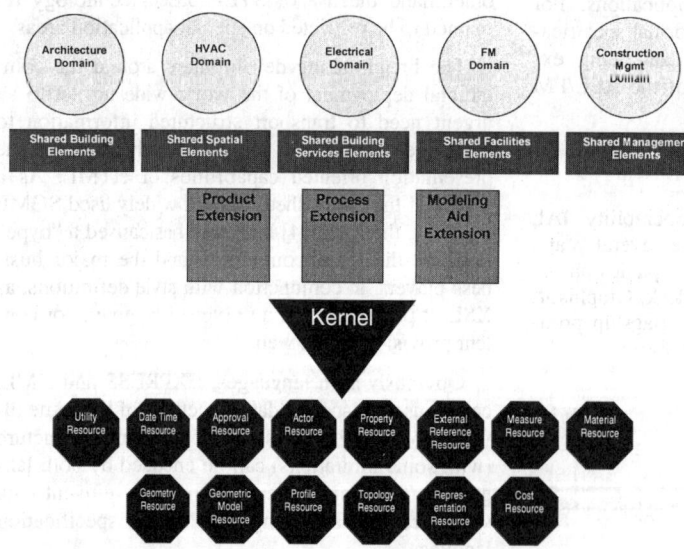

Figure 2 IFC 2x architecture

1.2.2 *IFC content model*

The scope of the IFC model combines support for various domains, such as architecture, building services, facilities management, construction management, and others.

The specification of IFC includes several aspects for the description of the total object information:

- the project model, including the logical and spatial structure of the building,
- the building element model, including the semantic definition of elements and their connectivity,
- the representation model, including the multiple geometric representation of the building,
- the content model, the dynamic assignment of styles to provide content information to the elements

The interplay between the different sub models of the IFC model is governed by the IFC architecture. One of the goals of the IFC architecture is to provide a clear separation of such sub models for clean implementations. Therefore the content sub model is self governed and accessed by the element model via a clearly defined interface. For historic reasons, the content model has been referred to as the property set concept of IFC. Within the latest IFC 2*x* developments, the property set definitions have undergone a major quality improvement.

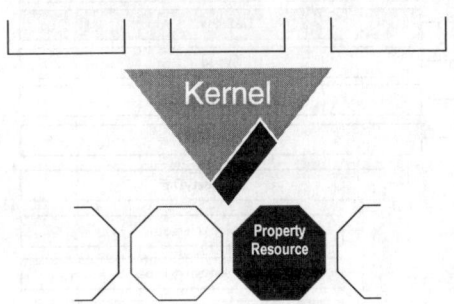

Figure 3 Content model within the IFC architecture

The content model is defined at the level of the resource layer and it is related to the object definitions at the central kernel of the IFC model. Figure 3 shows the areas within the IFC architecture

1.2.3 *Property set definitions in IFC*

The hierarchy of semantic object definitions (like the inheritance tree of object → product → element → building element → door) is clearly separated from the content or property set definitions. The link between both aspects of the IFC model is maintained using the well established objectified relationship concept. Figure 4 demonstrates the high level assignment of properties. By allowing the association of a single content definition to several object instances, a type – occurrence concept is supported.

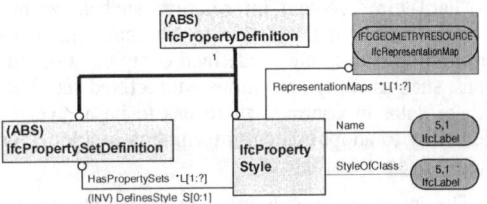

Figure 4 Definition of styles

The property style defines the combination of all property sets, applicable to an object plus the representation map, i.e. the multi-view block based geometric representation, which is common to all occurrences of this object type.

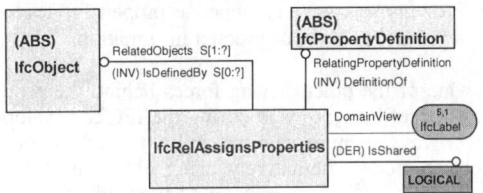

Figure 5 High level assignment of properties

The property set defines a generic set of properties, which gets its semantic meaning from the name string. In order to allow the correct interpretation, an additional agreement has to be established between the participants of the exchange. The IFC EXPRESS definition merely provides the container structure, which can be adapted to satisfy regional and local requirements.

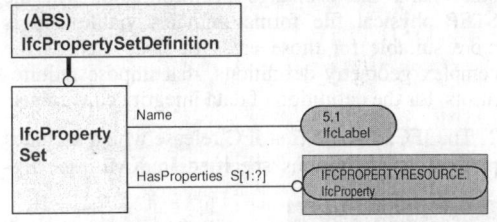

Figure 6 Definition of property set

Each property set defines a list of properties, which can be of type:

- simple property (name and value)
- simple property with unit (name, unit, value)
- enumerated property (name and enumerator)
- complex property (name and set of properties)

Again, the semantic meaning has to be derived from the name string.

1.2.4 *Reasons for property set definitions in IFC*

In the context of an international data exchange standard, is has been proven to be extremely difficult to "hard-wire" content information, such as manufacturer data or data relating to national catalogues or building codes, into predefined component definitions, such as entity definitions with a fixed set of attribute data. In contrary, there has to be a level of flexibility to adopt content information in a local, or industry segment context.

The incentives for defining property sets can be summarized as the following:
- Property sets enables dynamic extension of the IFC object model,
- Property sets allow for the aggregation of properties into any tree structure,
- Property sets can be attached to many object instances to support the type–occurrence concept,
- Property sets can be grouped as property styles to represent a library of product information.

One of the other driving forces behind the property set refinement cycle during the IFC2*x* development was the envisioned possibility to exchange the content model alternatively using either the STEP technology or the fast growing XML standard.

2 XML SUPPORT IN IFC2X

2.1 *Document type definitions of the IFC content model*

Since XML based data exchange appears to be most suitable for content information the IFC content model was chosen as the first candidate for using XML as an alternative exchange mechanism. On the other hand the exchange of information via the STEP physical file format remains viable and is more suitable for those areas of IFC, such as the complex geometry definitions, that impose requirements for the definition of data integrity constraints.

The IFC2*x* is the first IFC release which includes parts of its definitions specified in XML, see fig-

*X*tra
*X*tensible
*X*ML support

Figure 7 IFC2x logo

ure 7 for an explanation of the *"x"* within IFC2*x*. There are two DTD's provided:
- PSD (Property Set Definition markup language)
- PSML (Property Set markup language)

The PSD provides the markup language to define the property set structure and the significant name strings to capture the semantics. IFC2*x* property set definitions, either defined by the IAI international or by regional IAI chapters, are published as XML files according to the PSD. Essentially the PSD establishes the property set meta model in XML.

The PSML provides the markup language to exchange content information according to the property set structure. Each PSML exchange should refer to a PSD meta definition for the definitions of the content structure.

2.1.1 *Property set definition markup language*

The DTD of property set definition markup language, PSD [6], was designed to mirror the IFC R2.x content model and to add the slots necessary to capture the semantic definitions of the significant name tags. Included are the links to the IFC object model which can be defined, e.g. by indicating IFC object classes to which the property set is applicable.

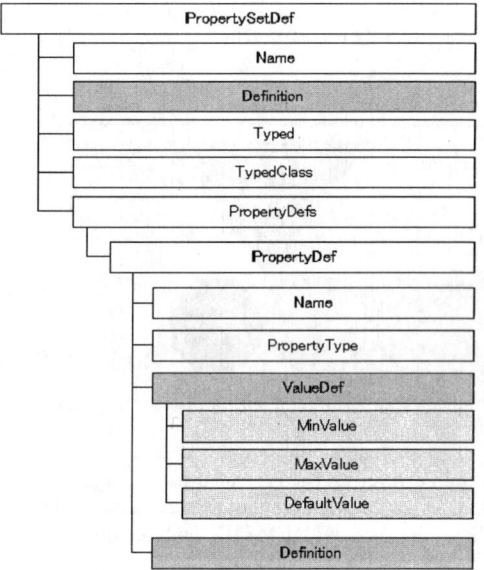

Figure 8 Structure of the PSD - DTD

The PSD defines in addition the definition element to property set and the individual properties and the value definitions used to constrain property values and to provide a default value for property initialization. Figure 8 shows the DTD structure (without the individual property definitions).

The delivery of IFC2x specifications include the xml files for all property sets, that are defined as part of the standard.

2.1.2 Property Set markup language

The PSML defines the exchange structure for property set content information. It is used to share product information defined according to:
- the overall IFC property set syntax,
- the agreements provided by an PSD xml file for this property set definition,

The tag set of PSML DTD consists of the following three major parts:
- Tag definitions for describing property style information, i.e. property style name, attached object type, contained property sets, etc.
- Tag definitions for describing property set level information, i.e. property set name, attached object type, contained properties, etc.
- Tag definitions for describing property level information, i.e. property name, value type, value, unit, etc.

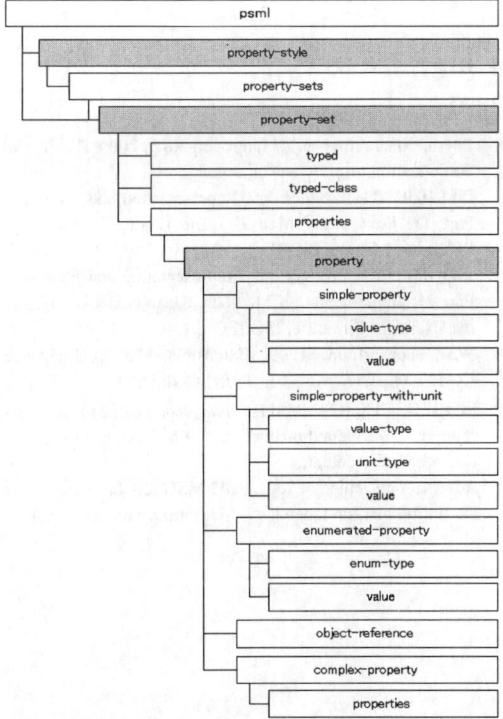

Figure 9 Structure of the PSML - DTD

The property set markup language tries to provide only a general property tree structure by entity definitions, whereas the special semantic of the content is provided by the "to be agreed upon" type name attributes. This allows for the property set markup language to be generic and for the provision of a general purpose XSL style definition.

Other initiatives, such as the affiliated aecXML activity, look into developing a well-formed XML exchange protocol by standardizing the content model through explicitly defined schemas reflecting special business transactions. This leads to a wide set of tag definitions and to the open question on how to merge a variety of these schemas. In contrary the PSD/PSML approach defines only a minimum of tags to capture the structure of product information, the interpretation is done by an interpretation of the name attributes.

3 PROVE OF CONCEPT

Two prototype implementations have been developed to demonstrate the functionality of the XML specifications and the link to the EXPRESS world.

3.1 Outline of XML to IFC conversion

Using the property set markup language and some software components, an XML to IFC conversion prototype was developed. The XML to IFC converter prototype consists of three major components:

1. XML document handling component:
   To access the XML document, the W3C defined the Document Object Model (DOM), which is a language neutral interface that allows programs to dynamically access the XML data. It is able to use DOM based component on many operation systems, i.e. Windows, Unix. The prototype uses MSXML component which is included in Microsoft Internet Explorer 5.

2. IFC data handling component:
   The IFC is based on STEP technology, therefore many STEP compatible development tools can be used to make IFC data handling component. In this conversion prototype, an ActiveX component is used for IFC data interface. The component has functions such as import/export IFC data file, creation/setting IFC object, and so on.

3. XML to IFC property set mapping program:
   First of all, the converter imports XML document which contains property set data by using DOM component. At this point, property set data is extracted on memory as a tree structure and ready to access by program, i.e. Visual Basic for Application. Second, IFC component exports the property set data into STEP physical file format according to the IFC content model.

The prototype shows that the linkage between XML and IFC can be implemented easily with only two components and a small conversion program. The next figure 10 shows the basic components:

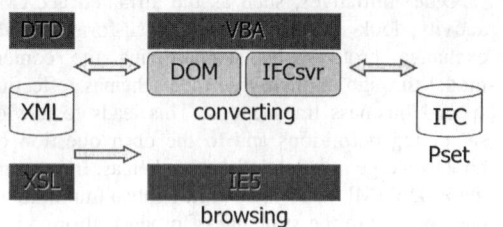

Figure 10 IFC P-set to XML P-set conversion

3.2 *Outline of XML property use in CAD*

Another prototype has been developed to demonstrate how PSD and PSML work together. The prototype is implemented in Visio and allows to read in an PSD xml file first to initialize an Visio object with the property set structure, declared within the PSD file.

The attribute set of the Visio object is created according to the standard PSD xml file for the IFC object type. The default value is used to initialize the attributes. After that, any PSML xml files, containing the actual product information, can be read in to assign the actual attribute values.

The figures show two stages, first after reading in the PSD xml file, second after reading in the PSML xml file.

Finally the internal model, consisting of the spatial information (story and spaces), the building element information (walls, doors, air terminals, etc.)

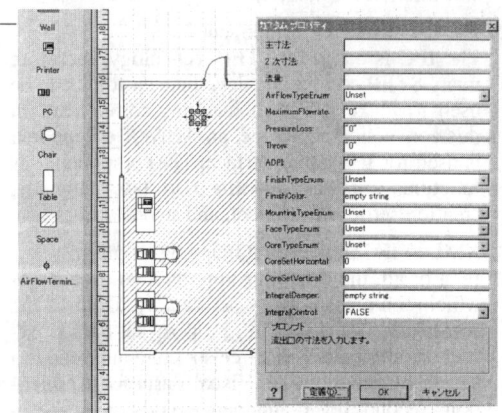

Figure 11 Screen shot after reading in the PSD file
(attribute list dynamically created and default values shown as given by the XML file)

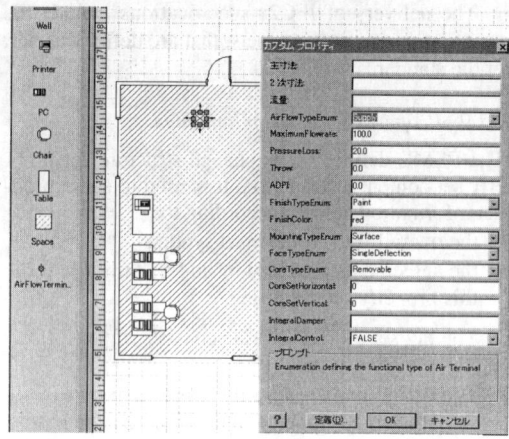

Figure 12 Screen shot after reading in the PSML file
(actual values are assigned to the attribute list, created in the previous step)

and the associated manufacturer properties can be stored in an IFC file, using the STEP physical file format.

4 REFERENCES

1 International Alliance for Interoperability, 2000, IFC Model Specification, http://www.iai.org.uk
2 ISO-10303-11: Product data Representation and Exchange Part 11: Description Methods, The EXPRESS Language Reference Manual, ISO/IEC, 1994
3 ISO-10303-21: Product data Representation and Exchange Part 21: Implementation Methods, Clear Text Encoding of the Exchange Structure, ISO/IEC, 1994
4 W3C Consortium, 1999, Extensible Markup Language (XML), http://www.w3.org/TR/REC-xml.html
5 ISO/WD 10303-28 Product data representation and exchange: Implementation methods: XML representation of EXPRESS-driven data
6 Adachi, Y.; Liebich, T (eds.), IAI MSG, 2000, Property Set Definition markup language (PSD), http://www.iai.org.uk

Concurrent engineering and cooperative work

Requirements for distributed engineering

M. Hannus & A. S. Kazi
VTT, Espoo, Finland

ABSTRACT: This paper presents some initial efforts from the recently started research project GLOBEMEN, whose focus is on defining and providing information and communications technology (ICT) support for dynamic virtual enterprises in one-of-a-kind industries. An exploration of the state of the art in ICT in terms of methods, models, and tools for distributed engineering is presented as a first step to identify the currently available functionality. Furthermore, an elaboration of the missing functionalities that will be harnessed /developed within GLOBEMEN is made. It is noted that the much-needed functionality of data/information exchange between heterogeneous systems was lacking. This is a core functionality that needs to be provided amongst others, to enable smooth operability of virtual enterprises in one-of-a-kind industries.

1 INTRODUCTION

Project oriented mode of operation is prolifirating in many industrial sectors as an attempt to provide increased flexibility and responsiveness. New ICT support is needed for dynamic, geographically and organisationally dispersed project teams, Virtual Enterprises (VE).

Currently available commercial ICT systems are primarily focused towards internal use within organisations and their associated supply chains.

This paper presents some approaches and early findings of the recently started GLOBEMEN project on ICT support in dynamic virtual enterprises in one-of-a-kind industries (GLOBEMEN 2000). The emphasis of the paper is on distributed engineering which is one of the key focus areas of the project (Fig. 1). In the area of distributed engineering GLOBEMEN, builds on the achievements of several EU projects, especially CONCUR, PROCURE, TOCEE, VEGA.

2 GLOBEMEN

2.1 *IMS and IST programs*

Global Engineering and Manufacturing in Enterprise Networks, GLOBEMEN (Jan 2000 - Jan 2003) is a research project within the global Intelligent Manufacturing Systems (IMS 2000) program. The project consortium consists of 23 organisations dispersed across different IMS regions: Australia, EU & Norway, Japan, Switzerland, and USA.

The EU participation is supported by the Information Society Technologies Programme programme (IST-1999-60002).

2.2 *Objectives*

The aim of GLOBEMEN is to create IT infrastructures and related models, methods and tools to support globally distributed and dynamically networked operations in one-of-a-kind industries. The focus is on inter-enterprise integration and collaboration in the three main facets of global manufacturing: sales and services, inter-enterprise delivery process management, and distributed engineering.

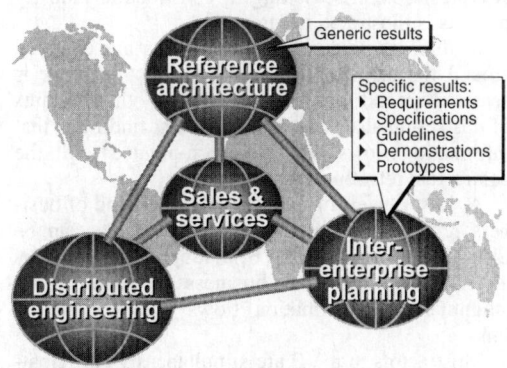

Figure 1. Focus areas of GLOBEMEN project

The vision of GLOBEMEN is that inspite of their apparent differences, various one-of-a-kind industries can be supported by fundamentally similar kinds of ICT tools. The project aims to find commonalities between various industrial domains and cultural environments and to suggest an integrating reference architecture. Different application systems can be developed based on the generic architecture. The potential is to magnify the global market for ICT solution providers while offering enhanced support to the industry.

For an overall description of GLOBEMEN see another presentation in this conference (Laitinen et al. 2000).

3 ENGINEERING IN VIRTUAL ENTERPRISE

3.1 *Virtual Manufacturing Enterprise*

Virtual Enterprise (VE) is a natural mode of business for many one-of-the-kind industries.

The GLOBEMEN project views a VE as a "temporary assembly of competences to exploit a business opportunity". It should be noted that competence, rather than capacity, is the key asset.VEs are established for the execution of projects from a network of companies who possess the necessary and complementary competencies. Once the project is completed the VE is dissolved. Several VEs may support the various life cycle stages of the product.

VE members are typically distinct enterprises while the VE itself is not a legal entity. In the manufacturing industry the VE is set up for business-to-business operations. The conditions are quite different from those of business-to-consumer operations. The VE setting is flavoured by the degree of formalism, continuity and trust between the members.

3.2 *Sample VE setting*

A typical enterprise setting in construction industry projects is illustrated in Figure 2 .

In this example information flows are centered around the architect while the contract network is centered around the client. Numerous other variants of this set-up also exist. An important finding is that information flows are often non-aligned with the contractual relationships.

At lower levels of the network other kind of business relationsships, such as supply chains, can be identified. Compared to VE, a supply chain is characterised by longer term business relationships and allignment of information flows with contractual links.

Most actors in a VE are simultaneously participating in many other concurrent VEs. Some VE members will not have previous experience about each

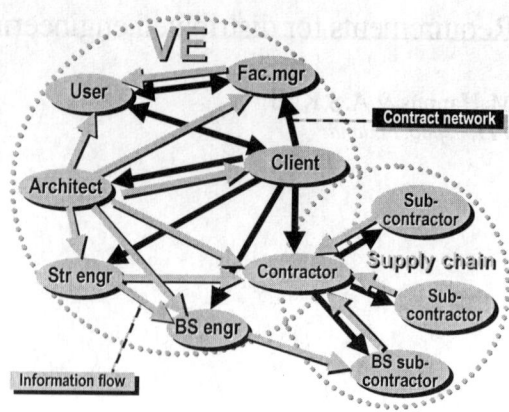

Figure 2. Example of typical enterprise relationships in a VE

other and may never operate together again. Communication in a VE happens broadly between persons of the VE participant organisations. In this kind of setting there is seldom a dominant actor who can define the rules of the game. Consequently the possibilities, and even the contractual power, to promote a common ICT infrastructure are quite limited.

As a summary, some of the main characteristics of a dynamic VE are found to be:
- temporary relationships,
- some VE members not known in advance,
- complementary competence provided by distinct companies,
- absence of a dominant actor,
- disparity of contract relationships and information flows,
- participation of actors in other concurrent VEs.

3.3 *Engineering in a VE*

Engineering activities in a VE are subject to the conditions described above (Fig. 2).

In an increasingly competitive envonment, engineering experts must be able to offer highly specialised services while being able to communicate on a high semantic level with previously unknown partners.

Consequently, we assume that the VE engineering staff has access to a variety of legacy tools within their own organisations. The persons will need to follow the working procedures of their own organisation while at the same time complying with the processes of the VE.

For the foreseeable future we expect that limited concurrency will be applied: information related to work in progress will be restricted. According to internal workflows and contracts in the VE, information to the VE members is only periodically released.

Key issues for engineering under dynamic VE conditions are:
- freedom to use legacy tools,
- standards for external communication,
- short set-up time and low cost of the common working environment,
- short learning time of common tools,
- protection of proprietary knowledge,
- division of responsibilities between VE members,
- periodical releases of information to VE partners.

4 DISTRIBUTED ENGINEERING ENVIRONMENT

4.1 *Basic system requirements*

The GLOBEMEN project is still in the stage of defining user requirements. However, in order to limit our scope we a look ahead at the possible system configurations which GLOBEMEN partners had in their mind. Figure 3 shows our initial anticipation how the previously described contradicting expectations on simultaneous differentiation and standardisation might be fullfilled.

In dynamic conditions it is mandatory to use a commonly available communication infrastructure, i.e. the Internet, to connect VE members. Each company has its own legacy tools for internal use. For external communication, available standards are used. Investments in legacy systems should be protected by isolating the mapping of internal data for external communication.

Once we found that the GLOBEMEN partners agree about this overall concept we proceeded into some more details. It turned out that the requirements of all partners (who participate in this specific work package of GLOBEMEN) could be related to the management of:
- documents,
- product models, and
- processes/workflows.

4.2 *Anticipated system architecture*

The resulting draft system architecture is depicted in Figure 4. At this level the architecture is in fact not specific to engineering and may suit other groups within GLOBEMEN, too. Therefore we provided the system components with generic labels:
- VE environment: A common data repository and sharing environment for VE members. It provides support for distributed groupwork and sharing of information using standard (or agreed) formats. The environment makes use of and controls modules for management of various types of data: documents, product and process models. These modules are likely to be commercial off the shelf

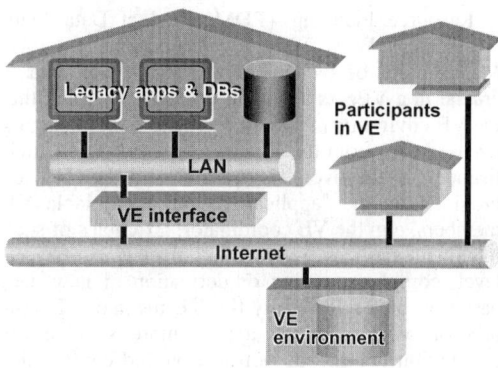

Figure 3. Connecting an engineering office to VE

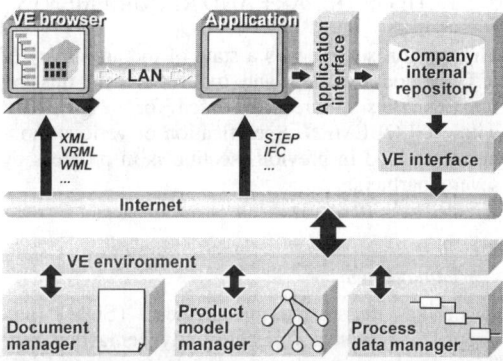

Figure 4. Anticipated system architecture for distributed engineering in VE

software products wrapped within the VE environment.
- VE interface: A company specific module which provides for communication between the company and the VE environment. This module controls transfer of released information to the VE environment.
- VE browser: A tool which enables users to search, view and retrieve data from the VE environment.

The following modules are internal tools of each company and as such, out of scope for development in the GLOBEMEN project:
- Applications: Any tools which companies use internally in performing their core business.
- Application interfaces: A service which provides applications an access to the company internal repository enabling users to store/retrieve data according to the workflow of the company.
- Company internal repository: A company specific information management system e.g. EDM (Electronic Document Management), ERP (Enterprise

Resource Planning), PDM (Product Data Management), WfM (Workflow Management).

Management of ownership and liablity of data is crucial in a VE. As a general principle we think that the VE environment should be focused on managing released data from the VE members. Possible "intelligence" of the overall system should therefore be provided by the "applications" of identifiable VE members, e.g. the VE coordinator. Examples of such intelligent services are workflow management at VE level, complex merging or derivation of new data based on data released by the VE members. Examples of services which appear more suitable for automation are change notification and conflict detection.

5 STATE OF THE ART AND REQUIREMENTS

This section summarises a state of the art survey of ICT methods, models and tools. Selected missing functionalities are recommended for development within GLOBEMEN. Specification of various modules described in previous section is in progress by several partners.

5.1 *Best Practice in distributed information management*

Electronic document management (EDM) and GroupWare systems are currently being deployed by technology leading companies and are mainly used internally or within supply chains. Inter-company use in project based co-operation is still un-common. Recently a growing number of related services have been introduced to the market by specialised ISPs (Internet Service Providers).

Because many companies do not yet have their internal EDM systems in place, there is a tendency that a common system is set up and shared among various project actors. This will obviously be a temporary solution. Once the deployment of these technologies proceeds further it will become necessary to shift focus onto interfaces between various company specific EDM systems.

The dominant communication practices in the industry are still based on traditional paper based information. The awareness about the necessity for ICT oriented practices for inter-enterprise communication is growing. Some guidelines exist about setting up and using ICT in inter-enterprise engineering projects.

5.2 *EDM / PDM*

Engineering Data/Document Management (EDM), and Product Data Management (PDM), are key enablers of information sharing and communication within a VE. The functionalities of these systems include: support for concurrent engineering, group-work and workflow management, document and product configuration management, access control, viewing and redlining/mark-up, version management. Both EDM and PDM technologies deal primarily with documents. PDM adds support for product configuration management, usually by linking of Bills of Material (BOM) and documents. Recently use of some systems has become available as online services via the web.

Claimed "out of the box" solutions still suffer from tedious system set up. The ability to share data from heterogeneous sources is limited by the availability and usability of standards.

Standards that are relevant for exchanges between different systems include: STEP PDM Schema (PDM Implementor Forum 2000), and OMG PDM enablers (OMG 2000b).

Many related EU and national projects have been identified and GLOBEMEN will attempt to learn from them (Hannus 2000).

Some effort will be spent by GLOBEMEN into migration from current document management technology towards product model management.

5.3 *GroupWare*

GroupWare (CSCW, Computer Supported Collaborative Work) tools allow people to work together collectively while located remotely from each other. These tools generally facilitate the sharing of calendars, collective writing, e-mail handling, shared database access, electronic meetings with each person able to see and display information to others, and other activities.

Different products offer different services as per their own definition of "GroupWare" in an attempt to satisfy user needs. Additionally, there are web-based solutions such as BSCW (Basic Support for Cooperative Work) which was developed under EU projects BSCW, CoopWWW, and CESAR. GroupWare functionality however, is not available as a one-point packaged solution.

Within GLOBEMEN, the intention is to provide a GroupWare environment using currently available technology.

5.4 *Product Data Technology*

Significant developments in this area have led to the formulation, adoption, and exploitation of standards such as the Standard for the Exchange of Product Model Data, STEP (ISO 10303) and Industry Foundation Classes (IFC). Until recently, only toolkits for developing file conversion software have been commercially available. First compliant commercial CAD tools have now started to emerge.

A key problem is that standard data models are constantly evolving. This instigates the continuous upgrading of software to remain compliant. Specific technologies, such as EXPRESS-X language and related software tools, are available for mapping between schemata (e.g. of different standard versions).

Gaps exist in terms of support for product data warehousing, exchange of partial product models, server-client architectures, model merging, etc.

Through exploitation and further development of the Express Data Manager (EPM Technology 2000), software for STEP model management, developments in GLOBEMEN will demonstrate a product model repository with model merging and partial model extraction capability.

A possible migration scenario is illustrated in Figure 5: develop a servlet which retrieves STEP files from a document management system and transfers them further to Express Data Manager for merging operation.

5.5 Application integration technologies/standards

Java is a platform independent programming language for developing Internet applications. Remote Method Invocation (RMI) is a Java-based protocol for client-server software communication over the Internet. The problem, seen with RMI is that it requires the use of identically defined class interfaces on both sides (client and server). Therefore it does not appear to be suitable for integrating systems of independent actors who only temporarily co-operate on a project. A standardised RMI API (Application Programming Interface) would be needed but is not available today.

Two key technologies/standards provide communication between heterogeneous systems in the VE

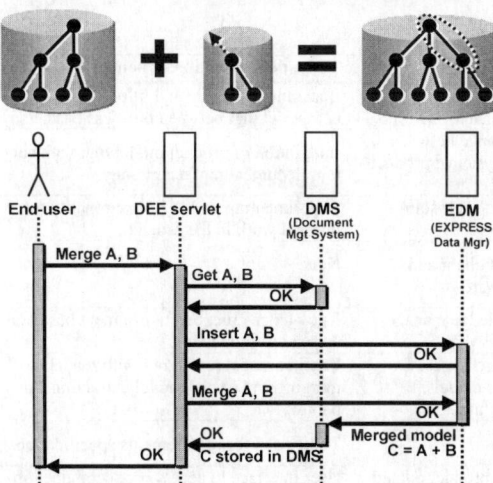

Figure 5. Merging of product models

through message brokering: Object Management Group's (OMG) Common Object Request Broker Architecture (CORBA) is a middleware specification (OMG 2000a). Microsoft's Distributed Common Object Model (DCOM) is a middleware implementation (Microsoft 2000). Similar concerns hold true for CORBA and DCOM as mentioned previously in connection with Java/RMI.

The GLOBEMEN project will consider potential areas for such standards for inter-enterprise system integration.

5.6 Knowledge Management

Knowledge management entails the capture, consolidation, dissemination, and reuse of knowledge in addition to the translation of new best practices to tangible programmable processes to be automated through ICT (Kazi et al. 1999).

Currently, knowledge management is not supported by packaged solutions, but needs to be built up through the combination of different "infrastructure" elements: open interoperable computing platforms, communication networks, knowledge creation analysis tools, external and internal content, collaboration tools, enterprise-wide and inter-enterprise-wide messaging, web content management tools, "push" and "pull" technologies, intelligent agents, case-based retrieval, portable documents, object databases, document management, process management tools, etc. State of the art tools today, facilitate concept classification to help identify knowledge, and then use either semantic, collaborative, or visualisation retrieval technologies to leech the knowledge from applications.

While there is evidence of support in terms of "process" knowledge management, there is a lack of solutions for "product" knowledge management. This is an area that will be addressed and developed within Globemen.

5.7 Virtual Reality

The Virtual Reality Modeling Language, VRML, is a language for describing three-dimensional image sequences and possible user interactions with them. Virtual Reality (VR) technology offers interesting opportunities for distributed engineering within and amongst the participants of a VE.

Versatile tools for viewing VRML models are available and provide the necessary basic functionality for integration with engineering applications. Current tools basically provide 3D object visualisation, while interaction with product related data is limited.

Within Globemen, a VRML based user interface will be used to access product model objects and related documents (Hemiö et al 1999).

45

5.8 The Unified Modeling Language

The Unified Modeling Language (UML) is a language for (software) systems engineering (OMG 2000c). UML provides a series of diagrams:
- Use Case: Describe functionality provided by a system to external interactors.
- Class: Describe the static structure of a system.
- Object: Describe the static structure of a system at a particular time during its life.
- Sequence: Describe a set of message/information exchange sequences.
- Collaboration: Describe entities involved in message exchange sequences.
- Statechart: Describe the behaviour in response to external stimuli.
- Activity: Describe the behaviour in response to internal processing.
- Component: Describe the organisation of and dependencies among different components.
- Deployment: Describe the deployment scenario of a system.
- Package: Describes the system at a high level in the form of interacting packages.

Added functionality comes from that fact that several software vendors (e.g. Rational, Visio) provide a built-in capability to generate C++, VB, or Java classes based on user drawn UML diagrams.

Currently there is a lack of a standardised process to guide system development based on UML. Within GLOBEMEN, it is expected to use certain diagrams for requirements capture. The diagrams currently being used include use case diagrams to illustrate business processes and sequence diagrams to capture the flows of information and data between different enterprises (project partners).

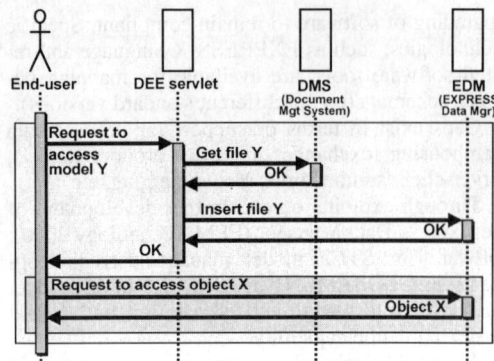

Figure 6. Remote viewing of partial models

5.9 eXtensible Markup Language

The eXtensible Markup Language, XML (W3C 2000), provides a universal format for structuring document content on the web. This facilitates the ability to not only define tags, but to also define the structural relationships between them.

Within GLOBEMEN, XML based product model data access over the web will be provided. Figure 6 illustrates a possible migration option: Servlet retrieves a model, a STEP file, from document management system and transfers it to Express Data Manager which then delivers single objects in XML form based on queries from the end-user/client.

5.10 Summary of the state-of-the-art survey

A summary of the observations made in preceding section is presented in table 1 below.

Table 1. Summary of Observations

Technology	Available functionality	Missing functionality	To be developed in Globemen
Best Practice in industry	Distributed document mgt; non-integrated product model mgt.	Distributed product model mgt between actors in a VE.	Basic infrastructure and pilot use for product model mgt between business partners.
Document mgt, PDM	Distributed document (file) mgt over the web.	Document exchange between heterogeneous systems; product model mgt.	Integration of product model mgt with current document mgt technology.
GroupWare	Workflow mgt, Internet conferencing, ...	Packaged off-the-shelf solutions/products.	Experimenting with VE functionality for internal work in the project..
Java CORBA DCOM	RMI (Remote Method Invocation) for client-server software interaction.	Independence of client and server side software.	None
Knowledge Management	Knowledge retrieval & structuring; process knowledge mgt	"Product" knowledge management.	Knowledge creation environment based on "product" knowledge.
Product Data Technology	STEP and IFC standards. First commercial CAD tools emerging. Conversion toolkits.	Product data warehouses. Exchange of partial models. Server-client architectures.	Product model repository with model merging and partial model extraction capability.
UML	Systems engineering and software development language.	Standard process	To be used for requirements specification.
Virtual Reality	Visualisation of 3D objects.	Interaction with product related data.	User interface to access product model objects and related documents.
XML	Structured documents	Product models	Product model data access over the web.

46

6 FUTURE WORK

Using this state-of-the-art survey, similar ones for sales and services, and inter-enterprise delivery process management in addition to industrial requirements specifications, work in GLOBEMEN will be co-ordinated and integrated into an ICT architecture for VEs. A reference model along-with associated implementation methodology and guidelines will be provided based upon which ICT support for virtual enterprise networks can be set up and operated through seamless inter-enterprise data and information exchange. The provided models and architecture will be generic enough to be usable in one-of-a-kind industries, standard bodies, and IT vendors. Industrial prototypes will be used for result validation and implementation demonstrated through selected applications. Finally, the findings will be synthesised, consolidated, and disseminated to the industry at large, standardisation bodies and ICT vendors.

7 CONCLUSIONS

ICT support to enable seamless data/information exchange and sharing is a key requirement for inter-enterprise collaboration in VEs. Initial observations from this study however indicated that amongst others, the functionality of data/information exchange between heterogeneous systems was lacking.

Future undertakings in GLOBEMEN are to be based on the aforementioned finding. After having identified the available technologies and associated limitations, feedback from industrial partners will be solicited to first identify their requirements and then furthermore functional specifications for distributed engineering environment development. Based on the study presented, available technologies and solutions will be exploited where feasible, and certain missing functionalities required for proper inter-enterprise collaboration developed to fulfil the needs of the industrial partners.

Developments in GLOBEMEN for distributed engineering to support inter-enterprise collaboration and information sharing will include: integration of product model management with current document management technology, addition of VE specific functionality to GroupWare tools, product model repositories with model merging and partial model extraction capabilities, VRML based user interface to access product model objects and related documents, and XML based product model access over the web.

REFERENCES

EPM Technology 2000. Express Data Manager:. http://www.epmtech.jotne.com/products/index.html

GLOBEMEN 2000. Global Engineering and Manufacturing in Enterprise Networks. *Intelligent Manufacturing Systems IMS*, project no 99004, http://globemen.vtt.fi

Hannus, M. 2000. Bookmarks to European and national research projects. http://cic.vtt.fi/links/euproj.html, http://cic.vtt.fi/links/euprojnt.html

Hemiö, T. & Salonen, M. 1999. Virtual Reality: Human Interface to Product Data. *In: Hannus, M., Salonen, M. & Kazi, A.S. (Eds.), Concurrent Engineering in Construction: Challenges for the New Millenium.* Helsinki: CIB Publication 236: 131-138.

Intelligent Manufacturing Systems 2000. http://www.ims.org

Kazi, A.S., Hannus, M. & Charoenngam, C. 1999. An Exploration of Knowledge Management for Construction. *In: Hannus, M., Salonen, M. & Kazi, A.S. (Eds.), Concurrent Engineering in Construction: Challenges for the New Millenium.* Helsinki: CIB Publication 236: 247-256.

Laitinen, J., Ollus, M. & Hannus, M. 2000. Global Engineering and Manufacturing in Enterprise Networks – GLOBEMEN. *ECPPM 2000.*

Microsoft Corporation 2000. DCOM: http://www.microsoft.com/com/

Object Management Group 2000a. CORBA: http://www.omg.org/corba/

Object Management Group 2000b. PDM Enablers (Ver.2). OMG Document: mfg/2000-01-02, 2000

Object Management Group 2000c. UML Resource Page: http://www.omg.org/uml/

PDM Implementor Forum 2000. STEP PDM Schema: http://www.pdm-if.org/pdm_schema/

World Wide Web Consortium 2000. Extensible Markup Language: http://www.w3.org/XML/

A framework for dealing with dynamic buildings

Jutta A. Mülle, Jens Nimis & Peter C. Lockemann
Institute for Programming Structures and Data Organisation (IPD), University of Karlsruhe, Germany

Manfred Hermann, Daniela Schloeßer & Niklaus Kohler
Institute for Industrial Building Production (IfiB), University of Karlsruhe, Germany

ABSTRACT: The design world of architects and engineers is changing. Costs arising during the whole life cycle of a building are being taken into account. Therefore, complex tools to support design decisions are coming up, and an increasing number of experts from a great variety of disciplines will have to cooperate in a more and more interrelated and sophisticated manner. Our hypothesis is, that an integrated framework based on the metaphor of a "dynamic building" as component-based, spatial model will bridge between the information technical representation and the "classical" building planning. All planning, cooperation, usage, and aging processes of the building life cycle will be reflected in such a "virtual" dynamic building, for which we are developing an appropriate framework.

1 INTRODUCTION

The planning of buildings comprises the whole technical life cyle of a building from the planning, to the usage with running, maintenance, and renewal phases, to the restitution of buildings. Because of the semantical complexity of this process, many experts from a great variety of disciplines will have to cooperate in a more and more interrelated and sophisticated manner.

We will meet these challenges in an interdisciplinary way in collaboration of architects and computer scientists. Our hypothesis is, that an integrated framework based on the metaphor of a "dynamic building" as component-based, spatial model will bridge between the information technical representation and the "classical" building planning. All planning, cooperation, usage, and aging processes of the building life cycle will be reflected in such a "virtual" dynamic building, that will provide for an intuitive access to the complex and interrelated information for all participants of the planning process.

In the following, we will first describe our basic approach to model and transfer a dynamic building with an integrated cooperative framework into practice, and then propose a system architecture to technically realize our vision.

2 REQUIREMENTS

Our approach aims to realize the metaphor of a dynamic building in order to achieve the following main characteristics:

- Integration of all design data along the whole life cycle of a building, i.e. an integrated product model. Our approach proposes a kernel model, that is based on a so-called n-dimensional design space with the three geometric dimensions and additional building-design specific dimensions [1].
- Allowing for comfortable integration of design tools on demand, that will be based on a dynamic view handling mechanism.
- Support of cooperation, i.e. a flexible cooperation model, that will be based on the workflows of the building life cycle [2]. The cooperation management has to be adaptable and flexible and should be integrated into the working environment.
- Management of Constraints, in order to handle vague and incomplete information, as it is typical for early planning steps. The grade of consistency and of completeness will become higher during the goal-oriented design process. Detecting consistency violation is also a basic mechanism for cooperation support.

3 BASIC APPROACH

To meet the requirements we propose a framework for an integrated planning platform for dynamic buildings. The main idea behind our platform is the metaphor of a dynamic building, i.e. a component-based, spatial model supporting the cooperation processes not only by the spatial dimensions, but additionally, by various other aspects of the building life cycle. These dimensions form a kernel to integrate the product and process model in a cooperative planning environment. The technical realization transforms the metaphor of dynamic building to basic elements, so-called dynamic components. They support dynamics in the evolvement of the design:

- along time, i.e. change of the components,
- along space, i.e. working in a distributed

environment, and
- along the life-cycle, i.e. change of the components during the life cycle time of the building.

The dynamic components will provide for a very flexible modeling as common basis of the design objects in the whole heterogeneous and distributed framework handling the life-cycle of buildings. In our work, we concentrate on the planning phase, but also from that point of view, we have to handle the later phases of the building life cycle. In future work, we will also analyse how our platform is suitable to support these phases.

Finally, the system architecture has strongly to support distributed and cooperative work using an integrating information base.

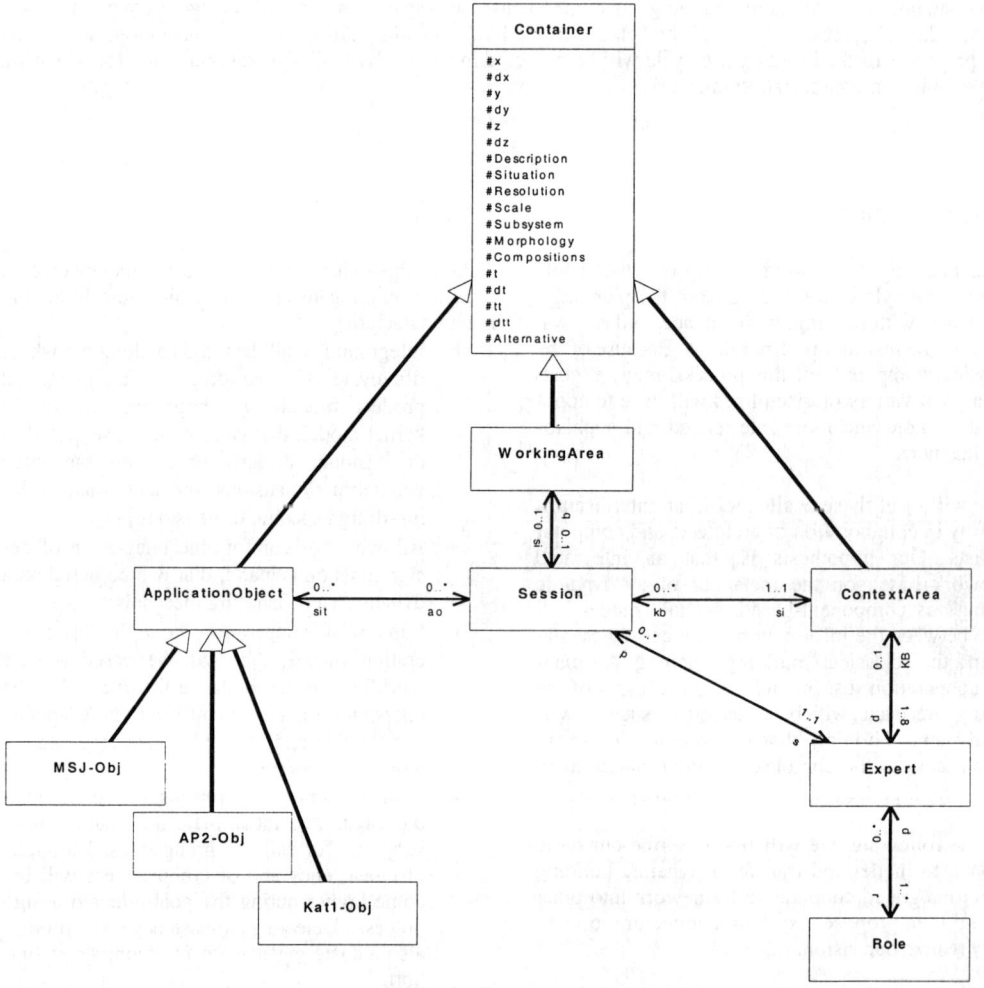

Figure 1.

50

4 DATA MODELLING

Integration is especially important in the planning environment, because there are a large amount of experts involved in a large project working with a great variety of application tools, highly specialized to the task they are supposed to work on. Each tool works on its own specialized data model so that integration means in a first step integrating these different models by the integration platform in a suitable an easy way.

Analyzing the characteristics of these different models showed in a first term the close relationship of the objects to the spatial area in the geometrical representation of the design and additional to other dimensions of an multi-dimensional design area. As illustrated in figure 1 we isolated these area-related properties from the objects to a so-called Container class, which represents an area in our n-dimensional

ContextArea, which supports a cooperation model and the workflow of the design process. It will be further described in the next section.

Each time an expert wants to work with our planning platform, he or she starts by specifiying the area in the design space, where he or she wants to work on, the so-called WorkingArea. In the following the expert will get connected either by an online connection to the system, restricted to the specified area or by a copy of all objects, overlapping the sepecified area. In the latter case a check-in/check-out mechanism will track the changes to these objects.

5 SUPPORT OF COOPERATION

Building design is an inherently cooperative task, where a lot of experts from different organisations and from various disciplines are involved, see also

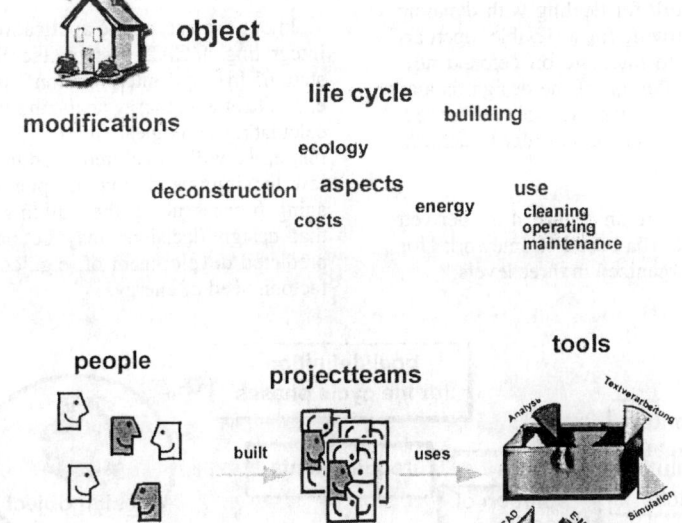

Figure 2.

design space.

This Container class is inherited by all ApplicationObjects, independent of the concrete application the object belongs to. This way in our integration kernel each object has a kind of proxy object, that makes it visible to all other applications via our integration plattform.

Various interaction schemes can now be implemented by the detection of overlappings, the so-called collisions of different objects, even from different applications in our platform.

The common properties modelled by the Container class are also inherited by the WorkingArea and the

Figure 2. Additionally to the above presented integration of data, there is the need for a more organizational support of cooperation.

The characteristics of a cooperation model in this context are
- Team-based cooperation.
- Mostly following an integrational as well as quality enhancing cooperation form.
- Supporting a cyclic and iterative process.
- Having a goal-oriented planning process.
- At most at the beginning, havingplanning process that is highly dynamic, handling vague and incomplete information, that

will basically affect on the cooperation model.

- Need to the management of objectives during the whole planning process.

Figure 3 sketches the relationships between the components of the planning platform in respect to cooperation support. The basis forms the common information base, which will consist of the design data and meta-information about the cooperation itself. Each participant will have a context and a dynamic information profile, that have to be handled explicitly in the framework. Conflicts will be handled by collisions of context areas and working areas of different experts. Our cooperation models also allows for defining rules how to handle conflicts between experts.

6 SYSTEM ARCHITECTURE

Realizing a framework for dealing with dynamic buildings requires to provide for a flexible, open architecture, that allows to integrate on demand new expert tools, for a distribution of the design data as well as of the clients, on which the tools are settled, and for cooperation mechanisms in order to achieve simultaneous engineering.

The system architecture in figure 4 is derived from an integrated Corba-based framework for CAD/CAM [3] and is organized in three levels:

- The storage level, where we have distributed repositories, potentially following different data models, e.g., object-oriented, relational formats. These repositories store the design data, in form of attributes, interfaces and methods for the tools at the application level.
- The middleware level, where we get the technical integration using the CORBA standard, and the semantical integration with CORBA services and special enhancements, i.e. the dynamic components and the cooperation models.
- The application level, where we will couple various design tools. Some tools will be embedded into a so-called integrated desktop that delivers a framework-specfic interface and allows for fully integrating the tool, other tools will keep their special interface, like the CAD tool, so that we get a looser coupling of these kind of tools.

In our project, on the application level, we are integrating a CAD system, the Bentley Micro-stationJ [4], and interpretational programs (IPs), e.g., a heat and energy analyzing tool, and a tool, calculating ecological figures [5]. In the second phase, we will add a prediction tool, that will allow to simulate during the planning phase the aging process along the building life cycle, so that design decisions may be based upon the predicted development of, e.g., costs, ecological factors, need of energy.

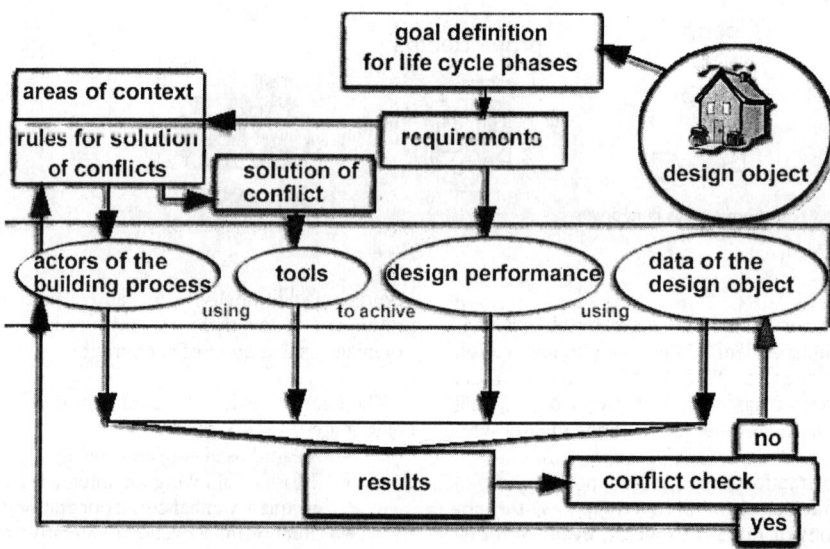

Figure 3.

52

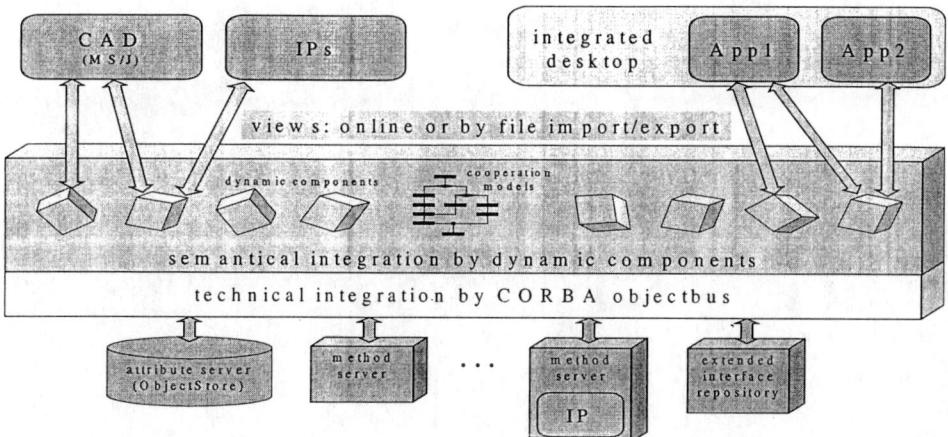

Figure 4

7 SUMMARY AND FUTURE WORK

In this paper we presented a framework for an integrated planning platform for buildings. It is based upon the metaphor of a dynamic building, a component-based, spatial model supporting the cooperation processes not only by the spatial dimensions, but additionally, by various other aspects of the building life cycle. These dimensions form a kernel for the integration of the product and process model in a cooperative planning environment. The basic elements are identified as dynamic components. The system architecture allows to realize distributed and cooperative work using an integrated information base.

Futher work will focus on the validation and refinement of the rough protoype system, especially concerning the cooperation model, the integration of the product model, and the view handling component. On the application level, i.e., the planning tools and interfaces for the expert users, we will concentrate on investigating dynamical simulation models during the planning phase to be able to make design decisions according to the potential development of the building during the whole life cycle, as, e.g., predicting the costs coming up during the aging process by renewal or maintenance requirements.

References:

[1]   P. Lockemann, J. Mülle, R. Sturm, V. Hovestadt: Modeling and integrating design data from experts in a CAAD environment. Proc. European Conf. On Product and Process Modeling in the Building Industry, Berlin 1994

[2]   U. Forgber, C. Müller: A Planning Process Model for Computer Supported Cooperative Work in Building Construction. IKM Weimar, 1997

[3]   G. Hillebrand, P. Krakowski, P. Lockemann, D. Posselt: Integration-Based Cooperation in Concurrent Engineering. Proc. of the Second Intl. Enterprise Distributed Object Computing Workshop, IEEE, San Diego, Nov. 1998, pp. 344-355

[4]   Bentley MicrostationJ, http://www.bentley.com

[5]   M. Hermann, N. Kohler, H. Koenig, Th. Luetzkendorf: CAAD System With Integrated Quantity Surveying, Energy Calculation and LCA. In: Green Building Challenge '98, Intl. Conf. On Performance Assessment of Buildings, Vancouver, Canada; October 1998

ISBA, a technical information system for the construction, operation, service and ageing management of nuclear power plants

H. Schillberg
Siemens Nuclear Plant GmbH (SNP), Erlangen, Germany

H. Drozella
INIT GmbH, Bochum, Germany

ABSTRACT: ISBA (i.e. an acronym based on the German, meaning "Civil engineering and plant layout information system") is a tool that was initially developed for the application in the civil engineering field at the planing of nuclear power plants. Today it has reached a state where it can provide information for the entire civil engineering and plant layout field, not only at the design stage, but throughout the entire lifetime of the plant. That covers the construction phase, where the basic information is created, the operation and service as well as the ageing management.

The system is built on an object oriented database. Information retrieval is gained via several viewers: a 3-D model or intelligent 2-D plan in CAD, or in a VR program. The latest viewers available today are internet browsers, which enable users worldwide to practice concurrent engineering by utilizing one central data pool.

1 INTRODUCTION

Civil engineering work and plant layout for a nuclear power plant demands a big amount of information which is produced by numerous engineers involved. It is a must to hold redundant data as small as possible in order to avoid e.g. the erroneous use of not up to date planning versions in the design process. In addition the communication of separate software systems through interfaces implies also the well known risk of information loss. For that reason the information system ISBA was developed as one central information pool. At the beginning it was the aim to provide in one system information related only to civil engineering. Today all information of interest may be managed concerning not only the planning and construction phase but also operation and service aspects as well as ageing phenomena of the plant. This information can belong to disciplines as e.g. structural design, routings and component arrangement with its corresponding execution and licensing documents as well as tracking of planning versions and planning chronology during the design and erection phase in a digital archive. Information concerning facility management and the documentation of ageing phenomena became a vital component of the information system.

The heart of the system is the global product data model in an object oriented database system combining the information of all areas mentioned above.

The building model is one essential element in ISBA. It is the user interface to reach the desired information. At the outset AutoCad was used as a "viewer", where the element of interest was picked with the mouse and the desired information was shown via a dialog box. Later on the viewers were supplemented by 3D Studio Viz of Autodesk, where the 3-D model could be viewed in virtual reality quality.

One big progress in the development of ISBA was the step into the world of internet applications. Today only an internet browser is needed to communicate and work with the system. With the additional shareware program "Cosmo Player" this communication may be proceeded with the 3D model.

So, in principle, engineering processes with customers and companies that are involved in a specific project may be communicated and handled worldwide via Internet using only a network browser.

One important aspect has to be respected and solved for this case, i.e. the security of data transfer and the access to company confidential intranets.

2 THE INITIAL PROJECT

ISBA was initially developed for the European Pressurized Water Reactor the so called EPR which is a common design work of NPI (Nuclear Power International), Siemens and Framatome in cooperation with EDF and the major German Utilities for a new generation of nuclear power plants.

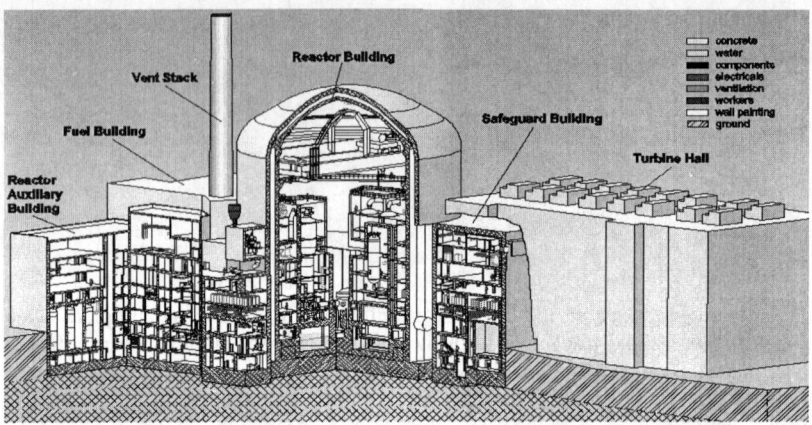

Figure 1. View at the EPR

This nuclear power plant shall be built in France in the next years.

It was intended to maintain information which is of global interest for civil engineering during project planning of a nuclear power plant and for its documentation following the project completion.

So in the basic stage it contained the following information:

- Geometric modeling of all components
- Graphical representation of all loads and their combinations
- Static system idealization
- Construction materials used
- Determination of construction costs as well as invitation to bid, contract award and invoicing
- Construction sequences for supporting scheduling as well as optimization of construction process and
- Object oriented storage of all planning documents and information.

3 FIELD OF APPLICATION

Primarily ISBA is an information and facility management system. The other important feature is the planning and construction tool which is used to create the building model. This model can be built during the planning phase for new plants as well as for existing plants on the basis of the 'old' documents.

The possibilities for the use of ISBA in a project are various and not limited to civil engineering aspects. Therefore its application was extended to the entire field of civil engineering and plant layout.

3.1 Planning and construction phase

ISBA is utilized to plan the civil structures. In this phase it is used as a CAD system which creates the 3D model of the building. In our historically grown practice this model normally is transferred to another CAD system where it represents a referenced building and where all equipment is planned.

This is the part of the construction tool in ISBA, but by creating the 3D building model an essential part of the information system is generated. The 3D building model is part of the global product data model in the object oriented database system.

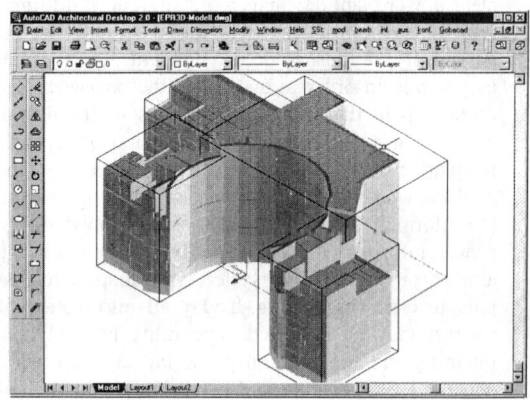

Figure 2. View at one part of the EPR building model

In order to complete also the information system with equipment which is planned in an other CAD system these elements have to be imported in ISBA. In order to produce not only graphical elements in the model a so called object manager was developed that gives such imported objects the same characteristics as objects created in ISBA themselves.

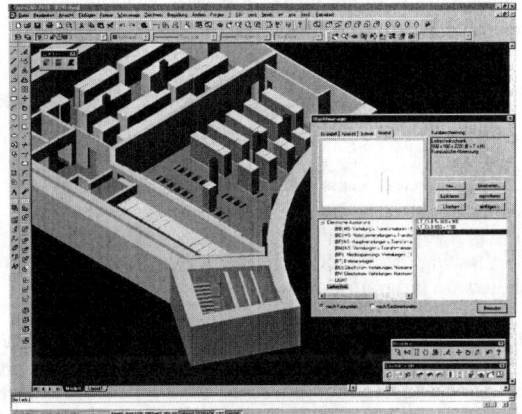

Figure 3.   Screen shot of a dialog with the 2D drawing

This is the graphical part of the system. System data are complemented when they are available.

On this basis drawings e.g. layout and load drawings are generated. These drawing do not contain only graphical information but also other data from ISBA. That can be loads, material quantities, operational data of systems and so on. As long as these drawings are still on the screen, they are absolutely intelligent and have direct access to the database.

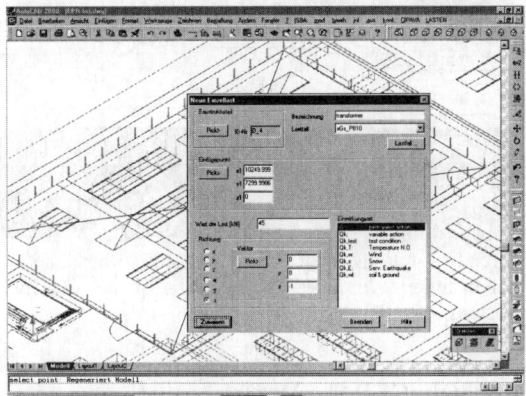

Figure 4   Definition of loads

Documents that are produced during the planning or the construction phase can be linked to that parts of the model to which they belong to. Also documents from the construction site such as protocols and quality reports can be stored in the system. All this information forms a digital archive which is available during the entire lifetime of the plant.

Figure 5   Planning and recording of switchgear cubicles

3.2 *Operation and service of a plant*

During the operation of a plant or preparation of service activities and revisions it is a big assistance for people who do not permanently work with an information system to retrieve the information wanted with help of a graphical model of the plant. They only have to be taken to that part of the plant where the information is desired from, click the respective object and obtain it via dialog boxes.

Figure 6   Pipelines, air ducts and cable trays with database reference

Also the management of areas is easily to be organized with ISBA. For revisions the placement of inspection tools of the various revision teams can be optimized. E.g.: in case of a research reactor the test equipment for tests that are performed in parallel can be optimized.

Figure 7.   Area management

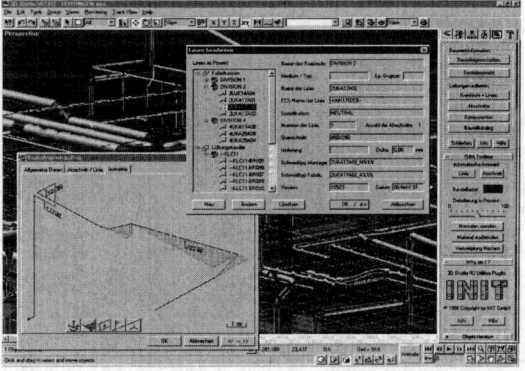

3.3 *Ageing management*

In the United States a clear tendency is to be observed to extend the lifetime of existing nuclear power plants up to 60 years of operation. This tendency will surely appear also in other countries operating nuclear power plants. This requires a strict documentation of the quality status of the plant. That means that the archive of the plant has to maintain its actuality and completeness over at least 3 to 4 generations of engineers and administrators. It is

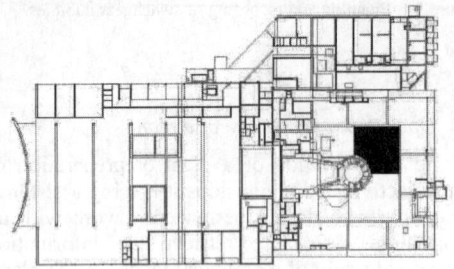

easily to be imagined that this is a difficult task when conventional archiving methods with original paper documents are the only way to fulfill this job.

Figure 8. Documents linked with the corresponding building object.

Information systems like ISBA provide an excellent possibility to create a digital archive. All documents are virtually linked to the objects they belong to. When some information is needed only the object or part of the building of interest has to be clicked with the cursor and either the information is taken directly from the screen or retrieved as paper document.

This archive also records all modifications of the plant with the corresponding states and documents.

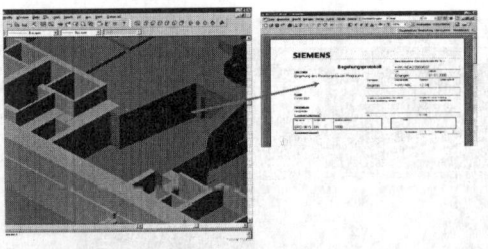

3.4 Internet communication

Nowadays internet communication gains more and more importance. A local server solution enables teams to work concurrently at one project. This approach is normally limited to one location. When more institutions or parties are involved in one project, then it is advantageous to use the inter-, intra- or extranet.

For this purpose a module was developed that enables the access to the central data pool with help of only an internet browser. The client can communicate with the system according to predefined rights. Either he can only read or read and write. It is also absolutely necessary to structure the system in such

a way that the rights for the access to specific information can be limited or denied.

The graphic display is either an intelligent 2D graphic, or a 3D model. When the 2D graphic is chosen then only an internet browser is needed.

Figure 9. Internet access with a browser

Viewing the 3D model in the browser requires the shareware program Cosmoplayer.

Figure 10. Internet access with the Cosmoplayer

Today in principle the communication and handling of engineering processes of all partners involved in a project is worldwide possible via Internet using only an internet browser. In this regard one significant aspect becomes very high importance. This is the security of data transfer and the access to company confidential intranets. Although there are

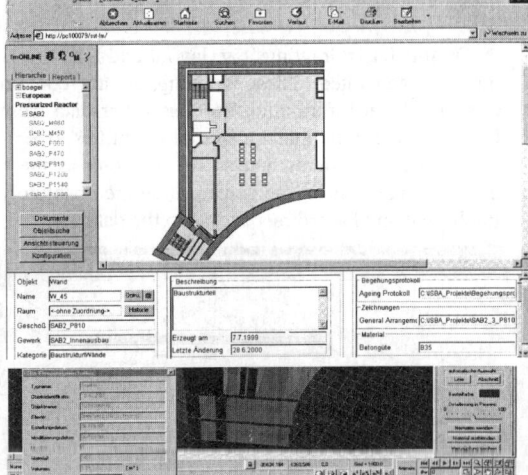

already solutions available, it has still to be done a lot in order to improve the handling of the security systems.

4 PRODUCT DATA MODEL

In the following chapters some information about the theoretical background is given.

4.1 Object-oriented product data modeling

Integrated information processing requires constant availability and exchangeability of all planning data. This affects not only the disciplines of statics, design and construction but also licensing and design modification processes, tracking of planning versions and planning chronology, especially for planning of large structures. A prerequisite for this is the implementation of a global product data model combining the information in the areas listed above and extending to the corresponding requisite degree of detail. The heterogeneous nature of the information requires a correspondingly flexible modeling technology, which is available in object-oriented modeling.

4.2 *Modeling of Structural Components*

Classification of the structural components is based on the corresponding function in relation to supporting behavior of the structure (walls, ceilings, beams, columns) and accounts for the desired information access based on common structural elements (buildings, levels, rooms, etc.).
The product data model utilizes the inheritance feature in the hierarchical structure. Information and component behaviors are inherited by the subordinate classes from the classes above. Proceeding deeper into the hierarchy involves more precise definition of the special class characteristics.

5 OBJECT-ORIENTED FORMULATION OF PROBLEMS SPECIFIC TO CONSTRUCTION, FACILITY MANAGEMENT

5.1 *The room problem*

In power plant planning, the room is the smallest unit with which geometric locations are indicated. For example, the location of a component is indicated with a room number. However, dead weight and loads under operating and faulted conditions are transmitted through foundations, hangers and supports to the surrounding building structural components. Ultimately the building structural components such as walls and roofs are required for dimensioning. Specification solely of the room number is insufficient for civil engineering processing. It is therefore necessary to establish the relationship between rooms and the surrounding building structural components.

The resulting problem is the classic problem of *many-to-many* relationships. This is demonstrated with a simple example in Fig. 4. Consider room R1, surrounded by the four walls W1, W2, W5 and W6. If wall W2 is considered, it can be seen that it separates two rooms from each other. Or, more generally, a room is surrounded by *many* building structural components, and a structural component separates *many* rooms from each other.

Many-to-many relationships cannot be realized without redundancy in database systems utilizing relational design models. Object-oriented database systems (OODBMS) enable the establishment of re-dundancy-free relationships with *associations* between objects. An association is a persistent link between two objects.

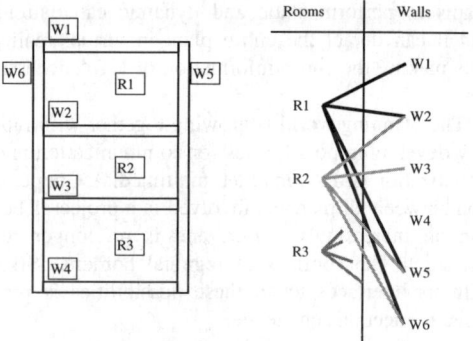

Figure 11.   The room-problem: representation of the many-to-many relation.

5.2 *The Room-Wall Problem*

Furnishing of a room is determined by its use and function in the plant. This results, for example, in different surfaces for a wall (Fig. 5). Modeling the product data by associations with attributes stores information in the object-oriented database which, although they are inserted for design reasons, already form the basis for determining quantities and costs as well as facilitating visualization.

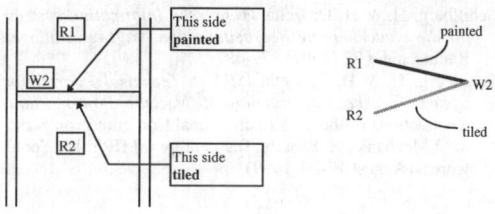

Figure 12. . The room wall problem: attributes of associations.

6 SUMMARY AND OUTLOOK

ISBA is a technical information system based on object-oriented technologies. It controls synchronous application of CAD and the database system based on an object-oriented product data model.

Initially ISBA was developed to provide an information system that supports the civil engineer to process the vast amount of load data which have to be considered in nuclear engineering. Today it can provide information for the whole nuclear power plant and is an ideal tool for facility management tasks. It is a tool for the determination of construction costs, it communicates with finite element programs to perform static and dynamic calculations and it can depict the entire plant in virtual reality thus making the entire information of ISBA accessible.

The planning world is growing together with rapidly developing possibilities for communication and visualization (ISDN, Internet, multimedia). Cooperation between all planners involved in a project is becoming increasingly closer, and is no longer restricted to the confines of regional borders. ISBA with its interfaces to all these possibilities is prepared to meet this challenge.

REFERENCES

Weber, B. & H. Drozella *Objektorientierung bei der automatischen Nachweisführung nach DIN 18800 (11.90).* Bauingenieur 68 (1993) 411-417

Stroustrup, B. *Die C++ Programmiersprache.* Addison Wesley (1993)

Richter, H *Symbiose von AVA und CAD - das digitale Bauwerksmodell.* AEC Report (1-95) 58 - 61

Objectivity *Objectivity/DB Reference Manual Rel. 3.81.* Objectivity Inc. Menlo Park CA (1996)

Schillberg, H. & H. Drozella *Objektorientierte Produktmodellierung für die Tragwerksplanung im Kraftwerksbau.* Darmstädter Massivbau-Seminar 'Multimedia im Bauwesen' Band 16. Freunde des Institutes für Massivbau der TH Darmstadt e.V.(1996)

Schillberg, H. & H. Drozella *Technische Informationssysteme für die bautechnische Bearbeitung von Kraftwerksanlagen* Bauingenieur 72 (1997) 479-487

Schillberg, H. & H. Drozella *ISBA, a Technical Information System for the Construction of Nuclear Power Plants* Transactions of the 15th International Conference on Structural Mechanics in Reactor Technology (SMIRT-15) Seoul, Korea, (August 15-20. 1999)

Product and Process Modelling in Building and Construction, Gonçalves, Steiger-Garção & Scherer (eds)
© *2000 Balkema, Rotterdam, ISBN 90 5809 179 1*

Distributed structural design using CORBA technology

M. Baitsch, K. Lehner & D. Hartmann
Department of Civil Engineering, Ruhr-University of Bochum, Germany

ABSTRACT: Structural optimization is a complex engineering task combining mathematical optimization methods with numerically intensive structural analysis to allow users to design highly efficient structural systems. In a distributed structural optimization application, separate server components can be created to provide basic services for mathematical optimization and structural analysis. These services can then be used by user-friendly client components that interact with design engineers. The Common Object Request Broker Architecture (CORBA) can be employed to implement distributed applications. In more detail, so-called process objects have been created that encapsulate the individual computational steps carried out in the process of design optimization in a completely uniform manner. In particular, it is shown how a network of various communicating process objects can be set up to implement the structural optimization process. This approach is exemplified by an example from the field of structural design, where NURBS are used to optimize a curved structure.

1 INTRODUCTION

Structural design in the classic sense is basically an iterative procedure requiring a wide variety of comprehensive engineering activities: An initial design is selected, evaluated and, based upon the insight gained during evaluation, continually improved until a satisfactory result is obtained. This affinity to strategies used for solving mathematical optimization problems provides a rationale for the formulation of structural design in the context of an an optimization problem. If it is possible to (1) describe the variable aspects of the product by a set of decision variables, (2) to define a design goal as a function of the decision variables and (3) to specify a set of constraints to eliminate unviable designs, then the time consuming and error prone process of manual iterative design can be greatly facilitated by employing formal optimization methods (Ramm et al. 1994). In this context, optimization is considered as an amelioration of the initial design rather than the attempt to find the best possible solution in a purely mathematical sense.

More formally, an engineering design problem can be transformed into a mathematical optimization problem by suitably defining an n-dimensional vector $\mathbf{x}$ of design variables, an objective function $f(\mathbf{x})$ and a set of m constraints $g_i(\mathbf{x}), i = 1...m$, that each map the design vector to a real number. Also, side constraints can be imposed as upper bounds u_j and lower bounds $l_j, j = 1...n$, on each component of the design vector. In turn, the mathematical problem can be solved by

determining the minimum value of $f(\mathbf{x})$ with the help of an appropriate optimization method, subject to the above constraints, i.e.:

$$\min_{\mathbf{x}}\{f(\mathbf{x}) \mid g_i(\mathbf{x}) \leq 0, \, l_j \leq x_j \leq u_j\}$$

$$i = 1...m, j = 1...n. \qquad (1)$$

Most mathematical optimization methods are well understood and readily available – the success of design optimization therefore usually hinges on the quality and accuracy of the product model, i.e. how the objective criterion and constraints adequately reflect the engineering properties of the underlying structural system.

2 DISTRIBUTED DESIGN OPTIMIZATION

In order to reduce the ensuing complexity associated with "real world" design optimization problems, it has proven to be a good idea to create individual modules each representing a particular aspect of the optimization process (Michalena, Scheffer, Fellini, and Papalambros 1999).

The task of solving the mathematical optimization problem involves the execution of a numerically ori-

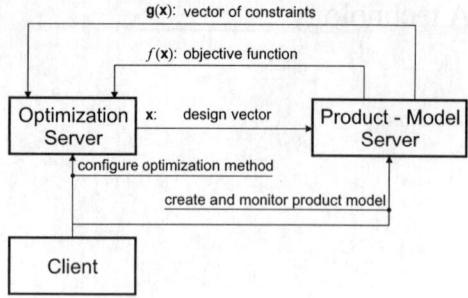

g(x): vector of constraints

f(x): objective function

| Optimization Server | **x:** design vector | Product - Model Server |

configure optimization method

create and monitor product model

Client

Figure 1: Distributed design optimization

ented mathematical algorithm and is completely independent of underlying engineering problem. The evaluation of the objective function and the constraints, on the other hand, are determined by the specifications of the engineering problem alone and are independent of the optimization method used. In a distributed, client-server based structural optimization application, it is therefore possible to create a separate server for general mathematical optimization (optimization server) that can interact with any problem-specific product model server, provide the servers adhere to the same interface specification. The optimization server and a product model server must be controlled by one or more client component, which represent the interface by which the user can employ the distributed system (see Fig. 1). The user-friendly client components basically allow users to input their design problems, monitor the optimization process and visualize final and intermediate results.

Following this decomposition philosophy, a CORBA-based software platform for design optimization has been established by the authors which will be presented in this paper with the focus on the product model aspect.

CORBA (Common Object Request Broker Architecture, (The Object Management Group 1999)) is an open, vendor-independent architecture and infrastructure that enables computer applications to work together over networks. Using the standard protocol IIOP (Internet Inter-ORB Protocol), a CORBA-based program can interoperate with another CORBA-based program for the same or another vendor and on the same or different computer hardware. Also, the operating system, programming language or network need not be identical for all components. CORBA ist based on the concept of object oriented modelling, i.e. CORBA applications are composed of individual units (instances) that interact remotely using remote procedure calls.

Communication between objects in CORBA is based on a so-called object request broker (ORB), a middleware technology that manages communication and data exchange on a network-wide, hardware and

operating system independent basis. ORBs promote the interoperability of distributed object systems because they enable users to build systems by piecing together individual objects. The implementation details of the ORBs are generally not important to the developers. In fact, the developers of a distributed CORBA-based software system need only concern themselves with application specific interface definitions between objects. The Interface Definition Language (IDL) is a technology-independent syntax for describing object encapsulations. It defines the interfaces in a CORBA application to describe all relevant services that CORBA clients can use and CORBA servers must implement. The IDL is only a definition language, i.e. it only describes what objects must be able to do, not how they do it. Therefore, to create a running application, the generic IDL specification must be mapped to the programming langue used to implement to client or server component.

An example of a CORBA-based application that implements a universal optimization service along with examples in structrual optimization is given in (Lehner, Baitsch, and Hartmann 1999).

3 PRODUCT MODEL SERVER FOR DESIGN EVALUATION

As depicted above, the most critical task in engineering optimization is to establish a model which is used to predict and evaluate the properties of the future product – the so-called product model. This model can include various submodels that address specific properties, for example:

- costs of assembly,
- structural safety,
- maintanance costs (heating, ventilation).

Obviously, these aspects can not be treated independently. In fact, for each aspect a numerical model must be established and, if necessary, somehow interlinked with the others. The treatment of each submodel requires basically three activities:

1. Prepare the input for the specific submodel. The input may partially consist of the output of another submodel.

2. Evaluate the submodel.

3. Interpret the results. This includes formulation of the objective and constraints for the design problem as neccessary.

A simple example from the field of structural optimization illustrates these steps (see Fig. 2. The design problem is to minimize the weight of a supporter, whereby the usual structural requirements (with re-

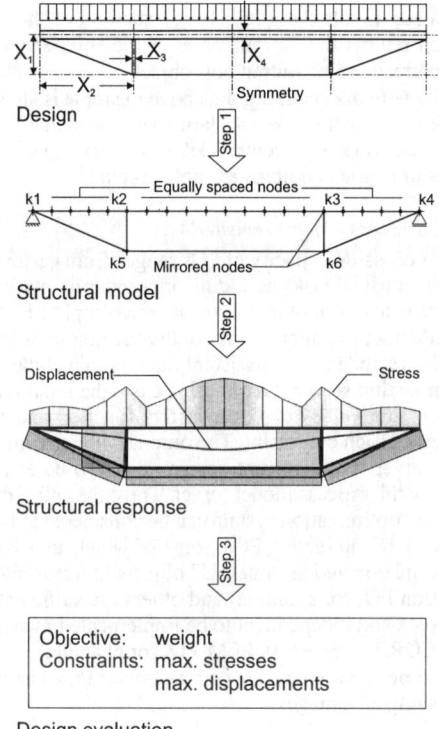

Design

Step 1

Structural model

Step 2

Displacement — Stress

Structural response

Step 3

Objective: weight
Constraints: max. stresses
              max. displacements

Design evaluation

Figure 2: Structural optimization of a simple supporter

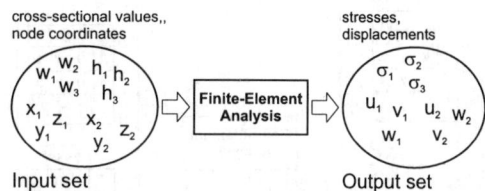

Figure 3: Mapping of numeric input on a set of numeric output

it is realized that the structural analysis transforms, in a well defined sense, a set of structural variables into a structural response represented by another set of variables.

Finally, in **step 3**, the results of the structural analysis are compared to maximum quantities (allowable stress, for example) according to the local design code. Results of this comparison together form the vector of constraints. Additionally, an objective function has to be identified.

This simple example can be extended in many different ways. Functions, or transformations, can be set up to perform all standard geometric operations. More complex POs can be created by including such powerful tools such as Non Uniform Rational B-spline (NURBS) that allow the design of complex, free form curves or surfaces with only a few key parameters.

3.1   *Process objects*

Obviously, all the operations in the above example basically perform the same task: A transformation of one multi-dimensional space into another (see Fig. 3). This insight holds even for the most complex computational operations carried out during structural design. The implementation of these mappings led to the development of so-called process objects (POs), which form the building blocks of the optimization problem description.

Since each submodel or simple transformation has its own input and output with a specific meaning, a semantic must be provided which allows the user to identify the desired quantities correctly. A simple way to do this would be to put each input and output in a separate array and define the semantic in the documentation. In the light of user-friendly interfaces, this is a poor decision.

More suitable is to use the semantic of input and output values to define additional structures, such as point vectors, hash vectors, matrices or vectors of scalars, to simplify the interfaces to POs. These structures are called ports and are accessible by self-explanatory strings (see Fig. 4). For example, a Finite Element PO has an input port of type point vector named "Points", representing the coordinates of all nodes and an output port of type matrix named "Displ", which contains the computed nodal displace-

gards to stresses and displacements) must be fulfilled. The design is described by two variables defining the position of the vertical members and the height of the system, respectively, and two cross-sectional variables. Together, these variables define the design vector. In this case a finite-element analysis tool is used to predict the behaviour of the structure under loading. In order to precisely evaluate stesses at intermediate positions of the top beam, some additional nodes are introduced. Therefore, the structural model is described by the nodal coordinates and the quantities defining the cross sections.

Dealing with **step 1** of the procedure outlined above, a key observation is that the number of potentially independent structural variables is larger than the number of design variables. A mapping from the optimization variables to the structural variables must be defined. This can be done by mirroring the two key-points $(k2, k5)$ on $(k3, k6)$ and by linearly interpolating the intermediate points between $(k1, k2)$, $(k2, k3)$ and $(k3, k4)$.

Step 2 is straightforward, it requires the solution of the governing mechanical equations. In some cases it will be necessary to employ more sophisticated analysis techniques like geometrically nonlinear analysis or the inclusion of time varying behaviour. In this step

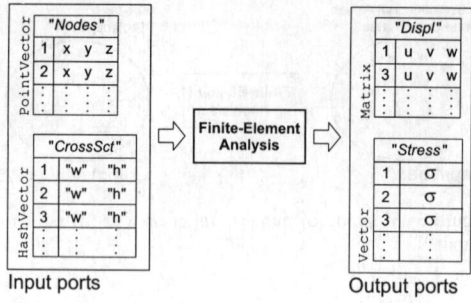

Input ports   Output ports

Figure 4: Identify input/output components by strings

ments. The access of the x-coordinate of the 5th node, for example, can therefore be easily handled by the input port. Methods to access ports simply expect a string to identify the port and return a reference to the target port, realized as an instance of a port object stored in the PO. Also, since each PO can have a variety of ports, it is possible to query the types of ports available at runtime.

3.2 Connecting process objects

By the introduction of the formalized interface for the POs, it becomes possible to treat arbitrary submodels involved in design evaluation in a unique manner. The possibility to connect POs to create a network representing all interesting properties of the structure is the next key issue.

In detail, it should be possible for any component of an output port object (as described in Sec. 3.1) to serve as input for any component in an input port object. This is done by employing so-called Connection-Objects (CO). Each output port object has a method which returns a specific CO for the requested component. The CO itself holds a reference to the output port object and, in addition, information about which component it refers to. The input port object, on the other hand, has a method by which a client can set a CO previously obtained from an output port. Internally, whenever the input port object needs its values

updated, it calls the getValue() method of the CO which returns the current value of the component in the corresponding output port object.

The network realizing the above example is shown in Fig. 5. For the sake of clarity, only a coarse view of the network is presented, details concerning the indices of connected items are not depicted.

3.3 Process objects revisited

The processing capacity of POs ranges from performing such trivial tasks as adding or comparing two real numbers to one another to executing a complete FEM-based structural analysis. Also, the the notion of POs can be extended in a consistent manner to include the optimization server itself. In this case, the input consists of the values of the objective function and the values of each constraint. The output of the optimization server is the current design vector to be evaluated by the process model server. Thus, the entire distributed optimization system can be considered to be a network of interacting POs, some of which are trivial and implemented as "internal" objects (a linear interpolation PO, for example) and others are sufficiently complex and independent to be implemented as separate CORBA servers (a FEM-PO, for example).

The network character of interlinked POs has the following advantages:

- POs can uniformly represent both elementary as well as complex tasks in the optimization process.

- POs allow comprehensive network connectivity.

- POs implemented as CORBA-based components can, in principle, be run on and accessed from any computer system on the Internet.

- POs can provide rapid implementation due to their inherently specific functionality.

In addition, existing (commercial) software can be wrapped into a PO.

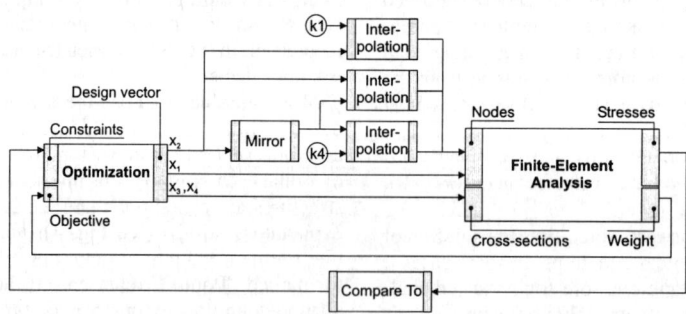

Figure 5: A sample network of POs

Table 1: Material properties	
Property	Value
Young's Modulus	$2.1 \cdot 10^5$ N mm^{-2}
Density	7850 kg m^{-3}
Yield stress	360 N mm^{-2}

4 APPLICATION EXAMPLE

The example in this section is based on a design problem which can be stated as follows: Minimize the weight of a curved girder where each of the 119 structural members has a circular-tube cross section (see Fig. 6) and is subject to stress and buckling constraints. Material properties are listed in Table 1. Since the arch is compression-loaded and expected to undergo large deformations the finite element analysis is fully geometrical nonlinear and uses an incremental-iterative solution procedure (Zienkiewicz 1991). Two different loadcases (one symmetric and one asymmetric) are considered.

Rather than using one design variable for each of the 61 nodes, the number of design variables can be significantly reduced by employing so-called NURBS-curves to describe the girder geometry. One NURBS-curve positions the center-line and a second width-curve is used to describe the distance of the top and the bottom beam from the center line. More precisely, the width-curve is used to scale the normal vectors of the center line which are then added to or substracted from the corresponding evaluated points on the center line. The portion of the PO-network that prepares the input for the FE-model is shown in Fig. 7.

The optimization problem consists of 16 design variables – 13 for the geometry definition and three for the cross-sectional values – as well as 476 constraint functions resulting from combination of stress and buckling restrictions with two load cases for each truss member. In addition, the total height of the system is limited to 15 m and the width of the arch is restricted to 2.1 m by imposing appropiate bounds on the design variables.

The optimization was carried out using a combination of the SQP-method (Schittkowski 1985) and evolutionary strategies (Grill 1997). The "optimum" design together with the structural response for the asymmetric load case is shown in Fig. 8. Detailed values for the design variables are given in Table 2.

5 CONCLUSIONS

Complex engineering processes such as structural optimization can be decomposed into a series of simpler, well defined computational tasks, where each task can be characterized by the way it transforms its numerical input to numerical output. Process objects (POs) have proven to be an excellent paradigm to encapsulate individual computational tasks, employing input ports, output ports and connection objects. Because of the general nature of PO interfaces, they can easily be set up to create an entire network of communicating POs. Thus, POs can be considered as reusable "building blocks" to construct solutions for numerically oriented engineering problems. With the help of suitable user-frendly clients, users can interactively create their own problem-specific PO networks.

By suitably defining the PO interfaces as CORBA specific IDLs, POs or a set of POs can be implemented as separate CORBA servers located anywhere on the Internet. Complex POs such as servers for mathematical optimization or servers to carry out structural analysis contain a large amount of technical expertise and knowledge requiring experts for proper maintanance. Thus, these types of POs can be implements as CORBA servers residing under the control of the software vendor and offering the software services to customers. Users of such POs or software services therefore need not concern themselves with

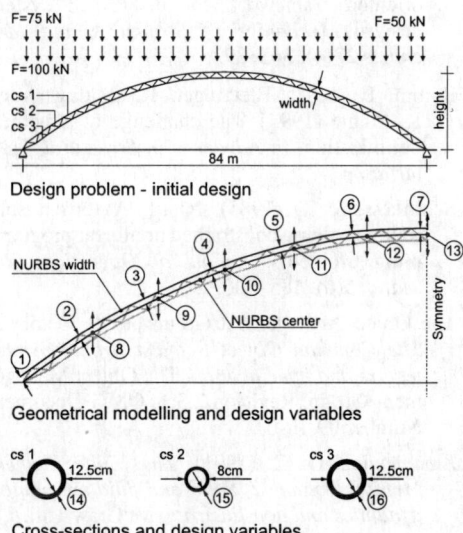

Figure 6: Initial design

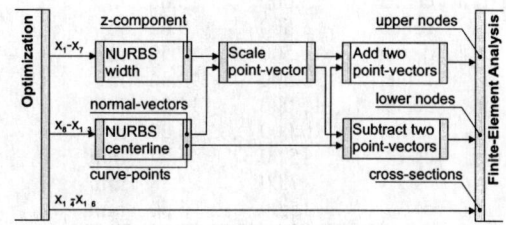

Figure 7: Network of POs (partition)

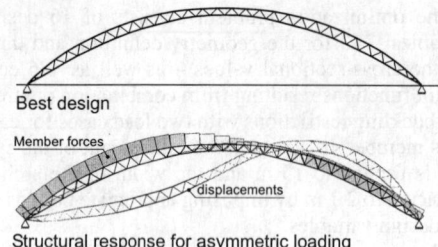

Best design

Member forces

displacements

Structural response for asymmetric loading

Figure 8: Optimized structure

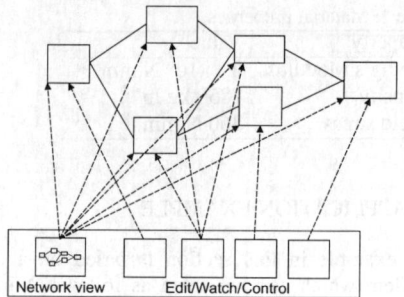

Network view   Edit/Watch/Control

Figure 9: Interaction of components

technical details such as installing proper library versions or frequent updates and bugfixes. As long as the IDL interface between software components does not change, any improvments (or bug fixes) a vendor incorporates into his software is completely transparent to the end user and does not required any action on his part.

For example, Fig. 9 shows a possible network of various server POs that are controlled by a series of client components. The network view client ("Network view" in Fig. 9) relies only on the general PO interface, whereas the other clients are specific to their server counterparts ("Edit/Watch/Control").

Using the well known Model/View/Controller (Gamma, Helm, Johnson, and Vlissides 1993) pattern, clients can be notified whenever a change in a corresponding PO occurs. Because the clients are updated whenever necessary, users can continually monitor the status of their optimization process. This is another example of the flexibility and generality that results when networked POs and CORBA servers are used to implement advanced engineering software systems.

Table 2: Design variables

Var.	Item	Initial val.	Opt. val.	Unit
x_1	width	0.600	0.100	m
x_2	width	0.600	0.716	m
x_3	width	0.600	1.050	m
x_4	width	0.600	1.050	m
x_5	width	0.600	1.050	m
x_6	width	0.600	0.646	m
x_7	width	0.600	0.745	m
x_8	center	5.000	5.08	m
x_9	center	8.500	8.62	m
x_{10}	center	11.500	11.59	m
x_{11}	center	13.500	13.36	m
x_{12}	center	15.000	14.86	m
x_{13}	center	15.000	15.00	m
x_{14}	cs 1	15.00	6.97	mm
x_{15}	cs 2	7.00	1.88	mm
x_{16}	cs 3	15.00	6.60	mm
Objective		13790	6514.290	kg

REFERENCES

Gamma, E., R. Helm, R. E. Johnson, and J. Vlissides (1993, July). Design Patterns: Abstraction and Reuse of Object-Oriented Design. In *OOPSLA Conference Proceedings*, Kaiserslautern, Germany.

Grill, H. (1997). *An object oriented programming system for discrete and continuous structural optimization using distributed evolution strategies (in German)*. Dissertation, Ruhr-University Bochum.

Lehner, K., M. Baitsch, and D. Hartmann (1999). A CORBA based universal optimization service. In *1st ASMO UK/ISSMO Conf. on Engineering Design Optimization*.

Michalena, N., C. Scheffer, R. Fellini, and P. Papalambros (1999). CORBA-based object-oriented framework for distributed system design. *Mechanics of Structures and Machines 27*(4), 365–392.

Ramm, E., K.-U. Bletzinger, R. Reitinger, and K. Maute (1994). The challenge of structural optimization. In *Advances in Structural Optimization*.

Schittkowski, K. (1985). Nlpql: A fortran subroutine solving constrained nonlinear programming problems. *Annals of Operations Research 5*(6), 485–500.

The Object Management Group (1999, October). *The Common Object Request Broker: Architecture and Specification*. The Object Management Group. Revision 2.3.1, OMG Document Number 99-10-07.

Zienkiewicz, O. C. (1991). *The Finite Element Method, Volume 2, Solid and fluid mechanics, dynamics and non-linearity*. McGraw-Hill.

Global engineering and manufacturing in enterprise networks – GLOBEMEN

Jarmo Laitinen
YIT Construction Limited, Helsinki, Finland

Martin Ollus
VTT Automation, Finland

Maria Anastasiou & Roel van den Berg
Baan Development BV, Barneveld, Netherlands

ABSTRACT: The GLOBEMEN project aims to create IT infrastructures and related tools to support globally distributed and dynamically networked operations in one-of-a-kind industries. The aim is to guide and encourage the industry and IT vendors to develop and adopt improved IT infrastructures by combining the views and requirements of various industries. The project will demonstrate a functionality, which offers IT vendors attractive market opportunities for product development satisfying the needs of various industries world-wide.

The measurable and tangible objectives of the project are to: (1) define industrial requirements and (2) compliant architectures for globally distributed product life cycle management, customer support, project and manufacturing management in the virtual enterprise, (3) implement proof of concept in industrial prototypes, (4) demonstrate core features of the architecture, and (5) promote deployment by IT vendors, manufacturing industry, academia and standardisation.

1 INTRODUCTION

The manufacturing in the 21st century will consist of virtual enterprises combining distributed core competencies and teams in customer solution provision co-ordinated by means of a flexible and globally networked IT infrastructure. Advanced collaboration between main contractors and SME-suppliers and the application of modern information technology will allow the effective co-ordination, communication and collaboration across geographical and organizational borders.

Manufacturing today is fragmented into numerous industry domains operating in geographically dispersed locations and supported by isolated and inflexible IT infrastructures. At present flexible information exchange requires that all the participants use the same kind of systems or that specific interfaces have been built to enable it. The management of the information in dynamic relationships becomes difficult. However, these relationships are common in one-of-kind manufacturing. The effective co-ordination of the distributed operations requires also the distribution of the responsibilities and management and developing new inter-enterprise co-ordination processes.

The GLOBEMEN project aims to define a reference architecture, required tools and infrastructure to support globally distributed and dynamically networked operations in a range of manufacturing industries. By combining the views and requirements of various industrial domains involved across the globe the project intends to guide and encourage the industry and IT vendors to develop and adopt improved IT infrastructures. The project aims to identify industrial requirements, define reference processes and demonstrate functionality to fulfil the gaps between existing systems and the inter-enterprise co-operation needs. The project results will offer attractive market opportunities to IT vendors for product development to satisfy the needs of various one-of-kind industries world-wide.

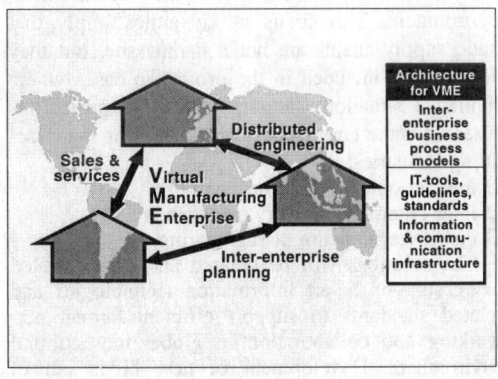

Figure 1: Global manufacturing management in the virtual manufacturing enterprise (VME)

2 OBJECTIVES

Based on the described vision and aims the objectives of the project are to:

❏ define *architectures* for globally distributed life-cycle management, delivery process management and engineering in the virtual enterprise.

❏ implement proof of concept in industrial *prototypes*,

❏ *demonstrate* core features of the architecture, and *promote deployment* by IT vendors, manufacturing industry, academia and standardization.

Tangible outcomes of the project will be:

1. Reference *architecture* for flexible distributed one-of-kind manufacturing enterprising system. This includes identification and specification of:

❏ virtual enterprise models with organizational and information views including resources, capabilities, product and activity configuration and their incorporation

❏ supporting tools and infrastructures (IT systems, guidelines, etc.)

❏ reference processes for distributed communication, collaboration and co-ordination, focusing especially on inter-enterprise processes

2. *Specifications* of required methods, functionality & tools.

3. *Guidelines* on implementation of virtual manufacturing enterprises.

4. *Demonstrations* of target system features and usage scenarios.

5. Industrial *prototype* implementations.

3 FOCUS

The focus of the work is on inter-enterprise information exchange and control. Consequently, intra-company IT infrastructures and stand-alone applications, however advanced, do not belong to the project. The target industrial business environment is dynamic customer focused networking in the global environment. This focus on dynamics imply that static supply chains are not a main issue, but they may still be included in the project in cases where similar IT situations are anticipated. The project focuses on three core business processes of manufacturing industries:

❏ interaction with customers and users,

❏ inter-enterprise delivery management,

❏ distributed concurrent engineering.

The solutions will be created and demonstrated using state-of-the-art information technologies and related standards to support efficient human networking and collaboration in global multicultural environment. Development of new IT is out of scope. Hence, the main contribution of the project will be in the integration of the technology to support inter-enterprise collaboration. The selected functionality specified is tested with pre-competitive prototypes by the software vendors. The requirements and architecture will be defined from the end-user, industry point of view. However, this will be done with support from vendors of core IT within and outside of the consortium.

The European partners represent one-of-a-kind production within the following industries: ship-building, construction and delivery of power plants and telecommunication systems. The Inter-regional partners represent additional industries such as aerospace, metal cutting, machinery, process plant, food processes and electronics.

4 STATE OF THE ART SURVEY

The project started with a first survey of the current practice of inter-enterprise business processes and information and communication technology (ICT). This technology is seen as the enabler for current and future virtual enterprise concepts and functionality. One aim of the survey was to help to clarify the scope of future work in the project. The scope of the survey includes topics such as tools, methods, and models, and can range from as wide as reference architectures to middle ware technology. The information in the survey will be used and elaborated during the project based on industrial needs and the proceeding of the work. What is presented below are the preliminary results in a nutshell.

The survey looks at developments from the point of view of integration. In the past decades advances in information and communication technology have driven the integration of activities in and between enterprises. These developments can be described and classified with figure 2. This figure shows a framework for inter-enterprise integration. Each of these layers will be briefly discussed below. In the state of the art analysis, which is part of the GLOBEMEN deliverables, the levels in the framework are discussed in more detail.

1. Standards for physical integration

Naturally physical integration is needed to facilitate co-operating applications and enterprises. Recently wireless integration has rapidly grown in importance. Relevant standards at this level of integration are WAP, MAP, and Ethernet.

2. Syntactical standards for application integration

The type of integration that concerns the actual software systems is called application integration here. Application integration is concerned with the usage of ICT to provide interoperation between enterprise resources. Application integration is split up in two parts. The lowest level is syntactical integration, for integration at the level of "form". Syntactical standards support that sources and messages have a

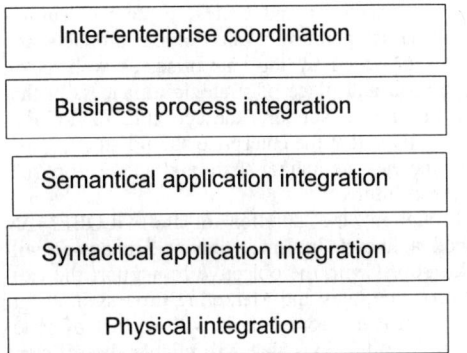

| Inter-enterprise coordination |
| Business process integration |
| Semantical application integration |
| Syntactical application integration |
| Physical integration |

Figure 2 Framework for Enterprise Integration

similar format. Standards in this area are Java, XML, Corba, DCOM.

3. Semantic standards for application integration
The higher level of application integration lies at the semantic level, the level of "meaning". Semantic integration should result in the fact that the output of applications is meaningful to other applications. This type of integration abstracts from the technical details of software implementations. Examples of standards at this level are EDIFACT, STEP and BizTalk.

4. Modeling languages for business process integration
Integration in virtual enterprises requires that a common understanding exists about the shared business processes in the broad sense. Modeling languages are needed to make these business processes explicit. Main areas are modeling of data and modeling of processes. Examples of standards at this level are IDEF, Petri nets, UML, ER, LOTOS, SDL, VDM, Z, B, χ. Time is generally incorporated in these modelling languages, but distribution in space usually is not.

5. Inter-enterprise co-ordination
In business-to-business e-commerce the trend is not so much towards one electronic market, but instead to several electronic (supply) networks, each one made up of closely co-operating business partners. In business-to-business environments Internet technology is applied in an extra-net fashion to bring people to your site who are already on your side. In those situations e-commerce applications could in fact be seen as a sign of trust between business partners, a token of commitment to each other. In a sense, the application follows the mutual trust. But like a "brick and mortar" retailer cannot automatically transfer his reputation in traditional business to the Internet, the mutual trust between business partners cannot automatically be "ported" to the e-commerce application. Thus dedicated effort at the level of inter-enterprise co-ordination is needed, e.g.

☐ Partner selection
☐ Certification
☐ Inter-enterprise best practice definition

The framework in figure 2 is not specific to the era of "the new economy". But over the years integration itself has become more "open".

Until recently it typically concerned integration of two given systems within two given units (often within one organization), who wanted to share specific data and applications for given purposes in fairly stabile business processes. Without trying to make the challenge of that type of integration marginal we could classify it as deterministic integration: many aspects were known in advance. The business context of the integration, the players, their roles and their systems were known and fairly stabile. The integration solution was dedicated to a specific case. The emphasis in the integration effort was on the technical challenge of making the two information systems operate as one.

Currently most attention is devoted to a type of systems integration which is quite different, one based on much more standardized technology, enabling communication between information systems belonging to business partners not familiar with each other in advance. The technical integration challenge for companies nowadays is to create "openness" to systems which are not known, owned by partners who are not known, enabling co-operation in business functions which are not automatically streamlined across the partners. Compared to the first type of integration, the second type of integration is much less directed. It is more flexible to a wider variety of business applications. It is more opportunistic.

In the era of deterministic integration the information technology was dominating and highly influential on the business opportunities that could be realized with the information infrastructure. In terms of figure 2 this translates into an emphasis on the three lowest levels of the framework, or even the first two.

In the opportunistic type of integration the technology plays much more a supporting role to unleash the intellectual potential that can be realized through smart forms of co-operation. The two lower layers are becoming more and more standardized and a commodity, leaving more room to creatively exploit the benefits of open technologies. Thus the emphasis has shifted towards levels three and four (business process modelling). Inter-enterprise coordination mechanisms became much more important.

5 APPROACH

The approach of the GLOBEMEN project is to address three main aspects of manufacturing:

❏ Sales & services in dynamic networks: Management of distributed sales, services and operation, maintenance & renewal support
❏ Inter-enterprise delivery process management: Inter-enterprise resource planning and dynamic supply chain management, integration of planning and manufacturing.
❏ Distributed engineering: Product and process engineering in a distributed global environment, document and product model management, group-work support.

The development within each of the three application areas is done in parallel based on the requirements and needs of the specific business processes in question. Applications will be developed and demonstrated supporting the business of the involved industrial companies. The industrial driven work covers:
❏ Definition of industrial requirements,
❏ A compliant architecture,
❏ Specification of required methods and tools,
❏ Preparation of guidelines to support industrial implementation and deployment,
❏ Demonstration of proposed architecture and implementation of industrial prototypes.

The three business activities are working in close collaboration with a parallel integrating activity. The aim is to develop a generic architecture for describing and supporting the business activities of a global virtual manufacturing enterprise network. Based on industrial requirements specifications the work will be co-ordinated and integrated into an IT architecture for enterprise networks and virtual enterprises. This will include a reference model and associated methodology or guidelines, which will guide the set up and operation of IT support for virtual enterprise networks. In the work equal weight is given to human and organizational issues and to technical issues. The solutions will support efficient human networking and collaboration in a global multicultural environment.

The models and architecture will be generic enough to be usable in all one-of-a-kind industries, directly usable by standards bodies and by IT vendors. The results will be validated through industrial prototypes, and implemented in selected applications. The validated results will be consolidated and disseminated to the industry, standardization bodies and IT vendors.

6 CASE EXAMPLE

An example of the industrial prototypes is in the sales and services area, and more specifically in the Sales and Marketing process. The process aims to facilitate an enterprise's understanding on market changing directions and trends, customer requirements, market perceived value of the products and services provided by the enterprise, as well as on competition and place of strategic alliances. In that way, the process supports the activities of defining the objectives that the enterprise should attain, in order to increase its market share and achieve optimal success in future.

Throughout the duration of the GLOBEMEN project, a knowledge management framework will be developed with the objective to support the execution of the Sales and Marketing process in an inter-enterprise environment. It will consist of practices and applications that will enable the efficient acquisition, use, organization and transfer of globally distributed implicit and explicit knowledge related to Sales and Marketing process.

A large amount of information and knowledge is required for an efficient performance of the sales and marketing process. The information and knowledge that is required for the execution of the process has to do with:
❏ Market segmentation, market changing directions and trends
❏ Competitors, customers and partners
❏ Innovation and trends
❏ Products and services
❏ Existing contracts and projects

In addition to the large amount of information and knowledge, there is an inherent complexity in managing this information. This complexity stems from the fact that various relationships and dependencies exist between the above mentioned different categories of information and knowledge. The information is globally distributed in different departments of the company, as well as in different organizations within its business network, thus creating difficulties in communicating and sharing the required information or knowledge. It is often the case that the required information resides in many different places and sources such as: the WEB, press releases, databases existing in the business network. In addition, the individuals involved in the process have important process – relevant knowledge, that could be helpful to their colleagues if it could be derived and be available to them in some way.

Based on the discussion above one can conclude that the knowledge management framework should provide the process participants timely access to the right information anytime, anywhere quickly and in a secure way. A summarized list of the requirements that the environment should fulfil is following presented:
❏ It should enable the storage of the required information and knowledge into a repository. The development of the repository should cater for the efficient management of its content, as well as for security issues.

- The users should be provided with efficient to access and update the required information. Moreover, the access to the repository should be controlled according to the role of each individual in the process.
- Since the information required to sales and marketing process is subject to irregular and unpredictable change, the process participants should be provided with functionality to search for valuable information in various sources, such as the WEB, CDs and local drives. The environment should enable the storage of the acquired information in the repository.
- The environment should ensure that all the users can quickly and easily obtain the information and knowledge that is relevant to their task and responsibilities. Moreover, the users should be facilitated to organise existing knowledge and information and view it from different perspectives, depending on their role in the process.
- The users should be provided with functionality to aggregate existing knowledge, synthesise different pieces of information and generate reports related to Sales and Marketing issues, such as reports on technology trends, reports on existing / new products and services.

A complete list of the industrial requirements can be found in the GLOBEMEN deliverables.

Throughout this case study, it is expected that a significant knowledge and expertise on how the Sales and Marketing process could be executed in an inter-enterprise environment with the support of new information and communication technologies will be acquired. Moreover, the practices and applications, which will be developed, are expected to promote learning and knowledge sharing in a virtual enterprise network, and in that way support the involved partners to further develop its internal and external knowledge awareness and responsiveness.

7 GLOBAL COLLABORATION

The project is a global collaborative research project within the Intelligent Manufacturing Systems (IMS) framework. The project is a kind of a "virtual research enterprise". The project has partners from four IMS regions (Australia, EU, Japan, Switzerland and USA). The consortium represents a wide range of different manufacturers of one-of-a-kind products and from all regions. The system vendors in the project will be the main exploitation channels for the results, which will be based on the work in the participating research organizations.

The main benefit from world-wide collaboration is that partners provide end-user requirements from different cultures, social and organizational patterns.

This will be of enormous value for the specification of the requirements of the IT infrastructure. The developed solutions will also be tested in different settings during the project due to the different environments in the regions. This type of bench marking within the project will increase the possibility for exploitation of the results into successful commercial products.

Specific knowledge from different regions will be used in the project. The enterprise engineering methodology in the GERAM reference architecture has been developed in Australia and the main persons behind the architecture are participating in the project. Methodology for operations support is mainly developed in Australia and Japan. Some Japanese partners will provide Supply chain management contribution. These partners have earlier already developed networking approaches in the area. A major CAD vendor from USA also contributes to the project. The utilisation of the competence of the inter-regional partners will enlarge the applicability of the solutions to be developed compared to e.g. a pure European project.

8 CONCLUSIONS

An IT architecture for inter enterprise collaboration in the one-of-a-kind industry will be defined and implemented and demonstrated in industrial prototypes by a collaboration where the end users will give the industrial requirements and evaluate the results. The research partners will mainly perform the development work together with the system developers, which also are the potential commercial exploiting partners. However, all partners will have their own benefit from the work.

The manufacturing companies (end-users of the systems) benefit from the methods and tools developed for efficient dynamic global enterprise networking. Most end users will develop solutions to some specific needs in their own business based on the architecture provided by the system developers and the research partners. Their prototypes will be implemented into normal operations. In this way they will have immediate benefit from of the results. This approach also ensures that the evaluation of the results relies on concrete industrial requirements. The end-users usually have the role of network managers in their customer deliveries. Consequently, they also act as channels for the technology transfer to their SME-subcontractors. This technology transfer will have a special emphasis in the project. It will start during the first part of the project with selected suppliers by involving them into the definition of the requirements and solutions.

In each IMS region there are several hundred companies with similar global networking activities as the end-user partners in this project. All these companies are potential users of the outcomes in this project.

The software providers in the project operate in different fields of production information systems both in the areas of process and product data management and STEP-technology. They will exploit the new knowledge and experience gained in the project to develop next generation systems, which provide better support in an inter-enterprise dynamic environment. In this respect both the requirements and gaps identified and reworked to specifications as well as the demos presenting the scenarios and the prototypes developed and tested are used by them. The advances in the products of the software providers make the results commercially available for practical implementation of external companies (with delay).

The project results support the strategic aim of the research institutes: to develop their scientific and technological knowledge to be able to serve both the industrial end-users and software providers. The national and international research and development projects of the research institutes are effective channels for technology transfer to external companies. The research institutes thus exploit and disseminate the results, both requirements, specifications, methodologies and demonstrations in future projects supporting external companies, both end-users and IT providers.

REFERENCES

http://globemen.vtt.fi/, Homepage for the IMS GLOBEMEN project

http://extranet.vtt.fi/gm21/, Homepage for the IMS project: Globeman21, Enterprise Integration for Global Manufacturing in the 21st Century.

Bernus, P & I. Nemes 1999 Organisational Design: Dynamically Creating and Sustaining Integrated Virtual Enterprises. *IFAC World Congress*

Bernus, P 1999. What, Why And How To Model In The Enterprise? *Proc. Int.Conf. on Enterprise Modelling*

Larsen, L.B., C. Kaas-Pedersen & J. Vesterager 1998. Creating a generic extended enterprise management model using GERAM – presentation of the IMS/Globeman21 project. *First International Symposium on Concurrent Enterprising, ISoCE'98, 4-6 June, Sinaia, Romania, pp. 45-55.*

Product and Process Modelling in Building and Construction, Gonçalves, Steiger-Garção & Scherer (eds)
© *2000 Balkema, Rotterdam, ISBN 90 5809 179 1*

Outsourcing of CAD production: The CaribCAD project

Meijnardt Scheers, Valerio Curti & Elisa Pandolfi
Department of Civil Engineering and Geo Sciences, Building Engineering, Delft University of Technology, Netherlands

Godfried Augenbroe
College of Architecture, Georgia Institute of Technology, Atlanta, Ga., USA

ABSTRACT: The CaribCAD project develops the technology for outsourcing CAD production work over the Internet. Applications range from being able to regenerate analogue drawings in electronic formats to the supporting routine of communication patterns between globally collaborating design firms.

The structuring of CAD production work deserves special attention, as it is the key to managing production effectively and performing QA at the right moments

The broad objective was to develop the technical basis, human capacity and protocols for the distribution ('outsourcing') of CAD workloads from engineering companies in Europe to specialized CAD bureaus in Developing Countries. The project brought together technologies from the fields of the World Wide Web, co-operative engineering, workflow management and data sharing.

The method of approach is applicable to purposes in the field of remote architectural- and engineering education as well.

In this paper most attention will be given to the outsourcing and to the electronic regeneration of the paper drawings of existing facilities

INTRODUCTION

The acronym CaribCAD stands for: a Co-operative Approach to the Realisation of Internet-Based CAD. The project was sponsored by the EU in its fourth framework (INCO-DC). The objective of the project was to develop the technical basis, human capacity and protocol required for distributing (outsourcing) Computer Aided Design/ Drafting (CAD) workloads between engineering companies within Europe and specialised companies in Developing Countries.

The project sought to achieve this objective by adopting a broad perspective that would possibly enable Developing Countries to become participants in future global co-operative engineering community efforts and to instigate competition between EU firms.

Many partners from Europe and Developing Countries co-operated in the project: co-ordinators from The Netherlands and the Dominican Republic, universities and private firms from the United Kingdom, The Netherlands, the Dominican Republic and Guyana.

Two pilot projects were set up:
- Outsourcing of CAD-production;
- Joint venturing in overseas projects.

This paper will focus on some aspects of the remote outsourcing of CAD production.

OUTSOURCING DEFINED

The fact that, up until now, no adequate definition for Outsourcing has been found in any of the relevant literature is an indication of how new the phenomenon is. The following definition is therefore proposed: **"It is the contracting, outside of an organisation, of services that are traditionally performed in-house."** This involves commissioning an external supplier to perform certain activities for an organisation, such as information planning, system development, or maintaining and operating information systems. This may include transferring staff and resources to the supplier. Nowadays, the proliferation of web technology has created more opportunities for distributing work to remote locations and for managing this work asynchronously time-wise and space-wise.

TEST OF TRADITIONAL OUTSOURCING

Some tests established what traditional outsourcing of CAD workloads would actually be like and examined the results when compared to CaribCAD preliminary goals and issues.

Several tests were then set up. The present document

particularly addresses just two of them, explored in co-operation with a Russian and an Indian firm.

A paper-based drawing was copied onto three A4 pages, scanned and sent.

Information about the required task was not requested by, or given to the two performers neither about standards, nor about procedures.

The results were utterly different:

- The Russians recognised that even when laid out on different A4 sheets, the drawing constituted one representation (Figure 2.) while the Indians drew every single A4 sheet separately…(Figure 1.)
- The Russians understood that an architectonic drawing is composed of different elements and so they represented each element by using different layers and colours, though only graphical colours were used. The Indians drew each and every element in the same layer and colour.
- With the present limited assignment, it turned out to be easier to check and modify the Russian drawing than that of the Indians.

In these tests there were some usual traditional out-sourcing process problems that resulted. From the information left on the final drawings, or omitted from them: e.g. useless construction lines, ordinary errors, details not drawn, etc. it has been concluded that the drawing method used should have been an overlaying one. Nevertheless, the presence of extreme deviations and the inaccuracy of the dimensions all suggested that a superficial redrawing process might be used, instead of what at first was considered to be an overlaying process.

Some conclusions:

The tests give an idea of the inadequacy of traditional processes when it comes to redrawing existing paper drawings but they are not exhaustive with respect to the subject in general.

For engineering companies a traditional outsourcing process leads to loss of time where checking, adjusting and cleaning the digital drawing is concerned because of the project requirements, the limitations of the applications used, and the problems created by drawings. The conclusion is that the system is not suitable for improving quality, or for reducing costs and execution times, unless the client:

- Defines the necessary conventions;
- Defines the different requirements for each project;
- Defines the required level of presentation;
- Instructs and tests the performer beforehand.

DIGITIZING METHODS

Due to experience gained in the traditional outsourcing test, where drawings were digitised – and results were unreliable – the different methods of producing a digitised drawing from a paper drawing were tested.

In fact there are three ways:

1. Using the scanned drawing to make an overlay;
2. Vectoring the scanned drawing by using software;
3. Redrawing on the basis of written dimensions.

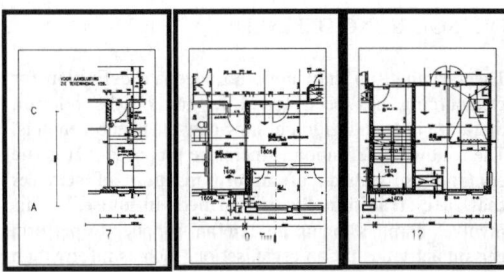

Figure 1. Indian products

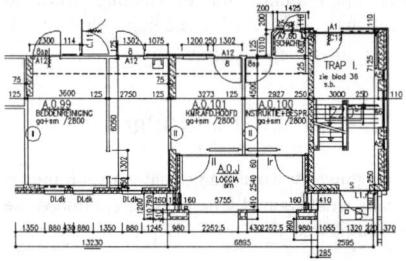

Figure 2. Russian product

(Figures 1. and 2. not reproduced on the same scale)

Overlay:

In an overlaying process the draftsman draws over the scanned drawing directly onto the screen. He does not have to look at the given dimensions (which are explicitly drawn). This means that if the hand made drawing is not precisely drawn or if the number of manipulating processes is very high or if, finally, because of the methods used by programs to represent raster pictures, the drawing dimensions are incorrect then the overlaid drawing will also be incorrect.

It is easy to conclude that if one is to obtain a precise and accurate drawing with an overlaying conversion then the original drawing must be precise and accurate as well. As this is not the case, though, in most architectonic practice and due to the fact that texture segmentation remains a fundamental issue in image analysis, pattern recognition and computer vision, overlaying still does not result in a precise and reliable digitised product.

Professional vectoring:

Software programs that perform automatic vectoring were subjected to testing, by outsourcing a number of drawings to professional companies used to working with

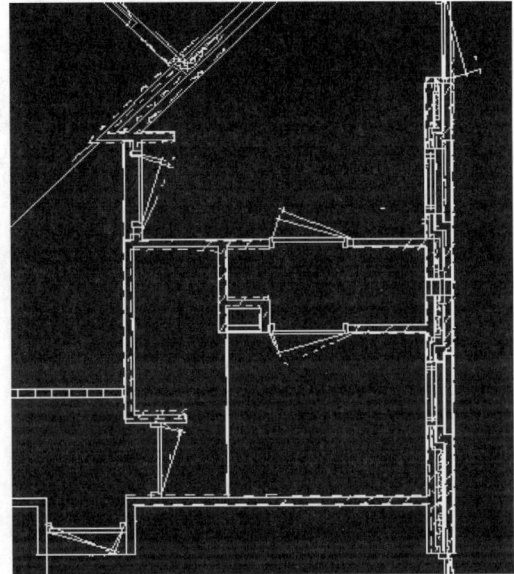

Figure 3. Professional vectoring of a paper drawing

such tools. The aims of the tests were to evaluate the state of the art in new ways in order to completely automate a CAD workload and to examine the results.

The image obtained (Figure 3.) should be self-explanatory and give an idea of what the problems are:

- The final process can never be reliable due to all the reproductions that a drawing has been subjected to which means that the original will more than likely not be to scale;
- The precision/imprecision of the final CAD is based on and depends very much on the precision of the paper-based drawing;
- The paper-based drawing has to be vectored according to its scale but it has to be transformed to a 1:1 scale for further uses in CAD programs;

Thus, there are:

1. Difficulties in handling the final CAD;
2. Difficulties with checking the correctness of the drawing;
3. Difficulties in checking its adequacy;
4. Loss of time in conjunction with preventive or subsequent despeckling and calibrating;
5. Loss of time because of the further processes of:
    - Extra OCR text recognition;
    - Purging of the hatching which normally results in incorrect elements;
    - Connecting all the single vectors in order to compose continuous lines;
    - Giving every single vector its final layer, colour, thickness, etc.;
    - Redrawing all the pieces of hatching;
6. The need to define very strict formal converting methods for every kind of drawn element.

7. The need to define very detailed methods in order to check the quality of the drawings.

Conclusions: Due to the fact that texture segmentation remains a fundamental issue in image analysis, pattern recognition and computer vision, vectoring is still not possible; so for engineering companies automated vectoring leads to a loss of time in in-house pre- or post-processing which is more time-consuming than actually redrawing.

Redrawing based upon written dimensions:

Setting up a complete new drawing based upon the written dimensions readable from the printed or plotted scanned image happens to be less time consuming and very accurate compared to the other two methods of digitizing.

This method is strongly recommended.

REQUIREMENTS ANALYSIS AND APPROACH TO CAD PRODUCTION

In order to enable successful outsourcing it is necessary to answer questions resulting from the tests; also by looking closely at various professional practices and at new CAD software developments. A requirements analysis therefore was made up. According to the analysis proposing the following items was a way to approach CAD production.

Digitised drawing applications:

1. Develop a system that is independent of the possible future uses of CAD;
2. Define the most economic way of performing. Aim, for instance, to have less of the mundane in-house work done by highly-skilled designers so that such valuable manpower resources can be deployed in highly skilled and really creative tasks, and in order to bring DC partners, step by step, closer to Western standards;

Digitised drawing:

1. The redesign should be, as far as possible, independent of the mistakes of the originals;
2. Check the drawings as little as possible without making real checks on paper, but by comparing the simple and minimal elements with each other;
3. Check the drawn elements using the same easy self-referring quality assurance procedure as that used by clients;
4. The drawing should be composed in a way that is closest to the client's, while remaining on the system of cross-referencing, since that allows all the previously quoted advantages to be benefited from;
5. Theoretically, libraries should be quite well defined, but depending on the office practice - not all the

offices already have well organised and complete libraries of standards, symbols, procedures, methods, etc. - this could be done step by step while outsourcing.

Transmission and communication:

1. Communicate by making use of a functional and well-assorted combination of easy, user-friendly, effective and almost instant tools, such as the new generation of communication tools available on the WWW at no expense. Also use normal e-mail and more traditional medias; definitely in a continuous way, at least at first or at key moments;
2. Depending on the tool used there will only be misunderstandings or delays but no loss of data, except in special cases: e.g. attachments bigger than the amount set up and then accepted by the local mail server/program, crashes in the servers etc. Using synchronous systems for key information or systems that are safer and with which you can directly check the results would avoid most of the problems;
3. Cross- referring (X-refs) each drawing to the small, atomic elements composing it, does effectively reduce the size of the CAD elements to the possible minimum;

Quality level:

1. All the quality assurances should be reduced to the necessary minimum;
2. One should view as acceptable the representation, as it is or as it is built, of what is given in the drawings themselves without forgetting that certain information can easily be extrapolated by normal architectonic means;

THE SYSTEM OF EXCHANGING FILES

The "**depot**", term proposed by TUD, refers to the system of organising and subdividing a CAD drawing into the smallest possible elements and creating final drawings by reassembling those fragments (Figure 4.)

The Depot (Figure 7.) is a sort of "supermarket" of drawings where the cross-references composing the final drawings, as well as the final drawings themselves, are stored. It is useful to be able to create new drawings starting from the idea that the elements can be divided into generic, almost generic, more specific, specific and very specific categories. It amounts to a way of showing the state of the art in CAD techniques by introducing the concept of Presentation Levels, which is used as a better alternative to scales. It is also possible to return to the scales later, at the moment of printing.

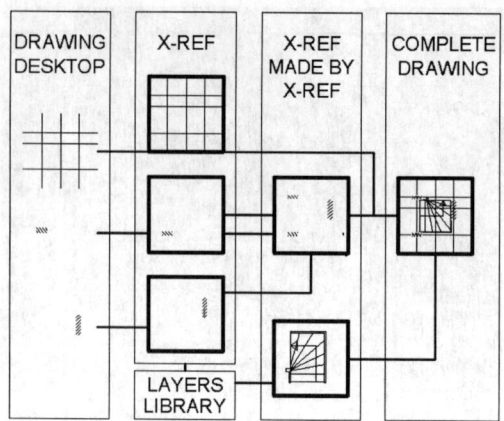

Figure 4.

REMOTE MANAGEMENT OF CAD PRODUCTION[1]

The CaribCAD project has adopted emerging workflow management technologies to allocate CAD production tasks to a remote provider, whereas the production is monitored and managed by the local client. Experience in other areas of outsourcing has shown that proper distant management capabilities and agreed enforceable Quality Assurance (QA) procedures are the key to successful execution of outsourcing jobs.

For the type of CAD production work described in this paper, an open partnership approach was chosen and the approach to the inherently rigid project execution will be briefly described. (Figure 5.)

As the right part of the figure indicates, open partnerships may be supported by template models for given types of outsourcing projects. The template model does not typify the partnership but the type of project. The requirement is

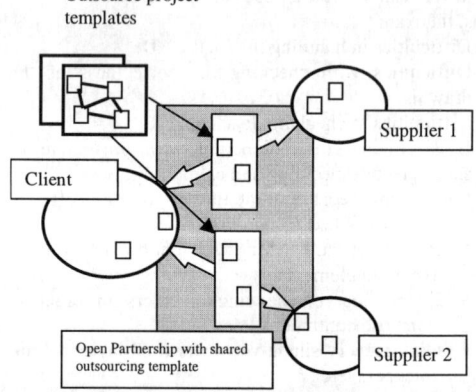

Figure 5.
The characteristics of open outsourcing partnerships

76

that templates are defined so that they can be reused in similar projects with different clients and different suppliers.

As described in the previous sections, the most important aspect of CAD production outsourcing is the quality assurance (QA) of the generated CAD files. One of the major challenges of project management is that of supporting the instruction, testing of comprehension, timely error notification and enforcement of client quality demands.

The groupware approach: a common tool set.

When discussing the general requirements of an 'outsourcing package', it is necessary to mention that tools must operate in a collaborative group environment, allowing group member specification, authority, security settings, etc. All group members are connected by e-mail, and have (permanent or dial-up) access to the Internet for file transfer and web browsing. The remote users should have only minimal specialised software, typically a WEB Browser and a mail tool. The tools must allow easy specification of a tailored 'project information repository' enabling all group members to have flexible access to all project documents. The repository must support versioning and setting of user permission for browsing and editing. An adequate group ware solution must be selected to meet these requirements.

The tool set must allow the generation of a project-specific central task repository that enables process building and enactment through the assignment of tasks to group members and the monitoring of execution. The model underlying the 'workflow execution and support' should have a graphical representation transparent to all users.

On the basis of these requirements the CaribCAD team implementation adopted a three-layered approach to the creation of a collaborative working environment,

- Communication;
- CAD file and Document Management;
- Collaboration and Work Flow.

For the first two layers, the strategy was adopted of creating an Intranet within the Internet building in the Microsoft Exchange Server environment. This allows each remote partner to access their group-ware environment through a simple Web browser interface using standard HTML protocols while at the same time retaining high levels of sophisticated development on the server side. The end result is complete support for mail, shared calendars, databases and tasks with no sophisticated software configuration on the client side. This is important when reliable and quick deployment to new remote users is essential. The third layer will be described in more detail below.

Workflow management:

Group management is based on formal process models, i.e. a 'neutral' representation of the workflow and document flow, capturing the procedural aspects of task execution and their associated documents. The model defines roles, tasks, task owners, inter-task dependencies, documents, privileges etc. The model starts with top-level tasks and decomposes and refines them until the level of atomic drawing activities is reached. The template model for the CAD regeneration outsourcing contains the following three top-level tasks:

1. Provide instruction (exchange and reach agreement on instruction set);
2. Do performance test (a test designed to ascertain worker skill level on the supply side and to reach consensus on QA procedures);
3. Conduct an actual project, starting with the negotiation of the contract, and ending with the delivery of the CAD files.

After each task has been refined to an adequate level of granularity, all operational and dynamic issues, actors, deadlines, authorities, resource allocations, etc. are taken into account. The resulting WF-Model defines precisely what is done, by whom and when. The next stage is then to cast the workflow model in an adequate workflow engine to provide the necessary run time support for the enactment of the process. CaribCAD has tested different process management tools for this purpose and it has been found that different partnership settings and different process templates require different workflow paradigms. PILOT1 was conducted with Keyflow™, which uses a message-based paradigm.

System architecture:

The 'project space' component comprises a data repository built on top of the Microsoft Exchange Server which provides a storage mechanism for all objects such as documents, CAD files, tasks, mails etc. together with the relationships between these objects. The object model holding this data is based upon the Collaborative Data Objects Model and the interfaces to these objects are compliant with the MAPI standard. The following briefly describes how the CaribCAD environment operates.

Workflow Models are developed using a graphical design tool that is embedded inside Microsoft Outlook, the output of this tool is a specially formatted e-mail message that is stored as a workflow template that can be executed many times. Figure 6. illustrates a simple Keyflow generated workflow from Pilot 1. Each node in the workflow has a role and a time scale, together with the data required for its performance.

The workflow is initialised through a WEB forms interface, which supplies the relevant data for that particular operation; actual performers substitute the roles, then CAD files and other documents are attached. Once executed the state of the flow is monitored the

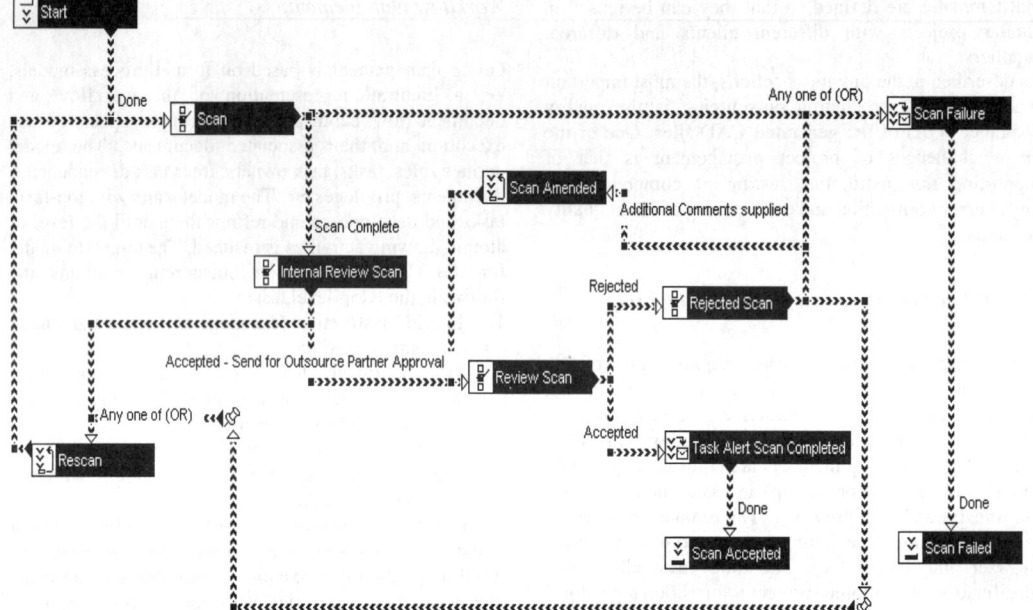

Figure 6. Typical workflow template.

WFM server which is a separate application. As each node is processed, mails and tasks are placed in users task lists on the server. A separate application monitors these task lists and as tasks arrive it sends a notification mail to the user's normal (local) mailbox. When a user receives a notification they use the WEB browser to remotely access the CaribCAD Pilot project server and view their tasks and associated data. These tasks are rendered into HTML on the server using Outlook WEB Access and other server side scripts. This allows quite complex information to be presented on the client's browser and it limits the amount of Internet traffic required. For example, the user can browse and select information from range of CAD files without downloading the entire file. This is achieved by means of server-based applications, which translate the graphical content of the CAD file into HTML.

One of the aims of the CaribCAD project was to establish a number of proven cases of successful CAD outsourcing and to develop an outsource 'package' for their support. One of the significant results is a set of re-usable models for significant classes of outsourcing projects.

In follow-up commercialisation efforts these models could be hosted on a dedicated outsourcing web site, where the client chooses the appropriate template, initialises it and deploys it with the accepted remote bidders for his 'auctioned' outsourcing job. It is important to note that the workflow model is set up in such a way that a rigorous selection of potential providers is supported, involving tests of their CAD proficiency (top level Tasks 1 and 2) before they can be considered potential business partners.

FINAL CONCLUSIONS

Advantages of the CaribCAD system:

From the tests described above (India-Russia) what became apparent is that there are some inherent problems involved in setting up a traditional outsourcing process. Thus, with reference to the problems already exposed, the method developed by CaribCAD at the TUDelft that is based on a "Depot of elements" (Figure 7.) concept, represented as single cross-references or as a combination of cross-references of the same element in all its locations the results will be:

1. Fewer mistakes than with straight outsourcing systems, thanks to the method of redrawing every single element, as computed by the control-dimensions and thanks to the control-dimensions usually being quite correct;
2. Greater ease when it comes to checking the correctness of the drawing, thanks to the system of cross checking small cross-references with other small correct cross-references and the final drawing(s) itself/themselves;
3. Easy drawing check adequacy; by correctly using/checking layers, entities, and cross-references the process/drawings then become immediate and self-denouncing;
4. Greater speed with:
   - Understanding an unknown drawing because of being able to use a quite exhaustive Depot with all its symbols,

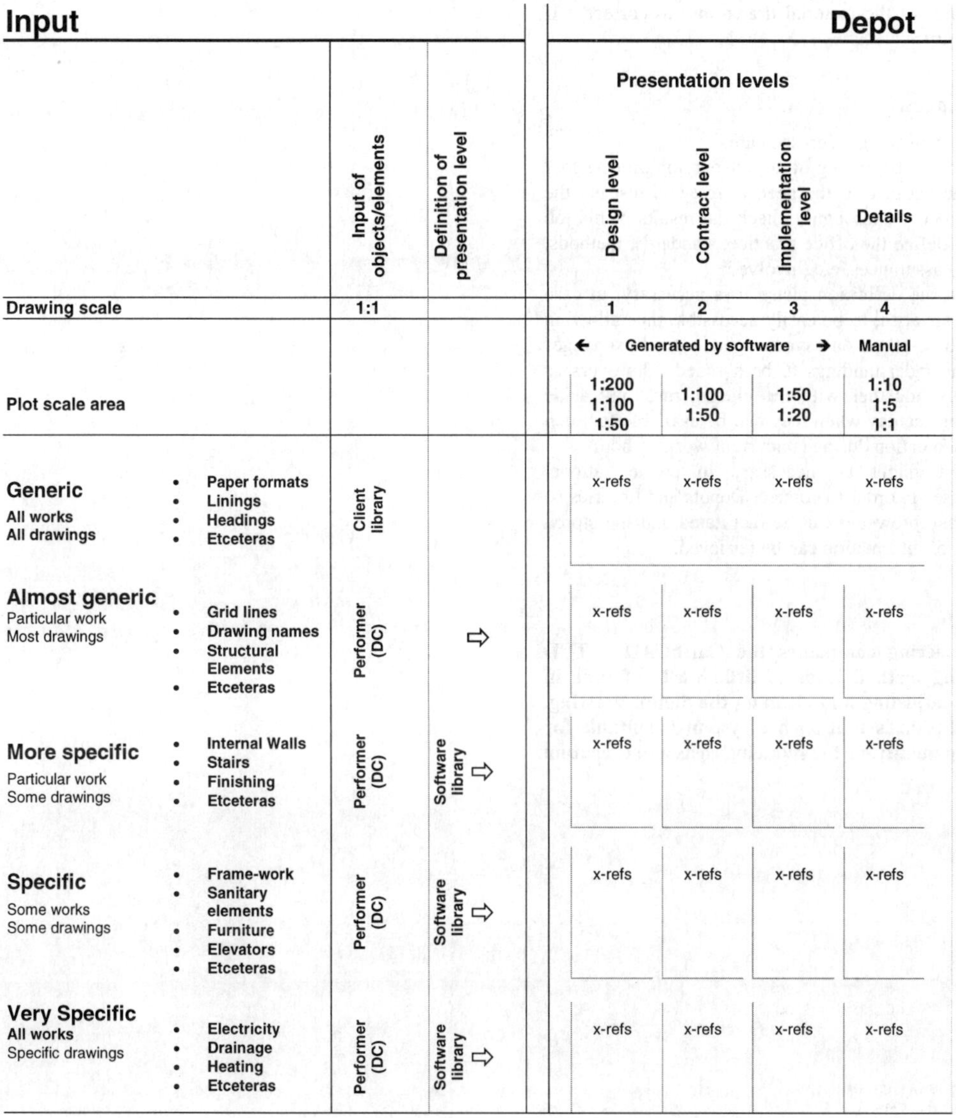

Input

Depot

		Input of objects/elements	Definition of presentation level		Presentation levels Design level	Contract level	Implementation level	Details
Drawing scale		1:1			1	2	3	4
					← Generated by software → Manual			
Plot scale area					1:200 1:100 1:50	1:100 1:50	1:50 1:20	1:10 1:5 1:1
Generic All works All drawings	• Paper formats • Linings • Headings • Etceteras	Client library			x-refs	x-refs	x-refs	x-refs
Almost generic Particular work Most drawings	• Grid lines • Drawing names • Structural Elements • Etceteras	Performer (DC)	⇨		x-refs	x-refs	x-refs	x-refs
More specific Particular work Some drawings	• Internal Walls • Stairs • Finishing • Etceteras	Performer (DC)	Software library ⇨		x-refs	x-refs	x-refs	x-refs
Specific Some works Some drawings	• Frame-work • Sanitary elements • Furniture • Elevators • Etceteras	Performer (DC)	Software library ⇨		x-refs	x-refs	x-refs	x-refs
Very Specific All works Specific drawings	• Electricity • Drainage • Heating • Etceteras	Performer (DC)	Software library ⇨		x-refs	x-refs	x-refs	x-refs

Figure 7. The "Depot". Boxes at right contain X-refs; a column of X-refs composes a drawing.

- Referring to a library of different "How to" suggestions, constructed over the years and by drawing conclusions from the replies given to the performer's questions.
- Instructing new employers without losing too much working time;
- Retrieving missing information that cannot easily be guessed or explained through simple communication with the client or by using fast communication programs and tools;
- Representing only the defined information and asking for the ill-defined;

- Checking small and handy pieces of drawing: cross-references;
- Perhaps adjusting and lightly cleaning up wrong pieces of drawings.

5. There is no need to define very strict formal representation methods. That is only necessary when the final drawing has to be 100% correct.

6. There is no need to define very detailed methods for checking the quality of the drawings because by checking the simple parts of a drawing, the cross-references, it becomes easier for draftsmen unfamiliar with the project to understand the real

meanings of the material drawn and to correct it if necessary.

Disadvantages:

The only disadvantages foreseen are:
1. At the very beginning of an outsourcing process loss of time because of the need to test and instruct the performer to thoroughly check the results of his job and to define the office practice, standards, methods, quality assurances, etc. involved;
2. Still in the definition phase it is necessary, in case problems arise, to be easily accessible thus allowing time to be saved on asynchronous e-mail exchanges and misunderstandings to be avoided. The overseas partners, together with the clients, must put aside times in the day when they can be available for such communication during concurrent working hours.
3. That it might be necessary to create rational databases in order to connect Depots and libraries so that easy browsing can be facilitated and the stored pieces of information can be retrieved.

Outcomes:

For engineering companies the CaribCAD – TUD outsourcing method leads to little waste of time in checking, adjusting and cleaning the digital drawing. The conclusion is that such a system is suitable for improving quality, or for reducing costs and execution times!

---

[1] Augenbroe, Godfried, and Lockley, Stephen (1998) CaribCAD: a technology to outsource CAD production work. In *Proceedings of ECPPM*, (ed. R. Amor), BRE, pp. 29-36.

Human centred concurrent engineering

Prototype internet desktop for engineers

Tomo Cerovsek & Ziga Turk
University of Ljubljana, Slovenia

ABSTRACT: In the recent years more and more engineers got a comfortable access to the Internet. A European Union's 5th Framework project - Intelligent Services and Tools for Concurrent Engineering (ISTforCE) - is focusing on how this could change the way they work. To date, the Internet was a communication platform (email) and a source of information (Web pages) but was not used as a place where actual engineering work would get done. One of the goals of ISTforCE is to design a services platform through which engineers at a given design or consulting company will access the services on the Internet and collaborate in real time. In this paper were exploring a special angle of that idea. We claim, that is possible to create infrastructure on which real construction companies and virtual teams of construction companies can rent and customize services on a project by project basis and where providers of engineering services can market their products. This infrastructure is considered a vertical service that an ISP (internet service provider) offers to its engineering clients, just like they offer email accounts and some space for Web pages today. We believe that such infrastructures will be central to future integration of the profession. In the paper we analyze the key requirements towards such a service, present very basic product and process models that are required to implement it, and present a functioning prototype. In the conclusions we speculate how this would change the ways of communication and collaboration in construction and how much more elaborate product and process models contribute to enhanced functionality.

1 INTRODUCTION

The use of computers have changed dramatically the way in which the AEC professionals work. Productivity is higher, projects are better, time required for design is reduced, costs and constructability issues can be predicted more precisely, etc. In this chapter we give a short historical overview that will illustrate evolution of the use of computers from tools, via communication media to working platforms.

1.1 From tools to platforms

The role of computers has gone through various stages. Negroponte (1970) speculated that their role could be that of tools, assistants or media. First small discrete tools helped to perform repetitive calculations or just giving values that were not listed in the tables the computer was really nothing else but a tool, like a calculator or ruler. With increased capabilities integration of various tools and exchange of information between them became possible. Machines have actually changed the way that engineers think. Almost all calculations are performed on PC. Younger engineers are helpless without electricity powering their PCs and the new generation of architects cannot express their ideas without 3D modeling software. Even older experienced engineers, who sworn to Markus tables a few years ago are using analysis packages and Excel to perform the calculations.

Until recently, the Internet technology has been used at a rather basic level. But explosive growth of AEC Internet resources (Figure 1) and increasingly complex services attract users. After years of research (overview in Vanier, 1997) just on time delivery of building codes is commercially available (TenLinks Inc., 2000). Manufactured products and materials specifications can be found on the Web (CONNET, 1999), email is replacing fax machines, mobile terminals (computers, phones, WAPs) are found useful on the building site. Electronic documents are replacing paper while governments are preparing suitable laws (Baker & McKenzie, 2000) that make digitally signed and encrypted documents equal to traditional paper based documents with hand signature. Internet and distributed working environments are taking important part in commercializing engineering content on the Web.

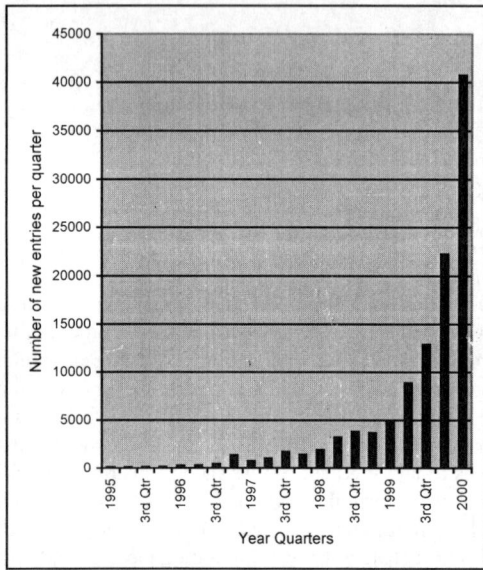

Figure 1: Age of AEC resources on the Internet.

1.2   Problem statement

We believe that in the near future the growing number of services on the Web, Internet's communication functionality and the tools on the workstation will merge into a seamless platform where engineering work would get done; both tasks performed individually, as well coordination, negotiation and other collaborative activities. Requirements for a platform that would support this include:

- it should be open, so that any service or tool, Web or workstation based, could be integrated in it.
- it should be customizable to persons: several AEC professionals with different personal and professional requirements will be using it.
- it should be customizable to projects: each construction project may require a different IT infrastructure
- it should be scalable: companies with different IT infrastructures are entering AEC projects.

The platform should be usable both on modest as well as state-of-the-art equipment.
- it should be extendable: it should not impose on the project management to adopt any predefined product or process models. Instead it should allow them to define a tailor made solution that fits each project best.

The ISTforCE platform will attempt to satisfy all the above requirements. In this paper we focus on the last one.

1.3   Key idea

It has often been noted (e.g. Techolz, Armathwaite, xxxx) that construction products, process and virtual companies are unique. Unlike mass production industries, construction is producing in lot series of one. Instead of joining the struggle to define "standard" data structures and procedures into which any construction project could fit, we propose to build an infrastructure where it becomes extremely easy to create a customized project support infrastructure on the Web. This is infrastructure would be created by the Chief Information (or Internet) Officer (CIO) of the project on request from the chief project manager. The infrastructure is created by using various templates, services and tools, which are provided by the platform-platform.

A prototype of this platform-platform - a plaform where platforms can be built, is presented in this paper.

1.4   Methodology

We have used loop-type rapid prototyping model as methodology to:
- Explore possibilities of enterprise computing,
- Help to define essential features to meet the requirements and needs of end user,
- Help to evaluate existing solutions and identify their pitfalls,
- Identify of possible solutions,
- Identify of actors and use-cases.

UML (Booch et all, 1998) was used as the visual modeling tool for the development of the software.

1.5 About the paper

In Section 2 the basic components are identified. In Section 3 we have focused on the actors, use-cases and architecture. In Section 4 a platform prototype is presented.

2 MANAGING ON-LINE COLLABORATION

In general we can divide components to those supporting teamwork, and those supporting individual's work. Both can be supplemented with external services offering information exchange. Following subsections are divided to major components/areas that enable functional collaboration according to the statement above.

2.1 Document management

Document management system is one of the most important components that enables basic collaboration and exchange of design information. Project documentation includes very complex data-types that go far beyond simple integer, numeric, decimal, date, time, and string data types. These are integral data-types such as time series, images, collections, presentations, html pages, text based documents and especially spatial data-types such 2D, 3D CAD drawings, geodetic surfaces and so on. Some advanced document management systems use functions that are a part of kernel with expert classes, user defined functions, functional indexes, methods defined by the data-types and parallel processing architecture.

Typical features: security based on roles, customizable descriptions and notification, versioning management, off-line distribution, viewing without applications with readling, audit trail. Multiple interpretations of the documents Integration of external features are desirable as well. With rather simple implementation of document management system with multiple presentations we are able to meet basic requirements.

2.2 Communication components

Major issues of communication components are not just the means that enable communication between team members. Capturing the interaction and negotiations between team members is important. Types of communications that can be captured are of course strongly related to the use of available technology. Participants can communicate with textual representations, audio/video or application sharing

that can include graphics, text, video and audio. Some traditional excepted tools for communication such as telephones, fax machines may not be excluded from this type of components. Capturing and streamlining communications is crucial for improvement of the efficiency of the team and knowledge discovery. Despite the fact that text based communication is the easiest to implement, not all possiblities are used. Asynchronous (e-mail and e-mail discussion-mailing lists, discussion forums) and synchronous communication via chat application should be stored in team activities repositories. Simple recording of on-screen activities while videoconferencing, application sharing and telephone talks should be saved in document management system.

2.3 Time Management

Due to the fact that workflow in construction industry is quite undefined and priorities are frequently changed during design activities, a good time management is required. Wastes of time of course depend on working habits of team members and their perception. Time management components should make it easier to detect required frequency of meetings. Time management components should include: Integration of personal and team calendar, task assignment, planning of meetings and appointments, to-do lists, and an analysis of the workflow.

2.3.1 Human resources management

Underestimated value of human resources against the importance of IT technology is giving an impression that IT can manage relationships between team members automatically. Latest technologies that enable concurrent engineering are useless if there is no communication protocol. Managing available persons according to their skills, personal qualities, keeping all members satisfied and busy is far to complex to be carried by someone who does not have feelings. On-line collaboration tools can just make it easier to detect problems, push forward driving forces and remove restraining forces. In geographically distributed environment it will be extremely important to create comradeship and other social values. Components that enable time based review are important but components that enable evaluation of the content should be even more important with integrated workflow.

2.4 Custom Databases

Custom databases provide internal collections (provisions, codes), reporting forms, RFI(xxx), risk identification, etc.

2.5 On-Line Design tools

Analyzing the widely used structural engineering programs we learned that nothing actually changed in the last 20 years. Input file is in ASCII format and analyses are executed from command line. Just tedious typing of input parameters for fixed-column format was replaced with graphical user interface via mouse clicking. After the model is prepared pressing the button do the typing of input parameters in the command line. Realizing the fact that preparation of the model and interpretation of the results takes most of the time and analysis itself just seconds, we propose new business model that will fit into current opportunities and can be implemented immediately. The idea is to move analyses programs from the PC to the server and give pre and post processors away for free. Proposed model would be in benefit of both users and software vendors. Vendors could charge on per project basis, for certain period, by processor time, by a number of runs or combination of those. That would be very interesting for users as well (don't need to buy expensive software, no installation guides, central storage of inputs and outputs, will be able to use specialized program just once) and vendors' worries about security issues and support tasks would be reduced. With such organization, software vendors or their certified representatives can control the use of their software completely. They will be able to focus on high performance of the program for particular machines and better utilization of hardware and software (high performance computers, parallel processing, etc). Non-linear analysis, which is becoming widely used still need lots of time and that could be in benefit of users as well. On-line help desks and expert exchange would help in adoptance of such services.

2.6 External supplementary services

Various external services such as manufacturers and product services, virtual libraries, technical information services, professional news, weather reports, expert exchange, etc.

3 USER CENTERED COLLABORATION FRAMEWORK

3.1 Identification of actors and objects

Well-defined roles of participants in the project are crucial for successful team communication and collaboration. We would like to clearly identify actors that will be specific for the new working environment. Especially in geographically distributed business to business environments integration of various systems can be very complicated task. Organization towards unified IT infrastructure, which will bring most from IT, is far more to complex to be carried out by one of the AEC professionals without additional IT background. The importance of such personnel, called CIO (Chief Information Officer), was introduced by Gartner Group and exceeds the fashion type buzzword. It is logical consequence that new type of professionals with strong AEC and IT background will be required. So the role of Construction IT experts will excel current status and that will allow participating professionals to focus on their primary activities and not on IT. Following is the generalized list of the actors that will take part in the system and are the most typical:

- CIO: Chief Information (Internet) Officer should cover most of the Construction IT issues. CIO consolidates, and streamlines IT systems. CIO should provide technology vision and leadership for developing and implementing IT initiatives that create and maintain leadership for the enterprise in a constantly changing and intensely competitive marketplace. Collaborative planning processes can be complicated task while managing not just internal team communication but as well external. Evaluation of enterprise initiatives and overall coordination for initiatives IT infrastructure and architecture running as well as ensuring ongoing investments are made. Sourcing make vs. buy decisions relative to outsourcing vs. in-house provisioning of IT services and skills. Partnerships establishing strategic relationships with key IT suppliers and consultants. Technology transfer provide enabling technologies that make it easier for to communicate between al participants as well as increase revenue and profitability. Training provide training for all IT users to ensure productive use of existing and new systems. Skills Needed: Strong business orientation (broad experience in our industry sector managing IT or related activities a plus (i.e., consulting or vendor in our industry). Demonstrated ability to bring the benefits of IT to solve business issues while also managing costs and risks. Skilled at identifying and evaluating new technological developments and gauging their appropriateness for the business. Ability to communicate with and understand the needs of non-technical internal/external participants. Ability to conceptualize, launch and deliver multiple IT projects on time and within budget. Ability to mesh well with the existing management team by being a good listener, a team builder and an articulate advocate of their IT vision. Personal Qualities: Superb leadership, communication and interpersonal skills; an ability to function in a collaborative and collegial environment; sensitivity to others; high integrity and intelligence; excellent judgment; a conceptual thinker strategic and well as pragmatic; and an ability to generate trust and build alliances with co-workers.

- Project Manager: Project manager is responsible for delivery of the final product (project) within the available time and budget. That must be achieved through available human resources and carefully planned use of IT. He is responsible to keep the track on project activities and major milestones. Coordinates and streamlines teamwork indicates wastes of time, possible conflicts. In the constant pressure of time of the participants in the team, he should be responsible to evaluate the quality of the solutions
- Platform Provider: Provides main infrastructure integrated into selected system.
- Service Provider: Provides a service and send request for registration
- Platform Administrator: Integrates available solutions and service into the platform that enables required level of interoperability between other services and components. Platform administrator also maintains and keep track of changes of existing, already registered services. He is also responsible for implementation of changes on request of CIOs.
- User: User is primary one of the AEC professionals. Besides usual IT skills required for design, he should not be occupied with the use of IT that enables on-line collaboration. He/she must be focused on assigned tasks and exclusively professional issues. Users can be as well general contractors, suppliers, etc.

3.2 Identification of use-cases

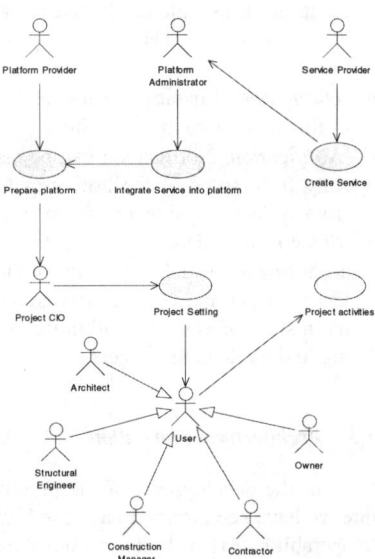

Figure 2: Main Use-Case view of the system

- *User registration.* With registration process user creates new profile and registers to the central database of users. The registration is required for the general use of the service. User profile contains personal information, additional business information, and logon information. Based on personal profile system also enables features such as WAP interface, desirable way of notification and additional professional references and skills. When the user completes registration he gets his own log file repository, he can choose weather he wants to be listed on public directories and what kind of information are allowed to be published. User privileges are assigned based on the role in the specific project. User is identified with unique ID and in the near future digital certificates will replace such identification so that it will be possible to identify user across different services.
- *Project Setting.* Project setting registers the project into the database of projects. Each project has clearly defined project workspace with available components, human resources and supporting services. CIO performs project setting on the request of project manager. Project mangers also defines special requirements. Project settings can be stored as templates for later use within other project. Template includes selected set of compo-

nents, services, team specifics and external connections.

- *Project activities:* Inludes operations on documents such as upload, download, view, change, delete, describe, discuss, compare and notify about.
- *Discussion.* Enables posting and reviewing of available discussion on any item.
- *Notification.* Notification can be send automatically from some of the use cases to the user or from system to system or user to user.
- *Show history.* Displays the history of any object according to available methods of the object. Project manager and team leaders are allowed to see the history of any user within the project according to the role in the project.

3.3 Architectural foundation

In the development of the prototype architecture we have been trying to achieve high level of interoperability, extensibility and transferability at service integration level. At user level our approach should offer customizability and adaptability. Architecture is composed of various layers (Fig 3) .

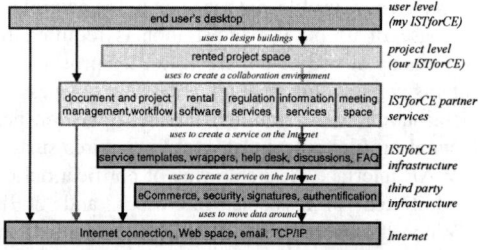

Figure 3: Architecture of the collaboration platform.

Such organization allows to integrate various services on user or project level. A user can rent several design tools for personal use as well as can use other design tools and services within particular project. Project space consists of services that are integrated into the platform using different templates. Templates are supported by e-commerce essentials, security issues. Details of the architectural foundation are described in another paper (Turk and Cerovsek, 2001).

4   PROTOTYPE SESSION

Web design and dynamic delivery of the content are one of the most difficult tasks that will help in adoption of the service. Providing smooth navigation through the project pages according to users privileges and the role in the project is crucial towards our opinion. Improved communication and collaboration by means of web-mediated tools can be achieved by detailed analysis of the user interface. Common impression of available systems on the Internet is that organization of the contents is too information centered instead of user centered. Working on several millions of Euros worth projects under pressure of time, budget and team members, demands well structured simple pages with the most important information. Each professional or even each user has its own original attitude and aims.
Internet technology is changing on daily basis and is becoming very sophisticated. Java, JavaScript, vector graphics, numerous plug-ins and controls enable almost anything that one can imagine. And for commercially available services it is quite tempting to show high tech capabilities to attract the user. The problems of the use of such technology are: *inconsistency of URL and content.* (Users often cannot find the page that they were looking at and is stored in their bookmarks. Using plug-ins, http protocol looses one of the most useful characteristics by giving a page another dimension that actually contains information.), *design takes control over the content.* It is not very rear that pages are over designed and content is hidden behind numerous clicks an windows.

In our prototype we are considering several issues related to dynamically delivered web content to the end user, especially for the front pages. We have implemented the Page Gather algorithm introduced by Perkowitz, M., Etzioini, O. 1998. Page synthesis is a very complex problem and as defined by authors it addresses following issues that seems to be trivial but they are not:

1. What are the contents of the page?
2. How are the contents ordered?
3. What is the title of the page?
4. How are the hyperlinks on the page labeled?
5. Is the page consistent with site's overall graphical style?
6. Is it appropriate to add the page to the site? If so where?

Availability of different user interfaces depends on the role of the user in project. After user is registered to central database (Fig 4) he can start using the system. Within the user interface he is able to maintain personal as well as project-related information and activities.

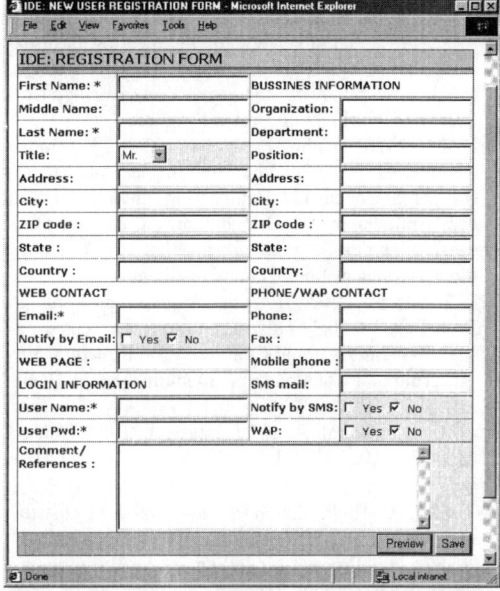

Figure 4: Registration form

User is able to customize his working environment

The end user environment should allow users to work concurrently on multiple projects from a central place.

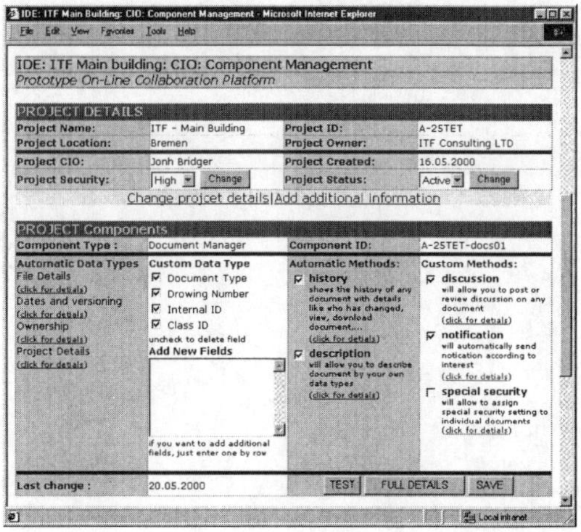

Figure 5: CIO interface showing component management layer.

We believe that

Figure 6: WAP access to project data

5 CONCLUSIONS

Trends show that web-based applications will cover activities starting with pre-project planning, through design, cost and scheduling analysis, supply chain management, to maintenance and finally demolition. It is hard to believe that there will be a service that will cover all these aspects. We believe that an inclusive platform of platforms, presented in the prototype could enable exchange of information that would be of benefit to vendors and users. Today, Internet companies are integrating various systems by acquiring companies offering highly specialized services (such as directories or document management functions) in order to dominate the market. They force the users to use their component which is integrated into the their platform, even though it is not the best one. We believe the platforms should be opened.

Another issue we discovered while playing with the prototype was a potential of the information overload caused by digital information exchange. With paper material, it was hard for the sender to prepare it (draw or type it), expensive to move it around (using mail or courier services) and it also took time by the receiver to understand it. Digital material is easy to create and distribute, but just as hard to understand or use. The bottleneck is just one and it is at the receiver's end.

The provider of collaboration services on the Web accumulates log files of what and how the users of his service were doing. Undoubtedly, the owner of the project data are the design and planning companies. But what about the log files? The log files store the process that the service was supporting and the service provider can, by some analysis and data mining syntheses best practice examples and implement winning process models. Are they allowed to do so? Service providers should offer clear log file policy towards AEC teams and privacy terms regarding tracking user activities within the project.

ACKNOWLEDGEMENT

The presented work is supported by the European Commission, in the frames of 5th framework ISTforCE project. The contribution of the partners in this project as well as that of funding agency is gratefully acknowledged.

REFERENCES

Baker & McKenzie, 2000. Global E-Commerce Law, at http://www.bakernet.com.

Björk, B-C, 1999. Information technology in Construction: domain definition and research issues, in C.J. Anumba, editor, *Computer-Integrated design in Construction*, SETO, London, United Kingdom.

Cerovsek T. & Turk Z., 1999-2000. Evaluation of enterprise solutions for AEC Industry, on-going work.

Fenves, S.J., 1998. Towards personalized Structural Engineering Tools, in I.Smith, editor, *Artificial Intelligence in Structural Engineering*, Springer Verlag, Berlin, Germany.

Fruchter, R., 1998. Internet-based web-mediated collaborative design and learning environment, in I.Smith, editor, *Artificial Intelligence in Structural Engineering*, Springer Verlag, Berlin, Germany.

Furnes, T. A., & Barfield, W. (Editors.), 1995. Virtual environments and advanced interface design, Oxford University Press , New York.

ILPF - Internet Law and Policy Forum., 1999. Survey of International Electronic and Digital Signature Initiatives, http://www.ilpf.org/digsig/survey.htm.

Jacobson, I., 1992. Object-Oriented Software Engineering. Addison Wesley Publishing Company, USA.

Negroponte, N., 1970. The Architecture Machine; Toward A More Human Environment. MIT Press, Cambridge, MA.

Perkowitz, M., Etzioini, O. 1998. Adoptive Web Sites: Automatically Synthesizing Web Pages, *American Association for Artificial Intelligence*

TenLinks Inc., 2000, A Guide to Find Building Codes, Regulations and Standards, at TenLinks home page, *http://www.tenlinks.com/civil/ce-codes.htm*

Turk, Z., 2000, Introduction to ISTforCE, at ISTforCE - Intelligent Services and Tools for Concurrent Engineering home page, http://www.istforce.com/.

Turk, Z., 1997. A Framework for Engineering Information Technologies, in K.S. Pawar, editor, *International conference on concurrent enterprising*, University of Nottingham, United Kingdom.

Turk, Z., R. Wasserfuhr , P. Katranuschkov , R. Amor , M. Hannus in R.J.Scherer, 1997. Conceptual Modelling of a Concurrent Engineering Environment, in C.J. Anumba in N.F.O. Evbuomwan, editors, *Concurrent engineering in construction, Institution of Civil Engineers*, London, United Kingdom.

Turk, Z. and Cerovsek, T, 2001. A Prototype Portal to Web Based Collaborative Engineering, accepted paper, First International Conference on Structural Engineering and Construction, Honolulu, Hawaii, January 24-26, 2001.

Booch G., Rumbaugh J., and Jacobson I., 1998. Unified Modeling Language User Guide. Addison-Wesley, USA.

Wasserfuhr, R., 1998. Shared Process Model for Distributed Co-operative Engineering, in R. Amor, editor, Product and Process Modelling in the Building Industry, BRE, United Kingdom.

Product and Process Modelling in Building and Construction, Gonçalves, Steiger-Garção & Scherer (eds)
© 2000 Balkema, Rotterdam, ISBN 90 5809 179 1

Towards a personalized concurrent engineering Internet services platform

R.J.Scherer
Technische Universität Dresden, Germany

ABSTRACT: Engineers are individuals. Engineers usually participate in different projects in parallel. Therefore they need a personalized workplace with easy access to data, information and knowledge. A generic Internet-based workplace easily configurable on demand, equipped with intelligent tools and basic concurrent engineering (CE) system capabilities is sought. Such a personal CE system must be capable to manage multi-project participation in different virtual enterprises with different client-server systems. The workplace tools should also allow easy access to rental engineering services on the net, such as special engineering software or design code repositories. It has to provide e-commerce and e-payment including proper authorization according to the employee status of each engineer. The workplace must be personalizable to individual working styles and sustainable over long time. Such a workplace is like a service platform. The platform should be capable of plugging into servers of different virtual enterprises. For this, a new plugging-in technique is developed, which will also provide access to the electronic market place and support e-commerce as a business model. The user-relevant information like information and knowledge management data will be kept on the platform, whereas mass data storage should be outsourced.

1 INTRODUCTION

In recent years much effort has been made in the development of multi-tier client-server systems in order to manage project information more error-free and, extended by information logistics components, provide information to the right person at the right time and allow and support concurrent engineering, i.e. co-operative, collaborative and simultaneous working [1, 2]. These systems are now extended by agents in order to outsource functionality, make it generic and re-usable in the form of services. However, in building construction practice only the electronic document management components are applied nowadays, sometimes extended by some workflow functionality. They support collaborative and co-operative working, support red-lining and to some extent also version management. On the research level, workflow is highly developed and allows the users a dynamic interaction with the system in order to adapt the project workflow to any unforeseen impact on the project progress. Project activities and user activities are synchronized permanently by the system and the system provides each user with a transparent view of activities of the whole project, which allows the user to decide upon the

sequence of doing his tasks in a way optimising the project progress [3]. Usually the user downloads and uploads information from the system by hand, sometimes the system does it for him when he starts and finishes a work task. All these systems have one thing in common. They are project-centred. They are designed to serve one project and support team work in a virtual enterprise [Fig. 1]. During a running work task the user is self dependant for the information management on his private work place.

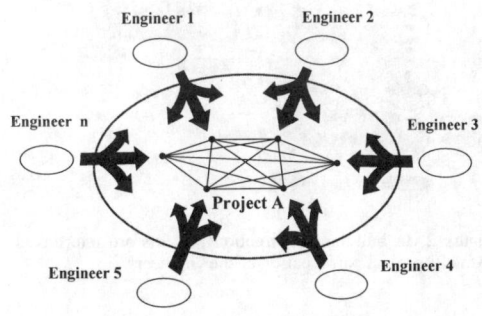

Figure 1. Project-centered virtual concurrent engineering team

This kind of working, which may often occur in some engineering disciplines, i.e. that a team of engineers is working for one and only project, is very seldom in the building construction and civil engineering industry. There, due to the high-grade granularity of most of the projects and companies, the short project time and because for each project a new team of companies is established, i.e. long-term virtual or extended enterprises do not exist in building construction – coordination of one team around one project is not sufficient at all, for both optimising the project and serving the engineer. Each engineer is usually involved in several projects in parallel. This means a project is not independent from others but all projects are interlinked with each other via the engineers to some extent [Fig. 2]. Therefore projects are currently hampered by ad hoc changes of human resources provided and the engineers are hampered by the different software technologies prescribed by the different project, which they have to learn in order to use and participate in the project.

The work in building construction must be seen project-centred as well as engineer-centred. The objectives are therefore twofold. First, a workplace is needed which serves the engineer's individual needs through easily accessible engineering services and through a personal information and scheduling management system with a long-term stability. Second, these personalized services has to be complemented through a project management system, which manages the cross project coordination of work tasks and human resources.

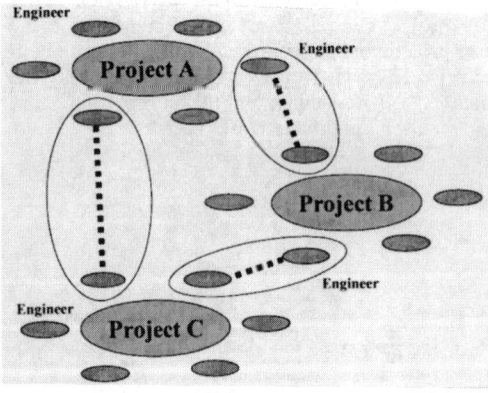

Figure 2. In building construction projects are interlinked by multi-project participation of the engineers.

2 OBJECTIVES FOR THE CONCURRENT ENGINEERING SERVICE PLATFORM - CESP

The work space for an individual engineer needed can be considered an assembly of different smart services. These intelligent services are to be embedded into a platform that provides a framework for the services and a standardized generic interface to the outside engineering server world. This means an appropriate methodology as well as an appropriate technology has to be developed.

The methodology has to cover such aspects like the principles of human interaction with technology, the role of technology, the model of engineering design and an engineering ontology for formulating requests in the engineering terminology and receiving system responses in an engineering way of understanding, i.e. properly arranged tables and graphics, like technical drawings and VR representations highlighting technical aspects of behaviour and work progress. The technology has to contain a meta-model to properly handle and integrate the components, adapters, server specifications, transactions and interoperability methods and a meta-model that allows to plug-in any kind of server, as long as the server follows some given specifications. This offers the user independence of the client-server system of a particular virtual enterprise. He can participate in any virtual enterprise without purchasing or learning new software systems. Moreover, he will now be able to participate in several projects that are running in parallel and/or different virtual enterprises at the same time, use the same design and management tools in his work space and manage and co-ordinate different projects and their data in his personalized manner. He remains with his own familiar tools in his familiar working environment, i.e. that the working space will get sustainable.

The concurrent engineering service platform has to offer the user a working environment with the following properties:

1. Independence. The interoperability services and the meta-model based framework of the platform should make the user independent of any kind of server in the different CE worlds in which he would like to participate in. The services should allow him to attach to any kind of server, to map the data into his unified form and to properly organize his multi-project dependent tasks with his personal planning service. An information logistics management system is needed, which will support him to keep track of the proper information flow.

2. Individuality. The platform should allow the user to carry out his work according to his individual preferences and abilities. It should provide him with services and tools which have individually configurable and adaptable interfaces. They will give him the freedom to choose the sequences of doing his work according to his individual preferences and they will keep assisting him in handling complex data and information structures inherent in the CE systems.

3. Capability. The CESP will increase his personal capabilities of doing sophisticated and specialized engineering work as well as of doing daily routine work more efficiently by providing him with the just-in-time knowledge related to engineering and about the IT system. This can be achieved through knowledge-based assistants, knowledge-based services and easy access to external engineering services and knowledge repositories (e.g. code libraries), which are nowadays already available on the global e-market place.

4. Sustainability. Due to the independence of the platform and due to the adaptability of the tools and services installation and configuration the user is able to accumulate his investments in the platform services according to his personal individual kind of working. The independence of the platform will shield him from the permanent evolution of the CE systems, i.e. new versions, interfaces, semantics and paradigms, which considerably increase the system's capability, of course, but on the other hand decrease his personal working capacity, because otherwise he has currently to learn and adapt them. This is of utmost importance when humans often have to work in different environments, as it is the case in building construction.

5. Lean. Data warehousing, safe transaction techniques, long-term archiving and data security are techniques, that are already widely developed and services are going to be provided by specialized companies and general data management service providers. These services are expected to be outsourced in order to keep the individual CE services platform lean from such non-individual services, which are more profitable to rent and focus the platform on the logistics and management of the information and knowledge he needs.

3 SERVICES

According to their functionality the services can be grouped into four different categories,
1. interoperability and information management services,
2. knowledge-based services,
3. e-commerce services,
4. personal user support and help services.

A particular service may be grouped into one or several categories. For the prototype to be devel-

oped [3], eight different characteristic services are selected (Fig. 3). These are:
5. Interoperability Services - IOS
6. Knowledge Based Model Access Service – MAS [5]
7. Knowledge Based Design Assistance Services – DAS [6]
8. Knowledge Based Code Checking Services – CSS
9. Remote User Specialized Rental Engineering Services – RES [7, 8]
10. Technology Support Tools for E-commerce Services - ECS
11. Training and Online Human Support Service – TOS
12. Personal Data Management and Planning Service –PPS

From these services each even-numbered one belongs to two categories. These are the services placed at the four corners in Figure 3. On the left edge of the platform in Figure 3, the interoperability and information management services are shown, on the top edge the knowledge-based services, on the right edge the e-commerce services and on the bottom edge the individual support and help services.

4 PLATFORM FRAMEWORK

The modelling framework of the service platform will be based on the modelling framework developed in ToCEE [9] and accordingly extended. This framework has five layers:
1. The Meta model layer sets the basic principles of the whole modelling paradigm. It serves for the formal definition of all allowable basic and user-defined data types as well as for the

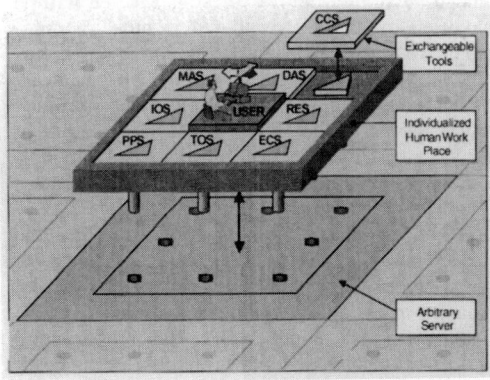

Figure 3. Concurrent engineering Internet service platform – CESP.

93

generic definition of object classes (called "concepts" on the Meta model level), containing a set of attributes describing its state and a set of operations defining its behaviour.

2. The Kernel model layer defines in three schemas the high-level generic concepts, which are common to all lower level models representing Product, Document and Process related information: (1) TC_IfcKernel, the ToCEE adaptation of the IFC Kernel schema, (2) TC_Communication_Model, which contains information on the system itself and the information processes in it, and (3) TC_Model_Population which has a similar purpose as the SDAI dictionary model (ISO 10303-22) [11], but is extended to include important for CE environment meta model information.

3. The Neutral model layer presents the concepts for each modelling perspective, i.e. Neutral Product Model, Neutral Document Model, Neutral Process Model, Neutral Contract Model and a common high-level Conflict Model. These neutral models are implemented respectively in each data management server of the CE environment.

4. The Aspect model layer further specializes the Neutral Product Model for the selected domains in building construction by defining aspect models, e.g. for architectural, structural, HVAC and geotechnical design.

5. At last, the Application model layer contains the native models of diverse applications used in CEE.

In addition to these five model layers, a set of independent resources are available to all other models for referencing – except for the application models, which however can copy constructs that are found useful.

The ToCEE modelling framework is an extended version of the IFC version 1.5.1. It will be modified in such a way that it is upwards compatible to the IFC version 3.0, which is under current development by the IAI [10]. The ToCEE meta model contains information about a product model stored in the product model repository of the system. This is needed in order to enable individual processing of local models for each actor together with a shared product data space for cooperative work, as well as for supporting the needed model interoperability features. The use of meta model information in the product model data server is based on implementation of the SDAI dictionary model (ISO 10303-22) [11] with specific extensions added to enable the operability of CE. It contains constructs for the concept entities *Model*, *Model Schema*, *Mapping* and *Mapping Schema*. Through the meta data a model "knows" not only the entities it contains, but also

who is the user who created it, which users are allowed to read/write/modify the data, is the model checked out (i.e. locked for changes), which is its up-to-date version, what mapping specifications are associated to it and for which other models, what is the base reference model to be used for consistency checking etc.

4.1 Plug-in method

The integration of the different services, servers and clients is achieved by a formal specification of the functionality and a semantic integration of the components. There, the data structure of the data which can be manipulated and stored are described. The specification consists of a set of EXPRESS-C specifications, an extension of the EXPRESS language, which is part of ISO 10303 STEP [11]. Each server plug-in implements a disjoint subset of the global concept registry. Concepts are classes in the sense of object-oriented programming but additionally they know about their instances and they are available at runtime. Services and servers are servers and clients as well depending on the particular communication event, whether they are sending a request or receiving a response. Therefore some of the components of the CESP can be both. All components which offer server services to the platform are registered in the platform server registry.

4.2 Platform architecture

The platform is a kind of middleware but enhanced with project knowledge, actors, responsibility, etc., which allows real information logistics. Whereas in ToCEE information logistics was concentrated on project, in ISTforCE information logistics is primarily focussed on the platform user and secondarily on the various projects the user is dealing with. Nevertheless the platform is nothing else than a client-server system for which the ToCEE 4-tier client server architecture [12] will be used as the start-up architecture. For one of the services, the knowledge based model access service, the client-server system will be enhanced by agents [5]. All platform servers are clients as well, because they store only the data, which were downloaded by the user for some while from an external project server. Therefore we do have two plug-ins [Fig. 3]. The external server platform plug-in and the services platform plug-in. Both are formally identical and are described in EXPRESS-C as mentioned above.

4.3 Interoperability

For interoperability, we have to distinguish between system interoperability and model interoperability. System interoperability will be achieved by the UPRL (uniform process resource locator) method developed for ToCEE [5]. UPRL extends the standard WWW addressing technique towards co-ordinated concurrent access to shared object models. It allows client applications to manipulate server-side documents or objects, based on a formal interface definition, which is expressed in EXPRESS-C. An UPRL is defined as URL plus project semantics, like person, work task, document, object and view plus methods, like notify, get, download and modify. Each object request with an UPRL produces an information container [9] as response output which can be used by a client application to work with the data contained in the container.

Model interoperability which may also be called functional interoperability means the ability to support at run-time the model and data transformations needed in the information exchange and sharing between the components of the integrated environment. Such transformations are necessary for example when changes in one local aspect model, source model, have to be propagated and checked against the constraints of another local, discipline-specific aspect, the target model, e.g. architectural spatial model to structural model. This is achieved with the help of methods like *model mapping*, *model matching*, *consistency checking* and *model merging*, which are available as prototype methods from the ToCEE project.

Model mapping means the use of special specifications and methods for the transformation of the modelling objects of one schema to another. Model mapping involves the conversion of one modelling representation (source) into one or more other modelling representations (target) without awareness of the context, i.e. already existing local data in the target. Model mapping is realised in the following way:

1. The *mapping specifications* defining the equivalences between the different data models are maintained explicitly in the product data server.
2. The *mapping operations* are performed by using these mapping specifications, invoked through calling the server functions 'map' or 'mapTry' for the appropriate data models.

A mapping operation always creates new data. In principle, these can be a new model schema, new object classes in an existing schema and complete or partial transformations of object instances belonging to one model into object instances of another model. Each mapping operation must be able to deal with the cardinality cases 1:1, 1:N and M:N.

4.4 Data storage strategy

The platform is intended for storing only the data, which are notices, versions and alternatives developed by the engineer for seeking solution for his engineering tasks. For data, which are not currently used only the UPRL address is stored in order to provide the user with newest data version at any time. For his private data also an external server provided by an Internet provider may be used in order to reduce data storage solely on the storage of UPRLs. This means that data storage will be reduced to information management data, only. Product data views as well as documents will be described as objects, containing their meta data as attributes. In conjunction with the information logistics component a personal management of these design data is possible. The data from external servers are kept via the information service by reference and can be downloaded on the CESP on demand. For data storage third parties' servers can be used as well due to the extended URL reference mechanism. The personal planning system will be complemented by a remote storage service, which offers users the ability to store their personal data in a reliable high-end security storage infrastructure and access them at any location. This will free users from individual backup and recovery tasks and offer them a platform for life-long personal information management, while security standards like SSL will ensure that a state-of-the-art level of trust is available for such an information bank.

4.5 Cross project activity management

The personal activity planner will be interlinkable to the workflow systems of the several projects of a user, allowing to merge the different workflow data of the user to an homogeneous activity schedule, keeping track of proper authentication and authorization. At the first look, the personal planning system will come as a messaging tool, allowing users to send messages, but additionally messages can be linked with tasks. With the help of an interactive graphical interface the user can rearrange tasks, can break down his tasks in smaller units, publish their work load on a shared calendar, give them priorities and outline dependencies with the other tasks. Tasks are modelled as objects.

The interlinkage of the platform workflow system with workflow systems of all the projects the user is involved in gives him transparency to the activity schedules of all projects and he will always obtain an actual worklist served by all projects. This allows him to properly decide upon the priorities of his tasks and on the contrary he will be properly monitorable by the project managers of each project, too.

ACKNOWLEDGEMENT

The support of the Commission of the European community during 1995 through 1999 under contract no. ESPRIT-20587, ToCEE, which allowed us the develop the multi-tier client-server system, which is applied her as the start-up system, is acknowledged. The ongoing support (2000 through 2002) of the CEC under the programme IST, contract no. IST-1999-11508 is very much appreciated and will allow us to develop the above outlined Concurrent Engineering Service Platform.

REFERENCES

[1] Scherer R.J. 1998. *A Framework for the Concurrent Engineering Environment*. Proceedings of the 2nd European Conference on Product and Process Modelling in the Building Industry, Amor R., Scherer R.J. (ed.), Building Research Establishment, Watford, Great Britain.

[2] Storer G. 1999. *VEGA Final Report*. Taylor Woodrow, London, Great Britain.

[3] Wasserfuhr R., Scherer R.J. 2000. *Distributed Management of Co-operative Design Processes.* Proceedings of the conference Construction Information Technology 2000, Gudnason G. (ed.), Reykjavik, Iceland.

[4] Scherer R.J. 2000, Fact sheet of the EU project IST-forCE, Dresden University of Technology, Institute of Applied Informatics in Civil Engineering, Dresden, Germany.

[5] Scherer R.J., Katranuschkov P. 1999. *Knowledge-Based Enhancements to Product Data Server Technology for Concurrent Engineering*. ICE 99 Den Haag, Netherlands.

[6] Scherer R.J., Gehre A. 2000. *Approach to a Knowledge-based Design Assistant System for Conceptual Structural System Design*. Proceedings of the 3rd European Conference on Product and Process Modelling in the Building Industry, Gonçalves R., Scherer R.J. (eds.) Balkema Rotterdam, Netherlands.

[7] Mangini M., Protopsaltis B. 2000. *E-Commerce: A New Frontier for Engineering Software and Services*. Proceedings of the 3rd European Conference on Product and Process Modelling in the Building Industry, Gonçalves R., Scherer R.J. (eds.) Balkema Rotterdam, Netherlands.

[8] Červenka J., Pukl R. 2000. *Modelling of Building Structures in the Web*. Proceedings of the 3rd European Conference on Product and Process Modelling in the Building Industry, Gonçalves R., Scherer R.J. (eds.) Balkema Rotterdam, Netherlands.

[9] Scherer (ed.) 2000. *A multi-tier Client-Server System for Concurrent Engineering*. Final Report of ToCEE, Dresden University of Technology, Institute of Applied Informatics in Civil Engineering, Dresden, Germany.

[10] International Alliance for Interoperability (IAI); Liebich, T., See, R. (eds.). *IFC Object Model Architecture Guide*. Final, Oakton/Virginia: PDF file, 1999.

[11] ISO 10303 -1, -11, -21, -22: *Product Data Representation and Exchange* - Parts 1, 11, 21, 22, ISO TC184/SC4, Geneva., 1994-96.

[12] Wasserfuhr R., Scherer R.J. 1997. *Process Models Supporting Information Logistics in Concurrent Engineering for the Building Life Cycle*. Proceedings of the International Conference on Concurrent Enterprising, pp. 141 - 152, , K.S. Pawar (ed.), Nottingham, Great Britain.

Product and Process Modelling in Building and Construction, Gonçalves, Steiger-Garção & Scherer (eds)
© 2000 Balkema, Rotterdam, ISBN 90 5809 179 1

Testing of building structures in the web towards virtual labs

J. Červenka & R. Pukl
Červenka Consulting, Prague, Czech Republic

ABSTRACT: Design of modern building structures needs advanced tools based on the current knowledge about structural and material behavior. These innovative tools should be available due to correspondingly innovative communication technologies, i.e. Internet and e-commerce. Design engineer should effectively rent or buy professional services and software needed, including training, user support and knowledge database. During the EC funded ISTforCE project an on-line virtual testing laboratory for concrete structures will be developed. It should enable virtual simulation of the real structural behavior via Internet. This laboratory will be based on SBETA-ATENA software provided by Červenka Consulting and it will be plugged in to the Concurrent Engineering Service Platform (CESP) that will be developed by the ISTforCE consortium.

1 INTRODUCTION

Computer simulation of real structural behavior enables to create a virtual model of the civil engineering structure, subject it to designed conditions and investigate its response. It brings a new quality in structural design and development. It can give an answer to many complex structural problems, but it also poses new questions and opens new undeveloped areas. One of them is a possibility of an Internet-based access to the available simulation tools and an effective on-line user support.

During the EC funded ISTforCE project (Intelligent Services and Tools for Concurrent Engineering (CE)) a Concurrent Engineering Service Platform (CESP) will be developed. The focus of this platform is an engineer working in the CE world. It should support its individuality and provide him with various services and tools. One of the Rental Engineering Services (RES) included in the CESP platform will be an on-line Virtual Testing Laboratory Service (VTLS) for concrete structures. It will enable virtual simulation of the real structural behavior via Internet. This laboratory is under development by the authors, and its main features and components are described in this paper. The virtual test laboratory will employ SBETA-ATENA software provided by Červenka Consulting.

Figure 1 View of an offshore oil platform

2 SIMULATION OF STRUCTURAL BEHAVIOR

Computer simulation that is used in the VTLS is based on a finite element numerical model of structure, constitutive model of material behavior and non-linear solution techniques. It is an extension of the linear finite element method into the range of non-linear behavior, with emphasis on exploitation of rich sources of knowledge from material engineering.

The authors are involved in the development and application of the commercial software SBETA-ATENA for non-linear analysis of concrete and reinforced concrete structures. Over a hundred users around the world are using the program for design, research and development. It serves as a tool for assessment of the load bearing capacity of designed or existing structures, for investigation of structural damages and failures, or for developing new structural systems (Margoldová et al. 1998).

An example of such investigation is the computer simulation of an offshore oil platform (Fig.1). An offshore platform is a large complex structure requiring employment of advanced design methods. Failure of one construction detail can lead to the total collapse of the structure, which happened several years ago. Reasons for this failure were explained using SBETA nonlinear finite element analysis.

The investigated platform was a large cellular reinforced concrete structure, its section is shown in Fig. 2. During construction it undergoes submerging for deck mating and then it is again elevated and positioned in the site. The tri-cell walls must be designed to resist the water pressure due to the submerging. The analytical model for SBETA was a symmetrical section of the tri-cell wall (Fig. 2). Two various reinforcement setups has been compared – with short and with long transversal bar (Fig. 3). The same structural detail has been investigated also experimentally by to a Norwegian engineering company. Loading due to water pressure (Fig. 3) has been increased up to failure of the tri-cell wall. The failure in SBETA simulation has been caused due to a combination of normal and shear forces.

The case with the short bar failed at the water depth of 70 m. The failure bypassed the short

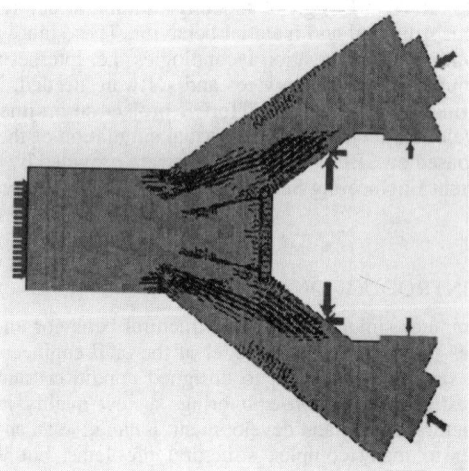

Figure 4. Crack pattern for the analysis with short bar

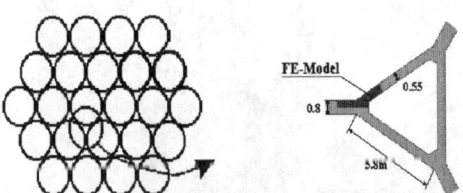

Figure 2. Plane section of the offshore platform and detail of the tri-cell section with the analytical model

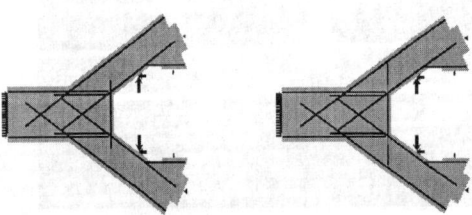

Figure 3. Water pressure loading scheme and reinforcement setup with the short (left) and the long (right) transversal reinforcing bar

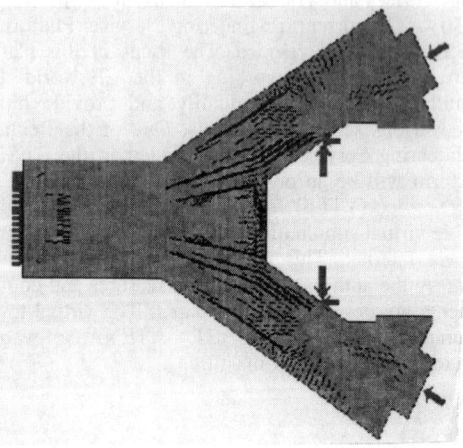

Figure 5. Crack pattern for the analysis with long bar

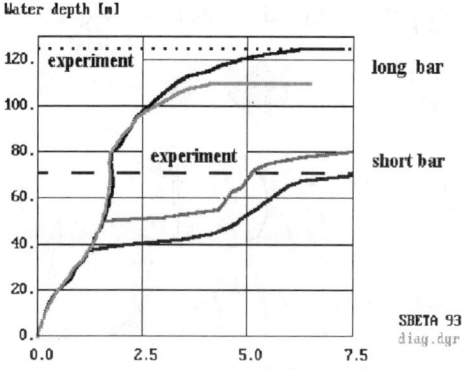

Figure 6. Comparison of critical water depths from SBETA analysis and experimentaly obtained values.

reinforcing bar and was mainly due to the development of a diagonal shear crack. The crack pattern is shown in Fig. 4. The case with the long bar failed at the water depth of 110 mm. The failure was due to combination of reinforcement yielding and concrete damage (Fig. 5).

The short bar, used in the structure instead of the long one, decreases the shear resistance of the tri-cell wall substantially. This mistake initiated the platform collapse. The same conclusions resulted form the above-mentioned experimental investigation. Experimentally obtained critical water depth and load-displacement diagrams from the SBETA simulation for both cases are compared in Fig. 6. Multiple curves for the SBETA simulation results from analysis with various finite element meshes (coarse and fine).

Another useful area for the application of non-

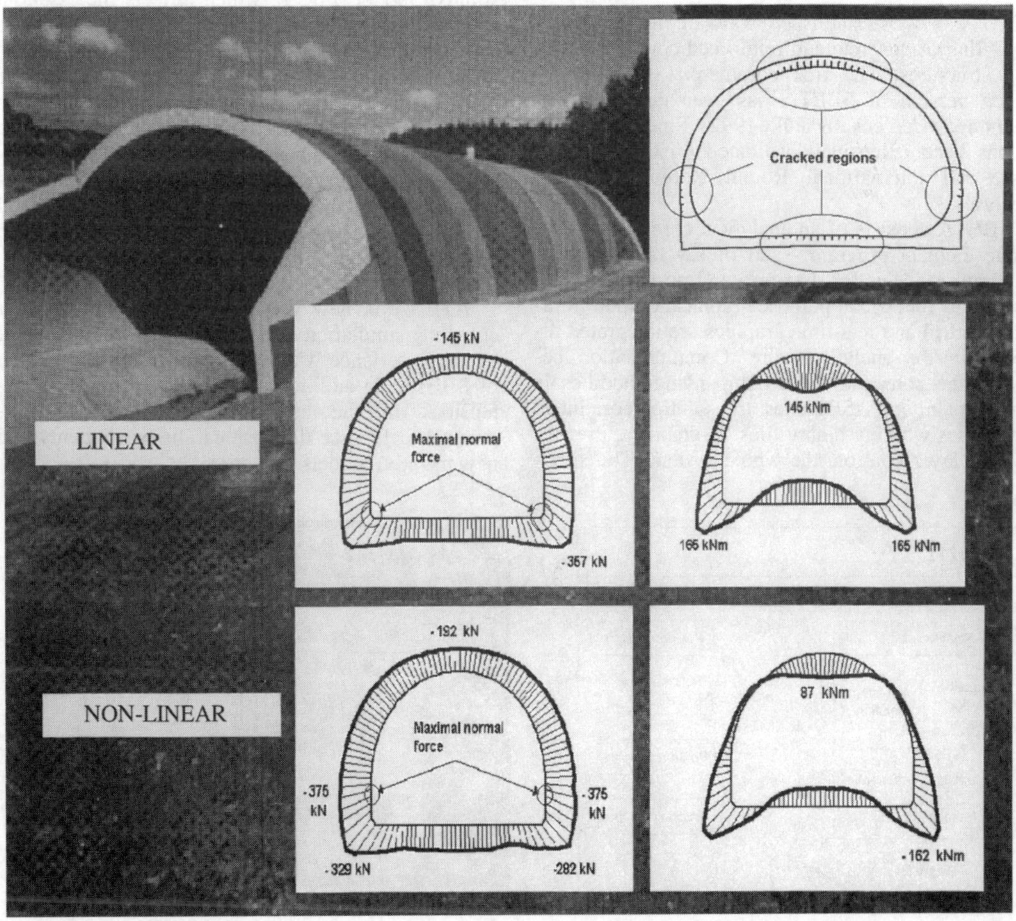

Figure 7. Example of tunnel analysis showing the reduction of moment forces due to non-linear analysis.

linear analysis is the design of reinforced concrete structures. Only non-linear analysis can solve the conceptual problem in the current design methods that assume non-linear material behavior for cross-section check but not for the calculation of internal forces. In reality the internal force distribution is strongly effected by the non-linear material behavior in the statically indeterminate structures. In some cases this redistribution of internal forces can be favorable for the construction as it is demonstrated on the second example, which is a small-prefabricated tunnel under a highway communication. As it is documented on Fig. 7, there is a favorable transfer from moment forces to normal forces. This can be utilized in the design phase of the structure to decrease the amount of necessary reinforcing, which can induce substantial savings of the construction cost.

3 SBETA-ATENA SOFTWARE

SBETA is well-established software for computer modeling of concrete and reinforced concrete building structures. The first commercial FORTRAN-based version of SBETA has been developed 10 years ago (Červenka & Pukl 1992). Since this times it has been extensively validated on experimental data and international Round Robin prediction analyses.

SBETA consists of an analytical core – nonlinear finite element program - and of several additional program modules for data pre- and post-processing, as well as for special purpose graphical output. Solution control and real-time graphics are integrated directly in the analytical core. Communication between the core and the accompanying modules is based mainly on ASCII files, the solution core internally uses working binary files. A graphical user interface layer controls the whole system. The struc-

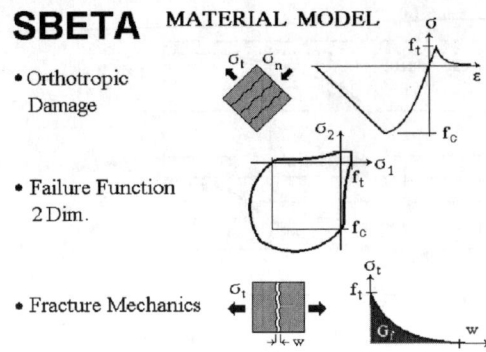

Figure 9. Overview of the concrete material model

ture of the SBETA system is schematically drawn in Fig. 8.

SBETA solution core is based on nonlinear constitutive model (Fig. 9), which reflects the essential features of the material behavior (Červenka 2000). In the case of concrete these features are brittle behavior in tension resulting in crack propagation, and the confinement effect under multi-axial stress states resulting in stress-dependent strength. The continuum mechanics model is based on the smeared crack concept and the damage approach. Objective solution is assured due to the employment of the crack band theory. Furthermore, a wide range of tools has been developed in order to support the features of reinforced and pre-stressed concrete.

ATENA is new generation of the software for computer simulation of concrete structure. It is based on the experience with development and application of SBETA. In addition, it offers broad range of extensions in the analytical core such as axisymmetrical, three-dimensional finite elements and new material models.

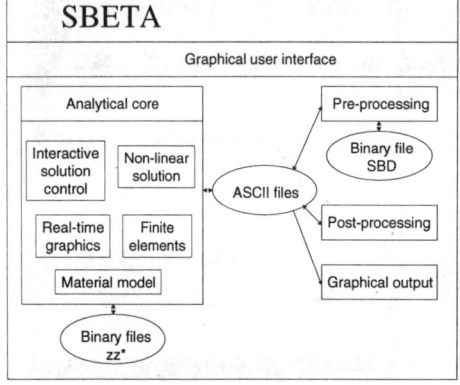

Figure 8. Scheme of the SBETA system

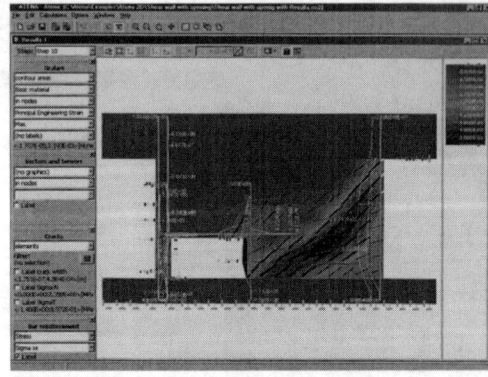

Figure 10. A typical outlook of ATENA interface during post-processing.

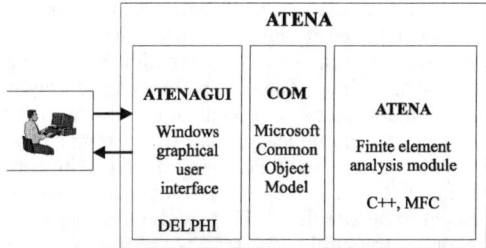

Figure 11. Layered concept of ATENA.

ATENA is conceptually object oriented and based on the MS-Windows environment. User-friendly graphical interface offers an excellent support to the various projects can access and use. The right figure indicates how an external service is part of a personalized engineer's platform, where it can be accessed when necessary.

The virtual laboratory performs advanced numerical simulations of reinforced concrete structures as it is shown on examples in previous sections of this paper. The virtual lab service can be used in almost all stages of the construction process. In the design phase it is a useful tool for checking the design by enabling more precise calculation of structural load carrying capacity or service state deformations. During the construction stage, it is an excellent tool to explain unexpected crack development or to evaluate an effect of various unexpected conditions that may occur on the construction site. The simulation of real structural behavior can be also utilized during the maintenance of the structure, where it can be successfully applied for assessing the influence of material deterioration or various accidents on the structural performance.

The structural simulation cannot be a fully automated process. Human interaction is necessary for the following tasks:

- Definition of additional input parameters that are necessary for the non-linear analysis and are usually not part of standard architectural data. Here a communication between a client and a consultant is necessary since consultant has the necessary technical knowledge but needs specific problem information from the client.
- Human communication and intervention is needed to define the objectives of the simulation.
- Communication is also important after the simulation is completed in order to clarify and explain the obtained results to the client.
- Human support is necessary for the quality control process, before a final report is sent to the client.

However, it is important to identify repetitive tasks that can be automated and eliminate the dupli-user during all stages of the nonlinear analysis. Layered structure of the program employs COM-based communication between the analytical FE program and the graphical user interface, which integrates pre-processing, post-processing and real-time interactive graphics and control. Typical outlook of the graphical user interface can be seen in Fig. 10. The layered concept of ATENA, shown in Fig. 11, enables various concepts for the Web implementation.

4 VIRTUAL TESTING LABORATORY

A platform will be developed during an ISTforCE project that will support a civil engineer working on several different projects. This Concurrent Engineering Service Platform (CESP) will incorporate the IFC based product model for data interoperability, workflow management support for balancing the workload, knowledge based access to product model data and knowledge based design assistants. The platform can be customized according to engineer's needs and preferences. Various services can be registered at the platform in order to support the engineering work. The virtual testing laboratory is one of these services. Its relationship to the CESP platform and to an engineer working on various projects is schematically shown in Fig. 12. The left part of the

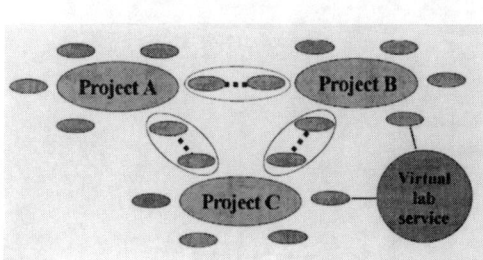
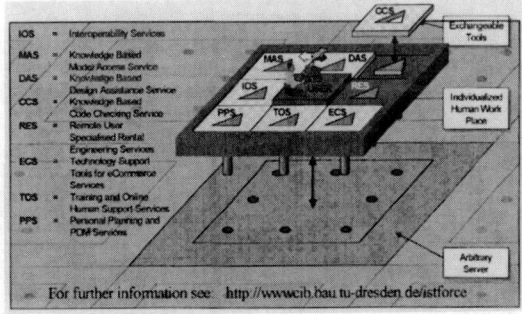

Figure 12. The scheme of the ISTforCE platform and its support for multi-project participation.

figure indicates the relationship between engineers working on different project. In this scenario, the VTLS service is a tool that various engineers from cation of work. The following actions in VTLS can be easily automated:

- Transfer of structure geometry and its conversion into a numerical model.
- Generation of additional input parameters for the nonlinear analysis. This can be automated only partially, and human intervention will be needed for verification and quality control.
- Generation of reports and animation sequences from the performed simulations. Re-

ports consist mostly of figures and description of parameters and settings that were used in the analysis. All this information has already been defined. Thus the majority of the report content can be generated automatically.

A typical communication and data exchange between the client and VTLS is depicted in Fig. 13. The input for the virtual lab is a description of the problem to be simulated. This can be a whole structure or a structural element. The geometry of the structure can be defined in the format of a STEP physical file containing Industry Foundation Classes (IFC) objects. The VTLS development is based on

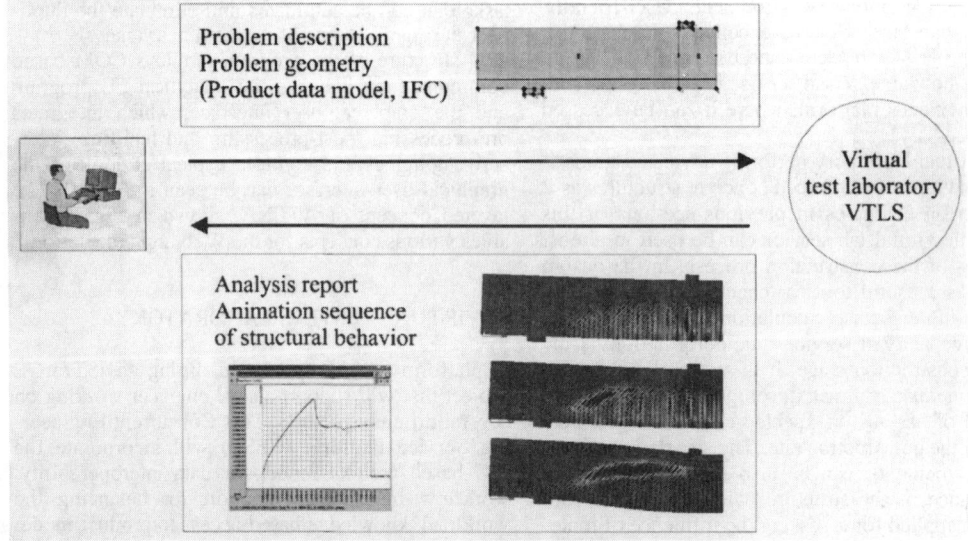

Figure 13. Typical data exchange between an engineer and VTLS.

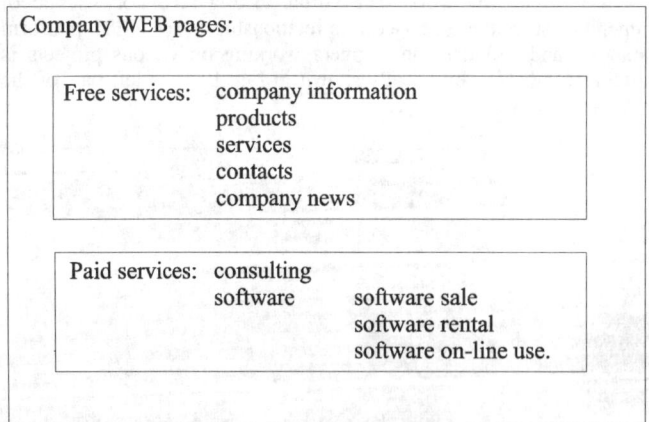

Figure 14. Main structure of VTLS WEB pages.

the IFC 2x standard. The STEP physical file can be directly imported into the finite element software, where the IFC data is transformed into the internal geometrical representation. Since the IFC version 2x does not contain any classes for the description of boundary conditions or actions, they will have to be inserted manually by the VTLS support stuff. The additional data will be identified based on the supplemental description that has to be provided by the client or through human communication. In order to support the document exchange and interaction between the consultant and the client VTLS will contain a rich set of support services. The individual support services are shown in Fig. 15. The figure shows individual components of the VTLS service that are used to support the client consultant communication during a simulation project. Before a prospective client can utilized the VTLS, he needs to obtain an authorization to access the service. After a service is consumed, invoice is generated and sent to the client. An authorization is required only for ac-

cessing the paid services. The virtual lab will be accessed via an internet browser and certain services and information will be available free of charge (Fig. 14). The other paid services will include:
- On-line software purchase.
- Software renting.
- Remote software execution. This module is developed by other partners in the ISTforCE consortium, and therefore is not discussed in this paper.
- The client-server technology implemented in ATENA (see Fig. 11) enables also another form of remote program usage that may be advantageous in certain cases. In this approach, only the graphical user interfaces can be downloaded to the client computer, and the non-linear analysis can be executed remotely on the VTLS server. The communication between the graphical interface and the solution module will occur via DCOM. This approach may be advantageous when the client is famil-

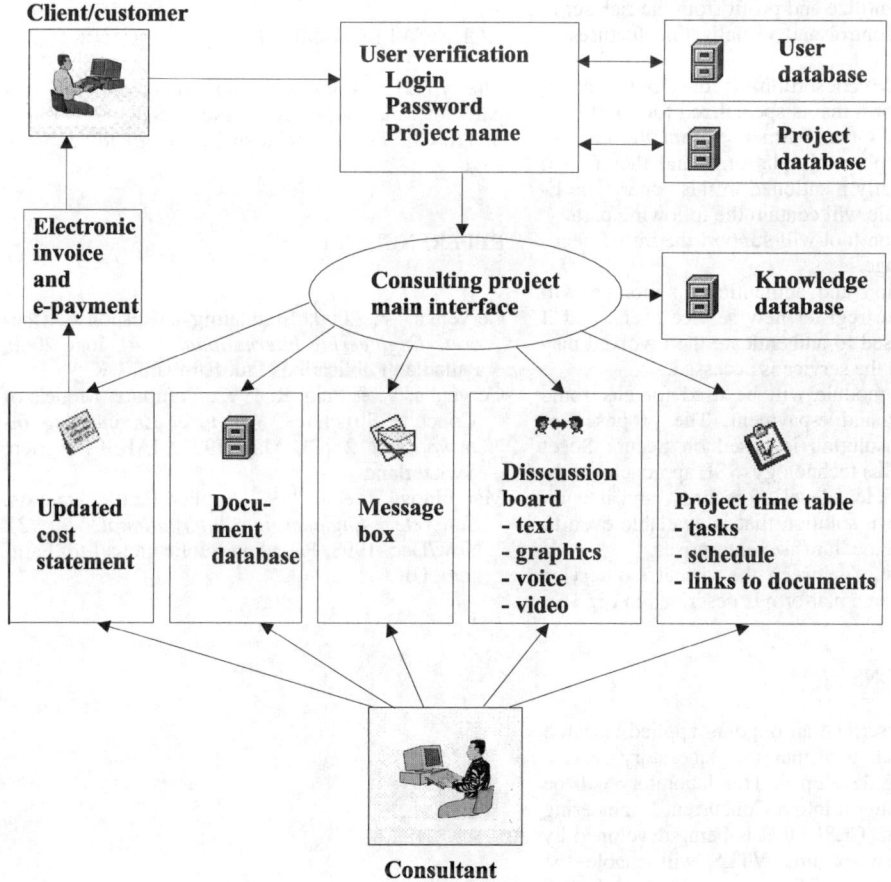

Figure 15. Individual support elements for the VTLS consulting service.

103

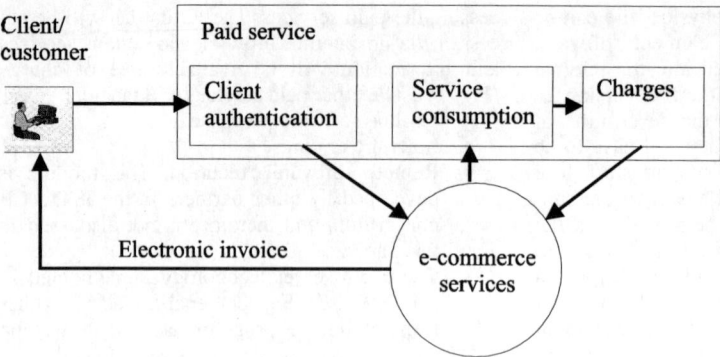

Figure 16. The relationship between VTLS and e-commerce services.

iar with ATENA program, but does not have the necessary computer power at his current location. Other advantage is that in this scenario both the client and the consultant may control and observe the analysis, and both of them can utilize and profit from the rich set of runtime control and visualization features of ATENA.

The ISTforCE consortium is developing an e-commerce platform that is specialized for consulting service charges. Other partners within the consortium are developing this platform, and therefore it will be only briefly mentioned in this paper. The E-commerce module will contain the following parts:

- Negotiation tool will support the initial negotiation stage.
- Registration and authentication module will be used to register new service users, and it will be used to authenticate the involved parties when the service is accessed.
- Payment module will be used for electronic invoicing and e payment. The proposed e-payment solution is based on Secure Socet Layer (SSL) technology. SSL approach exhibits a sufficient level of security, and it is an inexpensive solution that is available even to small and medium size enterprises.

The relationship between the virtual lab service and the e-commerce platform is described in Fig 16.

5 CONCLUSIONS

This paper described an ongoing applied research project, in which a virtual test laboratory service (VTLS) is being developed. This laboratory will be available as a plug in into a Concurrent Engineering Service Platform (CESP) that is being developed by the ISTforCE consortium. VTLS will enable the renting of advanced non-linear analysis of reinforced concrete structures on the Internet.

The VTLS utilizes the modern concept of the ATENA software. Its solution engine and material models are based on the extensively validated program SBETA.

ACKNOWLEDGEMENT

The VTLS development and implementation is a part of the EC-project No. IST-1999-11508 - ISTforCE. This financial support is greatly appreciated.

REFERENCES

Červenka, V. 2000: Simulating a Response. *Concrete Engineering International 4 (4),* June 2000, Palladian Publications Ltd, Farnham, UK.

Červenka, V. & Pukl, R. 1992: Computer Models of Concrete Structures. *Structural Engineering International 2 (2),* May 1992, IABSE, Zurich, Switzerland.

Margoldová, J. et al. 1998: Applied Brittle Analysis. *Concrete Engineering International 8 (2),* Nov./Dec. 1998, Palladian Publications Ltd, Farnham, UK.

Virtual enterprise and e-Business

Product and Process Modelling in Building and Construction, Gonçalves, Steiger-Garção & Scherer (eds)
© 2000 Balkema, Rotterdam, ISBN 90 5809 179 1

Configuration management in large scale infrastructure development

Th. P. J. van Rijn, H. van de Belt & R. H. Los
TNO Building and Construction Research, Delft, Netherlands

ABSTRACT: Large Scale Infrastructure (LSI) development projects such as the construction of roads, railways and other civil engineering (water)works is tendered differently today than a decade ago. Traditional workflow requested quotes from construction companies for construction works where the works to be performed were completely specified by others. Changes such as the introduction of Design-Construct contracts and Build-Operate-Transfer contracts also change the role of the construction company. The construction company becomes more of an information manager than a "concrete pourer". With Configuration Management the information will be managed more explicitly. Therefore the use of Configuration Management in Large Scale Infrastructure offers another way of working supporting the new roles of construction companies. Using Electronic Document Systems (EDMS) is a first step towards the full implementation of Configuration Management, which eventually will result in object management.

1 INTRODUCTION

The current way of working in Large Scale Infrastructure (LSI) development is still paper based. Drawings are exchanged in great numbers usually concurrent with the actual execution of the work. To co-ordinate the workflow with the information flow represented in drawings remains a difficult task. The workflow is not always performed as successful as can be, resulting in failure cost in the order of 10 – 15% of the overall construction costs. Another issue is the lengthy process-time to distribute a drawing, evaluate it and correct it. The latter problem is being addressed by the introduction of electronic document management systems, which work over the Internet or on a dedicated Intranet. Experiments have been conducted in the Netherlands with such set-up, using a dedicated ASP (Application Service Provider). The experiments show that apart from document management there is also a need to control the workflow.

A second development is the growing insight that managing the design and construction of complex structures in design-construct projects cannot be done paper based when the paradigms of "Configuration Management / System Engineering" are to be applied. These paradigms require an object-oriented approach in the sense that information about the physical objects are being controlled instead of their representation. To this end the project management for the construction of the Dutch HSL (High Speed

Line), a railway between Paris and Amsterdam introduced a hierarchical object tree approach. The Dutch part of the project sum amounts to 4 billion €. The approach, in which objects are managed using formal configuration management and change control, are implemented in a commercial-off-the-shelve PDM system.

This paper will give more insight in both developments and a proposal for the design and implementation of the mentioned object management system for Large Scale Infrastructure projects. Reported is the Configuration Management approach for the Dutch HSL and construction projects in the utility sector. These projects are industrial construction projects in which TNO acts as consultant and is involved in the implementation of the ICT environment of the projects.

The second part of the paper will address our vision on Intranet based object management systems. It will address the foreseen legal boundary conditions and technical issues involved.

2 OLD AND NEW WAYS OF WORKING

2.1 *Traditional tendering*

For many years the Large Scale Engineering (LSE) industry has been working in a traditional way, following the cascade model for product development.

In traditional tendering a construction company bid to realise the work on the basis of a set of documents and drawings which specified the product to be realised in detail. Design decisions, which had been taken in earlier stages of the process, were uncoupled of the consequences in the realisation process. Consistency of product design with the production of a LSE object was at the least, very weak, but very often non-existent. Of course this way of work leads to miscommunication, complex realisation management and many sorts of rework. Often LSE objects were realised with longer duration and extra cost than originally expected.

In traditional ways of working paper is the only means of communication. Considering the size of the Large Scale Infrastructure (LSI) projects of today it becomes obvious that an enormous amount of paper is being generated during all stages of a LSI project. Responsibility and accountability (if used at all) is entirely based on paper documents being distributed by regular mail or facsimile. For drawings the only official way of communication is the set of calques with appropriate signatures which give a representation of the object under development, as electronic information still does not have any official legal status.

Various participants each make numerous drawings with their own drawing procedures and agreements. In complex projects different cultures come together and new ad-hoc agreements, methods and procedures are invented. Each discipline then uses, stores and manages its own drawing files according to own rules and standards without interaction with other disciplines.

2.2 *Focus on Large Scale Engineering*

The previous description of traditional working methods holds for many building and construction (BC) projects, not only LSE struggles with it. This paper however does not address the complete BC industry. It is focused on the development of Large Scale Engineering projects, and infrastructure projects in particular. Delineating it still more, we focus on Large Scale Line-Infrastructure (LSLI). LSLI has a specific complexity which not only depends on the product and associated technology, but also on the many stakeholders involved, each with its own priorities. Especially the numerous governmental regions, through which LSLI passes, increase the complexity, because of its differing rules, regulations and different interpretation of legislation. LSLI projects can then be characterised by:

- "High" investment cost;
- Long duration but with programme urgency to secure earliest economic benefits;
- "High" environmental impact considerations;

- Technologically and logistically demanding, which leads to:
- Requiring multi-disciplinary input from many organisations, in turn leading to:
- Use of a consortium of organisations for execution of the project;
- Optimisation of lifecycle value based on function requirements and associated conditions.

Even small improvements of above characteristics will lead to tremendous benefits, not only to client and contractor, but also to the general public.

2.3 *Emerging of new ways of collaboration*

New ways of collaboration exist today and are emerging. All aim at improving the effectiveness of the total process, from initiation until operation, maintenance and even demolition. Careful steps are taken towards Life Cycle Management and the consideration of the complete life cycle of a product.

In Design-Construct contracts principals try to liase with a single actor who is held responsible for both the design and the construction of an object. Key to this type of collaboration is that the construction company is able to influence design decisions in an early stage. Design decisions, which may have large consequences to the realisation work. Construction experience and knowledge is used during the design to solve issues and to ascertain that construction work is feasible without having to investigate expensive solutions as a consequence of earlier design decisions.

In Build-Operate-Transfer contracts a contractor accepts the responsibility to deliver a product based on a formal specification and accepts to operate the product for an agreed period of time. At the end of the agreed operation period the product is handed over to another operator (which may or may not be the client or its users).

In the Private Finance Initiative (PFI, PPS in Dutch) private organisations in co-operation with government or other public organisations initiate the financing of Large Scale Infrastructure development. In the Netherlands several initiatives are underway, e.g., development of A4 Delfland, where a consortium proposes to develop an office and public services part on top of a planned track of the A4 between Delft and Schiedam. Incentive for private partners is the possibility to deliver an exploitable office park. Incentive for public organisations is the possibility to obtain financial support from commercial organisations without which the development of necessary infrastructure would be delayed on budget grounds. For the general public it is important that the use of scarce land in the Netherlands will be better optimised by multiple use.

The Alliance model followed in the AMOI method establishes an early relationship between all relevant stakeholders, e.g., client, contractor, designers, specialists, facility managers, users. AMOI is a Dutch acronym for Asset Management of Underground Installations. Through an interactive collaborative approach based on creating win-win situations for all participants, concepts for complex products may be described and visualised early in a project. The essence being that clients actively participate in the process and project, each stakeholder being responsible and accountable for parts for which they can exercise influence and control.

Risk bearing construction management is a model where a single contractor carries out complete inception, design and construction management, also bearing the risk for the resulted product. Risk may be financed from expected process improvements and associated economic benefits.

3 EXPERIMENTS WITH EDMS

3.1 *EDMS description*

EDMS generally stands for "Electronic Data Management System" or "Electronic Document Management System". In the context of this article we use the latter meaning. Of course it is the data in the documents that are of interest. The term document may be used in a broad sense and include letters, faxes, reports, contracts, but also drawings, calques, spreadsheet files, etc. A general definition, which is, used in the Dutch Half-Time project is: "An EDMS is a system to store and manage digital documents." Half-Time is a 3 year project TNO and the Hollandsche Beton Groep NV (HBG) are collaboratively executing. Aim of the project is to reduce the realisation time of building and construction projects by 50%. Building time is defined as the time the Construction Company can exercise influence and is hence depending on the contract form. The project started in October 1997. The project is partly funded by the Dutch Ministry of Economic Affairs.

When entering documents in the EMDS a set of information is attached to enable version control and management. Information may include origin, author, timestamp, version, etc."

An EDMS is usually based on a client-server architecture, where the client manages the interaction with the user on the one side, and the interaction with the server on the other side. The server attends to document management tasks. An EDMS server takes care that documents are stored in a database. Controlled document check-in into the database is done by authorised users, and a number of standard attributes are filled in automatically. To consult the document the user is only able to view it, never to change anything. Users with sufficient authority however, who are allowed update access to documents, do controlled creation and editing of documents. All activities performed by users are registered and kept in the database. In this way a history of document access and change is kept.

3.2 *Relation to Workflow Management*

An EDMS in itself only supports the storage, retrieval and historic use of electronic information. It does not do anything to support the management of the process and the associated workflow of information objects (documents such as drawings, contracts, etc.). In order to enable improved process management it is necessary to control the progress of work and workflow using a workflow management system. Workflow management is explicitly describing processes in terms of activities, responsibilities, resources, predecessors and successors in combination with a computerised system to monitor the flow of work and the state of affairs of associated documents. A workflow management system monitors the workflow it does not execute the steps. Many EDMS-es provide workflow management functionality.

3.3 *Implementation in Half-Time project*

The Half-Time project addresses a wide variety of built objects but focuses not specifically on Large Scale Infrastructure projects. In the Half-Time project several pilots have been carried out to investigate the use and benefits of EDMS, with and without associated workflow management. In order to quickly introduce EDMS in the detailed design phase of projects, a simple workflow was taken within a clearly scoped part of the design process. The design activities investigated comprised the routing of drawings between installation engineers and structural engineers. Four different consultants and the main contractor exchanged drawings using an EDMS with the associated workflow. The participants were so enthusiastic that they prolonged the use of the system after the pilot was concluded.

4 CONFIGURATION MANAGEMENT

"Configuration is the process of identifying and defining the items in the system, controlling the change of these items throughout their lifecycle, recording and reporting the status of items and change requests, and verifying the completeness and correctness of items". Many definitions exist of Configuration Management. This one is taken from the IEEE Std-729-1983.

Configuration management (CM) denotes the method, which enables management of specification, design and realisation of complex products. It supports and enables tracking and tracing of the current state-of-affairs in design and realisation of physical products. CM consists of a set of activities, which in conjunction form the methodology: configuration structuring, configuration identification and formal change request procedures. The essence of Configuration Management is in establishing an explicit way of working, i.e., carefully denote what to do to enable a clear decision making and retracing of steps. As such CM is one of the three main aspects of project control, besides time (duration) and money (cost). CM consists of three parts: Configuration identification, Configuration control en Configuration status accounting (CM = CI + CC + CSA). In addition specification and design will be evaluated and reviewed by experts, to check for consistency and completeness and to verify that the CM method has been followed. (In some descriptions of CM the latter review steps are excluded).

4.1 Configuration Identification

Configuration Identification describes the system structure, the nature of its elements, their identity, and gives access to each item version. Each information object (document) needs to be uniquely identified in a pre-described way. A clear description needs to be made of what types of configuration objects (specification documents, product model objects, etc.) are being managed and what information is contained. A configuration includes hierarchical relationships between configuration objects. Identification includes the registration of changed versions, and the state of objects and documents, e.g., In progress, Verified, Released.

Another aspect to be explicitly registered is the Configuration Baseline. A configuration baseline is the minimal set of information which determines the product during its life cycle. The baseline does not contain any redundant information and consists of an identifying name and a set of references to objects or documents, which describe the configuration at that moment. The baseline is also a necessary part of the change process. Depending on the moment of establishing the baseline one may speak of "Specification Baseline", "Concept Design Baseline", etc.

4.2 Configuration control

Configuration control organises versions and changes to system items while keeping coherency and consistency on the complete system. Configuration control is established by the introduction of a Formal Change Request procedure. If a baseline is established changes to the configuration, which de-

mand an update of the baseline, need to be requested using a request for change. Usually the change requests are gathered and evaluated and decisions made by a formal change control board. A change control board may reject changes, accept changes or forward changes to a next management layer.

In this context the use of an EDMS tool is a specific implementation of configuration management.

5 CONFIGURATION MANAGEMENT AT DUTCH HIGH SPEED LINE

In the Netherlands the Dutch part of the High Speed Line South (HSL Zuid), a high-speed train link from Amsterdam to Paris, is being developed based on Design-Construct contracts. In the pre-tender phase Dutch Railways developed a reference design in order to support the use of Design-Construct contracts.

The purpose of the reference design is to enable evaluation of the offers for construction of sections of the HSL Zuid. During the development of the reference design an important management instrument used is configuration management. One of the activities carried out was the introduction of a Formal Change Procedure (see *Configuration Control*).

Another important activity in configuration management is determining and identifying how a configuration can be described. At HSL Zuid it was decided to introduce a neutral object tree to register the physical objects and its changes. The neutral object tree is a model of a single project, without a reference model or type-model. Some characteristics are:
- An Object Tree is an instance model describing the objects of a specific project;
- An Object Tree is a decomposition tree using "contains", or "consists of " relations.
- An Object Tree supports a minimum set of relations

For further details refer to (Nederveen 2000) en (Nederveen & Tolman 2000).

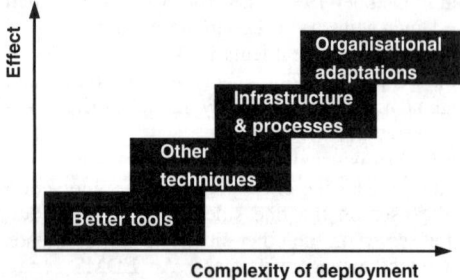

Figure 1: Relation between complexity of deployment and effect of measure

6 A NEW VIEW ON THE TOTAL ORGANISATION

Different and new ways of collaboration also have consequences for the organisation of the work. It means project and process risks are taken by other participants. Eventually stakeholders who are able to do something about risk aspects will become owners of the risk and will have both responsibility and accountability for risk management of specific risks. This also means that relationships will need to be nurtured and long-term relationships will emerge based on this same shared responsibility enabling the introduction of Life Cycle Management and Supply Chain Management. Relationships will be based on trust in stead of "just-in-case" behaviour.

Organising the work in building and construction remains the major goal. Based on changes in collaboration the organisation of the work needs to also reflect the changes. One of the characteristics of building and construction, the project where a number of organizations temporarily works together to realize a product, will need to adapt based on changing ways of collaboration. Explicit determination of roles, communication and information sharing and exchange will become the basis for success of new collaboration and contract forms.

It may be obvious that organisations need to change. Organisations however seem reluctant to approach the necessary changes. One of the reasons being that the complexity of the deployment and effort to be put into changing an organisation will be large. On the other hand the benefits will be substantial as well. The picture shows the relation between the effect obtained and associated complexity and hence risk of changing ways of working.

6.1 *Roles of players*

That new ways of collaboration will change the participation of organisations in Large Scale Engineering may be apparent from the Dutch VISI project.

Figure 2: The track of the Dutch High Speed Line

Source: Website of Dutch Ministry of Transport, Public Works and Water Management

VISI is an acronym for *"Voorwaarden scheppen voor invoeren standaardisatie ICT in de GWW-sector"*. In English "Establishing pre-conditions for the introduction of ICT standardisation in civil engineering." Its objective is to develop a generic model for digital communication, which will enable partners in large scale and complex infrastructure development schemes to organise communication and cooperation in a flexible way. This, in turn, will significantly reduce costs and duration of development of large infrastructure works. VISI has proposed a framework for standardisation and execution of communication. The framework will describe the process on the boundaries of identified roles. The responsibilities of each role will be specified including the required communication. The information exchange is defined by messages. The contents of all messages will be reflected in a common information model.

6.2 *Information exchange and sharing*

The upcoming of Internet has tremendously increased the flexibility with which people are able to communicate. The introduction of new ICT applications, which are based on the Internet and which support the possibility of information integration, exchange and sharing will increase the way people are able to communicate. Information sharing based on agreed communication messages is a prerequisite for collaborative ways of working to be successful. If not present in a project, how can one communicate and ascertain that the same things are meant, that the same drawings are used, that everyone involved knows the current state of affairs?

7 THE FUTURE: OBJECT MANAGEMENT SYSTEMS

Will Object Management Systems be the future to enable and support improved management and control of complex LSLI projects? What steps need to be taken to come to an Object Management System? The introduction of computer based tools and the use of electronic document management will be steps into the future. Steps which will lead to object management and object management systems.

7.1 *Transition from paper based information exchange to object sharing*

Object sharing can only be done on an electronic basis. Only when information is freely available and accessible the concept of sharing can be realised.

An EDMS using its workflow management features is a step forward towards electronic object sharing. Numerous different types of electronic documents might be controlled using an EDMS. Workflows are defined for day-to-day activities. A document might eventually evolve into a set of information about a certain object.

Subsequent steps need to build on improving communication and in particular making agreements followed by standards on communication. Both form and content of communication messages must be fully understood by parties who communicate. This means that structuring of information will become more and more important in the coming years. Structuring information needs to start on a project basis where the benefits will be clearly visible, not on a meta-level where no immediate benefits are visible and where project details cannot influence the structuring itself. Extensions to a meta-level to derive product-type models might be based on a multitude of projects.

7.2 Object trees are a structuring mechanism

Documenting a complex product, one starts with creating a list of the components that can be found in the product. After a while, especially with complex products in a complex process, the list has grown to an extent where it no longer is useful. People cannot retrieve the information, cannot find a specific component, etc. Next people start to define a structure for the items in the list. The structure is based on some relation between the items. One can think of the "is a type of / can be a" relation, or the "is composed of / is part of" relation or any other relation one chooses in order to structure the design of a complex infrastructure product. In the HSL Zuid project the decomposition of the product was selected as the structuring relation. An object tree is a structuring mechanism which supports determination and identification to describe a configuration. At HSL Zuid it was decided to introduce a neutral object tree to register the physical objects and their changes. Section "Configuration Management at Dutch High Speed Line" gives some more information.

An object tree can simply be extended with all kind of attributes, each describing something specific of the product. The hierarchical nature of an object tree suggests to use the same mechanism for registration of requirements, cost, planning, risk, etc. and to relate those to the objects in question. Adding responsible actors to an object one can determine interfaces between actors. Managing the interfaces is a major task in the risk management of a project. Especially at interfaces, where people from different disciplines and from different organizations interact, problems arise in the execution of the project.

7.3 E-commerce also needs structuring mechanisms

In E-commerce information is made accessible to different actors in a pre-defined way. It can also be viewed as another type of information sharing. In the Building and Construction industry E-commerce and business-to-business communication over the Internet is progressing. It gives opportunities to electronically communicate with suppliers in a clear pre-defined way during the complete duration of a project and the life cycle of supplied off-the-shelve products. A next step can be the composition of products to be ordered using business-to-business communication over the Internet.

The European partially funded eConstruct project (Tolman & Böhms 2000) is concerned that information and communication tools are not being developed to resolve problems arising during the inception, design, realisation and exploitation of international construction projects like high speed rail links and inter-European state highways. Consequently eConstruct aims to develop, implement, demonstrate and disseminate a new Communication Technology that can be used over the Internet to support meaningful communication across the borders of the European member states.

7.4 Barriers

7.4.1 Legal aspects
Investigations need to be carried out into the legal aspects of electronic information exchange and electronic information storage and retrieval. Different responsibilities need to be regulated and accountability needs to be determined.

7.4.2 Trust
Building and construction is an industry, which was based on institutionalised distrust. Trust is not something that you can turn on and off, but has to be deserved. To try to establish a trustworthy relationship in the BC industry, partners in the relationship need to be of equal worth. With that it is meant that partners need to respect each other and want to create a win-win situation where both gain and loss are split between partners on the principle that each party is equal and profits and losses are distributed to the ratio of their participation in the project.

7.4.3 Who is the winner
The question, which pops up every time innovations are proposed, is: "Who benefits from the innovations, changes, etc.?" During the process of innovation introduction, the innovator invests with the ex-

pectation of a reasonable return on investment. At the same time innovation is aimed at qualitative and quantitative improvement of products and economic improvement of the process. Eventually the client will benefit. On the other hand the innovator gains a head start and may obtain a larger portfolio based on improved quality of both product and process, shorter duration and lower price.

8 CONCLUSION

Investigations in the Half-Time project showed that distribution through electronic exchange in stead of regular mail caused the physical distribution time to be eliminated. It also showed that explicit and up-to-date knowledge of drawing state, i.e. who was working on a drawing, improved the clarity and progress of the process. Project leaders seemed eager to use the workflow functionality since the system electronically tracks progress of drawings.

Although the Half-Time project does not explicitly focuses on Large Scale Infrastructure, the conducted experiments give reason to believe that the introduction of an EDMS will assist in the migration towards the use of Object Management Systems.

Introducing Configuration Management is mainly an organisational issue. Tools are already available to support Configuration Management. In LSI projects the rule to "keep it simple" will lower the threshold to actually implement and use such tools in projects. Filling in an object tree as the basis for configuration management will in subsequent projects be further investigated and tested. Object trees might in future be complemented with the definition and filling in of requirement trees, cost trees, risk trees, etc.

When in future e-commerce and business-to-business communication will be fully used the need for information structuring and configuration management will grow, resulting in the arrival of distributed object management systems over the Internet.

REFERENCES

Belt, H. & Willems, P.H. 1999. *Half-Time R32310: Subproject 3.2, IT-support of processes, Inventory and selection of an EDM system.*

Nederveen, G.A. 2000. *Object Trees – Improving Electronic Communication between Participants of Different Disciplines in Large Scale Construction Projects,* PhD-thesis, Delft University, ISBN 9013619-3.

Nederveen, G.A. & Tolman, F.P. 2000. *Neutral Object Tree Support for Inter-discipline Communication in Large Scale Construction,* ECPPM 2000.

Tolman, F.P. & Böhms, H.M. 2000. *Electronic Business in the Building and Construction Industry: Preparing for the New Internet,* CIB Conference 2000.

VISI 1999. *Communication in projects, challenges for the sector infrastructure, Final report (in Dutch) of the VISI Investigation Phase,* LWI.

CARUSO: Customer care and relationship support office

Hubert Baumeister & Piotr Kosiuczenko
Institut für Informatik, Universität München, Germany

ABSTRACT: Customer Relationship Management (CRM) is an inherent business strategy for companies big and small. The technology has reached a point where it is truly enabling the way enterprises manage their customer relationships. The goal of the EU funded project CARUSO is the design of a software toolkit that facilitates the building and maintaining of high quality business-to-business and business-to-customer relationships. CARUSO is designed to allow a multi-dimensional way of looking at markets, customers, suppliers, products, personnel, internal and external information, communication and action flow. This will be accomplished by the following core features: front-office application builder with customer care and marketing desk, basic technologies comprising a general communication server, intelligent information, document and contact access, unified messaging, and a customizable user interface. Emphasis will be put on exploiting existing tool packages as much as possible. The CARUSO toolkit is targeted at European Small and Medium Sized Enterprises (SME) and allows them to optimize their business operations to the mutual benefit of both the supplier and the consumer.

1 INTRODUCTION

Currently European SMEs are affected by major changes in global economics. In the US millions of jobs were lost in the 80ies which led to the metamorphosis from a product(ion) oriented to a service oriented market by which millions of new service jobs were created. A similar development is now going on in Europe.

It has been realized that gaining new customers is much more expensive than keeping the current customers. Companies are fighting for the same customers, but, at the same time, customers are finding that it is very easy to switch from one company to another. The financial impact of customers disloyalty can be immense. A recent Harvard University study reported that in many companies a five percent improvement in customer retention could increase profits by 85%. Therefore, it is no wonder that companies are searching for ways to reduce customer turnover. Recent surveys have also shown that poor service or inattention is the cause of 65% of customers leaves (VanLaeken 1999).

Thus, keeping customers or increasing their loyalty can be achieved by focusing on their needs. As a result, companies are trying to improve the quality of customer interaction and the service of customer requests, starting with the very first contact, and on throughout the sales process to the service and support provided after the initial sale or service.

Customer Relationship Management (CRM) is an inherent business strategy used to achieve this goal. This relatively new concept influences strategy, business processes, as well as information systems in many companies. CRM and a high quality supplier relationship are essential success factors in a highly competitive global marketplace.

In the past, business relationship management was both cost and time intensive. In particular SMEs, with naturally limited resources, were not in a position to carry out broad-scale marketing programs or administer their supplier relationships in a effective way. Now, technology has reached a point where it is truly enabling the way enterprises manage their business relationships.

The market trend is to cover the definition of CRM with a single integrated platform. Island solutions like customer care, support, and contact management cannot fulfill the CRM demands. For example, Siebel (Siebel 2000) bought Scopus Technologies to integrate customer service technology in their products. Other companies have signed co-operation agreements or developed modules, interfaces, respectively in joint efforts. Having their origin in backoffice applications such as accounting, purchasing, and production, Oracle (Oracle 2000), BAAN (Baan 2000), or SAP (SAP 2000) follow with their strategy this path to integrate the missing building blocks for customer relationship.

The EU-project CARUSO (Customer Care and Relationship Support Office) aims at providing European SMEs with a tool package that provides the flexibility of individually built applications tailored to their specific needs in the field of customer and supplier relationship management. It is also going to facilitate business strategies which will help to understand and anticipate the needs of an enterprise's current and potential customers. Another objective of CARUSO is to provide the technology that helps to grow customers into a position of equivalent partners who will pro-actively influence the life-cycle of goods and services including areas such as pre sales, marketing, post sales, and maintenance. Thus customers will gain a more direct influence on the nature and quality of products and services offered including the provision of all relevant information to them.

2 THE CRM CHALLENGE

The ultimate goal of CRM is to attract and retain customers and increase the profits. CRM is a complex process that requires, on one hand, a redesign of current business processes and, on the other hand, integrated IT support. The key factor is how well an organization manages its customer relationships from the first contact through the sales process, customer service, and ongoing customer retention activities.

In recent years, call centers have gained popularity as being cost-effective and efficient. Organizations are now realizing the critical importance of every customer contact and the potential of the call center for customer relationship management strategies. The proper application of call centers can improve the overall quality of customer interaction while streamlining customer requests and orders. In addition, call centers are nowadays increasingly responsible for business interactions that are being conducted through alternative emerging communications channels, such as e-mail, internet, fax, voice mail, pager, etc.

One of the most urgent challenge facing call centers today is the fact that it is becoming increasingly difficult and costly to recruit, train, and retain qualified call center agents. Technology is a key to help reduce the learning time and costs of call center employees. Moreover, technology can help inexperienced representatives deliver much better customer service.

Another important issue is equipping agents with the necessary empowerment and competence to allow him or her making proper decisions without unnecessary call escalation. Call escalations are usually very time and resource consuming and decrease the customer satisfaction (Cusack 1998); therefore it is desirable to avoid them as much as possible. One way

to achieve this is implementing some form of unified agent desktop application to give call center agents the information they need to respond quickly and accurately to customer questions and requests. Another, more important way, is to integrate the agent desktop system with knowledge-bases and back office systems as well as with the company's business strategies. This will help save time, for example, by automatically retrieving customer account information from the corporate database and displaying it on an agent's desktop.

A growing number of call centers are also considering the use of sales configuration to help automate the sales process. These are often provided by rule-based engines which help to translate customer needs into sales opportunities and, at the same time, simplify the selling of complex products and services. By the means of sales configuration technology, companies can rapidly roll out new campaigns. This technology also allows inexperienced agents and even new hires to present very complex products and services and to interact with the company workflows.

All the technical means can not and should not eliminate the personal relationship with the customers. The ability for call center agents to view the information about the customer can help maintain a more personal interaction. Current script generators enable call center agents to create personalized, one-to-one correspondence based upon the customer's profile and information gathered during the contact so that the company presents one face to the customer (Schmid and Bach 2000).

Another emerging challenge is the fact that call centers must be able to address multiple contact channels including phone, fax, postal services, internet, e-mail, voice mail, etc. In fact, many call centers are developing into multimedia communication centers within the next year. Companies are integrating their call centers with their Web pages to enable customers to help themselves as well as to schedule callbacks or initiate on-line chat sessions with customer service representatives. This allows customers to use the way they prefer to contact and interact with companies. Therefore, organizations are looking for unified messaging solutions to help manage the flow of interactions across the various communication channels.

One of the key aspects of CRM is that it is centered around the customer and not around the departments of the company. Usually this implies that the business processes that are needed for dealing with a customer cross the boundary of single departments or business units. It is crucial that business workflows in a call center can be integrated with the back office workflows, and that the workflows can be modified appropriately if needed. Workflow automation software allows for directing and monitoring work that goes outside of the call center to assure completion or tracking progress. This helps reduces fulfillment time of new product orders and allows call center agents to be better informed of the current status of a customer request.

To effectively deal with customers, a CRM system needs to store customer profile information in addition to a complete customer contact history, including any documentation that is associated with the customer. This can increase revenues through improved cross-selling and up-selling capabilities, and, moreover, they help companies improve their understanding of buying patterns and customer preferences as well as the targeting of their marketing efforts (Berry and Linoff 2000).

However, maintaining this information can pose considerable problems because usually this information is spread throughout the different IT systems of an organization, and it exists in a range of formats. Further, to be effective, it is important that only the information necessary to deal with the customer's current issue is retrieved, while unnecessary information is suppressed.

The American market research company AMR Research estimates the total size of the world-wide CRM market at more than $2.5 billion in 1998, and growing at more than 50% a year. This does not include software from vendors who incorporate CRM functions in their core products. The major players in the CRM market segment are Siebel, Vantive, Clarify, Point, Applix (APPLIX 2000), Corepoint, and IMA (cf. (MSI 2000)). Remedy offers a solution for call centers with very powerful workflow engine (Remedy 1999). Newcomers from the USA originating from internet based applications, like Pivotal, Firebond, Upshot, and Vignettetry, try to break into the CRM market segment now. The market leader in CRM, following AMR, is Siebel Systems founded only a few years ago. Siebel is the largest player in the market, a position they have achieved by focusing on large accounts and not on SMEs. Siebel's deals are typically in the millions of dollars. Its closest competitor is Vantive; the two have about 25% of the market.

The CRM market in Europe is still immature. Outside of large corporations little has been done, and even there the market is scarcely beyond its infancy. New companies have started to move in which indicates that the market has reached the critical growth phase. These companies include SalesLogix, Onyx, Pivotal, Firebond, Upshot, and Vignette. Their CRM systems mostly evolved from existing contact management systems and internet based applications for small- to medium-sized businesses. One problem with these systems is that they address the needs of the US market rather than the needs of European companies.

A lot of the currently offered software packages fulfill only part of the CRM demands. SFA (Sales Force Automation) is the predecessor of CRM. They focus on sales and marketing application, primarily for the sales force (help desk, complaint management, telemarketing are gradually also offered by these packages. There are several players at the moment like Siebel, Remedy, Scopus, Heat, Clarify, Vantive, Point, Cincom, and many others.

Usually the costs for adapting CRM software to the business processes of a company is much bigger than the costs of the software itself. This is the reason why currently the major players address the Fortune 500 enterprises only. The software packages represent a major investment. For example, a Clarify solution is a million project and only suited for a minimum of 50-100 agents. On the other hand, in particular newly funded small companies require highly integrated and powerful instruments to establish a solid customer base.

3 PROJECT OBJECTIVES

Building and ensuring high quality of supplier consumer relationships is the prime objective of this project by means of a Customer Care and Relationship Support Office (CARUSO). CARUSO provides management and control tools to monitor and improve European SMEs relationship to customers.

The goal of CARUSO is the generation a software tool-kit that facilitates the building of highly scalable front office applications to maintain high quality consumer-supplier relationships. Emphasis will be put on exploiting and integrating existing tool packages as much as possible.

The installation of many CRM systems is very expensive because specialists are needed to adapt the core system to the needs of the company. This makes the use of CRM less suitable for SMEs. CARUSO addresses this problem by providing an easy to use application builder, which allows the rapid generation of front office applications tailored to the individual requirements of a company. A script developer is integrated into the application builder to provide conversation scripts to aid and guide the agent. The script developer offers immediate modification options to respond to, for instance, specific campaign requirements

Since many of the relevant data dealing with customers are spread among the IT systems of a company, CARUSO provides interfaces to the company's Enterprise Resource Planning (ERP) systems and other installed software applications, and provides interface options for the most common used data bases. In particular interfaces to back office systems allows the design of front office applications that provide a unified access to the back office func-

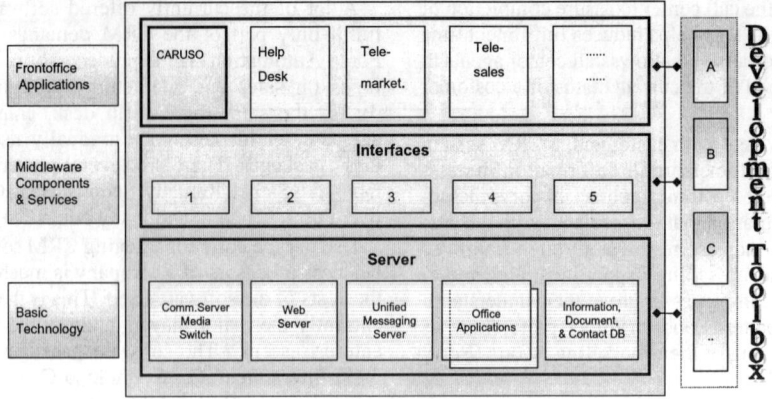

1	General Communication Server	A	Application Builder
2	Office Application Interface	B	Script Developer
3	General Data Base Interface	C	General GUI
4	Unified Messaging Interface		
5	Other Middleware Interfaces		

Figure 1: CARUSO Architecture

tionality. This is achieved by representing databases, ERP systems, and back office applications as software components using standard middleware technologies like CORBA, COM/DCOM, or Enterprise Java Beans (EJB).

To protect the companies investment into their Private Branch Exchange (PBX) systems, CARUSO makes use of the installed basis of the PBX by providing a Computer Telephone Integration (CTI) module which is compatible to nearly all available European PBX systems, offering in addition the functionality of Interactive Voice Responder (IVR), and Voice-Mail-Server as well as Power-Dialer.

Another aspect targeted by CARUSO is the integration of different communication channels, like phone, e-mail, fax, web, etc. in a Unified Messaging System, which allows consistent communication with the customers in a variety of ways.

CARUSO provides adaptation of front office functions based on customer profile and contact history by means of a dynamic graphical user interface (GUI), and skill-based call distribution taking into account the agent's tasks and expertise. This allows a more efficient interaction with the customer by providing the help desk agent with only the functionality needed to resolve the customer's issue. Further, the interaction will be guided using information about solutions to frequently occurring problems stored in a knowledge base.

In addition, the CARUSO toolkit has the following crucial features:

- full support for workflow, document, and contact management
- web interface for the help desk agent which allows rapid dissemination of adapted or new front office applications, and a similar interface to the one used by the help desk agent can be used for the customer (self help)
- a solution that is geared to European standards and requirements rather than US', in particular multilingual support
- a management information and control system that allows to monitor customer satisfaction and the effectiveness of the CRM processes, and provides data mining functions to identify customer behavior patterns.

4   THE ARCHITECTURE OF CARUSO

The architecture of CARUSO consists of three layers: the front-office layer, the middleware layer, and the basic technologies layer (cf. Fig. 1). The front-office layer contains the front-office applications which are designed and customized by using the tools of the development toolbox. The development toolbox contains, among others, the application builder and the script developer.

The basic technology layer consists of a general communication server; office applications; information, document, and contact databases; and a unified messaging server. The communication server manages various communication media (e.g. telephone,

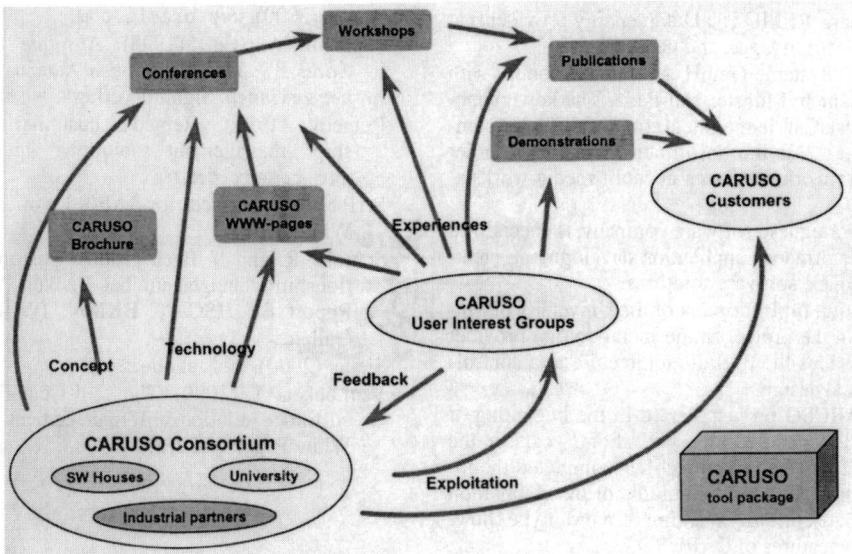

Figure 2: Dissemination

text, data, fax, e-mail, WWW, video), and is prepared not only for Voice over the internet protocol (IP), but also for IP call center functionality.

The task of the middleware layer is the integration of the communication services with the other basic technology components and the front-office applications. The middleware layer provides a uniform access to customer data and history stored in the various databases of an enterprise, like contact databases and ERP systems. CORBA, EJB, and COM/DCOM are used for the middleware components and services.

The component based approach together with the development toolbox makes it possible to extend CARUSO with new interfaces to basic technologies and allows the generation of customer care applications tailored to the individual enterprise requirements. Further, it is possible to start with only a few components and add new components as needed.

The overall design of CARUSO follows an iterative object-oriented approach based on the Unified Modeling Language (UML) (Booch et al. 1999) and the Unified Process (Jacobson et al. 1999). Within the last few years the UML has become the standard notation for object oriented modeling. The UML is a diagrammatic notation for modeling object-oriented software systems. Class, component, and deployment diagrams model the static aspect of software systems, while Use Case, interaction, and activity diagrams are used to model the dynamic aspects.

The UML can be used with any object-oriented software development method; it itself does not constitute a development method. For CARUSO we have chosen the Unified Process.

Within one release cycle, the Unified Process distinguishes 4 phases: inception, elaboration, construction, and transition. The goal of the inception phase is to establish the business case, the goal of the elaboration phase is a project plan and a sound architecture, the goal of the construction phase is the final system, and the goal of the transition phase is to deliver the system to its end users. Within each of the 4 phases the basic workflows business modeling, requirements, analysis and design, implementation, test, and deployment are executed, possibly several times.

The advantage of this approach is that at the end of each iteration the result is a running system which allows for immediate feedback. Immediate feedback is important because it is very hard, if possible at all, to gain a relatively precise and complete specification of a system as complex as a CRM system in one big step. Only by testing successive versions of the system, missing and inappropriate functionality can be discovered.

5 CONCLUSIONS

In this paper we have given an overview of the EU-funded project CARUSO. The project is pursued by a group of companies that cover the role of end user, technology provider, software developers, and methodology provider.

The role of the end user is played by REMU who wants to apply CARUSO to maintain relations with

119

its customers. REMU is a Dutch energy provider that supplies electricity, gas, and heating.

DataCall Systeme GmbH is a software house with sites in Munich, Münster and Paris. The key competence of DataCall is the integration of different communication media into information systems in order to facilitate work processes at multi media workstations.

SFI, a Portuguese software company, is specialized in high-performance application development, creating tailor-made software solutions.

The Institut für Informatik of the University of Munich acts in the project as the methodology provider which develops the overall architecture and controls the technical design.

The CARUSO project started in the beginning of 2000 and is scheduled for a period of 2 years. In the moment of writing the project has completed the inception phase and is in the middle of the elaboration phase. We are currently working on a prototype showing the key features of CARUSO.

Among presentations and demonstrations on tradeshows, distributing press releases and publishing best practice reports, an user interest group will be used to exploit and disseminate the results of the project (cf. Fig. 2). The group will consist of industrial and institutional partners who will use and test CARUSO. After completion of the final tool, these partners will provide an adequate reference as a basis for the further dissemination of the results. The initial user interest group will consist of SMEs from Europe. It is planned to enlarge this group towards the final prototype phase to build the basis for an early and effective dissemination.

More information on CARUSO can be found on its web site caruso.isd.pt.

REFERENCES

APPLIX (2000). www.applix.com. White Papers.

Baan (2000). www.baan.com.

Berry, M. J. A. and G. S. Linoff (2000). *Mastering Data Mining: The Art and Science of Customer Relationship Management* (2nd ed.). Wiley.

Booch, G., J. Rumbaugh, and I. Jacobson (1999). *The Unified Modeling Language User Guide*. Object Technology Series. Addison Wesley.

Cusack, M. (1998). *Online Customer Care: Strategies for Call Center Excellence*. Quality Press.

Jacobson, I., G. Booch, and J. Rumbaugh (1999). *The Unified Software Development Process*. Addison-Wesley.

MSI (2000, March). New remedies for marketing exec insomnia. Published by MSI, www.msiconsulting.com.

Muther, A. (200). *Electronic Customer Care: Die Anbieter-Kunden-Beziehung im Informationszeitalter* (2nd ed.). Springer.

Oracle (2000). www.oracle.com.

Papmehl, A. (Ed.) (1998). *Absolute Customer Care: Wie Topunternehmen Kunden als Partner gewinnen*. Signum Verlag.

Remedy (1999). Remedy customer relationship management solutions architecture. www.remedy.com.

SAP (2000). Services for SAP/R3. www.sap.com. White papers.

Schmid, R. and V. Bach (2000). Customer Relationship Management bei Banken. Technical Report BE HSG/CC BKM/4, IWI-HSG, St. Gallen.

Siebel (2000). www.siebel.com.

VanLaeken, T. (1999). The Call Center's role in customer relationship management. Cincom White Paper, Vol 2(5).

Preparing furniture industry to be competitive in the advent of the electronic business

R. Jardim-Gonçalves, R. Tavares & A. Steiger-Garção
Università Nova de Lisboa, Facultat Ciências e Tecnologia, Departamento Engenharia Electrotécnica, UNINOVA, Instituto de Desenvolvimento de Novas Tecnologias, Portugal

M. Borràs & I. Gresa
AIDIMA, Benjamín Franklin, Valencia, Spain

ABSTRACT: Nowadays there is not so much work done in the field of Standards for Product and Process Data Modeling for the furniture industrial sector. This paper presents the research and development work that has been performed in the scope of the funstep Interest Group (FSIG) initiative, funded by the European Commission, towards a Standard Application Protocol for furniture product and decoration project data. This issue is now considered essential to push this industry, mainly composed by SMEs, to be competitive in the global market in the advent of the electronic commerce.

1 INTRODUCTION

Nowadays there is not so much work done in the field of Standards for Product and Process Data Modeling for the furniture industrial sector.

In the advent of the industrial exploitation of Internet, and especially in the areas related with the business-to-business electronic commerce activities, the existence of such standards is now considered essential to push this industry to be in the state of the art. Mainly for the SMEs, which represent more than 90% of this industry, this is a requirement to be competitive in the global market.

During last years, Research & Development work has been done in this field in conjunction with R&D institutes, Sectorial Industrial organizations and companies operating in this furniture related business, developing models, platforms and toolkits to support the furniture industry in this way.

The funStep Interest Group (FSIG) is a non-for-profit international association, open to anyone interested to be member, and funded by the European Commission, where its members have interests to cooperate, to follow the research and developments, and to get at first hand the results of the work on this subject. Now FSIG has 144 members (75% are from industry), from 18 Countries of 4 Continents.

One of the principal objectives of FSIG is to support the development of an International Standard model for product data exchange and electronic services. This work is being done in midst of the CEN/ISSS and ISO TC184/SC4 community, represented there by some FSIG members participating as delegates in its meetings.

To reinforce this relationship, a CEN/ISSS Workshop was set-up, and in parallel, a formal A-Liaison was already estab-lished between FSIG and ISO TC184/SC4. The general architecture of the funstep model includes the Standards ISO10303 (i.e., STEP) and ISO 13584 (i.e., PLib), and already achieved the ISO New Work Item stage (ISO-NWI), and is registered as part 236 of ISO10303: "Application Protocol: Furniture product and furnishing project data". The International Stage (ISO-IS) should be achieved within 2 years.

2 THE FUNSTEP INTEREST GROUP

The funStep Interest Group is devoted to the promotion of the development and implementation of data exchange standards and electronic commerce in the furniture industry, and more specifically those data formats originated by the ESPRIT 22056 funStep project based on the industrial requirements and open to collaboration with other working groups or standardisation organisations (ISO, IAI, CEN, etc.).

The funStep-IG was created in October 1998 with the support of the European Commission. The main aims of the funStep project is the improvement of the communication between furniture manufacturers and their customers; retailers, allowing the exchange of furniture products, catalogues and furnishing projects based on the ISO TC184/SC4 standards (e.g., STEP, PLib) between heterogeneous systems.

Due to the interest arisen, the scope was extended to cover all kind of products and relevant aspects such as electronic commerce.

The project puts together furniture manufacturers, retailers, suppliers, designers, industrial federations, software vendors, R&D centres, Universities and Standardisation Bodies world-wide from Spain, Germany, Italy, Portugal, France, UK, The Netherlands, Greece, Belgium, Sweden, Finland, and without funding; USA, Chile, Argentina, Australia, South Africa, Switzerland and Canada.

FSIG is structured by Chapters. Each Chapter corresponds to one language and is coordinated by a Chapter Coordinator. This is the way found to facilitate the involvement of the industry, since most of them are not English speaking, having its Chapter Coordinator as the interlocutor for the General Management and Technical Boards.

Since the beginning of the funStep-IG several activities have been carried out in each participating country to fulfill the scope of the project; presentations in international Fairs, Congresses, technical meetings, harmonisation meetings, mailings, newsletters, visits to companies, requirements collection, development of a Furniture Dictionary (10 languages) to be exploited in the Internet, etc.

One of the aims of the initiative is to keep closer the industry needs, so, about 75% of its membership is composed of furniture manufacturers, retailers, suppliers, designers, industrial federations or software vendors specialised in furniture industry. The other 25% of the Group consists of R&D centres, Universities or Standardisation Bodies.

ISO TC184/SC4 has approved the New Work Item project for development of the Product data standard in the furniture sector proposed by AIDIMA and UNINOVA with the support of funStep-IG Chapter co-ordinators: It is coded as Part 236 of ISO 10303 – Application Protocol: Furniture product data and project data.

In the schedule, a draft of the ARM is foreseen by September-October 2000. Implementations are expected by February 2001 (ISO meeting in Portugal). Working meetings planned on March, April and June 2000, including harmonisation meetings with other standards; Building and Construction (AP225), Automotive (AP214), Plib (ISO 13584), etc.

Under the umbrella of the funStep-IG, the funStep-AP project, within the initiative ECOM-IS "Electronic Commerce Open Marketplace for Industry Sectors initiative", was launched in November 1999 to speed up the standardisation official process of the specification for the data exchange in the furniture sector at international level.

The Consortium (furniture manufacturers, software vendors specialised in furniture, R&D centres, STEP providers) relies on the support of the European Standardisation Body CEN/ISSS Workshop "Furniture product and business data" and the European standardisation bodies in order to form a unique front in harmony with international organisa-

Figure 1 – Portal of funStep Interest Group at http://www.funstep.org

tions and the Subcommittee ISO TC184/SC4 close to the project aims.

The membership to the funStep-IG is free and open to everybody willing to collaborate or to receive updated information. Furniture manufacturers, retailers and software vendors are specially invited to join this initiative.

The official portal of FSIG is at http://www.funstep.org (Figure 1). There, it can be found all information related with this initiative, including a repository of relevant documents, automatic join form to be member and the "Furniture On-Line Dictionary". Figure 2 depicts a snapshot of it.

This dictionary, developed as one initiative of the Portuguese (UNINOVA) and Spanish (AIDIMA) Chapter Coordinators, have its terms translated for each language of each chapter including: terms, descriptions and synonyms. Its main objective is to make easier the communication among partners speaking different languages in the furniture industrial sector.

Organized in groups of "subjects", the Dictionary allows an easy access to the data depending on the scope of the query, e.g., office furniture, STEP related terms, etc.

This dictionary is active in the sense that users, if they have permission to do it via password, can update and manage the terms, adding or updating terms and automatically asking via Email to the other chapter contacts to translate it to their native languages, keeping the dictionary complete and updated.

The architecture of the dictionary relies on the top of a commercial RDBMS, with a program layer specific to provide the required functionalities for an automatic and easy access and management using an Internet browser.

Figure 2 – Snapshot of the "On-Line Furniture Dictionary"

3 THE FUNSTEP AP AND ITS RELATIONSHIP WITH CEN/ISSS FUNSTEP WORKSHOP AND ISO TC184/SC4

The acceptance of the ISO Community of the funStep Application Protocol was realised as an ISO TC184/SC4's PWI (Preliminary Work Item), as a first stage for towards an International Standard.

The PWI resolution was voted and accepted by unanimity of the ISO community participant members during the ISO meeting (Bad Aibling - Germany) in June 1998, and the funStep models were registered as Part#236 "Furniture product and furnishing project data" of the ISO 10303 Standard.

During the ISO TC184/SC4 meeting in Melbourne (February 2000) the result of the balloting of the second stage of the standardisation process (New Work Item - NWI) of "Furniture product and furnishing project data" was officially announced, and approved. Now the third stage is in process and the Committee Draft document should be ready in the beginning of 2001.

This work close to the ISO TC184/SC4 community is of a great help in disseminating the funStep project from the point of view of industry and implementors (software houses) though the models will be guaranteed as an international standard (ISO – IS).

Also, in the scope of the CEN/ISSS funStep Workshop, several CEN Workshop Agreements (CWA) will be published related with the standardization work of the funStep model.

4 THE FUNSTEP'S APPLICATION ACTIVITY MODEL

An IDEF0 Model is a complete, concise and consistent description of a system developed for a particular purpose. It is focused on the concepts of system modeling employing both natural and graphic languages.

IDEF0 is a public domain modelling system and is the one recommended by ISO TC184/SC4 for description of the Application Activity Models (AAM) for the Application Protocols developed there.

After a previous analyzing the context of the funStep objectives, IDEF0 was considered suitable for the aims and thus adopted.

In this context it defines:

- System as a set of interacting components with relationships among them.
- Modeling as the act of developing an accurate description of a system.

From an IDEF0 perspective, a model focuses on either system activities or system things.

IDEF0 is a static modelling paradigm that represents a system as a network of inter-connected activities. It uses a mix of graphics and natural language to capture and communicate process details. By adding information in a glossary processes are described.

The IDEF0 models are hierarchical. They start with a single activity at the highest level. This activity is then decomposed into sub-activities on the next page, and so on.

In the funStep case, and following the IDEF0 methodology, the purpose of a model was achieved answering the following set of questions:

- What are the manufacturer's requirements?
- What are the retailer's requirements?
- What are the software suppliers requirements?
- What kind of information to exchange?
- What relationships to consider?
- What kind of product to model?
- ....

focused on the following perspectives:

- home furniture manufacturer
- kitchen furniture manufacturer
- retailers
- big retailer
- software supplier
- Architect, Designer

4.1 IDEF0 - Diagram syntax and usage

An IDEF0 diagram contains boxes and arrows, which boxes represent activities of the system being modeled and arrows connect boxes together and represent interfaces or interconnections between the boxes.

Boxes are rectangles, and represent a function or an active part of the system, and so are named with

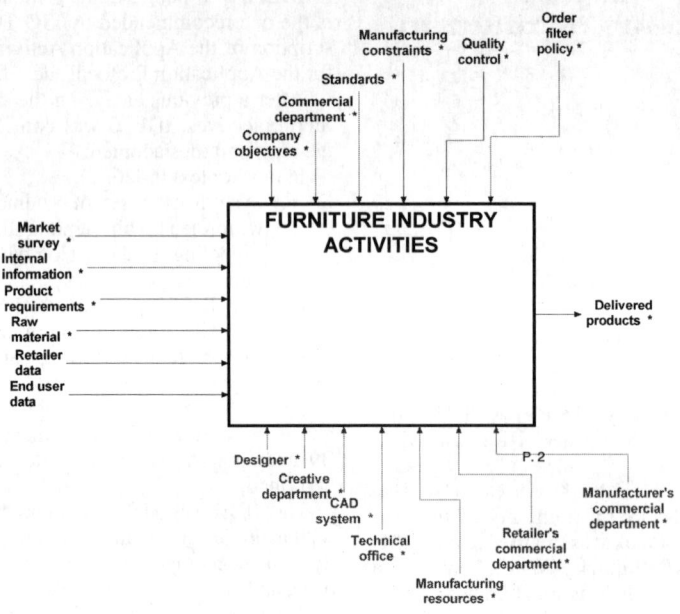

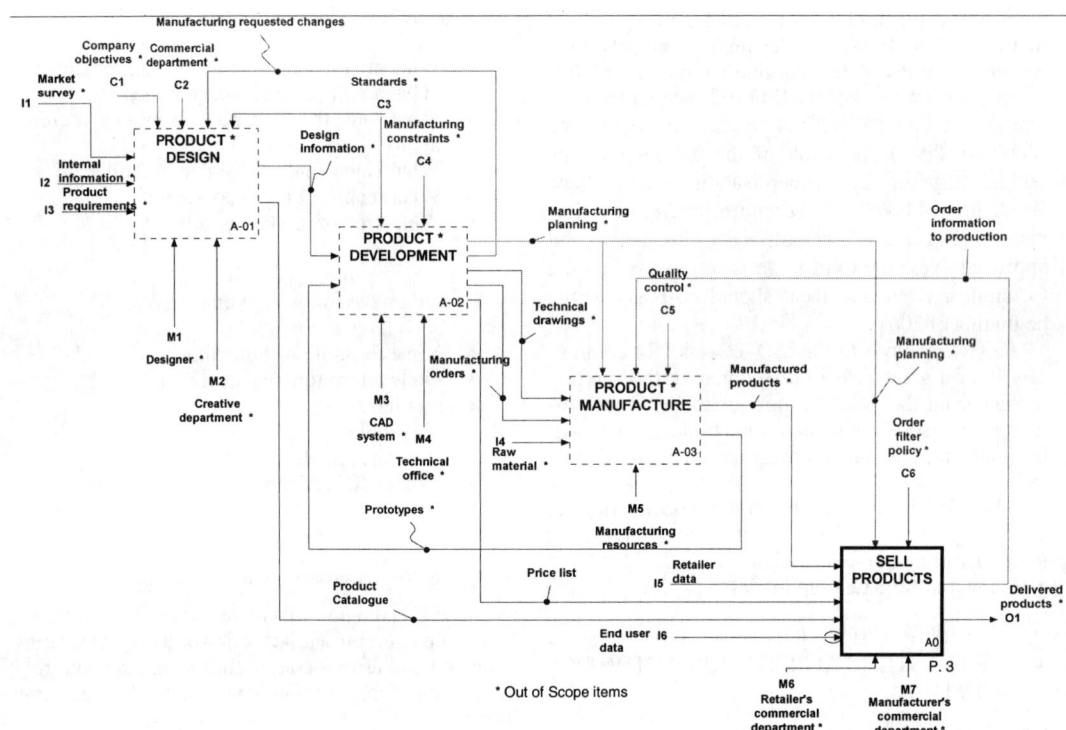

* Out of Scope items

verbs or verb phrases. No fewer than 3 and no more than 6 boxes should appear on any one diagram.

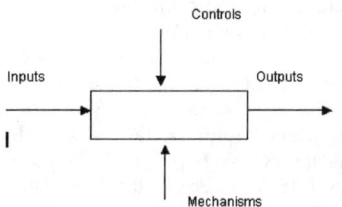

Boxes are placed in the diagram according to their relative order of importance, their "dominance". Usually the most dominant box is placed in the upper left-hand corner of the diagram. Dominance can be thought of as the influence one box has over other boxes. Also a digit placed in the lower right-hand corner of each rectangle can indicate dominance.

Arrows are single lines with arrowheads at their ends, they represent "things", and so they are labeled with nouns or noun phrases. Arrows are collections of things, they can have multiple tails (sources) and multiple heads (destinations).

A diagram is usually redrawn several times, creating versions. IDEF0 uses a diagram configuration control schema, built upon chronological numbers or C-numbers to distinguish different versions of the same diagram from each other. These codes are put in the lower right-hand corner. It is constructed from the author's initials and a unique sequence number.

A system is represented by a single box and his arrows to define the boundary around the system. It is called "Context Diagram".

The process of structured decomposition first breaks the one-box boundary of the model into a single diagram having from 3 to 6 boxes, then breaks one ore more of this boxes, and so on. The title for each diagram is taken verbatim from the box it decomposes.

IDEF0 identifies each diagram in a particular model by what is called a "node number". It has the form: model name or abbreviation, slash, the capital letter A (for Activity diagram),a hyphen, and zero, for example, MP/A-0. The node number for the diagram that decomposes the boundary is the same node number without the hyphen, MP/A0. All other node number are formed by taking the node number of the parent diagram and appending to it the number of the box that is being decomposed.

C-numbers are also used to link downward diagrams. In the Context area the author draws a tiny square for each box on the parent diagram, shades in the box that the diagram decomposes and place the c-number of the parent diagram near the shaded box.

IDEF0 has a notation that allows identify and validate arrow connections between diagrams. It is a

encoding Schema called ICOM (Input Control Output Mechanism)

By following it, one creates a set of implicit off-page connectors which can be quickly changed when boundary changes.

4.2 The modeling Process

IDEF0 is a true methodology because it integrates an interactive process for model development, configuration control notation for models, a diagram reference language and a model activation language. Its organization is centered on the specific roles people play on IDEF0 projects. They are:

- experts are information sources
- authors develop diagrams and models
- a librarian coordinates the written exchange of information
- readers review and validate data models
- Technical Review Committee approves a model for usage

4.3 The funStep AAM

The application activity model consists of a structured graphical representation of the main activities developed within the application scope and the information, materials and resources of such activities. IDEF0 is the methodology used for the development of the Application Activity Model.

The model to describe has to be with the relationship among the furniture manufacturers and the commerce (retailers, major retailers and private customers).

The model refers to the Product definition to use in the Project model in order to allow the exchange of Product Libraries and Orders (decorating projects) including graphic information.

The purpose of the AAM model is the "Identification of the more important activities that are carried out in the manufacturer-customer relationship, in order to identify the most important information exchange between them".

The aim of IDEF0 model is to provide a global vision of the activities to be performed in the system, the information flow and the required material.

The diagrams shown in annex contain an extract of the funStep IDEF0 AAM model information. ICOM (Input, Output, Control, Mechanism) with asterisk are out of the scope of interest of the system but are kept to give a global environmental view of the system.

4.3.1 Activity: Product Design (A-01)

This activity has as objective the definition of the requirements of a new product covering an analysis of the today-situation of the company and its products. It also has into account the related standards.

Commercial and Creative departments are involved and sometimes external designers are contracted for the occasion. From a conceptual point of view, in the product design several aspects must be kept in mind such as manufacturing, assembly and recycling.

The Design information is generated for the following phase, product development and also the Catalogue. During development process, the design task receives feedback from the development task until the product gets stable.

4.3.2 Activity: Product Development (A-02)
This activity of product development allows to document technically the design producing technical drawings. Sometimes prototypes are required for production.

CAD/CAE systems make easy the documentation for the manufacturing phase. Once the design is approved the price lists are prepared.

There is collaboration between the Technical office, product designer and the Manufacturing department in this phase

4.3.3 Product Manufacture (A-03)
This is an activity of material and data processing, its objective is to obtain the final product with the requirements of quality defined.

It begins at the end of the previous phases with the input of raw material, base material for the final product.

This activity consists of production programming, process planning and product manufacturing.

Technical drawings are the information generated during design and developing.

Manufacturing resources are human resources (CNC operator, programmer, etc.) and physical resources (CNC machinery, CAD/CAE systems, tools, etc.) necessary for product manufacturing.

The order information corresponds to the data in the order document asking for the product manufacturing, with delivery date, terms of payment, ship conditions, etc.

4.3.4 Activity: Sell Products (A0)
The product, when produced and controlled, is shipped to the Retailer as indicated in the shipping orders.

The activity of selling Products -pieces of furniture- (called in previous versions Manufacturer-Customer relationship) has been modeled with the activity named "Sell Products", and it stands for that period since a catalogue is sent to the Retailer until the purchase confirmation is received by the manufacturer from the Retailer. This purchase confirmation will make that the manufacturer launches a production order.

"End-user" refers to the client of the Shop, the person who is purchasing the pieces of furniture, the furniture buyer.

"Retailer" refers to the Shop but it can be a private customer sometimes (end-user as customer). Furniture manufacturers are their providers.

"Manufacturer" refers to the furniture manufacturer, who is supplier of the Shops or private customers (direct sale). Retailers are clients of the manufacturers

"CAD-deco" refers to the Computer Aided Design for furnishing in the Shop or at manufacturer's.

"Special-case-conditions" when an Order includes special products, they have a technical construction proof and an special budget calculation that is sent to the Shop with the delivery date to be accepted.

"Manufacturer's notification" an acceptance or rejection notification to the Shop.

"Order" usually sent via fax to the manufacturer from the Retailer. It consists of 3 parts basically:
· Retailer's management data; address, name, etc.
· Design; the view plans needed to describe the project completely
· Article description; quantity, code, description, price, etc.

When the Project contains special products (products that are not catalogued products) it can add sketches indicating dimensions, notes, etc.

A. Activity Inputs:
· Retailer data: They comprise the name, code and address data from the Retailer's
· End user data: They comprise the name, code and address data from the end user
· Price lists: They are the fees with all the article data; name, code, dimensions, price and a draw and other characteristics
· Product catalogue: It consists of a collection of photographs and other indications to show an extract of all the possibilities of the manufacturer's products
· Manufactured products: Products ready for shipping following the order indications and data

B. Activity Outputs:
· Order information to production: Once the order from the Retailers' has been filtered and everything has been checked, this notification with the order data starts all the Production process for the products to be produced, packaged and shipped.

C. Activity Controls:
· Order filter policy: This is the set of restrictions and controls for an order to be introduced in the manufacturer's system. It is credit policy and sometimes management decisions. Each manufacturer company uses this criteria in a very personal way. It is the set of technical and credit filters an order has to pass before the rejection or acceptance process.

· Manufacturing planning: This is the control that allows the selling product activity to take place, including information about delivery and expedition

D. Activity Mechanism:

· Retailer's commercial department: Personnel responsible at the shop of running the selling, create the furnishing project with the manufacturer's information and prepare the order to the manufacturers'

· Manufacturer's commercial department: Personnel responsible of the distributions of the commercial information to their clients and filter and compile the orders information.

5 PERSPECTIVES OF FUTURE WORK

The European Union (EU) is the first furniture producer and the first consumer world-wide. The furniture industry is one of the most important branches in the EU. The total production in 1998 was 74000 Millions of Euro, with a growth rate of 4.1%. This industry represents almost 4% of the European internal gross product.

The furniture sector employs about one million people, and it is structured on a wide basis of SMEs. There are 8800 companies with over 20 people, employing a total of over 600000 workers. There are 80000 companies with less than 20 workers, with a total of 300000 people.

The future work intends to keep this trend of the Furniture industry within the emerging digital economy and smart enterprises through the use of modern information technologies and standards, promoting the co-operation for consensus, standardisation and interoperability for data exchange and business-to-business e-commerce in the furniture and related sectors. It is envisaged a direct impact of these activities to over 3000 companies during nest 2-3 years.

To achieve that, the following issues should be set up:

1.- The progressive development of an interactive process of communication, dialogue and active dissemination of standards for data exchange and electronic commerce in the furniture industry using Internet, in order to achieve consensus in the Furniture sector and accelerate the standardisation process influencing the technical developments in the sector.

2.- To co-ordinate the evolution and extension of the data models –nowadays at ISO NWI stage- based on the industrial requirements -from software vendors and users-, and the results validation based on the definition of real test cases.

3.- The management of co-ordination of all the activities and new initiatives with an intensive use of Internet and videoconference in order to reduce costs.

4.- Manage liaisons already arisen with other workgroups in related sectors (shipbuilding, architecture, heat & sanitaryware, ...) and new ones to come.

6 ACKNOWLEDGEMENTS

We would like to thank all the members of funStep Interest Group that as collaborating and providing the required work for results and success of this initiative, and specially the Chapter Coordinators, that have been establishing the essential link between the general objectives of funStep and the industry in their countries.

In particular, we acknowledge the European Commission by the financed budget and its support and trust in our ideas and developments.

7 REFERENCES

http://www.funstep.org

http://www.nist.gov/sc4

http://www.cenorm.be/isss

e-Business and e-Commerce

Electronic communication in the building and construction industry: What will eConstruct bring about?

Frits Tolman
Technical University of Delft, Netherlands

Michel Böhms
Netherlands Organization for Applied Scientific Research (TNO), Netherlands

David Leonard & Jeff Stephens
Taylor Woodrow Construction, London, UK

ABSTRACT: The paper presents the aims and goals of a recently started new European project (IST[*] 10303) titled eConstruct, from electronic Business in the building and construction Industry: Preparing for the New Internet. The project focuses on the development of Internet communication technology that can replace many of the traditional communications encountered in current building and construction practise. The opportunity follows from the coming introduction of the so-called "Next Generation Internet" (NGI), the Internet for structured information exchange using new technologies like XML (eXtensible Mark-up Language) and related standards. This paper describes the project's scope, purpose, baseline, how eConstruct relates to other initiatives in more detail and discusses the first results of the project.

[*] Information Society Technologies – The European Fifth Framework programme

1 INTRODUCTION

Electronic business (eBusiness) will become the standard means of doing business. The question is not whether, but when. Electronic business in the building and construction industry is no exception.

One of the first steps towards eBusiness is to develop a suitable electronic communication technology that serves the particular needs of the international building and construction industry.

Open meaningful electronic communication in building and construction is not yet supported. The most advanced communications concern commercial transactions, using EDI, and the exchange of electronic technical drawings using proprietary exchange formats like DXF and DWG. Exchange of product models, the subject of ISO-STEP, is in our industry not coming off the ground. IAI-IFC is doing better but a lot of work and convincing still has to be done.

In recent years Internet showed both its appeal and its weaknesses. Today almost all the professionals in our industry are connected to the Internet. Everybody surfs around over the web and uses e-mail for informal, unstructured communication. Also the downside is clear for everybody:
- Internet cannot be trusted.
- It is insecure, slow, occasionally not working.
- Not always used for business communications.

Internet (Intranet, Extranet) is in principal ideal for the building and construction industry since it is cheap, widely available and not too difficult to use. The problems with the current Internet, which is unstructured, sometimes slow and insecure, will be overcome by the Next Generation Internet (NGI). The NGI will be structured, faster and much more reliable and will provide services as security, digital signatures and much more.

Developing a Communication Technology that is suitable for the whole variety of building and construction companies, ICT vendors and individuals, requires a small and dedicated team involving Industry and R&D groups from different member states.

The eConstruct [econstruct][1] project partners are:

Role	Company Name	
E	Taylor Woodrow Construction Ltd.	(UK)
E	Betanet SA	(EL)
R	TNO Building and Construction Research	(NL)
E,S	Stichting STABU	(NL)
R	Delft University of Technology	(NL)
R	Centre Scientifique et Technique du Bâtiment	(F)
S	Nemetschek AG	(D)
S	EPM Technology AS	(NO)

Roles: E: End-user, R: R&D, S: Software Vendor

[1] square brackets denote references

Besides these project partners eConstruct also co-operates with so-called Collaborative Partners, i.e. external organisations that have similar or related goals and that are willing to trade some of their effort for early information and influence. Our first Collaborative Partners are: Construction Information Systems and CSIRO in Australia.

2 WHY XML?

The main drive behind eConstruct is the new opportunity following from the NGI. XML is a protocol that extends the capabilities of HTML, which only permits objects to be tagged for the purpose of formatting an object within a browser. XML allows objects to be tagged with properties that define the data contained within each object. XML object tagging can occur directly inside an XML message or it can be contained in an external document called a DTD (Document Type Definition). Unfortunately DTDs have very restricted facilities for strong data typing, which is fine for document-oriented communication but not for data-oriented communication. In the near future "XML Schema" will replace DTDs and help to solve the problems with data typing, therefore enabling more meaningful communication.

XML is very different from EDI technology because XML messages contain, or reference, properties that define and describe the data contained within the message. EDI messages, on the other hand, are very long strings that contain data, however, it is up to the sending and receiving parties to know the exact format of the message and parse the data values accordingly. As such XML represents a much cleaner and more open and standardised method of handling interface messaging. It is not surprising that XML is threatening the existence of the current EDI infrastructure.

XML supports the development of industry or application specific languages called "XML vocabularies" that are tailored to a particular industry sector, or application type. These vocabularies are used to define the concepts that are clearly understood by the users. By using these concepts, instead of unstructured data, XML messages can be made content rich, secure, fast and potentially able to replace existing traditional ways of communication.

3 eCONSTRUCT

One of the many projects looking into the XML future is the European eConstruct Project that started in January 2000. The goal of eConstruct is to provide the building and construction industry with a suitable communication technology that supports most of the communications related to the procurement of materials, components, assemblies, documents, systems, services and equipment (also) over the national borders.

What eConstruct will not do is duplicate existing EDI initiatives who will probably also adopt XML themselves. The project will also not address work on the commercial and contractual aspects that are already covered by several other developments.

Initially eConstruct will focus on object identification and definition, i.e. "What makes a column a column?" If we want to communicate in the European building and construction industry we have to have one common vocabulary or "lexicon" of concepts that can readily be understood by people and computers, and that can be translated into the national languages and classifications in use in each individual country. This vocabulary is called "building and construction eXtensible Mark-up Language" (bcXML).

Next eConstruct will develop tools that support different languages and multiple classification systems. Tools that use bcXML-based communication will enable electronic business between participants in European and international projects.

Finally eConstruct will develop a cluster of three related bcXML-based applications that support the supply chain and the execution of large-scale projects involving partners from multiple member states, and – hopefully - are of enough interest to boost its application in industry.

The first applications of bcXML will be eCommerce type of applications, buying and selling over the Internet via some virtual market place. Think for a start about the stage before the actual buying and selling takes place, i.e. the 'shopping' phase. For example:

- You need a new boiler and want to search the Internet to find the product and/or supplier of your choice, or
- You need to paint your house and want to search the Internet for a company who might be able to do the painting.

In these situations EDI is not too much involved and the open Internet seems ideal.

When you require the procurement of non-standard components, assemblies, systems and services that need some additional design input - which is what building and construction projects are all about - shopping and such in a virtual marketplace suddenly becomes much more difficult. In those cases we have two options. Option one is to bridge the gap with the design/engineering world and link to their Product Data Technology (PDT) standards and systems. Luckily associated standardisation initiatives in this area like ISO-STEP and IAI-IFC are in the process of being able to support XML. Option two is to extend the current eCommerce process that is largely based on article numbers with a more parameter-based type of communication, i.e.

buy a concrete beam not out of a catalogue but, for example, specified with a required height (say 57,5 cm), width and length.

Both types of communication will be explored. Also the role of 2D and 3D graphics (a picture says more than thousand words) like VRML will be explored. Consider for example buying ceramics or furniture of your taste. Hard to express without graphics wouldn't you say?

4 RELATED PROJECTS

As always information gathering is one of the first steps to take. Who is who in XML? Who is doing what? And which developments are most promising?

The European eConstruct project is not the only project of its kind. The following related initiatives (among many others) are particularly interesting: LexiCon, aecXML, XML/EDI, GEN and ebXML.

4.1 *LexiConExplorer*

The LexiConExplorer is a tool from the Dutch STABU Specification organisation for capturing building and construction semantics. The development started in the Brite-Euram CONCUR project. Capturing BC semantics has been tried over the years in the form of Classification and Coding systems. In the electronic age new forms of classification arise. One of these new forms is the LexiConExplorer with its filling.

Developing the LexiConExplorer and filling the LexiConExplorer is heavily related to eConstruct goals. We were therefore lucky that the STABU became one of our partners. However the STABU is not only active within eConstruct, but also in other developments, like in a co-operation with representatives of building and construction in Norway. Also the STABU is active in ICIS, the international classification 'mantle' organisation.

The most interesting aspects of the LexiConExplorer and its content are that it can serve as a super classification system itself and that it can support different languages and classifications. With its structure and simple but clear definitions it allows many to many mapping of concepts between many different views.

4.2 *aecXML*

aecXML [aecxml] is an initiative of Bentley Systems in the US recently brought under the North American IAI umbrella. The scope and purpose of aecXML is roughly the same as bcXML. One exception is that Europe is much more fragmented than the US and the eConstruct emphasis on object identification and definition is not as pressing. Though for the time being both projects co-exist it has been decided that we will share the largest common set of elements as possible.

4.3 *XML/EDI*

The XML/EDI group is working on the merging of XML and EDI. One action is the development of common EDI directories and on-line repositories for global application in different domains. eConstruct will be participating in the building and construction industry.

Also on the national level eConstruct will be complementary with EDI or EDI-like developments, like for example HCP-BOUW (NL), GAEB (DE) and CITE (UK).

4.4 *GEN*

Another interesting initiative is the creation of GEN (from General Engineering Network) electronic marketplaces for global engineering companies. GEN is already working for several years and the first commercial results are available. GEN supports the required concept translation mechanism, but (still) in a closed environment. GEN (and particularly GENIAL) is concentrating on Electrical, Mechanical and building and construction. Several GEN results are already rewritten in XML, or will be so shortly. eConstruct will be co-operating with GEN in the domain of the building and construction industry.

A first eConstruct pilot project focusing on buying and selling of Prefab Concrete Elements in will use some of the GEN results.

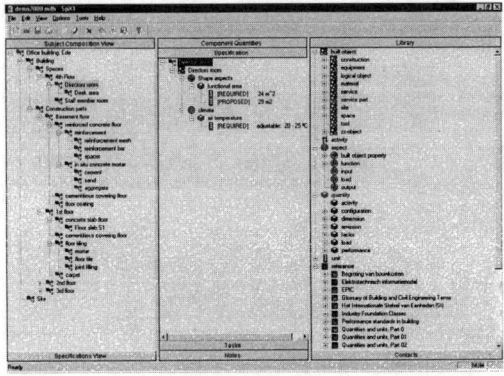

Figure 1 – The LexiConExplorer User interface

4.5 ebXML

ebXML, from Electronic Business XML, [ebxml] is a very interesting initiative of UN/CEFACT and OASIS intended to provide an open XML-based infrastructure enabling the global use of electronic business information in an interoperable, secure and consistent manner by all parties.

The United Nations body for Trade Facilitation and Electronic Business (UN/CEFACT) [un-cefact] and the Organisation for the Advancement of Structured Information Standards (OASIS) [oasis] have joined forces to initiate a world-wide project to standardise XML business specifications. UN/CEFACT and OASIS have established the Electronic Business XML initiative to develop a technical framework that will enable XML to be utilised in a consistent manner for the exchange of all electronic business data. Industry groups currently working on XML specifications have been invited to participate in the 18-month project. A primary objective of ebXML is to lower the barrier of entry to electronic business in order to facilitate trade, particularly with respect to small- and medium-sized enterprises (SMEs) and developing nations.

All the big commercial players in eCommerce, maybe except Microsoft, are present and strongly participating. Very interesting in the ebXML approach is that the project covers many layers of abstraction, starting at the definition of Business Processes and Electronic Markets and ending with specifications for message envelopes and such. By doing so a complete set of ebXML compatible standards and software will arise. Moreover these standards not only focus on EDI as it was, but more on something that can be called *document oriented business* (where the documents are XML documents that can readily interact with both humans and applications). As the approach also involves related technologies like Electronic Document Management and Workflow Management this clearly seems the wave to surf.

One additional reason for putting (part of) our hope and money on ebXML is that it is open and competes against an initiative by Microsoft, called BizTalk. Though Microsoft plays a VIP role in BC, it should not be our only alternative.

5 RESULTS

Though the project is only nearly five months old there are already some results coming out. The most obvious result is the fact that now we are sure that developing an XML vocabulary for building and construction is the right way forward and that our vision of bcXML and its perceived role is indeed feasible in the given timeframe. Many other industries are doing the same as we are. Often those industries

join the ebXML effort because they hope to avoid falling in Microsoft's hands by having to rely on BizTalk (an MS initiative for inter XML vocabulary communication).

We also learned from the GEN initiative that the concept of on-line transformation of both language and classifications are feasible. The first eConstruct pilot project will demonstrate simple buying and selling of Prefab Concrete Elements both nationally and over the national borders.

Also we now clearly see that bcXML will be based on XML Schema, the successor of Document Type Definitions (DTDs). The main problems with the DTD format's restricted expressiveness are overcome in XML Schema. The fact that XML Schema is not yet fully operational is for eConstruct no problem. However we hope and expect that the standard and some tools will become available in a couple of month.

What we also have done is to develop a 'vision' of the bcXML architecture. Basically we want to distinguish the bcXML ontology schema, that describes a set of relevant building and construction concept classifications both in a language neutral format and in a set of country specific formats, and the bcXML object schema, that describes the relevant building and construction project concepts both in an actor neutral format and in a set of actor specific formats.

A recent idea, still under consideration, is to add a third bcXML schema that describes the transaction side of the problem. The bcXML transaction schema will group concepts like Company, Participant, Business Process, Contract and such in one schema that can be easily mapped on, instantiated from, or replaced by, the ebXML Business Process Model concepts. This way we will be able to re-use the ebXML structure, Core Components and related software once it becomes available.

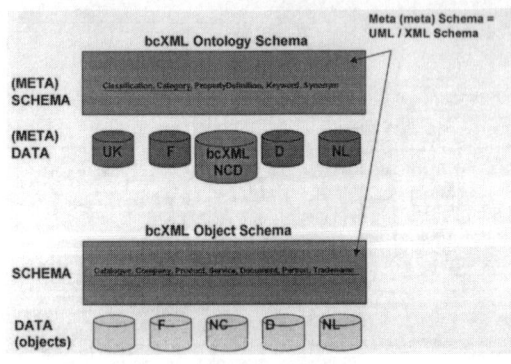

Figure 2 – bcXML Architecture

6 CONCLUSIONS

The European eConstruct project will develop a Communication Technology "bcXML" suited to the specific needs of the European building and construction industry. BcXML will be made available to the building and construction industry and submitted to suitable standardisation bodies and umbrella organisations.

BcXML will be able to support new document oriented construction processes, where the dumb unstructured documents are replaced by smart structured documents.

Communicating not only the *representation* but also the *meaning* of data makes all the difference. Communicating representations always requires human interpretation and transformation even if the purpose is to input the content in a database, or software application. If the human factor can be brought outside the information exchange process, the speed and quality of such communication can be greatly improved. If additionally the representations can be seamlessly translated into different national languages that conform to local classification standards, co-operation between European building and construction companies is enabled.

The eConstruct project is picking up speed. The next few month we will develop, implement and demonstrate a small but realistic part of bcXML. This part will solely focus on Prefabricated Concrete Elements (PCEs) and their environment (i.e. the description of structural parts made-up of assemblies of PCEs, the materials required for the production of PCEs, the required services like transport, assembly, etc.). The end user partners will create PCE libraries following different approaches (i.e. article numbers, parameters, etc.) that will be used in a (faked) construction project played by another partner. The bcXML ontology, object and transaction schemata will be instantiated for the PCE example. Yet another partner will implement gateways to PDT. Software will be tailored to the needs of the demo. External interested parties will be invited to play a role.

Though the eConstruct project will focus a number of applications of bcXML, many other applications are possible and many market opportunities will arise. As eConstruct is largely an open project, interested parties who are also willing to contribute to our projects goals are advised to become members of the Co-operating Partner Group.

REFERENCES

[econstruct]   www.econstruct.org
[aecxml]   www.aecxml.org/index2.htm
[ebxml]   www.ebxml.org
[un/cefact]   www.unece.org/cefact/
[oasis]   www.oasis-open.org

E-Commerce: A new frontier for engineering software and services

Mauro Mangini
Geodeco S.p.A., Bogliasco, Italy

Byron Protopsaltis
FIDES DV-Partner GmbH, Munich, Germany

ABSTRACT: Enhancing the capabilities of engineers, offering them new ways of working by means of the most advanced technologies, can be achieved in two different ways. Providing them the tools to accomplish their tasks or offering them the services of completing such tasks for them. This paper discusses the concepts and tools that will be developed within the framework of the ISTforCE Project: an environment to provide remotely engineering software and services.

1 INTRODUCTION

1.1 *The Current Situation*

Electronic Commerce opens the door of the global market to providers of goods and services. This perspective allows even small and medium enterprises to get in contact with potential customers and suppliers all over the world widening enormously their commercial possibilities.

The sale of goods through Internet is already a common habit in the Information Society and is growing at a very fast pace. However only a few kinds of services are already sold through Electronic Commerce. They are low-level services that require limited interaction with the customer, such as providing information stored in databases etc.

For engineering services at present too many problems are still open. Among others, engineering companies willing to sell their services on-line would need to provide "intelligent" interfaces able to interact with the potential users understanding their needs and advising them to select the most appropriate actions. Moreover an acceptable level of quality should always be guaranteed on the results.

1.2 *The Perspectives*

The new concept we want to introduce and demonstrate with the ISTforCE project is that in the near future engineering services could be sold through Internet (which is not possible today) by means of the new technologies we are going to conceive and develop. This concept, which in our approach shall be implemented for the engineering field only, can be eventually broadened to "higher level" services of different kinds to be provided by means of Elec-

tronic Commerce. The goal is to open a new way of working even for professions with high technological content in the Information Society.

Enhancing the capabilities of engineers, offering them new ways of working by means of the most advanced technologies, can be achieved in two different ways. Providing them the tools to accomplish their tasks or offering them the services of completing such tasks for them. Engineering software on rental basis on the net is an important aspect to broaden the capability of an engineer. It can provide him the necessary tools to solve a particular problem, provided that he has the necessary skill to use such tools. Moreover the renting of software frees the engineer to buy, install and maintain the software.

To be even more effective and immediate, we can offer directly the service of solving engineering problems. Offering engineering services through Internet will also allow very small high specialized enterprises to be stronger recognized in the engineering community enabling them to provide their knowledge and services to a large number of potential clients.

There will be therefore the possibility for the user to select:

1. Direct on-line service where he can interact directly with the software able to solve its problem and can get at the end of the session the desired solution;
2. Remote services where he is helped by an intelligent system to provide all the data necessary for

the engineering service, which will be then executed through human intervention.

Within the framework of the ISTforCE Project an environment to provide remotely engineering software and services will be implemented. A Product Model for consistent product data exchange will be established, and intelligent browsers implemented for data recognition and manipulation.

2 A NEW WAY OF WORKING

2.1 *Motivation*

European engineering companies, able to provide high quality engineering services, could highly benefit from the perspective of selling remotely their services and be able to reach potential clients even in distant countries. Despite the integration efforts in industry standards, differences among European countries are still significant concerning approaches to civil engineering design.

Moreover, specific technical capabilities are not always available to solve problems that are peculiar of countries where the construction will be executed but not of the country where it will be conceived and designed. For instance North European design offices having to deal with complicated seismic related issues for contracts abroad, need often the support of foreign companies or institutions having greater experience in the field.

The open market for engineering services within the European Union allows for instance an engineering company to bid for a project in other European countries where geological or climatic conditions are completely different and consequently also design solutions (i.e. seismic problems, ice engineering, etc.).

It is therefore evident the need for a new technology able to help the engineering companies willing to perform their services at a global level, to overcome the difficulties of lack of specific knowledge for problems peculiar of different regions of the planet.

From already completed ESPRIT Research Projects (COMBI, TOCEE, etc.) the basic technology for performing the design on a distributed basis has already been investigated, solutions have been proposed and prototypes have been implemented.

What is still missing is the possibility to find on-line the depositary of the missing knowledge or capability able to interact with the virtual enterprise performing the design on an automatic or semi-automatic basis.

2.2 *Domain*

The domain that will be addressed within this project is the market of engineering consulting services where companies offering their consulting services could perform design specific analysis to be used by any other subject in the global network who will pay for this service a fee. Other engineering related service providers like surveyors or environmental experts could also offer their services using the Electronic Commerce technology allowing the project manager to quickly set up the project requirements.

The tools that will be developed will concern the specific market of engineering services but the concepts will result strictly general and applicable to different kind of services with high technological content. It is possible to conceive in the near future for instance on-line high level services for medical care, legal consulting, fiscal problems and so on.

2.3 *Advantages*

The advantages of this approach are:

- lower project costs due to the increased productivity, and to savings of the stand-by and communication costs;
- shorter lead time due to automated processes;
- increased competition;
- higher efficiency;
- increased project monitoring possibilities;
- possibility to always select the best actor available on the market;
- possibility to deviate very quickly the design process from one actor to another.

3 THE CONCURRENT ENGINEERING ASPECT

3.1 *Broadening the Design Team*

In this view a new form of Concurrent Engineering at the global level could be established where also the commercial aspects of the problems are taken into account. Up to now the Concurrent Engineering approach was only envisioned within a restricted pool of companies or subjects willing to cooperate on a specific project. By means of the new technology the Concurrent Engineering approach will be possible by using external engineering services covering specific aspects of the project through Internet speeding and up consistently the design and construction processes.

From a previous Esprit Project (ToCEE) a new way of working with a Concurrent Engineering approach was investigated. Different aspects of the collaborative as well as co-operative working have been considered and tools have been implemented to put those concepts into practice. The scope though was restricted to a pool of players. A few engineering companies that agrees to work together on a project, sharing the information and the tasks to be carried out on the basis of a central management.

It is necessary now to improve and broaden the scope and methods by introducing the technology of Electronic Commence for all kind of engineering related services going, from specific design consulting up to field operators. This will allow the central management entity to play without restrictions in the global market of engineering services needed for the different phases of the project from feasibility study up to final design.

3.2 *Procurement and E-Commerce*

There are many services needed during the course of a project that could be provided by means of Electronic Commerce. For instance all services related to environmental conditions and loading such as:

- Topographic surveys;
- Geotechnical site investigations;
- Geophysics and geo-environmental investigations;
- Laboratory testing;
- Environmental impact assessment;
- Archeological investigations;

For the solution of such problems and for others, companies could be available to offer their services through E-Commerce. These companies will be able to receive requests for offers by the project manager, compile proposals, propose prices and finally perform the service.

3.3 *The Automatic Engineering Service Provider*

An autonomous entity available in Internet and able to interact with users seeking engineering services will be conceived and implemented. The architecture is shown in Figure 1. This entity will contain "intelligent" interfaces able to interact autonomously with the user not only guiding him through menus and buttons but also asking him questions concerning data availability, special needs etc. and offering as free of charge service a check on the effective need for the requested service.

A knowledge-based system will implement this function. If the user (or better say the customer) decides to order the engineering service then all the gathered information will be transmitted to the Central Design Office (see Figure 2) where the consult-

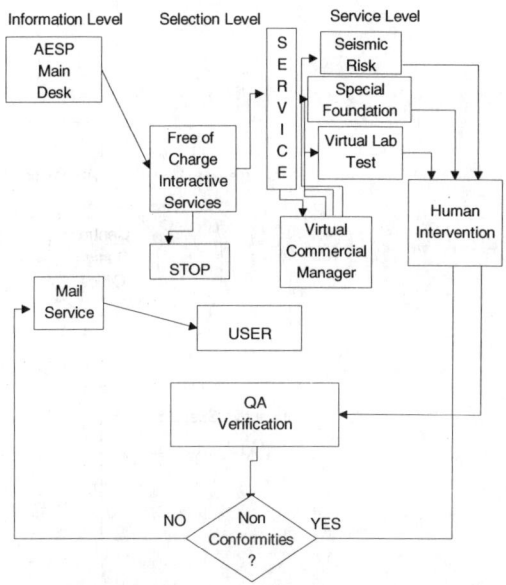

Figure 1. AESP Architecture

ing service will be performed with human intervention. A quality verification will be performed to ensure correctness of the results. A report will be then delivered to the client electronically signed by the author (the engineering consultant) who takes therefore full legal responsibility on the correctness of the results.

In the AESP there will be three different levels of interaction:

1. *An information level* where the user acquires information about the services available with detailed information (downloadable also in pdf format) about the methods that will be used and the obtainable results. This will be achieved through a Help Desk.

2. *A selection level* where the user will be able to select the service he wants to use and start a first free of charge interaction with it. This possibility will provide the user with additional feed back about his needs helping him to decide whether he actually needs the service or not. These features will also be very useful in attracting visitors that may use the free services and spread the information about the AESP services in the engineering community.

3. *A service level* where the user will be prompted with the interface for actually using the selected service. A negotiation will take place by means

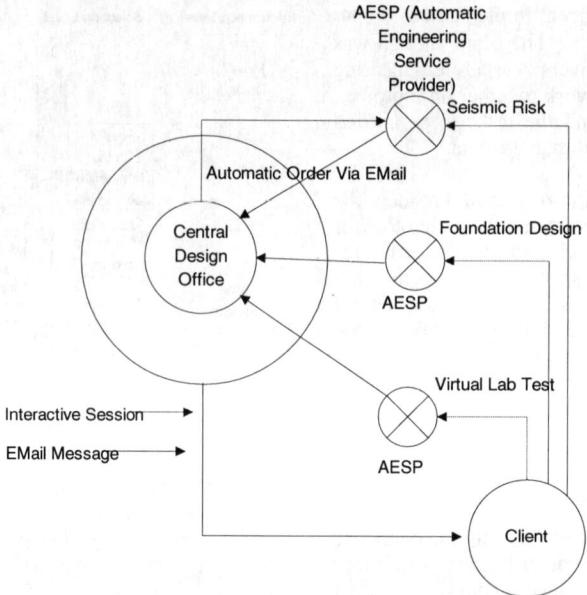

Figure 2. Interaction of the AESP with the central design office

of a negotiation tool and finally the consulting service will be ordered.

Three engineering services will be available through the environment:

- *Seismic Risk Assessment;*
- *Special Foundation Design;*
- *Virtual Lab For Concrete.*

These three services will require no direct interaction between the user and the software performing the analysis.

Human intervention will guarantee proper results checking and quality verification before release of the official results.

The Seismic Risk Assessment Services will be provided taking into account the local information concerning the seismic catalogues and geological conditions. State-of-the-art risk methodologies will be applied. The user will only need to input the coordinates of the site where he wants to build the structure and the available information about the soil characteristics.

The system will perform a complete seismic risk analysis, starting from the seismic catalogs available for the area, and taking into account also the local geological conditions, and provide to the user the likelihood of a given peak ground acceleration to be exceeded.

3.4 *The Software Rental Service*

An *Engineering Software Rental Service* will be implemented where object oriented applications for geo-technical design will be offered on rental basis.

This type of applications will be able to deal with different design codes and calculation methods and to support highly interactive work. They will contain iterative numerically intensive (finite element, kinematic Element method) as well as non-intensive computation methods. Decision making for the selection of the computational method and modelling via parameters within the computation will be implemented.

The applications are:

- Design and dimensioning of Excavation Walls;
- Excavation planer;
- Ground Failure;
- Stability checks;
- Slip Circle.

Delivery of applications on the Web requires centralized application delivery capability, which should be scalable, manageable and secure. It must also offer predictability in terms of performance and cost.

140

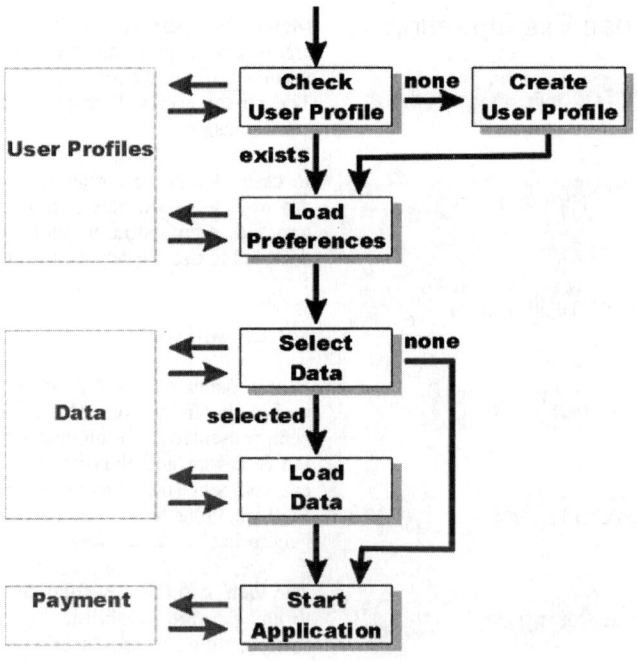

Figure 3. Profiling with user preferences

Management Services offered by commercial packages like the Citrix MetaFrame© or the Windows Terminal Server© with extensions will enable the packaging and publishing of applications to servers. Citrix optimizes additionally data transfer to a minimum so the user can work without performance reductions with the available bandwidth.

Users will be able to access all applications they need including the latest Windows-based applications and Java applications, from their desktop.

The set of applications can be delivered to customers without adjusting client system configurations. The total cost of application ownership can be reduced because users will be charged according to their actual demand. Users are able to access the latest update of the applications through their personal computers.

All applications look, feel and perform as if they are running on a local desktop. A flow chart diagram for logging in to the application server and run a problem (access on selection level) is shown in Figure 3.

3.5 *License Models for local Installation*

In case of downloading an application, Software Licensing will be offered. The software licensing will

include password protection, encryption and will be computer dependent.

3.6 *Electronic Licensing*

Electronic license distribution will account a considerable portion of the total software license acquisition and distribution market in the next years. More optimistic forecasts predict 100 percent market penetration by 2008. An electronic distribution and management of licenses will be tested within the project. This includes automatic license file updating and charging.

3.7 *Electronic charging for the software rental services*

Every application will include a cost calculation, which will depend on time of logging in, logging out, the type of application, the features used and the disk space occupied for data and results.

Applications will be coupled with a paying system, which will accept credit card charging and will also be able to send invoices. The license fees that will be charged will be based upon actual usage. Every invoice will include an itemization list with all features/programs accessed. This list will make the cost charged transparent to the customer.

Automatic License File Updating

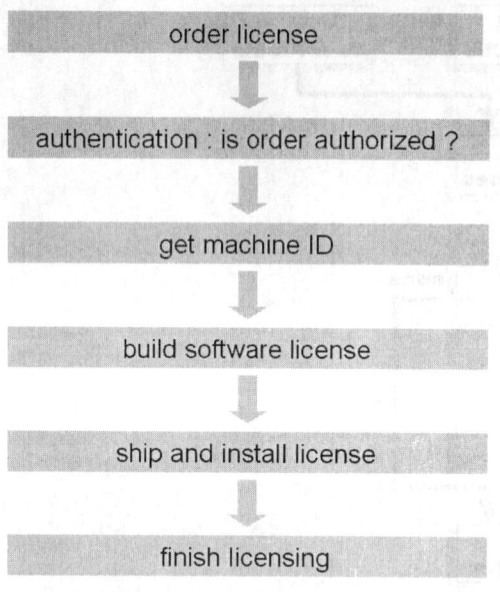

order license

↓

authentication : is order authorized ?

↓

get machine ID

↓

build software license

↓

ship and install license

↓

finish licensing

Figure 4. Automatic License File Updating

The aim is to offer rental software services through a classical Internet provider company. All applications will be installed and updated on the hardware of the Internet provider. A secure interactive administration tool will allow accounting operations like inserting, deleting customers, maintaining passwords, assigning access rights etc.

The prospect customer for the software rental services can be acquired in two ways according to the following schema:

- by usual ways like advertisement or marketing actions. The customer receives a password that allows him to access the system;

- customer is registering himself online through Internet. In this case a password for accessing the rental service is given to him after identification.

In both cases every time a customer uses the service, a file with "consumed units" is updated and the payment module is informed. In regular time intervals an itemization report per customer will be generated containing number of accesses, time and disk space consumed etc. This report can be viewed by using the administration tool so the revenues can be followed.

Many customers are not willing to be charged by credit card. In this case an itemization report and an invoice will be generated and sent to him by e-mail. The customer can then remit the amount by usual bank transfer.

In case of invoice requests, if a customer does not pay after a certain time period, then by using the interactive administration tool, the customer can be disabled to use the service further.

4 CONCLUSIONS

In this paper part of the work currently under way within the framework of the ISTforCE Project has been presented. The module for Rental of Engineering Software and Services will be part of a greater environment (the Concurrent Engineering Software Platform) that will provide to civil engineers a new generation of on-line services.

The idea is to conceive and develop an environment to enhance the capabilities of civil engineers by either providing the engineering software to solve specific technical problems on demand via Internet, or providing them directly the engineering services they need to perform their design tasks without limitations due to lack of knowledge or technology.

A system to provide engineering software on rental basis to be used remotely without any local installation will be implemented to enlarge the availability of specialized analysis software.

An automated environment able to interact with the engineer, checking his needs and allowing remotely deployed human consultants to solve his problems will be implemented.

5 REFERENCES

Mangini, M., G.Varosio, E.Parker 1994, "COMBI: KB Tool for Foundation Design", First European Conference on Product and Process Modelling in the Building Industry, Dresden Germany, October.

Mangini, M., H. M. Sparacello, 1999, "A Concurrent Engineering Approach to Geotechnical Design and Construction", Second International Conference on Concurrent Engineering in Construction, Helsinki Finland, August.

R. Wasserfuhr Shared Process Models for Distributed Co-operative Engineering.2nd European Conference on Product and Process Modelling in the Building Industry Watford, UK, 19-23 Oct 1998

R. J. Scherer, R. Juli, W. Reinecke, R. Wasserfuhr Towards a Concurrent Engineering Environment in the Building Construction Industry. Proceedings of the XIIIth FIP Congress 8-10-May 1998, Amsterdam

R. Wasserfuhr, R. J. Scherer. Process Models for Information Logistics in the Concurrent Building Life Cycle. International Conference on Concurrent Enterprising, Nottingham, Oct.1997

R. J. Scherer, R. Wasserfuhr, P. Katranuschkov, D. Hamann, R. Amor, M. Hannus, and Z. Turk. A Concurrent Engineering IT Environment for the Building Construction Industry. ESPRIT Conference "Integration in Manufacturing", Dresden, Sept.1997

Process re-engineering and e-business models for efficient bidding and procurement in the tile supply chain

J. Zabel, O. Peters & F. Weber
Bremen Institute of Industrial Technology and Applied Work Science (BIBA), University of Bremen, Germany

V. Steinlechner
Swietelsky Bauges.m.b. H., Innsbruck, Austria

ABSTRACT: This paper describes the first results of the European project e.bip (IST-1999-10710), which aims to innovate the bidding and procurement processes of tile layers by establishing a new broker service in the supply chain of the tile industry. The paper presents the case study of an Austrian tile layer. The main drawbacks of the as-is situation as well as the requirements and envisaged to-be processes are outlined. Furthermore, the basic concept of the e.bip architecture is introduced based on the analysis of different e-business models. The objective of the e.bip broker service is to support the co-operation between tile manufacturers, wholesalers and building contractors as well as provide a virtual marketplace to carry out electronic negotiation, ordering and invoicing and increase the availability of product characteristics.

1 INTRODUCTION

The bid preparation and procurement processes of small and medium enterprises (SMEs) are still predominantly based on traditional approaches, tools and communication channels (Segev et al. 1998). Additionally, re-using bidding data in subsequent procurement and order processing tasks is usually badly supported. Although the adoption of Internet and Web-based technologies is currently weak, the general attitude of SMEs towards its use is positive (Gebauer et al. 1998). Case studies confirm the potentials of e-commerce in procurement (Eyholzer 1999, Gebauer & Färber 2000). The companies benefit from time and cost savings, increased flexibility, improved inter-organizational information sharing as well as the access to new markets. Major obstacles to e-commerce for SMEs are the investment costs, the lack of adequate tools, difficulties to change the traditional organizational systems, missing standards for the value chain as well as security objections (Chappell & Feindt 1999, Segev et al. 1998). However, in order to face the pressure of the international competition, SMEs need to take advantage of the potentials of integrating bid preparation and procurement activities as well as applying e-commerce technologies within these processes.

The e.bip project aims to innovate the bidding and procurement processes of tile layers by establishing a new broker service in the supply chain of the tile industry based on existing solutions and standards. The broker service represents a mediation system increasing the availability of product charac-teristics, supporting the co-operation between tile manufacturers, wholesalers/traders and tile layers as well as providing a virtual marketplace to carry out electronic negotiation, ordering and invoicing.

This paper presents the results of analyzing different e-broker business models as well as the findings of re-engineering the business processes by means of the Baldauf case study, an Austrian tile laying company. The objective of these activities was to find the best e-business concept meeting the requirements of today's value chain in the tile construction sector.

The following chapter outlines the approach applied for re-engineering the business processes of the e.bip industrial partners. Chapter 3 provides a brief overview of tile value chain. The as-is model as well as the to-be processes of a tile layer are exemplarily described in chapter 4. Chapter 5 introduces the state-of-the-art for business models in electronic commerce and procurement. Before the paper is concluded, chapter 6 highlights the basic concept of the envisaged e.bip broker architecture.

2 APPROACH

The e.bip project applied a classical process re-engineering approach. Starting point was the analysis of the current business processes at all end user's sites. The current as-is processes were captured and modeled in order to obtain a detailed mutual understanding about the business activities of the different actors in the tile value chain. By analyzing the drawbacks and capturing the user requirements, a model of the to-be processes was developed defining the target that has to be reached within the e.bip project. In parallel, broker architectures and case studies of existing e-business solutions have been evaluated. The synthesis of to-be model and study result represent the basis for the specification of the overall e.bip architecture.

An important principle in the e.bip project is the use of iterative and participatory development approaches in all project phases. The as-is processes and their drawbacks have been captured by carrying out interviews with the e.bip industrial partners and distributing questionnaires. With regard to the user participation, a very easy modeling method had to be selected. Therefore, the models of the as-is and to-be processes were defined using the IDEF0 approach (Colquhoun & Baines 1991, Mayer 1990).

3 TILE VALUE CHAIN

This chapter briefly introduces the tile value chain by defining the main actors and outlining their relations and the traded tile categories. Furthermore, it is defined, which actors, relations and tiles are addressed by the e.bip project.

3.1 Actors

The value chain in the tile industry consists of different players namely manufacturers, stone traders, retailers, wholesalers, building contractors and customers. We defined their roles as follows:

Manufacturers are the producers of ceramic or natural stone tiles and act as sellers only. *Stone traders* sell complete solutions to their customers. The material needed is either bought from manufacturers or produced by the stone traders themselves. *Retailers* sell the products obtained from wholesalers to customers or tile layers. *Wholesalers* act as business intermediaries between manufacturers/stone traders and building contractors/ceramic retailers. The main added value provided by the wholesalers is the provision of logistics. *Building contractors* buy tiles from the wholesalers and retailers as well as sell them and their additional services to customers. Building constructors comprise tile layers, architects, building companies and engineering companies. *Customers* are the final consumers of tiles or tile products.

Companies from all above-mentioned roles – with the exception of customers – are present in the e.bip consortium. Consorzio per la Zona Industriale Apuana (CZIA) is a consortium of stone traders and ceramic and natural stone tile manufacturers in the Massa and Carrara region of Tuscany (Italy). TIBA is a Portuguese wholesaler and retailer of products and materials for the construction sector. Stadlbauer is one of Austria's leading wholesalers and retailers of building material, tiles and Do-It-Yourself-cater. Baldauf (Austria) deals in tiles and stones and has an own purchase department and independent work force of stone layers thus acting as a tile layer and building contractor.

3.2 Relations

By considering the number of sellers and buyers on a market for a certain good at a specific time (cf. Vetter & Pitsch 2000, Feld & Hoffmann 1999), most relations between members of the value chain are 1:1 relations respectively brokered 1:1 relations (wholesalers act as business intermediaries between manufacturers and building contractors). Building contractors sometimes use invitations to tender (n:1 relations) to maximize their returns. However, auctions (1:n relations) and stock market scenarios (n:m relations) are currently not applied in the tile value chain.

The e.bip project focuses on the business-to-business relations between manufacturers, wholesalers/retailers/traders and building contractors. As tiles are mainly purchased by color and texture and most customers need, in addition to the tiles, also the tile laying service, there is only a small market for online sales in the business-to-consumer sector.

3.3 Goods and their Characteristics

In the tile industry, many different types of tiles can be distinguished. The two most important categories are ceramic and natural stone tiles. Their characteristics can be summarized as follows:

- The specific value (price per unit of weight) of the product justifies long transport routes across several borders.
- Tiles are fashionable articles with short life cycles. This forces companies to either hold a very small stock (leaving little choice to consumers) or to take the risk of devaluation of merchandise due to de-fashioning.
- Tile production processes cannot be standardised. Both the varying qualities of the raw material (for stone and marble tiles) or the specifics of each firing process (for ceramic tiles) lead to variations in colour, dimensions, and quality, making it impossible to guarantee a large quantity of a exactly similar product.
- Different product identifiers for similar or identical items and the missing standardisation for construction material characteristics in Europe often leads to communication problems within the supply chain.

The envisaged prototype of the e.bip network will primarily address ceramic and natural stone tiles, but will be open for other tile products and materials.

4 PROCESS ANALYSIS AND RE-ENGINEERING

This section gives a short presentation of Baldauf and its major business activities. The findings from the analysis of the as-is and to-be processes at Baldauf are briefly described.

4.1 Company Profile

Baldauf Fliesen & Baustoffe GmbH (Baldauf Tiles & Building Materials Ltd.) is a small Austrian enterprise whose key activity is the trade and laying of tiles and natural stones within the regional market. Its customers are public institutions, commercial and industrial enterprises as well as private households. All tiles and natural stones are obtained either directly from the producer or manufacturer or from wholesale companies in Italy, Austria or Germany.

Baldauf has an annual turnover of about 6.5 million ECU and about 30 employees. Baldauf is a 100% subsidiary of Swietelsky Bauges.m.b.H., Linz (Swietelsky Builders Ltd.), one of the biggest building companies in Austria. Baldauf has two branches in Tyrol and Vienna. The Tyrolean branch, where the e.bip system will be implemented, is organized as an independent profit center directly reporting to Swietelsky. The profit center idea is so much part of the company's strategy (of both Baldauf and Swietelsky) that it is carried down even to the work force at the construction site.

Marketing its products, Baldauf is confronted with the problems of rising competitive pressure of an expanding market and the customers increasing expectations. To meet these challenges successfully, solutions have to found to safeguard competitiveness, to reduce costs inside the company and thus to increase a successful commercial activity.

4.2 As-Is Situation

This section introduces the top level processes of Baldauf (cf. Fig. 1) and highlights the major drawbacks and bottlenecks by focusing on bid preparation and procurement.

Generally, it was identified that many business processes are still predominately based on paperwork. Furthermore, fax, phone and face-to-face communication are the major means of interaction. Electronic data exchange is used scarcely and much information has to be re-typed. This is time-consuming and represents a source of error. Except from bid preparation, the activities at Baldauf are not well supported by the applied software systems.

For the whole bidding process, the e.bid application is used (Weber 1999, Krömker et al. 1997). This system also contains databases with customers and suppliers. Two additional applications are applied for order processing, stock keeping and invoicing as well as for the cost calculation of invitations to tender. The three used software systems are not integrated. As consequence, data have to be manually re-entered and the current status of an order can not be easily retrieved.

```
A0 Laying and trading of tiles and natural stones
   A1 Preparing bid
      A11 Processing of inquiry
      A12 Developing of technical solution
      A13 Calculating of bid
      A14 Compiling and distributing bid
      A15 Pursuing of bid
   A2 Receiving order
      A21 Taking over bid data
      A22 Adapting bid data
      A23 Compiling order confirmation
      A24 Sending out order confirmation
      A25 Archiving order confirmation
   A3 Procuring material
      A31 Selecting range of goods
      A32 Capturing needed material
      A33 Commissioning of material in stock
      A34 Ordering material
      A35 Capturing delivery
      A36 Paying material invoices
   A4 Processing order
      A41 Preparing laying work
      A42 Carrying out laying work
      A43 Documenting laying work
   A5 Invoicing
      A51 Collecting invoice information
      A52 Compiling invoice
      A53 Sending out invoice
      A54 Archiving invoice
   A6 Post-calculating
      A61 Comparing bid and processed order
      A62 Analyzing deviations
```

Figure 1. IDEF0 node tree of Baldauf's as-is processes.

The bid preparation process (A1) starts with processing of a customer inquiry, followed by the development of a technical solution, cost calculation as well as the compilation and distribution of the bid documents. Finally, the bid pursuing activities are carried out. At Baldauf, efforts have already been undertaken to increasingly include the supplier into the bidding process to minimize costs and handling time by introducing the e.bid system. However, the as-is analysis revealed that the entire bidding process is efficiently supported, but hindered by missing access to up-to-date product information (cf. A3 Procuring material). Price comparisons as well as searching for specific tiles are currently very difficult due to the lacking standardization of the product numbers and the difficult identification of the tiles with a specific tonality (cf. chapter 3.3).

By receiving an order (A2), the bid data of the e.bid system are taken over and adapted to the client's modification requests. Currently, the bidding data cannot be transferred electronically into the order processing system. In general the customer immediately gets an order confirmation as soon as he places the order. This confirmation is based on a bid or on sales negotiations and achieved as paper document.

After sending the order confirmation, the procurement process (A3) is triggered. Baldauf keeps a permanent stock of ceramic tiles and natural stones, borders, bars, bindings and oven materials for the sales department as well as for the processing of laying work. The following considerable problems have been identified.

The access to information about adequate product ranges (e.g. new products, their characteristics, prices and manufacturers) is one of Baldauf's main problems. Today, Baldauf's tile assortment is restricted to the products of a few Austrian and Italian manufacturers/wholesalers presenting regularly new collections at Baldauf and providing their information material (catalogues, brochures, price lists). All standard products are captured in a local repository of the e.bid system. Due to the lacking access to up-to-date product data, the stored information are often obsolete. Currently, the master data (i.e. supplier, customer and product data) are only manually updated once a year.

Furthermore, the negotiation and ordering processes with manufacturers/wholesalers are protracted since predominantly traditional communication means are used. Furthermore, Baldauf has very restricted possibilities to distribute itself invitations to tender to a broad range of manufacturers and wholesalers.

For software reasons, different procurement activities have to be carried out twice, e.g. data of an order confirmation can not be copied into the procuring program. The data of delivered goods are currently repeated captured in three different forms. The procurement system insufficiently supports this task.

Additionally, marketing material is missing that could make the customers' decision easier and quicker, e.g. product samples for Baldauf's showroom and professional design proposals about the combination of different products and colors (tiles and borders / wall and floor).

Baldauf's chief engineer prepares, coordinates and supervises the processing of orders (A4), including e.g. all technical details, schedule and personnel planning. The accounting of the piece-wages for the tile-layers is performed manually based on periodical reports about the laying work. These report forms partly do not contain all necessary information and causes problems in the subsequent tasks.

The invoicing activities (A5) include the collecting, compiling, sending and archiving of invoices. The preparation of the overall invoice is rather complicated and time-consuming, because the partly hand-written delivery notes of goods and reports about laying work have to be manually collected and summarized. Furthermore, the sales invoices for material are sometimes prepared without having a corresponding delivery note. Then the stock has to be corrected additionally when the invoice is prepared.

Occasionally the bid prices are compared with the data of the corresponding invoice to analyze any deviations. Post-calculations (A6) are often carried out when the expected success is missing. Analogously to invoicing, the collection and evaluation of the different information, forms and reports is very time-consuming and insufficiently supported by the applied IT system.

4.3 Requirements and To-Be Processes

Figure 2 shows the top level of Baldauf's to-be processes. Planning the ideal solution, it was noticed, that the 'What' remained mainly the same at this level, but that the 'How' changed significantly.

Since the as-is analysis revealed that the applied IT systems for procurement, order processing, stock keeping and invoicing do not support the corresponding processes efficiently and are additionally not well integrated, the to-be situation has been modeled assuming an optimal software support. The requirements towards the IT infrastructure are indicated in the IDEF0 diagrams by the supporting means or mechanisms arrows entering the bottom of the process boxes.

Process-spanning, it was noticed that Baldauf requires an integrated system supporting the entire bidding, procurement, invoicing and warehousing processes. Additionally, electronic communication means for negotiation, ordering, invoicing etc. need to be introduced to increase the efficiency of the communication processes. The internal report forms should be partly substituted by electronic replacements and revised in accordance to ISO.

Concerning the bidding and procurement processes (A1, A3), it turned out that Baldauf as SME is unable to spend the personal resources necessary to maintain a tile repository with up-to-date properties and price information. Thus, the idea arose to outsource such a service (This was the initial motivation to initiate the e.bip project). The required broker service should give access to a large amount of product information and allow Baldauf and other buyers of tile products to search for specific tiles. Additionally, the broker service should notify the buyers about additional products of certain series as well as provide additional information about suitable combination of tiles (design proposals).

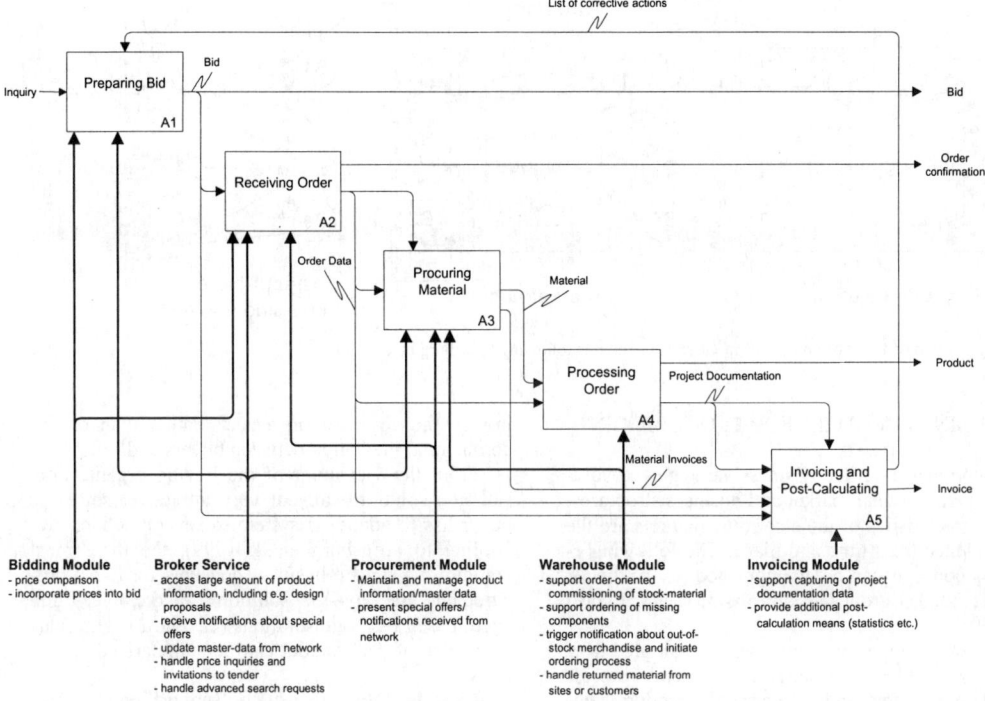

Figure 2. Simplistic model of Baldauf's to-be processes in IDEF0 notation (without controls).

Furthermore, the broker service should comprise e-commerce functionality to allow online ordering and negotiations. Also price comparisons, announcements of special offers and the possibility to distribute invitations to tender to the subscribed sellers of tile products should be supported. From the increased market transparency, the enhanced information access, and accelerated electronic transactions, Baldauf expects a considerable qualitative and quantitative improvement for all bidding and procurement activities.

For the process "Receiving Order" (A2), it is expected that the bidding data is transferred electronically to the procurement system and can be easily adapted for the generation of the order confirmation. The data of the order confirmation should be directly used for ordering merchandise within the procurement process (A3) as well as to check the stock, i.e. the system has to support identifying and commissioning of the available stock material and ordering the required components.

By processing an order (A4), improved IT-supported project planning and co-ordination is needed. Since this aspects are not addressed by e.bip, the specific requirements are not detailed in this paper. However, IT support for capturing returned material from sites and customers should be covered by the warehousing system.

Moving from as-is to to-be, the most significant organizational change concerns invoicing and post-calculation. Since the processes comprise similar activities for collecting order data, both as-is processes have been merged to one to-be process (A5).

For all orders with an order value of more than a defined threshold, the efforts required for the post-calculation will be captured directly with the invoice data to streamline the post-calculation activities. The data collection and entering process should be supported by the invoicing module. Facilitation of this activity is also expected from the partly substitution of paper documents by electronic reports and electronic invoices provided by the manufacturers/wholesalers. The invoicing module should provide advanced statistical functions to determine for instance a project-related list of piece wages, which is currently not possible.

The envisaged benefits of restructuring the current business processes and introducing the new software systems are mainly to increase efficiency of the bidding and procurement activities and produce a higher quality of the final offers as well. Also, due to the integrated system architecture and broker services, customer requirements can be handled more flexible. Thus, the overall customer satisfaction is expected to improve too.

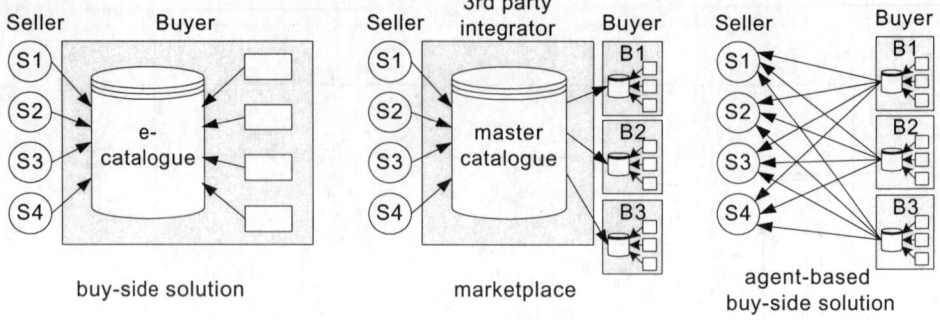

Figure 3. Common business-to-business procurement solutions following (Ginsburg et al. 1999).

5 E-BUSINESS MODELS FOR PROCUREMENT

The e-commerce and e-business strategies revolutionize the traditional market and business structures. Main characteristics of the electronic markets are the independence from time and place. The following e-business concepts can be distinguished (cf. Archer & Gebauer 2000, European Commission 1998, Eyholzer 1999):

Sell-side solutions are single seller, multiple buyer market hierarchies. The seller sets up and controls the sell-side content and maintains the product catalogue. *Buy-side solutions* are set up by a buyer to support the purchasing processes. In these systems, the buying organization is in control of the catalogue content and requests data from selected sellers/suppliers. On *Virtual malls*, third parties act as business intermediaries between sellers and customers and provide added value to the selling process by taking care of logistics, payment transactions, online catalogues, etc. *Matchmakers* are information intermediaries providing customers with links or information about where to purchase the desired products or services online or in conventional ways. *Marketplaces* are run by third parties to provide an online means for buyers and sellers to establish business transactions. In *Virtual Enterprises*, different companies join forces and share resources and policies or even IT systems and business processes. In *Inter-collaboration Models*, two or more parties belonging to different organizations collaborate to achieve a specific goal, like creating a product or service. Unlike a virtual enterprise, the degree of trust, integration and co-operation is rather low. The same applies to *Intra-collaboration Models*, where the parties are members of the same legal entity.

Sell-side and buy-side solutions are applied in direct business relations between sellers and buyers, whereas virtual malls, matchmakers and marketplaces represent the common brokered e-commerce concepts. Virtual enterprises and the inter-/intra-collaboration models are e-business concepts focusing on the co-operation aspects rather than on the commercial mediation between buyers and sellers.

From the viewpoint of the buying organization, sell-side solutions are not very attractive, since the buyer has to adapt his system to several sellers. According to (Ginsburg et al. 1999), the three most common types of e-business concepts for b-2-b procurement are buy-side solutions, marketplaces and agent-based buy-side solutions (cf. Fig. 3). The characteristics of each model can be summarized as follows:

Buy-side solutions usually support only a predetermined set of items and sellers. Thus, the market transparency is generally low and ad hoc buys are not possible. For the buying company, the solution is most flexible, but also very costly. Since the solution offers non-standard interfaces, sellers have to adapt their system interfaces to co-operate with the buyer. The model is applied when buyers and sellers consider their business relation as long-term.

Marketplace concepts can relieve buyers from the administration of the catalogue's content as well as facilitate the integration and market transparency through product data harmonization. An intermediary sets up the master catalogue and provides value-adding functions, like aggregating sellers' offers, matching of buyers and sellers as well as acting as agent of trust. The solution allows the buyer to access customized set of items and suppliers. The intermediary is responsible for subscribing sellers, managing of the catalogue's data as well as the design and functions of the system. Therefore, the solution is especially interested for companies with lower in-house IT infrastructure and know-how.

The third solution represent an *agent-based variant of a buy-side solution*. The buying company sends advanced software agents to generate dynamically an electronic catalogue. This model allows ad hoc buys between sellers and buyers. The solution provides maximum market transparency, but requires that all business partners apply advanced communication protocols and standards.

150

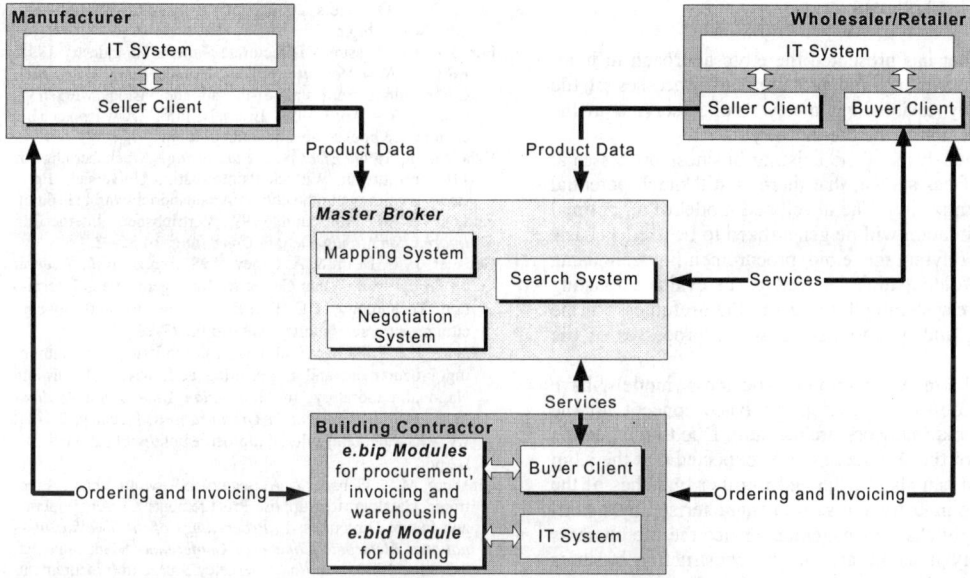

Figure 4. e.bip broker architecture.

6 THE E.BIP BROKER CONCEPT

A main requirement in the tile value chain is to cope with the different product identifiers. Thus, the non-brokered architectures, like the sell-side and buy-side solution, do not represent appropriate solutions for the members of the tile value chain. A pure agent-based solution is very expensive and requires advanced IT infrastructure and know-how. Since most companies in the tile industry are SMEs, an agent-based buy-side solution is not practicable in the tile value chain.

Therefore, a vertical electronic marketplace will be implemented within the e.bip project. The central service represents an electronic brokerage system ("Master Broker", cf. Fig. 4) that is accessed through the Internet. The broker service aims to increase the availability of product characteristics as well as provide the platform to carry out ordering transactions electronically between manufacturers, wholesalers/retailers and building contractors.

The Master Broker will integrate all product data provided by the sellers of tiles and unify the product information by providing a Mapping System to handle the problem of managing several product identifiers for similar articles. The Search System will offer advanced selection features for tiles, e.g. color matching mechanisms for finding tiles in a certain color or color range or checking whether a new series of tiles still has the same color as its predecessor series. Additionally, the buyers can initiate the Negotiation System by specifying conditions (e.g. deadline, optimal and maximal acceptable price) and

distributing these invitations to tender among the connected manufacturers and wholesalers/retailers.

Interested manufacturers, wholesalers/traders and building contractors can subscribe themselves to the virtual marketplace. The existing IT systems at the companies will be extended with external interfaces for carrying out business-to-business transactions with the electronic marketplace, the so-called "Buyer Clients" and "Seller Clients".

For the building contractors, an e.bip module for procurement and warehousing will be developed and the existing e.bid bidding application will be adapted to meet the identified requirements (cf. chapter 4.3) and take full advantage of the new broker service. A pilot implementation of this bidding, procurement and warehouse management system will be developed at Baldauf. The linked e.bid/e.bip applications provide the following features:

- The product data of the local e.bid repository can be automatically synchronized with the up-to-date information of the Master Broker.
- The system will support the information and workflow for all bidding and procurement activities by substituting paper documents as well as integrating communication means such as email.
- In case of a successful bid, the bidding data can be directly re-used for procurement. The e.bip procurement module will automatically check the stocks of the tile layer's warehouse and provide a list of required components for online ordering.

The e.bid/e.bip applications can be run with or without being member of the e.bip broker service as well as combined or separately.

7 CONCLUSIONS

The paper has presented the e.bip approach to innovate the bidding and procurement processes of tile layers by establishing a new broker service in the supply chain of the tile industry.

The analysis of the existing business processes at Baldauf has shown, that there is still much potential for optimization. The developed model of an optimal to-be situation will be generalized to be used as basis for specifying the e.bip procurement, warehousing and invoicing tools as well as to evaluate existing software system with respect to the usefulness for the bidding and procurement business processes of tile layers.

Furthermore, common e-business models have been outlined to develop the basic concept for the e.bip broker network architecture. Due to the generic nature of the architecture, it is expected that the e.bip solution can also be applied to other branches of the building industry and sectors characterized by a similar supply chain management. Since the tile business is a very dynamic area in the construction business with respect to the life cycle of product information, a successful implementation of the broker service in this sector represents an excellent test bed of the broker concept able to prove the transferability into other sectors.

At the time of writing this paper, the project has finished the analysis of the as-is situation as well as the modeling of the to-be processes and is beginning to start detailing the final architecture of the e.bip network. Future findings and results will be made available on the e.bip homepage on the World Wide Web (cf. Enns & Peters 2000).

ACKNOWLEDGEMENT

This work is partly funded by the European Commission through IST-1999-10710 project e.bip "Efficient Bidding and Procurement in the Tile Industry – Practical Trading Tools and Broker Services for the Exchange of Product Characteristics". The authors wish to acknowledge the European Commission for their support. We also wish to acknowledge our gratitude and appreciation to all e.bip partners for their strong support and valuable contribution during the various activities presented in this paper.

REFERENCES

Archer, N. & J. Gebauer 2000. *Managing in the Context of the New Electronic Marketplaces*. 1st World Congress on the Management of Electronic Commerce, Hamilton, Ontario, Canada, 19-21 January 2000. http://haas.berkeley.edu/~gebauer/publications/ArcherGebauer_WCMEC.pdf

Chappell, C. & S. Feindt 1999. *Analysis of E-Commerce practice in SMEs*, KITE Project. http://kite.tsa.de/bp_analysis.zip

Colquhoun, G. J. & R. W. Baines 1991. A generic IDEF0 model of process planning. In *International Journal of Production Research,* 29(11).

Enns, O. & O. Peters 2000. *e.bip homepage*, March 2000. http://www.ebip.net/

European Commission - Directorate-General III Industry 1998. *R&D for New Methods of Work and Electronic Commerce*, Contributions from state-of-the art and visions workshops Dec. 1997 – April 1998, Brussels. http://www.ispo.cec.be/ecommerce/books/consolidatedreport.zip

Eyholzer, K 1999. *Electronic Purchasing*. Arbeitsbericht Nr. 116, Institut für Wirtschaftsinformatik, Universität Bern. ftp://www.im.iwi.unibe.ch/pub/Arbeitsberichte/arbnr116.pdf

Feld, F. & M. Hoffmann 1999. Vertriebsnetz Internet. In *Information Management & Consulting.* 14:85-93.

Gebauer, J., C. Beam & A. Segev 1998. *Impact of the Internet on Procuremnt*, Fisher Center for Management and Information Technology, UC Berkeley. www.haas.berkeley.edu/citm/procurement/publications/arq-02-27-98.pdf

Gebauer, J. & F. Färber 2000. From Pilot to Practice: Streamlining Procurement and Engineering at Lawrence Livermore National Laboratory. In *Information Technology Applications and Management in Organizations*, Hershey, 2:1-23. www.haas.berkeley.edu/citm/procurement/publications/LLNL2.pdf

Ginsburg, M., J. Gebauer & A. Segev 1999. Multi-Vendor Electronic Catalogs to Support Procurement: Current Practice and Future Directions. In *Proceedings of the Twelfth International Electronic Commerce Conference*, Bled, Slovenia, 7-9 June 1999. http://haas.berkeley.edu/~citm/procurement/publications/Bled_final.pdf

Krömker, M., V. Steinlechner & F. Weber 1997. CSCCM - Efficient Bid Preparation for Supply Chains in the Building Industry. In *SMC '97 Conference Proceedings. 1997 IEEE International Conference on Systems, Man, and Cybernetics.* Hyatt Orlando, Florida, USA, 12-15 October 1997. Computational Cybernetics and Simulation. Institute of Electrical and Electronics Engineers - IEEE, New York. 4: 3585-3590. http://www.biba.uni-bremen.de/projects/csccm/results.html

Mayer, R.D. (ed.) 1990. *A Reconstruction of the Original Air Force Wright Aeronautical Laboratory Technical Report AFWAL-TR-81-4023*, The IDEF-0 Yellow Book, College Station, Texas, Knowledge Based Systems Inc.

Segev, A., J. Gebauer & C. Beam 1998. *Procurement in the Internet Age – Current practices and Emerging Trends (Results from a Field Study)*, CMIT Working Paper. http://www.haas.berkeley.edu/citm/procurement/publications/wp-1033.PDF

Vetter, M. & S. Pitsch 2000. Using Autonomous Agents to Expand Business Models in Electronic Commerce. In *International Journal of E-Business Strategy Management.* 1:207-213.

Weber, F. (ed.) 1999. *Efficient Bid Preparation in the Construction Industry - How to win more Bids with less Effort*. A Best Practice Report from the CSCCM Project, Computer Supported Co-operative Construction Management, ESPRIT Project 22828. BIBA Schriftenreihe (26), Aachen: Verlag Mainz.

IT in the early phase of construction projects

Entrepreneurial project preparation of building investment – The early recognition of pitfalls in a project limits costs and risks

Werner Gruetzner
SystemProject Werner Gruetzner München, Germany

ABSTRACT: Systematic and Strategic Preparation in the investment of a building investment (SPP) is still not very common within small and medium sized enterprises (SME). However, this first step of Professional Project Management (PM) helps the future client in making educated decisions and in investigating the project pitfalls in a very early, vulnerable stage of the project. The weak moment in a project is the "start" phase, where the biggest mistakes are made within the entire project process. Today, the project process with SME often starts with an architectural and planning process unsupported by clearly defined requirements of the entrepreneurs. Supported by a client coach and easily handled tools, the entrepreneur gets the overview of the entire project process before making any larger commitment. In the future, the entrepreneurial SPP shall be integrated naturally within the disciplines of a modern process system for building projects supported by electronic tools and Internet consultation.

CONTENTS

1 INTRODUCTION

Due to stronger competition and a time of grave structural changes, an entrepreneur will take every chance to emphasize his market position and net yield. But, while it is natural in technical fields to optimize results, the establishment of proper locations, optimal building plans, and plant design are frequently neglected with SME projects:

The systematic preparation of a project (SPP).

In the very near future, the SPP of the project pre-phase (PPP) will become much more frequently used by the SME's in Europe. Professional Project Management (PM) helps to manage the problems in the early phase of a project.

If a client neglects to include the SPP phase in the planning of his project, he will surely lose, as experiences has shown, from 10% to even more of his investment.

If you are considering the reconstruction or the extension of a factory or a new plant, there is every chance to get and to keep a project on line based on SPP.

Entrepreneurs, investors, banks investing in indus-

Fig. 1

trial and office buildings, planners and project engineers, would save themselves money, time, and frustration if they were able to foresee and therefore avoid problems. They are advised to ask for support in the earliest stage of a project discussion.

2. THE KEY TO SUCCESS

The key to the success of a project which best reflects the Owner's objectives, creating the highest possible net yield of investment, lies in the total commitment of the Owner, particularly to the *project pre-phase*, the most crucial phase, which frequently determines the ultimate success of the project. SPP minimizes the Owner's risks before he contractually binds himself to the investment, and before any consultants or planners have been contracted with, land or machinery is bought.

Once the planning and actual work on a project has started and, at that late point, expert advice is procured, the full benefits of that advice in regard to costs, time, function of the project cannot be fully realized. The success of every aspect of the enterprise is drastically reduced.

The problem which can be solved by SPP is to organize the link between the business plan of a company and its project as a subsequent logical measure. The main subject is to generate a „project plan" from the point of view of the client.

The SME-Owner, especially the first-time owner, can only proceed with highly developed but very easily handled instruments, with confidential professional coaching, and a step-by-step procedure of project preparation with stock-taking, situation analysis, site research, master plan concepts and strategic planning.

3.   TYPICAL FAULTS

3.1 *Practical cases*

The Owner (Client) has a lot of different aspects to deal with. 3 examples demonstrate the problems an entrepreneur might have:

Case 1:
The space does not suit to the requirements
In front of the factory owned by the fiber optics industrialist Mr. W., the recently delivered machines stand and can't get through the gate. Finally, after enlarging the gate to the required size, the machinery is brought in.

The configuration of the plant was precisely measured by the process engineers, but five of the machines fit into the required space, the sixth and seventh do not. Consequently, production can't start, delivery deadlines are not met, and an enormous slump in sales threatens the business.

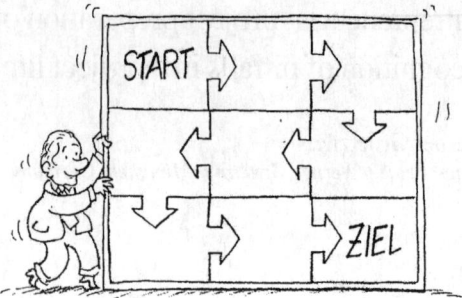

Fig. 2. *Start & Goal: Systematic project preparation*

Case 2:
Lack of clarity creates confusion
The Client of a sophisticated project asked his architect to develop proposals for varied planning solutions for his project idea ("vision"). The Client had not thoroughly thought out his requirements and had therefore not been clear in his instructions to the architect.

The architect designed what he thought would be the best plan, neglecting to consider many of the technical aspects.

After a long period of planning and a good number of contracts were awarded, the Client realized that the project did not meet his original vision. As a result, there were time consuming and costly disputes.

Case 3:
Faulty design do-ordination
A chemical production plant was ready to start operating. Shipping logistics were perfectly designed, but only 50 to 60 % of the output could be turned. What happened? At various stages of growth, the plant's pipes were designed with different dimensions, causing a bottle neck in the flow of products.

The system could have been altered, but space was limited negating that solution. Only a very expensive and sophisticated correction could be made but this was price prohibitive and, to make matters worse, the production had to stop, causing the plant to lose that anticipated income.

Scenarios like these are not rare.
If the entrepreneur does not call for professional help and ignores the pitfalls, they will become stumbling blocks for all involved.

3.2 *Typical pitfalls in a project*

Frequently, even if the projects are totally different, the pitfalls of a project and their consequences are very similar. Here are typical examples, by experience:

- The design of the project does not correspond to the projected business volume. Space requirements are over or underestimated. The reasons are, for instance: (too) great enthusiasm for the "beautiful plant" and how it appears to the public; overly optimistic expectation of turnover, false economy in space planning for personnel and machinery, or a lack of co-ordination with all participants.
- Major decisions made by the entrepreneur are based on inexpert advice. He has not educated himself in time to delegate tasks in a timely or productive manner. The project has not been defined properly. The "vision" of the entrepreneur has not been supported with sufficiently detailed requirements.
- Costs are out of control, over budget by 30, 40, even 50%. Unforeseen costs were not anticipated and threaten to burden the business. Effective control mechanisms are missing.
- Impatiently, an Owner insists on speed. He urges project progress without sufficient examination. In the hurry to begin, the project takes on a provisional character, avoiding the focus of a well prepared project schedule. A rushed start and bad organization leads to confusion and disappointment, especially at the moment when there is no longer a chance for sensible revision.
- The overall costs of building and running a plant are not considered. The building may be beautiful, but more costly than anticipated. Oftentimes, utility costs (services and maintenance) and financial charges were not adequately researched or budgeted.
- Energy supply, maximum use of the given space, service expenditures, and life span are just a few of the factors to be considered with the calculation of successive costs. Very often these factors are underestimated or not even considered in the client's investment planning.

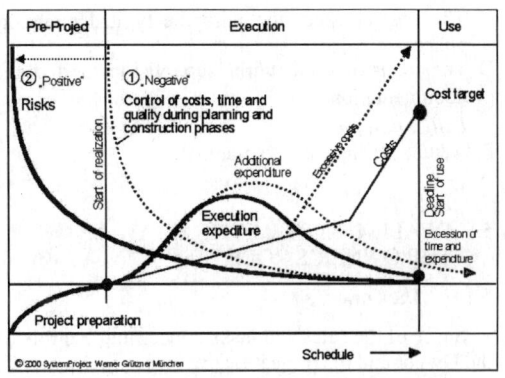

Fig. 4. *Project schedule: pre-phase preparation minimizes the Owner's risks:*

This list of "pitfalls" can go on and on; its variety amazes again and again. The specific stumbling blocks of a project can only be recognized by early, exact and individual examination.

4. PRE-PHASE PREPARATION

As the board of directors of a firm develops and decides its enterprise strategy, it has to determine the strategy for a construction project as well, keeping in mind the philosophy of the enterprise. There can be no successful project without a clear enterprise profile.

As the graphic (Fig. 4) shows, this Pre-Phase Preparation minimizes the Owner's risks before any consultants have been contracted with, land or machinery is bought.

At this stage ("start of realization") the Owner still has the unlimited freedom to say "Yes" or "No" to the project without the risk of significant loss.

Careful examination of the requirements peculiar to his project, and the thorough education of the Owner in all aspects of his endeavor provide him with the ability to select the most highly qualified expertise available to assist him in the planning phase.

The investment of a building enterprise is set from the first concept through to the actual use phase. The project schedule for planning and execution is well known, and is divided into 5 main steps:

Project pre-phase:

A The development and the thorough analysis of the enterprises strategy (objectives of the enterprise, project idea, "vision")

B The strategic project development and preparation for project commencement concepts

These two phases are most important to future costs savings.

Realization phase:

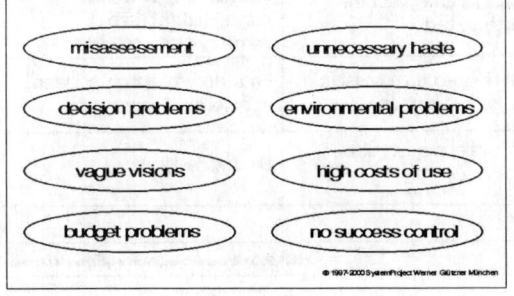

Fig. 3. *Typical project pitfalls*

C The project execution with the typical planning phases

D The construction, with supervision and final documentation

Utilization phase:

E Actual facility management.

5. DEALING SYSTEMATICALLY – THE OWNER'S ROLE

5.1 *The Owner's tasks*

Some of the rules and tasks concerning the role of the Owner and his project team:

- Definition of the overall objectives of the project in compliance with the medium and long term strategies of the enterprise
- Definition of the project targets
- Definition of the project organization including external consultants and integration of the future users
- Consequent project preparation and project pursuance with careful review and analysis of the situation, and the organization of internal communication and relationship

- Frequent and regular control of the client's targets with regard to costs, time, quality and function

Organization and control of the internal information and record system.

- Organizational planning before starting the building plan is the task of the Owner and is an exclusive responsibility which should not be delegated. Experience shows that management does not like to be overly involved with that (as seen in case 3).

But how shall the project participants learn the Owner's intention if not clearly communicated by himself? The complete and early organization of the Client's tasks gives him the chance to concentrate on his entrepreneurial goals. His direct engagement leads to direct savings.

5.2 *The Owner's support*

The more complex the task, the more the Client requires early professional partnership and support. If he wants to keep his investment under control in terms of finance, time and function, he must safeguard this support in the earliest stages.

No	Project Step	Major Themes	Sub-Themes	Activities
1	Business plan			
1.1	Entrepreneurial aims	Entrepreneurial analysis of situation and aims	Enterprise Strategy: situation in terms of facts, time, finances, rights & laws, areas	Analysis of the business plan
1.2	Project idea (vision)	From 1.1: conclusion concerning organizational & area alterations	First definition of aims for building (location, purchase, new facility etc.)	Alternatives (new project, extension, move, sale, abandonment)
1.3	Project strategy	„Business plan for the project"	Analysis of situation and definition of requirements, project organization	List of measures, time schedule, work program, decision "to do or not"
2	Strategic project development			
2.1	Project preparation	Project structure, action plan	Concepts: Operation, function, use, requirements, general development, organization	Stock-taking, research, tasks and job description
2.2	Purchase / sale real estate	Optimizing, validation	Decision criteria, analysis of the location, aptitude test	Research: rights of real estate, old disposed of harmful waste, contracts
2.3	Pre-project	Continuation of 2.1	Master planning, economic design, estimation, task description	Basic documentation, decision aids, ability of approvals
2.4	Document of decisions	Revised Project concept	Facts, data, scale, project guide (handbook)	Binding decisions
End of project pre-phase				
Subsequent: Phase of realization (3,4) and Operation (5)			© 1999/2000 SystemProject Werner Grützner München	

Fig. 5. *The Owner's pre-phase work: overview of the structure of themes (simplified)*

For the Owner's support with his tasks he cannot delegate, he gets a project guideline, professional structures, schemes and diagrams based on requirements both of the entrepreneurial aims and the building project tasks. This is fundamental for the subsequent planning phases.

An experienced Owner's Coach can offer all that is needed for proper project preparation. His expertise, along with a good working relationship between him, the Owner, and the other professionals on the project, can make the entire experience more pleasant and more cost effective. When brought in early enough in the project, it is his survey, provocative questions, and systematic analysis which give the Owner's confidence to make the right decisions.

In close touch with the client, the coach provides the important link between the Owner and external project support (project managers, planners, etc.). The coach co-operates with the client to prepare this team. To a certain extent, he "builds the bridge" from the Owner's "vision" to the realized project, free of subsequent project execution.

5.3. *The Owner's project team*

Corporate management establishes a small, internal project team for the entire duration of the project in which all fields relevant are represented and are supported by the coach.

This in-house project team is directly linked to the board of directors and shall be recruited from leading and decision-making members of the company acting independently.

5.4. *Owner's pre-phase documents*

The Owner's fully formed vision, arrived at with the help of expert input, is documented and adhered to by all those associated with the project, and in all phases and disciplines of the work. Still in the pre-phase, the Client states more precisely in all aspects what his project is like, asks for professional back up, and summarizes all results in documentation.

This becomes the framework for all subsequent project activities and the key concerning use, building, technique, costs, schedule, image factors (corporate identity), and project management.

It can be called the "What"-document - what the client expects from his investment. This document shall give unambiguous and detailed information concerning project targets, organization, basic concept of the plant and its technical requirements on process technique, infrastructure, mechanical & electrical installation system.

While the team and corporate management determine the individual project phases and decision processes, tight project management by the Owner is ultimate strategy for success.

A project guide - the "How"-document - concerns all works and relationships within the entire procedure and helps the communication among all participants.

6. THE INFORMATION PROCESS

6.1. *Information components*

Typical provisions within the Owner's project and information components comprise:
- Independent support of the Entrepreneur (Owner's coaching),
- Stock-taking (all information and documents relevant for the project),
- Situation analysis (subsequent to the enterprise's business plan),
- Site research and concepts (if necessary with survey),
- Master plan concepts and traffic concepts,
- Strategic planning of need and requirements (facts, data, finances),
- Strategic planning of project procedure (organization, schedule),
- Project studies, feasibility studies, project simulation,
- Provision of Owner's project key documents (e.g. handbook, project guide, key drawings),
- Provision of project documentation.

It will be helpful to discuss how the information system for project preparation can be emphasized as a step needed before starting the planning and realization of the project and optimized by software support in the sense of „project simulation" and „process modeling".

The SPP of the SME Owner shall be consequently integrated in the information process which usually starts with the architect's description of the building task. (Fig. 6)

6.2. *PC-programs*

In the early phase of project preparation it does not require more than simple PC-programs as
- Word (description of the project facts and data),
- Excel (estimation, quantification),
- Power Point (organization, functional relations) and a simple
- scheduler (bar time schedule).

They will be handled centrally by the client coach and allows easy exchanges of data within the client's project team

However, the stock-taking of information and documents relevant for the future planning works shall be provided in programs compatible to those used with the project planning later on.

The SPP creates general clarification of the requirements for the information techniques according to the project's need and usable by all who communicate with the Owner.

6.3. *Communication*

An SME entrepreneur has different requirements than those leading a large company which can afford to engage a PM office with an oncoming project. Therefore the communication in the PPP is different, and shall be economical and unsophisticated.

To meet this fact more effectively than with the present system of personal consulting and instrumentation - which is of basic need - an additional electronic system of consultation may be developed for the specific use by the SME Owner. This system can be provided in the Internet as a general scheme, patterning the Owner's tasks.

To save money and time (e.g. by travelling) a video conference tool designed for this special purpose may support the direct dialogue with the client.

6.4. *Security Risks*

The communication and data exchange in the Internet, as suggested before, does provide a problem of security which should be solved in the lest complicated manner.

A business or project plan, a finance or organization chart, are highly sensitive documents, containing the confidential information and data of the enterprise. They must not be transferred by e-mail without a secure protection system. Presently, a business plan should not be mailed electronically; but be transported with disks by a person in the confidence of the entrepreneur.

Though the Owner's project team is a small group, securing the knowledge and information of the company requires an intelligent technique.

7. COSTS OF SPP

The typical SME Owner does not expect to invest in expenses for consultation in the PPP.

But the SPP works individually to the Owner's requirements and at this stage is flexible to his specific need. The costs are limited and do not extend the investment.

However, the effect will be saving execution costs, shortening the planning time, and avoiding expenses for unexpected changes on a considerable scale.

8. CONCLUSION

Medium sized companies, communities, associations, institutions, project groups or development companies which choose not to have a permanent planning department, but who have need for expert

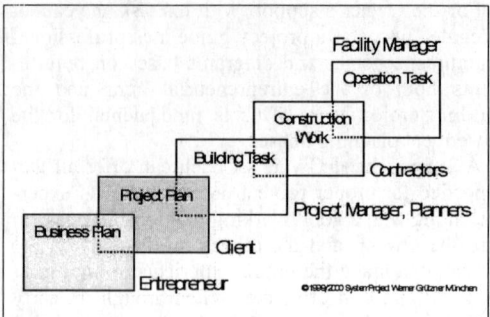

Fig. 6. *Information process*

development planning, are well advised to take advantage of this system of strategic project management, the "business plan" for investment.

By following the above steps, the SPP will result in a project more economically sound and of lesser risk. However, there is no scale of assessment at the beginning of a project. At the end of the project, the advantages are clear to see.

SPP provides optimal and cost effective procedures developed with experts in the field of project preparation in commercial and industrial building, and based on years of experience, to the client wishing to limit the possible pitfalls of his endeavor and striving to create a highly successful project.

SME Owners lack the experience of larger companies and consulting firms, so they will find it necessary to have concise,. Proven methodologies in the strategic planning of their projects at their fingertips. This can be made available through the assistance of computer technology and programs precisely developed for them.

ABBREVIATONS

PM   Professional Project Management
PPP   Project Pre-Phase
SME Small and Medium sized Enterprises
SPP   Systematic (& strategic) Project Preparation

FIGURES

Fig. 1. "Dealing systematically"*
Fig. 2. "Start&Goal"* –
      Systematic project preparation
Fig. 3. Typical project pitfalls
Fig. 4. Project schedule: pre-phase preparation minimizes the Owner's risks
Fig. 5. The Owner's pre-phase work: overview of the structure of themes
Fig. 6. Information process

(* cartoons by Josef Pretterer, Munich)

Product and Process Modelling in Building and Construction, Gonçalves, Steiger-Garção & Scherer (eds)
© 2000 Balkema, Rotterdam, ISBN 90 5809 179 1

Managing the brief effectively: The CoBrITe approach

Yacine Rezgui, Grahame Cooper & Peter Barrett
Salford University, UK

Dino Bouchlaghem, Mahmoud Hassanen & Simon Austin
Loughborough University, UK

ABSTRACT: The present paper gives a comprehensive overview of the CoBrITe project. First, the aims and objectives of the project are described, followed by a detailed definition and characterization of the briefing process. Then, an overview of the current technology implementations of the CoBrITe industrial partners is given. The paper also introduces five key areas that can promote effective briefing: Communication, Information capture, Information referencing, Information representation, and change management. Finally, the CoBrITe system demonstrator is presented. The project is ongoing and supported by the Link / IDAC program.

1 INTRODUCTION

A wide range of problems are encountered during the briefing stage of a construction project. There is little guidance and support for clients, whilst designers have difficulties both in capturing clients' needs and conveying conceptual design options to them. There is a central difficulty, associated with language, communication and the exchange of information between clients and design teams, which is now gaining widespread acknowledgement (Hassanen and Bouchlaghem, 1999).

The CoBrITe (LINK/IDAC UK funded) project argues that the construction industry has yet to exploit the potential of IT systems to assist both parties during this critical phase. This is in contrast to later stages of design and construction where computer-based techniques and systems are commonplace. The overall aim of the CoBrITe project is to improve the briefing process through more efficient and effective use of existing and emerging information technologies that can support client and design teams. This aim translates into the following objectives:

- highlight shortfalls and best practice by reviewing information management techniques and the use of IT during briefing;
- integrate recent/current research projects concerned with briefing and form a research network;
- access potential users' needs in terms of use and diversity in both small-medium and large pro-

jects, focusing on the five key improvement areas (Link IDAC 88);
- identify promising systems/products that could assist the Key Improvement Areas by conduct an audit of existing enabling information technologies (including interfaces);
- position the IT products within the framework of briefing and design processes in Level 2 of the IMI Process Protocol and the IAI Business Process Model, to inform their development;
- identify specific IT tools and methods for the five improvement areas;
- produce a prototype integration environment for the management of briefing and design information;
- deploy effectively the proposed approach by re-engineering the briefing process redesign (within the industrial companies involved in the consortium).

The present paper gives a comprehensive overview of the CoBrITe project, including an analysis of the briefing practices and information requirement. The paper also includes a description of the CoBrITe system architecture. The latter proposes a framework that integrates a set of proprietary and commercial software applications aimed at supporting the briefing process.

2 DEFINITION AND CHARACTERISTICS OF BRIEFING

A variety of definitions of briefing can be found (BSRIA 1990, CIB 1997, BRE 1987, BS-7832 1995, and CIRIA 1995). Consultants tend to consider the briefing process as a limited process with well-defined start and end to ensure records of changes in order to be able to claim fees for any extra work. On the other hand, clients prefer to consider the briefing process as extended until almost the final stage of construction to ensure that the final product meets their requirements and fulfil their objectives. Barrett and Stanley (1999) defined the briefing process as "the process running throughout the construction project by which means the client's requirements are progressively captured and translated into effect".

The agreed definition of briefing within the CoBrITe project is as follows: the process running throughout a construction project by which the requirements of the client and other relevant stakeholders are progressively captured, interpreted, confirmed and then communicated to the design and construction team. This definition is believed to be more suitable as it widens the customer base; emphasises the cyclic nature of understanding what is really needed; and delineates briefing activity (which must always involve deliberation of needs/requirements and therefore involve the stakeholders in some way) from the design activity which produces potential solutions in response to the brief (Hassanen and Bouchlaghem, 2000).

A comprehensive literature review of briefing practices in general (Hassanen and Bouchlaghem, 1999) and within the industrial partners in particular (Hassanen and Bouchlaghem, 2000) has been conducted. The latter has revealed the following main characteristics of the briefing process:

- Briefing involves a huge and wide range of initial/preliminary but crucial information/data from different independent sources.
- Briefing involves concurrent and collaborative work by different non co-located parties over the same information.
- Some of the actors, including clients, involved in the briefing process have little understanding and knowledge about buildings.
- All possible options should be comprehensively examined at this stage to ensure that no potential alternatives have been missed. However, and due to the short time allocated for this process, such examinations can not be conducted in depth or in detail.
- Many changes and revisions occur during the briefing stage; critical changes which affect the decision making, should be effectively reflected and notified to all relevant parties.
- Needs of the client have opposite impacts (especially in large projects) on the design attributes. Requirements need to be rated and ranked to identify the most important requirements to be fulfilled (in case of contradictions) and, hence,

maximise clients' satisfactions. This is a very complicated task if it is a large project.

3 STATE OF THE ART IT SUPPORT FOR BRIEFING

The commercial partners involved in the CoBrITe project have all implemented proprietary solutions to address the problems associated with the briefing process. Information technology is used extensively to represent information, with the "de facto" standard tools being Microsoft Office (Word processing, Excel spreadsheet, Access database, PowerPoint presentation software). In most cases this information, once captured and represented, is stored in a shared network directory. Although the use of document management systems for managing this electronic information is limited, several organisations are conducting trials to evaluate this type of technology. In addition, most partners make use of visualisation techniques. These range from presenting information in a schematic rather than tabular manner through 2D and 3D visualisation techniques and visual walk-throughs. Relational databases are also used to hold generic or common data. Project-specific information are then generated from these databases and conveyed into different forms, including room data sheets. This information can be reused in subsequent projects.

Due to legal and business considerations, none of the CoBrITe industrial partners use electronic media as their primary mechanism to distribute documents to other partners on projects, although awareness of the technology is overall fairly good. Despite the availability in electronic form of most information used and produced during the briefing process, paper-based information remains the principal foundation for communication. Organisations do use email to distribute information to other entities with whom they collaborate, for example in a partnering relationship, but hard copy information is still provided. It is also worth noting that electronic information is produced from a variety of de facto formats.

Furthermore, the CoBrITe have not yet adopted the modus-operandi of the so called "virtual enterprise" where most information is created, shared, and exchanged electronically. The face-to-face meeting is considered as an essential part of the brainstorming and requirement capture process. Several partners either use, or are considering using, proforma (either paper-based or electronic) in order to ensure that relevant information is captured for any stage of the process. These are based either on textual documents or on database technology. Also, the organisations involved in the project, with one exception, do not have a thorough recorded map of the processes involved in briefing, but these are, however, well understood on an individual basis. A document describing briefing and what is required from the brief exists in several cases, although these documents do not contain a comprehensive record of the briefing processes. This document is often out of date or obsolete, or not relevant to the organisation at the current time (i.e. is not used). It is worth mentioning that BAA believe that a good understanding of their business processes will produce efficiency gains in the briefing process. They are currently implementing IT based solutions to achieve their process driven approach to managing the brief.

It was also noted that most CoBrITe partners recognise the potential usefulness of building a corporate knowledge base but feel that the cost and complexity of implementing such a system would negate any benefit in the short term. There is a common feeling that current available commercial offers are not mature and reliable enough to undergo large scale implementation. Several organisations do, however, acknowledge the potential usefulness and expressiveness that this kind of technology would provide in the representation of information and in the communication of ideas or concepts.

4 PROPOSED FIVE KEY AREAS FOR TECHNOLOGICAL IMPROVEMENT THROUGH IT SOLUTIONS

Five areas of technological relevance, which may impact on briefing efficiency, have been identified within the CoBrITe project. These are communication, information capture, information representation, change management, and information referencing (see Figure 2). In addition, five general key areas for better briefing have also been identified in (Barrett and Stanley, 1999). The matrix in figure 1 relates the five general improvement areas for brief-

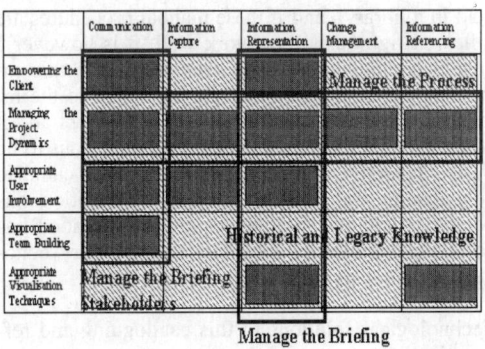

Figure 1. The CoBrITe Matrix.

ing (Barrett and Stanley, 1999) against the five areas of technological relevance to briefing. The matrix highlights four topics that can provide potential solutions to the briefing process, namely: manage the briefing stakeholders, manage briefing information and knowledge, manage the process, and manage historical and legacy knowledge. The latter are used as a guide for the development of the CoBrITe integrated infrastructure.

4.1 Key Area 1: Communication

Communication is the process by which information is exchanged between two entities. In most cases this will involve two identifiable individuals, but it also includes information exchange between individuals and organisations, or between two organisations. Communication issues in the briefing process impact on all other areas in the process. Information exchange in the briefing process is performed using several mechanisms; the most popular of these is the meeting, where individuals exchange information in a face-to-face manner. Other exchanges are performed through electronic media (email) or by exchanging hard copy information.

There are a number of requirements that any information technology solution supporting information sharing and exchange must fulfil. The process must be secure, it must conform to the ideals of non-repudiation (when a courier collects a document there is a written record of its collection and likewise for its delivery, and the same is expected for electronic information exchange). Additionally, the process should ensure that information is captured effectively. These communication mechanisms are well established as manual processes; there are several technologies available that would allow this communication to occur and its requirements to be met electronically. IT tools to assist the communication processes include: Email, Encryption, digital signatures, groupware solutions, document management systems, and workflow solutions. Other tools that may help communicate ideas, rather than hard information, include those related to visualisation.

4.2 Key Areas 2 and 3: Information Capture and Representation

The communication process is concerned with the exchange of information; this information must be captured and represented in order for it to be analysed and processed for the benefit of an organisation or a project. Information that has been captured and represented must be approved in some manner so that any errors can be corrected and misconceptions avoided, although this is likely to be classified as information management.

163

The mechanisms used for information capture in the briefing process are largely dependent on the processes undertaken to communicate that information. In meetings or face-to-face exchanges, the information is captured in the memories of the individuals present at the meeting. These memories may be assisted by minutes from the meeting, and occasionally a full typescript of comments made during the meeting will be produced. Note that legal cases are normally fully transcribed for completeness of the records, which may be an indication of how information could be captured for the briefing process. However, even a full typescript will result in the loss of some information, for example gestures or body language will not be recorded. Often visual cues will be used to illustrate points made at meetings, for example, picture boards or 3D drawings of existing buildings, and the inclusion of these in a project's information store so that comments made can be compared to the visual cue that triggered the comment should be considered.

The information capture process is more straightforward if the communication process consists of the exchange of documents or other paper and electronic information since the information is effectively captured and represented as an inherent part of the communication mechanism. Proforma can be used to help ensure that captured information is complete; these are not solutions in themselves, but act as an aide to the process. The capture of information and the representation of that information has been embodied in manual processes with differing degrees of success (depending on the capture and representation processes or methods used) for as long as humans have been communicating. Several technologies are now available that could improve these fundamental processes, many of which are already used by the commercial partners involved in the CoBrITe project.

Any technological solution or improvement to the information capture and representation processes must be able to record information in a manner that is understandable, i.e. as a written document or some other familiar structure that professionals can make use of. It must be accurate, so the information captured and represented reflects the information exchanged and precise, so that as much information exchanged as possible should be captured and represented. Tools that represent information should be easy to use (or they will not be used). As part of this ease of use, tools should be well integrated to avoid unnecessary duplication of information; related information is often formatted in different ways, but all information should be available regardless of its file format.

Information technology tools to assist in the representation of information are numerous, and widely used within organisations. These include word processors, spreadsheets, databases, and CAD. Mechanisms can be employed to record (capture) non-electronic information in an electronic form, such as scanning of sketches or hard copy documents. This information can then be stored on CD or DVD, and software exists which will take scanned documents and convert them to textual electronic data rather than as an image. The current voice recognition systems are likely to soon be able to produce typescripts from meetings. Other tools that help in the information capture process include data warehouses (new information and trends may be able to be extrapolated from existing legacy data), email (messages sent form a "permanent" record of information exchanged, therefore capturing that information automatically) and GroupWare. Once this information has been represented it may be linked using some referencing mechanism to allow the semantics of a project, which may span several information sources, to be followed.

4.3 Key Area 4: Information Referencing

All projects generate and use information; the complexity of the project will often dictate the amount of information captured, stored and managed. A large construction project can generate a massive amount of information that no individual can hope to assess and understand in its entirety. In order to help ensure that relevant information is not overlooked during the decision making process, this stored information should be referenced in some manner. Information is indexed by organisations so that the individual following the retrieval process can retrieve any required data with the minimum of effort. Like other processes in the general information management cycle, the reference and indexing of paper based records is well understood. People have been managing huge amounts of information using manual mechanisms for centuries (for example, cataloguing information held in a library), and if these manual procedures are followed rigorously they work well. It is however a hugely complicated task to cross reference this information and to check that all references to the information or from the information are correct and it is this process that allows errors and omissions to be made. Effective referencing of information can also help ensure that all the required information has been captured and that the required information has not been captured in multiple, contradictory documents.

Technological solutions to this cataloguing and referencing of information are an obvious application of information technology, which can iteratively process and index massive amounts of electronic information quickly and without error. Whether this technology would be useful in the context of the briefing process or not depends largely on the effectiveness of existing manual indexing and cataloguing mechanisms and on the amount of information that must be managed. Information technology for information referencing can be separated into several areas: tools that maintain references between documents, like a library catalogue; tools that create indexes of information; tools that can search these indexes and generate lists of appropriate information related to a search term or terms; and tools that allow sections of documents to be referenced.

Often this last category of tool (tools that allow sections of documents to be referenced) consists of the tools used to represent information, such as a word processor or CAD package. The Microsoft Office suite of applications performs this task adequately. Unfortunately these tools normally support referencing in a proprietary way, so references to document sections may not be available outside the tool the information was generated in. IT tools assisting the information referencing processes include: Index Servers, Hard copy to electronic copy conversion tools (OCR, Adobe Acrobat, etc.), Tools that represent information (such as Word Processors), Web technology (hyperlinking in HTML information).

4.4 *Key Area 5: Information and change management*

Information is not a static resource. As projects progress change to the stored information base that has been built up during the project are inevitable. Often these changes will themselves need to be recorded as new information for the project knowledge base. Sometimes why information has been changed is as important as the change itself. Projects that wish to record the reasons for change are likely to do so by adding a new piece of information into the project knowledge store, by perhaps updating a change log for a specific item of information. These manual procedures are again well understood and effective for hard copy if performed consistently within a project. Changes to information can often cause side effects; changes may affect related information and perhaps cause a cascade of changes within the collection of stored knowledge to become necessary. The manual process for updating related information could be very time consuming in a project with a large store of knowledge. Technological approaches to performing this operation are unlikely to be possible to implement if the cascade is to be done automatically; however, technology can help to identify changed information, the reasons for changes, and any related information that may be affected by a change. As an aide to the manual process information technology may help to ensure consideration of

all related information. Information, once created and stored, must be managed. It must be possible for personnel to access any information they need, and for authorised personnel to modify the information. Perhaps more importantly, information that should not be modified, or that should only be modified by a limited number of individuals, should be protected from unauthorised updates. In many cases organisations will wish to manage their stored information in a way that closely resembles their existing manual processes.

Any technological solution to assist with information management and change management needs to consider the security of information, "audit-ability" of changes and versioning of information to prevent loss of data. It is likely that this area of information technology will be very relevant to the briefing process. The technological areas addressing these problems include: document management systems, groupware systems, and workflow systems. The efficiency of an organisation's existing manual processes in the process of change and information management should not be underestimated or ignored. It may be that a technological solution that mirrors existing (manual) processes is the optimum solution for an organisation.

5 COBRITE PROPOSED SOLUTION

Following the above requirements and the review of the industrial partners briefing practices, as well as their current technology implementations, a web-based solution built around a shared workspace has been proposed. The latter makes use of BSCW for storing and archiving information. The shared workspace holds all information concerning the brief as well as its evolution. It is structured according to the four topics resulting from the analysis of the characteristics of briefing (as illustrated in Figure 1), namely: Communication, Folders, Process, and Legacy. The CoBrITe demonstrator authenticates users during the logon procedure. Users are then prompted with a list of projects to which they have authorized access. Users are then given access to a project shared workspace upon selection of a specific project. The CoBrITe demonstrator presented in the paper made use of the Wythenshaw Hospital project, provided by WS Atkins.

The Communication panel provides access to the details of any actor involved in the briefing process, and provides ways of communicating with the actors, mainly via e-mail or fax. The Folder panel provides a structured access to project information. The stored information include bitmap images (representing the planned building), text documents, CAD drawings, detailed spreadsheets, and structured

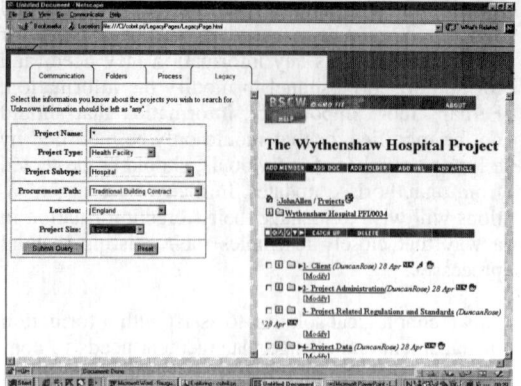

Figure 2. The CoBrITe shared workspace.

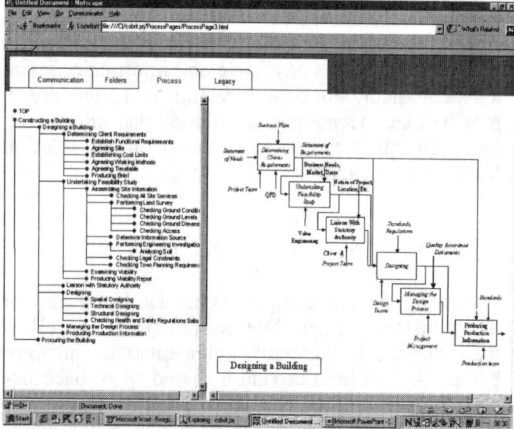

Figure 3. The CoBrITe shared workspace.

data stored in relational databases. The legacy panel provides structured access to corporate legacy information organized on a project-basis. Users can perform a search against a set of defined criteria, as illustrated in Figure 2.

The Process panel provides a process driven representation of the briefing lifecycle. The RIBA plan of work has been implemented for the purpose of the demonstrator. This panel provides also a process driven access to information. The IDEF0 methodology has been used to map the briefing process. Users can access information used as a control, mechanism, input or output to each briefing activity, as illustrated in Figure 3.

6 CONCLUSION

The paper presented a comprehensive overview of the work undertaken in the CoBrITe project. The main characteristics of the briefing process were discussed in the light of the current practice review with the industry collaborators. Five proposed key areas for technological improvement, with regard to IT implementation, are briefly covered. Finally, the CoBrITe demonstrator has been described. The latter provides a structured access to legacy as well as ongoing projects information and knowledge. The research team is now in the process of deploying the CoBrITe demonstrator within WS Atkins and Nuffield Hospitals in the UK.

7 ACKNOWLEDGEMENT

This research is funded by the UK government (EPSRC) and industry: AMEC Design and Management, W S Atkins, The Boots Company, BAA British Airport Authority, BDP Building Design Partnership, Currie and Brown, and Nuffield Hospitals.

8 REFERENCES

Barrett, P. and Stanley C. 1999. Better Construction Briefing, Blackwell Science Ltd, UK.

BRE. 1987. Better Briefing Means Better Building, Building Research Establishment Report by J. J. N O'Reilly.

BS – 7832. 1995. Checklist for Briefing – Contents of Brief for Building Design, the British Standards.

BSRIA. 1990. A Design Briefing Manual, BSRIA Application Guide AG11/98. Compiled by Parsole C.

CIB. 1997. Building the Team, Working Group 1, Thomas Telford, UK.

CIRIA. 1995. Planning to Build?: A practical Introduction to the Construction Process, CIRIA Special Publication by Potter, M.

Hassanen, M., and Bouchlaghem, D. 1999. Literature Review Report on Briefing Practices and Links with other Relevant Projects, CoBrITe Interim report.

Hassanen, M., and Bouchlaghem, D. 2000. Current Use of IT in Construction Briefing, CoBrITe Interim report.

Latham, M. 1994. Constructing the Team. Final Report of the Government/Industry Review of Procurement and Contractual arrangements in the UK Construction Industry, HMSO.

RIBA. 1967. Plan of Work, RIBA Handbook of Architectural Practice and Management. Published by the Royal Institute of British Architects, RIBA.

Product and Process Modelling in Building and Construction, Gonçalves, Steiger-Garção & Scherer (eds)
© 2000 Balkema, Rotterdam, ISBN 90 5809 179 1

Modelling with features and the formalisation of early design knowledge

J.P.van Leeuwen & B.de Vries
*Design Systems Group, Faculty of Architecture, Building and Planning, Eindhoven University of Technology,
Netherlands*

ABSTRACT: Creativity in architectural design requires a type of computer support that provides flexible structures in information models. Such models must follow the dynamic way that designers handle information, in particular during early stages of design. A framework is discussed that allows designers to define the formalisation of their own design concepts into so-called types of Features. The definition of these Feature Types can be done in a number of manners; three scenarios for this procedure are presented and discussed.

1 CREATIVITY AND DESIGN SUPPORT

1.1 *Dynamic nature of Design*

Architectural design problems are generally ill-defined, or 'wicked' as Cross (1984) characterises them. They require a very dynamic behaviour from designers, not only in searching for design solutions, but also in searching for the design problem (Coyne at al. 1991). Tasks such as analysis, synthesis, and evaluation do not occur in neat cycles, but designers tend to switch in a rather ad hoc manner between the different stages and tasks in design and often perform these tasks concurrently (Lawson 1990).

Obviously, during such a dynamic design process information is not treated as static data. Content and structure of design information is continually subject to change, which places a significant requirement on the development of computer support for design tasks. Formal information models for design must be as flexible and dynamic as the design process itself. They must evolve, as design evolves. The evolution of design information models has been subject of research by Eastman et al. (1995) and at the Design Systems group at Eindhoven University of Technology (van Leeuwen et al. 1995, 1997, 1998, 1999, van Leeuwen 1999). Similar issues are addressed also in the work by Ekholm & Fridqvist (1997, 1999) and Hendricx (2000).

Considerations on the dynamic nature of design lead to the following conclusions:

1. Design is a process of problem-solving and often concerns problems that are initially not well-structured.
2. Information related to design problems and solutions is dealt with in different ways depending on the approach of solving the design problem. Design involves creativity through combinations of these approaches:

a. Selection of an existing solution for a similar design problem. This involves matching information related to the problem and the existing solutions.

b. Creating a new solution for the design problem, involving the generation of new information that defines the solution.

c. Combining existing pieces of information in order to find new relations or structures in concepts and ideas that lead to design solutions. This involves analysis, re-interpretation, and re-structuring of existing design information.

d. Altering the design-problem in order to find a suitable solution. This means analysis, re-interpretation, and re-structuring of the information related to the problem, possibly even adding to, or dropping parts of the design problem.

3. Activities in design do not take place in a predictable order, the information dealt with in design activities cannot be foreseen: the content and structure of required information or generated information cannot be presupposed.

4. Individual designers, as well as the sector of design in the Building & Construction industry as a whole, are under constant development, with new knowledge, concepts, techniques, methods, products, materials, and styles emerging. Conceptual information models must evolve along with this development, in order to accurately represent the changing domain of design in B&C.

The above conclusions lead to the statement of new requirements for information models that are to support the dynamic nature of design. These requirements are denoted by the terms extensibility and flexibility, both ensuring the possibility for an information model to evolve along with the development of a design.

1.2 *Modelling with Features*

The technology of Feature-Based Modelling (FBM) has been developed initially in the area of mechanical engineering (Shah & Mäntylä 1995). As in the approach of Product Modelling (PM), FBM has started from the objective of generating semantically rich models of engineering data. Yet, the approach followed in FBM is different. PM approaches have developed conceptual models that represent the data structures to be used by their applications. These conceptual models largely aim at the later stages of design of the product, and are used to communicate the as-designed information between the various participants. Historically, the starting-point in FBM has been formed by geometry models, from which it was attempted to recognise the semantics of design. These semantics were then modelled using so-called Features. However, because much of the design information available during design cannot be recognised from geometric models, the design-by-Features approach was developed, where the semantically rich Features formed the primitives in building up the geometry. Combinations of both design-by-Features and Feature recognition joined the advantages of both approaches (DeMartino et al. 1994, Ovtcharova & Vieira 1995).

The result of this historical development from Feature recognition to design-by-Features, and eventually to the combination of both, has been that models consisting of Features are not, as in most PM approaches, predefined in large data structures. Features are defined as relatively autonomous entities of information that are given a position and relationships in the model only at design-time, not at the time of development of conceptual models. Also, the collection of Features available to designers is not assumed to be complete: designers can define and add their own Feature types to their collection of design tools. These characteristics of Feature-Based Modelling are very appealing to the dynamic architectural designer who is struggling with ill-defined design problems at the early stages of design.

Research on a Feature-Based Modelling approach for architectural design has led to the development and prototypical implementations of a theoretical framework with the following characteristics (van Leeuwen & Wagter 1997, van Leeuwen 1999):

1 Features are used to represent the semantics of a building design.

2 Features are the formal definition of characteristics or concepts of design.

3 Features are applied to multiple levels of abstraction of modelling the design (as opposed to the original FBM area, where Features are used only to describe the level of parts).

4 Features can be Generic Features, shared by the domain of architectural design, or Specific Features, which are defined for a particular view, e.g. a particular design style.

5 Designers can define Types of Features as the need to formalise a design concept arises.

6 Features form interrelated structures in a Feature model, using the relationships that are defined at the level of Feature Types, or by adding occasional relationships at the instance level.

7 Libraries of Feature Types represent bodies of domain knowledge. These libraries can also include instantiated data, mixed with the typological definitions.

Of these characteristics of the Features framework, issue number 5 (Designers can define Types of Features as the need to formalise a design concept arises) is discussed in detail in the remainder of this paper.

2 STRATEGIES FOR DEFINING FEATURE TYPES

2.1 *Concept identification*

Defining a Feature Type follows the decision to formalise a design concept. Therefore the first problem to address in Feature Type definition is: how to recognise and identify a concept? Two different points of view from which to approach this problem are discussed. The first point of view discusses how concepts can be acquired from sources of design knowledge. The second point of view presents the most common approaches in OO Analysis to classify a given knowledge domain. Both points of view must be considered in the processes of identifying concepts for the formalisation of design knowledge.

2.1.1 *Design domain knowledge and vocabulary*
The first point of view in the quest for concepts is taken from the body of knowledge in the domain of design. This knowledge, particularly in the complex discipline of architectural design, is not always readily available or easily accessible. Certain concepts in this body of knowledge are scientifically defined, such as the SI units (meter, second, Kelvin, Volt, etc.). These often are rather elementary concepts, which is to say that, in terms of information structure, they do not bare much complexity. Other, perhaps more complex, concepts may be defined in a less exact manner, but still be well conceived, such as industrial products of which all characteristics are

known and available from manufacturers. The terminology, used for these products and their characteristics, forms the basis for defining the Feature Types that are to represent this kind of concept. A third kind of concept is perhaps the most important in design, especially in early stages. These concepts form the core of architectural design theory and methods. They represent elements of design that can be either concrete or abstract; tangible or intangible; exact or indeterminate. For this kind of concept, the vocabulary of the design domain may be a suitable starting-point for their formal definition into Feature Types. This vocabulary, in architecture, is not formally defined either, but many terms have traditional meanings that are generally accepted.

The first consideration in the process of identifying a concept should therefore be whether a term exists that covers the potential concept. Terms are normally used to indicate the names of, e.g., systems, structures, products, materials, functions, organisational units, et cetera. An analysis of the way the term is used should be projected onto the concept being identified and reveal if the term actually represents that concept or not. If an existing, accepted term cannot be found, there are four possible consequences:
- The potential concept needs some adjustment to fit a term that is reasonably close to describing the concept.
- The potential concept covers a combination of multiple terms.
- The potential concept introduces a new term in the design vocabulary.
- Any combination of the three options above.

Whether or not new terminology should be defined involves a trade-off between aspects such as:
- *Acceptability of the concept in the design discipline*
  This may be an important issue when the concept serves, e.g., purposes of standardisation, regulation, or information exchange.
- *Desired or allowed level of ambiguity of the defined concept*
  Because new terminology, as opposed to traditional terminology, is not naturally known, its introduction may result in various interpretations of the term, which have to be verified against the concept's definition. Any ambiguity in the formal definition of the concept will then allow variance in the interpretation of the term.
- *Completeness and exactness of the definition*
  As a result of the previous aspect, the completeness and exactness of the definition of the concept cannot be based on knowledge that is inherently related to traditionally known terminology.
- *Uniqueness of the concept in relation to existing vocabulary*
  Using new terminology allows a concept to be

defined distinctly and independently from implicit meanings related to existing terminology. This can be a prerequisite when the uniqueness of the concept is to be stressed or when distinction from other concepts is necessary.

Closely related to design domain knowledge are the areas of design methodology and design theory. Design methodology, according to Roozenburg & Eekels (1995), is the science that studies the structure, methods, and rules of design. Design methodologies are either developed while focusing on the design process as a whole, or intended for specific domains or phases in design. An example of the latter, given by Roozenburg & Eekels (1995), is the morphological method, which relates characteristics and functions of a design with the variant components for that design in an array containing all conceivable solutions. For the identification of design concepts, it is interesting to look at the subjects used in specific design methods, especially those subjects that form an intrinsic part of the method.

A recently developed methodology for architectural design is presented as Generic Representations by Achten (1997). This methodology involves an approach to the identification of design content in architectural graphic representations. Its hypothesis is that graphic representations made during the design process imply the design decisions that are made. The research shows how it is possible to extract such design decisions from the graphic representations, by inferring the declarative knowledge embedded in these representations. The methodology proposed by Achten (1997) involves using generic representations, and the design knowledge acquired from them, as a model for procedural decision-making in design.

Many design methods, like the example given above, develop design aids, such as archetypes, design patterns, proportional or other measuring systems, rules for design schemata for instance for floor plans and elevations, and so on. These tools can be regarded as the design concepts that are applied in the context of the design situation at hand, using established procedures from the design method. The definitions of these concepts are not always clear and explicit but may involve implicit knowledge about the usage and meaning of the concepts themselves and of the procedures for using them in design. The formalisation of this kind of concept into a Feature Type requires that all relevant knowledge be made explicit, which may involve formalisation of other concepts and knowledge about concepts that have not been identified explicitly before. Classification strategies will help to identify these.

2.1.2 *OOA strategies for classification*
The term classification in Object Oriented Analysis refers to the task of the software engineer to identify

classes of objects in the domain for which software is to be developed. These classes then form the backbone for the design of procedures and data storage of that software. According to Sowa (1984) there have only been three general approaches to classification:

1 *Classical categorisation*
   The criteria for sameness of objects is formed by their properties: objects that have one or more properties in common belong to a category.
2 *Conceptual clustering*
   First, the conceptual descriptions of classes are formulated, then objects are classified according to these classes using a 'best fit' method.
3 *Prototype theory, or classification by example*
   The class is not defined conceptually, but by means of an example: a prototype. Objects are member of the class only if they sufficiently resemble the prototype.

In practice of Object Oriented Analysis, these approaches are combined and/or followed sequentially. Classification forms the main starting-point not only for the identification but also for the design of object classes. It supports the determination of structures of classes and of the structure of data and behaviour of these classes. As such, these approaches to classification are valuable also during the definition of Feature Types.

2.2 *Decisions in Feature Type definition*

Definition of a Feature Type is a procedure that is very similar to the definition of object classes in OO approaches for which many strategies and checklists have been described, e.g. Booch (1994). Aspects that need to be considered when defining a Feature Type are the following:

– *Bottom-up versus top-down*
   A top-down approach allows the designer to represent the logical hierarchies that are found in the domain of architecture, whereas a bottom-up approach stimulates the re-usage of existing Feature Types.
– *Typical versus non-typical*
   Is the information to be formalised typical for the concept, or does it merely concern a characteristic of a particular instance of that concept? This has to do with the reusability of the concept: when too much information is included in the Feature Type, then the reusability of the concept will be less: perhaps some less common characteristics should not be defined as part of the Feature Type, but modelled as instance level relationships for particular Feature Instances only.
– *Wide structures versus deep structures*
   Booch (1994, p. 140) discusses the subject of how to choose the inheritance relationships between classes of objects: deep inheritance trees tend to have classes that are less interdependent, but may not exploit all commonality; wide inheritance trees result in smaller individual classes, reusing other classes, but their complexity will be harder to understand. A similar problem exists with other relationships between classes, such as decomposition.
– *Presentation versus representation*
   This concerns the distinction between how a concept is presented to the user, e.g. on screen, and what actually comprises the concept. The latter kind of information must be modelled and stored and is used to generate the data for presentation.
– *Choice of relationship: specialisation, decomposition, association, or specification*
   For the definition of the relationships between Feature Types, these four relationships are available in the framework. Specialisation results in the definition of sub-types inheriting from super-types. The other three kinds of relationships are given a name by the designer to describe their role in the definition of a Feature Type.
– *Redundancy, completeness, and consistency*
   Although modelling a design information structure should aim at minimal redundancy and maximal completeness and consistency, it must be realised that the optimal configuration of information does also rely on aspects like reusability and practicability. Especially these two aspects will often justify certain levels of redundancy to exist in a collection of Feature Types. The pursuit for completeness should always be considered in the context and purpose for which a Feature Type is to be used and in relation with the amount of information that is likely to be available at the time of modelling or that designers are willing to provide. From the point of view of information management, consistency should always be pursued, yet in creative design, inconsistency may, to a certain degree, be acceptable. Moreover, the option to be inconsistent in dealing with information during design is often considered an important factor in creative processes.

2.3 *Scenarios for Feature Type definition*

Three distinct situations are recognised in which Feature Type definition may be initiated.

1 Feature Type definition from scratch
2 Feature Type definition from a prototype
3 Feature Type recognition.

2.3.1 *Feature Type definition from scratch*
The first situation in which Feature Type definition is initiated is when a designer (or e.g. an organisation for standardisation) decides to formalise a concept that has not necessarily been modelled in terms of Features before. The formalisation of such a con-

cept is started from scratch. For this approach, a procedure is described in this section, which guides a designer through the various decisions to be made when defining a new Feature Type. This procedure leads to a selection of the appropriate class of Feature Type and assesses the definition of all its attributes, possibly resulting in the definition of other Feature Types or the instantiation of Feature Instances.

The Figures 1 to 4 show the procedure for formalising a concept and defining a Feature Type. This procedure assumes that the designer has already identified the concept. It comprises four diagrams that guide the designer through a number of decisions regarding the contents of the concept. Diagram 1 starts with the determination of the primary nature of the concept, leading to the choice of the class of Feature Type that is to be defined. It distinguishes procedural, geometric, and constraint concepts from all others. This distinction may lead to the definition of a Handler Feature Type, a Geometric Feature Type, or a Constraint Feature Type respectively. If none of these apply, one is to proceed with diagram 2 of the procedure.

A Handler Feature Type requires the selection of the event that is to trigger the procedure defined by the handler. The parameters for this procedure need to be declared, meaning that the types are specified of the Features that can be passed as parameters to this procedure. The procedure itself must be defined, using one of the procedural languages made available by the design system.

In case of a geometric nature of the concept, a Geometric Feature Type is defined. This type requires selection of the parametric geometry that it represents. What kinds of parametric geometry are available depends on the geometry-modelling engine that is integrated with the design system. The selected geometry provides the types of the parameters that must be provided by instances of the Geometric Feature Type; these parameters are given from within the context of the geometry-modeller. The Geometric Feature Type further declares these parameters by providing the types of Features that can be passed as parameters.

For Constraint Feature Types, the type of constraint needs to be indicated, which depends on the availability of constraint solvers in the design system. Once the constraint type is selected, the parameters required by this constraint are known and the Constraint Feature Type further declares these as the types of Features that can be passed.

After the definition of any of the above Feature Types, there may be remaining aspects of the concept that have not yet been taken into account as parameters. If this is the case, the defined Feature Type is itself contained in a larger Feature Type, a Complex Feature Type, which must be defined next.

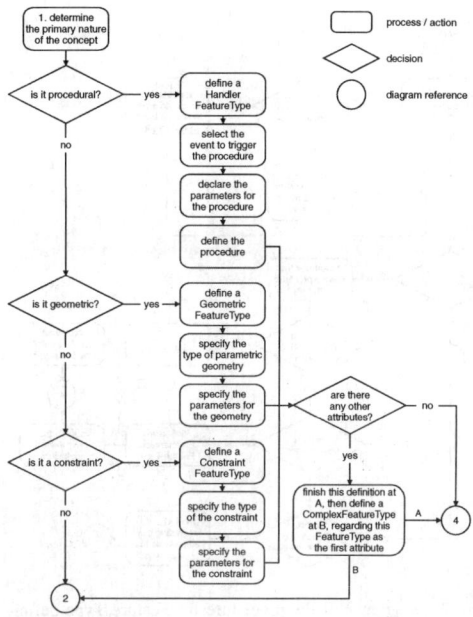

Figure 1. Diagram 1 of the procedure for Feature Type definition.

First, the definition of the current Feature Type is finished using diagram 4 of the procedure. After that, the definition of the containing Complex Feature Type can be started, which takes the Feature Type that has just been defined as its first attribute.

Diagram 2 shows how to proceed when the concept does not lead to the definition of Handler, Geometric, or Constraint Feature Type. First, all the attributes of the concept are listed that represent data to be stored by Features of this type. For each of these data-attributes, it is considered whether or not the attribute is relevant to the majority of occurrences of the concept. This is not necessarily a very clear decision, as the term 'majority' already indicates, more so because the possible occurrences of the concept may not come into view clearly at this time. Nevertheless, it should be questioned if the particular attribute really contributes to the concept's significance, or if it is relevant only for the one occurrence of the concept that the designer has in mind. If the latter is the case, the attribute should not be defined as a part of the Feature Type, but rather be modelled as a relationship at the level of the Feature Instances. Non-typologically defined relationships form a part of the Features framework that is not further discussed in this paper but elaborated in van Leeuwen (1999).

The number of data-attributes that are considered relevant for the Feature Type's definition are

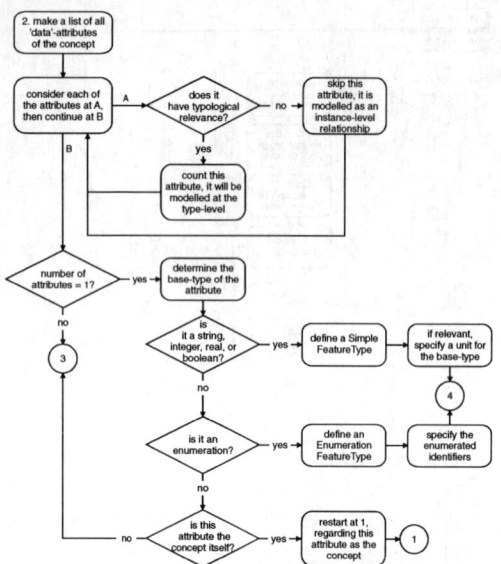

Figure 2. Diagram 2 of the procedure for Feature Type definition.

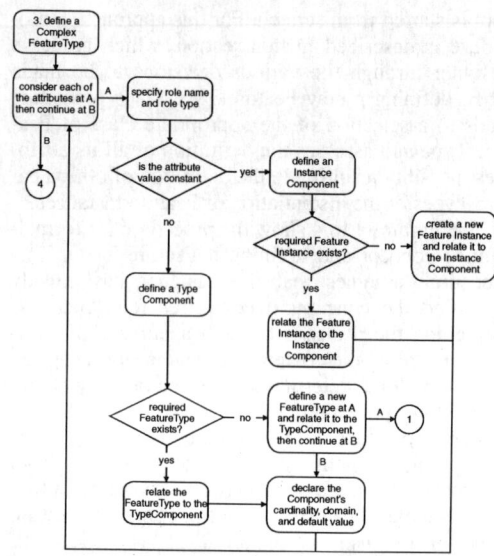

Figure 3. Diagram 3 of the procedure for Feature Type definition.

counted. If this number is exactly one, then the base-type of this attribute must be determined. For attributes that represent a string, integer, real, or Boolean value, a Simple Feature Type is defined. For attributes that represent an identifier chosen from a given list of identifiers, an Enumeration Feature Type is defined. If none of the above is the case, then the attribute itself represents a complex information structure, which must be represented by another Feature Type. Because this attribute is also the only data-attribute of the concept, it might in fact be that this attributes represents the concept itself. This is particularly true if no behaviour attributes are to be defined for this concept (see diagram 4), meaning that the concept exhibits no other characteristics than those represented by this attribute. Therefore, promoting this attribute to be regarded as the concept itself should be considered. If this is found to be the case, the procedure should be restarted, taking the notion of this attribute as the notion of the concept. Else, the procedure is continued at diagram 3, where the attribute will be the first and only attribute of a Complex Feature Type.

If the number of relevant data-attributes of the concept is greater than one, then a Complex Feature Type needs to be defined and the procedure is continued at diagram 3. This is also the case if no attributes are found relevant. This may appear to be an odd case, formalising a concept that has no characteristics, yet the mere existence of a Feature Type with a given name may be sufficient to represent a particular design concept at certain stages in the de-

velopment of a design or design theory. Perhaps later the content of the concept will become clearer and attributes will be added to the Feature Type that represents it. Also, behaviour attributes are yet to be dealt with, at diagram 4, which may give more meaning to the Feature Type being defined. Concepts with no data-attributes at all are modelled as Complex Feature Type that have no components.

If the procedure leads to diagram 3, this means that the concept will be represented by a Complex Feature Type. All data-attributes of the concept are to be defined as components of the Complex Feature Type, which are given a role name and role type (decomposition, association, or specification). For every relevant attribute of the concept the question must be answered whether or not the attribute has a constant value for all occurrences of this concept. If this is true, the Complex Feature Type will define an Instance Component, which is formed by a relationship to a Feature Instance. Possibly, this Feature Instance needs to be created.

For attributes with a value that varies for the different occurrences of the concept, a Type Component is to be defined for the Complex Feature Type. This is a relationship to another Feature Type, which, during instantiation, results in one or more relationships to Feature Instances. Possibly, the related Feature Type does not yet exist and must be defined in a new procedure started at diagram 1. For Type Components, the cardinality, domain, and default value must be specified.

After a component has been defined for each data-attribute of the concept, the procedure is con-

tinued at diagram 4 with the definition of the concept's behaviour.

The fourth and last diagram of the procedure for
defining a Feature Type adds behaviour to the type's
definition by means of adding event handlers. First,
a list is made of all the behaviour-attributes of the
concept being formalised. As with the data-attributes
in diagram 2, all those attributes are eliminated that
bear relevance only to certain instances of the concept and are not significant to the intrinsic notion
that the concept represents.

For each of the remaining behaviour-attributes,
the event is specified that will trigger the particular
behaviour, the event handler, of the instances of this
Feature Type. Next, it must be determined if the parameters that are to be assigned to the event handler
will be assigned in a similar manner for all instances
of the Feature Type, or if each instance will assign
the parameters in their own particular manner. If the
way of assigning parameters does not vary per instance, the parameter assignment can be done at the
level of the Feature Type, which results in relating a
Handler Feature Instance, containing the parameter
assignment, to the event handler. This Handler Feature Instance may need to be created in case it does
not already exist.

In the case of 'per-instance' assignment of parameters, only the Handler Feature Type can be
specified for the event handler. Again, this Handler
Feature Type may need to be defined if it does not
already exist. The actual parameter assignment is
done during instantiation, when an instance of the
specified Handler Feature Type is created.

After all the behaviour-attributes have been formalised into event handlers, the definition of the
Feature Type can be concluded by specifying the
domain for the instances of the type, and a default
value. The kind of content of both domain and default value depends on the class of Feature Type that
has been defined.

2.3.2 *Feature Type definition from a prototype*

The second scenario, Feature Type definition from a
prototype, is the situation where the designer ac-
knowledges a particular pattern of information,
modelled in structures of Feature Instances, as representing a particular concept that will recur during the
same or other design cases. The definition of a new
Feature Type can then be done on the basis of the
structure of Feature Instances that was modelled using relationships at the instance level.

The 'prototype' that the designer has built by creating this structure of Features is turned into a new
complex Feature Type that defines the relationships
as its components. From this point on, this scenario
follows a procedure similar to that of the first scenario, as described above.

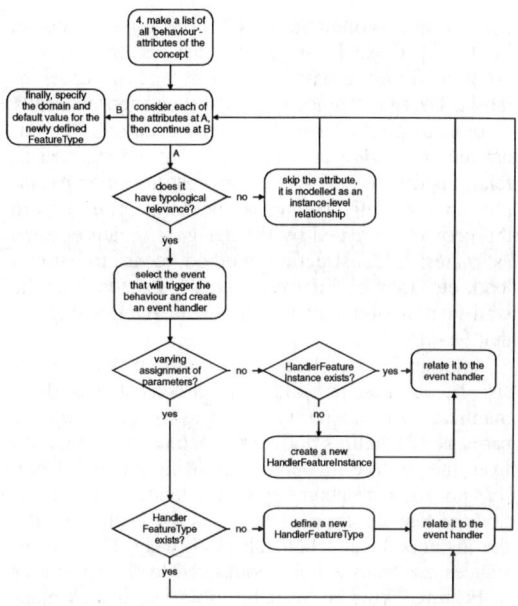

Figure 4. Diagram 4 of the procedure for Feature Type definition.

2.3.3 *Feature Type recognition*

The procedure of turning a prototype Feature structure into a Feature Type definition could also be initiated by a design system. Using pattern-matching
algorithms, a design system can search for recurring
patterns of Features and relationships at the instance
level. Once such a recurring pattern has been found,
it may be proposed to the designer as a concept of
design.

An important issue in the original area of Feature
model in Mechanical Engineering is Feature recognition. In that area, Feature recognition has the
meaning of recognising Features from a given geometric model. The geometry is analysed and
searched for patterns of geometry that match the
definition of known Feature Types. Once such a
match is found, the geometry can be replaced by an
instance of the found Feature Type. In this manner
the geometric model, which is poor in semantics, is
converted to a Feature model that provides all the
additional information necessary to, for instance,
manufacture the geometry of the designed product
with the available machinery.

Architectural design systems may well benefit
from a similar approach to designing elements of a
building. While the designer uses generic geometric
modelling tools, the created geometry may be analysed and interpreted as a structure of Features that
semantically enrich the geometry with detailed ar-

chitectural information. This approach is, of course, limited to those Features that can actually be discriminated on the basis of their geometric representation. Using inference methods, these geometrically recognised Feature structures may eventually be enhanced with additional Features that are defined as relationships to the geometric Features. For example, once a wall Feature has been recognised from the geometry created by the designer, Features such as material, construction method, cost, maximum load, etc. may be inferred from the existence of the wall Feature and added to the list of relationships of that Feature.

Another kind of Feature recognition that can assist the designer in building up a consistent and semantically rich design model, is to try and recognise patterns of Features not from a geometric model but from the Feature model as it is being created. Here, it is not the bare geometry that is matched to definitions of Feature Types. Instead, in the Feature model the instance-level relationships between Feature Instances are analysed and compared to the structures of Feature Types in available libraries. In this manner, a given constellation of Features that are interrelated by the designer during modelling at the instance-level, can be replaced by an instance of a Feature Type that has been found to define the same relationships at the type-level. This facility of the design system supports the designer in creating consistent models and adding knowledge to the model that is implied by the design actions. The degree of similarity between found Feature structures and the relationships in a particular Feature Type should possibly be variant, allowing the designer some freedom in using accustomed terminology and including cases that look similar to defined Feature Types. Mainly the latter may well appear to be a stimulant to the designer, since the system is now encouraging the creativity of the designer and helping the development of the design as it proceeds.

Figure 5 shows the procedure that is followed in case the user or the system requests a Feature recognition process to be executed. First a group of Feature Instances must be selected from which known Feature Types are to be recognised. Selection of this group can be performed entirely by the designer, supported by design system interaction or completely automatic by a design application. From the Feature Instance group the corresponding Feature Types can be determined.

In the Feature Type Library, some Feature Types are marked as a root type, namely those Feature Types that are considered a main architectural concept (e.g. wall, floor, space). The root Feature Type will be the objective of the recognition process. The question now is whether the selected group of Feature Instances contains an instance of such a root type: a root instance. In searching for a match between a possible root instance and the root types in

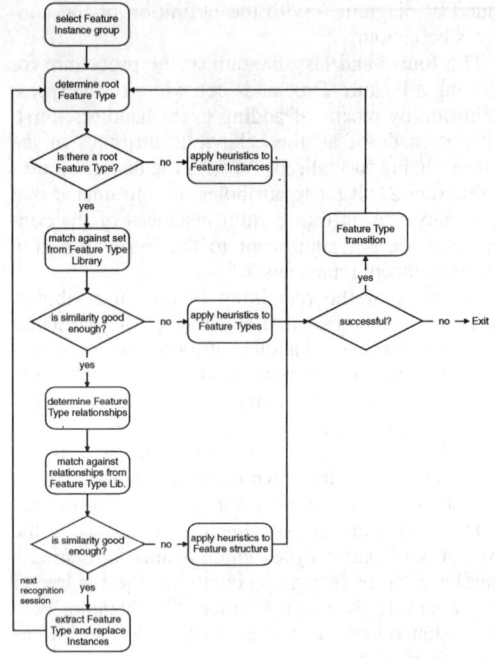

Figure 5. Procedure for computer aided Feature Type recognition.

the library, inheritance must be considered, meaning that a match is also made against sub-types in the library.

If this root instance cannot be identified, additional heuristics are needed to introduce an appropriate root instance. For example, four Feature Instances of an assumed Feature Type called Space Boundary could geometrically constitute a space. If this geometrical relationship is detected then a Feature Type Transition procedure is executed that infers an instance of the Feature Type Space. After that, the Feature Recognition procedure is restarted. Identification of the root instance in the selection is not necessarily a completely automated task; it can also be assisted by the user.

Once a root instance is found or inferred, the Feature Types from the selected group of Feature Instances are matched against the structure of the root Feature Type in the Feature Type Library. Not all Feature Types related to the root Feature Type in the Feature Type Library are necessarily present in the selected group of Feature Instances. The relationships between the instances are not yet considered in this stage of the recognition process.

Now the question is raised whether the similarity match is good enough. This decision can either be taken automatically by the system, using thresholds for the number and severity of missing instances, or in discussion between the system and the user. If the similarity is too low then additional heuristics are required, for instance a thesaurus of Feature Type names to detect possible cross-references between the used names. Feature Type Transition is executed and the Feature Recognition process is restarted.

If the Feature Type similarity is sufficient then the Feature Type relationships are determined from the selected group of Feature Instances. Considering the relationships in the Feature Type Library cluster, starting from the root Feature Type, they may:
- be absent in the selected group of Feature Instances,
- have a different typology (e.g. association instead of decomposition), or
- have a different role name.

First, a match is performed not taking these differences into account, just considering the topology of the structure. For this purpose graph matching techniques are used. Again, the successfulness of this match can be determined automatically by the system using thresholds or in discussion between the system and the user. Additional heuristics provide rules that can add or replace relationships in order to fit the selected Feature Instances in the structure of the Feature Type found in the Library. Since Feature Based modelling allows for describing a specific building concept in several ways, this process supports the conversion of different description styles to one generic style.

The Feature recognition procedure exits if one of the heuristics fails. At that point there are several possible results of the recognition process:
1 One or more root instances have been identified or inferred and the structure of instance relationships found between Feature Instances in the model has been replaced by an instantiation of the structure found in the Feature Type Library.
2 One or more root instances have been identified or inferred but a proper match of the relationships in the model to Feature Types in the library could not be made. In this case, the user can decide to use the relationships modelled at the instance level to define a new Feature Type: this is scenario 2 described in section 2.3.2 as Feature Type definition from a prototype. If a partial match could be made, then the user can alternatively decide to define a sub-type of the partially matched Feature Type.
3 No root types known from the Feature Type Library could be identified or inferred in the selected group of Feature Instances. Again, the user may decide to define a new type from the prototype instances, which in this case would also lead to the definition of a new root type.

3 DISCUSSION

The proposed strategy for formalising architectural design knowledge is in fact a design process in itself. It is the design of architectural design knowledge; design at a meta level. As such the meta design level process dangers from the same pitfalls as the architectural design process illustrated in the introduction of this paper, namely ill-defined problem, ad hoc process cycles, etc. The three described strategies offer a style guide for architectural design knowledge modelling based on Feature technology. FBM allows for describing a building concept in different ways using (slightly) different Feature models.

Applications, however, that will share FBM data require a predefined Feature model structure. Without this structure or additional knowledge it is impossible to extract information from the Feature model of a design. Therefore generic Feature Type libraries are needed that contain standardised Feature model structures. In that sense generic Feature Type libraries serve the same goal as standardisation efforts in product modelling (e.g. STEP Application Protocols, Industry Foundation Classes). In contrast with the STEP AP's, a generic Feature Type library is dynamic, it can be update anytime leaving the existing Feature Type structure unchanged. Secondly Feature Type libraries can contain Feature Instances also. This is especially useful in case of specifying supplier's information with a limited variable domain (e.g. the width of a door is either 800 mm, 820 mm or 840 mm).

Inconsistency and incompleteness is an inherent characteristic of Feature Based Modelling. This can be regarded both a pro and con of this modelling approach. In this respect, the following conclusions are drawn:
- Inconsistency and incompleteness is an elementary part of architectural design and thus a prerequisite for architectural knowledge modelling.
- Inconsistency and incompleteness are designer dependent. Apart from checking norms and standards there is no general rule a designer can count on for maintaining consistency. Also, incompleteness may be a target of a design process.
- FBM, as currently developed in this research, does not support any kind of strategy for maintaining consistency and completeness for a specific design part or design task. Future research must be conducted on this issue, as is done in the work of Eastman et al. (1997a; 1997b).

REFERENCES

Achten, H.H. 1997. *Generic Representations, An approach for modelling procedural and declarative knowledge of build-*

ing types in architectural design, PhD. Thesis. Eindhoven: Eindhoven University of Technology.

Booch, G. 1994. *Object oriented analysis and design*, second edition. Redwood City: Benjamin/Cummings Inc.

Coyne, R.D., Rosenman, M.A., Radford, A.D., Balachandran, M., & Gero, J.S. 1991. *Knowledge-based design systems*. Reading, Massachusetts: Addison-Wesley.

Cross, N. 1984. *Developments in Design Methodology*. Chichester: Wiley & Sons Ltd.

DeMartino, T., Falcidieno, B., Giannini, F., Hassinger, S., & Ovtcharova, J. 1994. Feature-based modelling by integrating design and recognition approaches. *Computer-Aided Design* 26(8).

Eastman, C.M., Jeng, T.S., Assal, H.H., Cho, M.S., & Chase, S.C. 1995. *EDM-2 Reference Manual*. Los Angeles: University of California in Los Angeles.

Eastman, C.M., Parker, D.S., & Jeng, T.S. 1997a. Managing the integrity of design data generated by multiple applications, The theory and practice of patching. *Research in engineering design* 9: 125-145.

Eastman, C.M., Jeng, T.S., Chowdbury, R., & Jacobsen, K. 1997b. Integration of Design Applications with Building Models. In *Proceedings CAAD Futures '97*: 45-59. Dordrecht: Kluwer.

Ekholm, A. & Fridqvist, S. 1997. Design and modelling in a computer integrated construction process, The BAS-CAAD project. In *Proceedings CAAD Futures '97*: 501-518. Dordrecht: Kluwer.

Hendricx, A. 2000. *A core object model for architectural design*, PhD. Thesis. Leuven: Katholieke Universiteit Leuven.

Lawson, B. 1990. *How designers think, the design process demystified*, 2nd edition. London: Butterworth Architecture.

van Leeuwen, J.P., Wagter, H., & Oxman, R.M. 1995. A Feature based approach to modelling Architectural Information. In Fisher, Law, & Luiten (eds.), *Modeling of buildings through their life-cycle*. CIB W78-TG10 publication 180, Stanford University.

van Leeuwen, J.P. & Wagter, H. 1997. Architectural design-by-Features. In *Proceedings CAAD Futures '97*: 97-115. Dordrecht: Kluwer.

van Leeuwen, J.P., & Wagter, H. 1998. A Features Framework for Architectural Information, a case study. In Gero & Sudweeks (eds.), *Artificial Intelligence in Design '98*: 461-480. Dordrecht: Kluwer.

van Leeuwen, J.P. 1999. *Modelling Architectural Design Information by Features*, PhD. Thesis, June 1999. Eindhoven: Eindhoven University of Technology.

Ovtcharova, J. & Vieira, A.S. 1995. Virtual prototyping through Feature Processing. In Rix, Haas, & Teixeira (eds.), *Virtual Prototyping - Virtual environments and the product design process*: 78-90. London: Chapman & Hall.

Roozenburg, N.F.M. & Eekels, J. 1995. *Product Design: Fundamentals and Methods*. Chichester: Wiley & Sons, Ltd.

Shah, J.J. & Mäntylä, M. 1995. *Parametric and Feature-Based CAD/CAM*. New York: Wiley & Sons.

Sowa, J. 1984. *Conceptual Structures: Information Processing in Mind and Machine*. Amsterdam Addison-Wesley.

Process modelling and management

Development of process modelling tools for the construction industry

K. Karstila
Eurostep, Helsinki, Finland

V. Karhu
VTT Building Technology, Espoo, Finland

B.-C. Björk
Royal Institute of Technology, Stockholm, Sweden

ABSTRACT: The methods used for modelling and planning of processes have in the construction industry so far often been of an ad-hoc nature. In the best of cases companies may have used formalised process re-engineering methods and tools, but a more common approach has been to use relatively simple, informal graphs. Researchers in the domain have favoured a technique called IDEF0. Use of this method in a number of R&D projects together with experts from industry have, nevertheless, revealed a number of weaknesses.

This paper describes interim results of the Finnish subprojects of the on-going international research project, Models for the Construction Process (*MoPo*). The project is a co-operation between Sweden, Finland and Slovenia and has an overall size of around nine man years. In the early phases of the MoPo project a lot of emphasis has been put on a study of the requirements that construction industry companies place on such tools and models. This study has been carried out in co-operation with design and construction companies, partly by carrying out analysis of real processes of the involved companies using the IDEF0 methodology, and gathering feedback on the methods "as they are in use", rather than confronting industry experts with more abstract models or processes that they are not familiar with. The experiences are reported in the paper.

In on of the work packages of the MoPo project a comparison was made of the features of six different process modelling languages. The data models of all of these were defined in the EXPRESS language and a synthesis method, called GEPM (GEneric Process Modelling method), was defined. A prototype application based on this model has subsequently been developed, on top of a groupware environment. The application database can be used for storing data which has been produced using software applications supporting traditional bar chart scheduling or IDEF0 models, and different types of views including object browsing can be derived from the database. The prototype has been tested using process data from a medium size office building project gathered in an earlier research.

In another work package a model and prototype tool (PIIP, Process Information Integration Platform) for construction process design, planning and management is under development. PIIP uses product data technology (PDT), familiar from the STEP and IFC activities, applying it to construction process modelling.

1 INTRODUCTION

The proliferation of IT use in construction (distributed CAD, document management and workflow systems, Internet etc.) has lead to a need for companies to analyse and document their current work processes more precisely than previously. The results of such analyses can be used for defining information management procedures for project teams as well as a basis for reengineering efforts that take into account the new possibilities offered by IT. The increased use of contractual arrangement other than the traditional bid-construct organisation, new management and the wide-spread introduction of quality systems also increase the need for advanced process analysis methods.

The methods used for this kind of analysis and planning have in the construction industry so far often been of an ad-hoc nature. In the best of cases companies may have used formalised process reengineering methods and tools, but a more common approach has been limited to relatively simple, informal graphs ("box-and-arrow" notation) and use checklists.

The more formal, different process modelling methods and/or resulting models can be categorised as follows:

- **Generic process description methods**, which typically are not specifically developed for construction process modelling only. Examples of this category are IDEF0-method (IDEF0 1993), PetriNets (Www-Petri Nets 2000), various kinds of flow charting or data flow methods. These methods are often based on a very limited set of concepts that have corresponding graphical symbols for developing models, which are represented and presented as graphical diagrams.
- **Construction process activity or functional models**. These models have been typically developed for specific purpose with specific viewpoint, using e.g. one of the above mentioned generic process modelling methods, like IDEF0.
- **Construction process information models**, which by using information modelling methods, like EXPRESS data specification language, describe and represent various objects, their attributes and relationships. The information model is then typically basis for implementation of computer software.
- **Construction tasks simulation models** which describe the dynamic aspects of some construction, typically mechanical, subprocesses or tasks of construction work, like earth moving of crane lifting. The goal of the simulation is to analyse the performance of the process, impacts of process disturbances on the queuing, performance etc.

A number of construction process activity models have also been developed as part of R&D projects, related to construction process re-engineering efforts. The IDEF0-method has been the prevailing method in these efforts. Sanvido's model (Sanvido et al. 1990) was composed to provide an open information architecture to support the provision of a facility. Karhu & Lahdenperä (1999) has developed a process model of current Finnish design and construction practice. The model of Zhong (Zhong et al. 1994) covers the overall building process although it is aggregated compared to Sanvido's model. Another model (Merendonk & Dissel 1989) is more comprehensive and includes many functional and conceptual schemes. More recently the UK Process protocol work, which is a joint venture between university researchers and practitioners, has defined a framework model for how the construction industry should work, and presented it using diagram techniques (Aouad 1998).

2 AIMS AND SCOPE OF THE MOPO-PROJECT

The MoPo-project (*Models for the construction process*, Www-MoPo) is an international collaborative R&D project which aims to develop methods and tools to support industry needs and requirements for construction process design, planning and development.

From an **industrial viewpoint** the project aims at developing methods, IT-based tools and reusable models for construction process analysis and planning, which could be deployed by companies and construction project teams in a relatively rapid time-frame. From an **academic viewpoint** the project aims at making an original contribution to construction management and construction IT research in the sub-domain of construction process modelling. This is an area which currently is relatively underdeveloped, and where lessons learned in other industries and other scientific disciplines could be studied and used as input to a synthesis of new knowledge.

The project is carried out by a multinational research consortium with participants from the three European countries: Sweden, Finland and Slovenia. The group is lead by Prof. Bo-Christer Björk at the division of Construction Management and Economics at the Royal Institute of Technology, Stockholm. The Funding comes from three national R&D programs (including IT Bygg och Fastighet 2002 in Sweden and the TEKES VERA programme in Finland) as well as from the industrial partners participating in the consortium. The planned project volume is around 9 man-years, spread over a period of two and a half years

The MoPo-project has been divided into two main phases: Phase 1 to do prestudies on earlier work in the subject area, to capture the initial industry requirements for process support methods and tools, and to develop early prototype tools to provide better basis for further in depth development. Phase 2 to continue and expand to more detailed requirements capture, and the development of methods and tools in selected areas based on the insight from the first phase.

In the following the three on-going subprojects being carried out in the Swedish part of the MoPo project are briefly presented. In the subproject ProFacil, an early draft of a model of the construction and facilities management process from an end users perspective has been developed (Björk et al. 1999). This model was developed using the IDEF0 method. The model has been evaluated by industry experts from a number of FM companies in order to provide feedback and directions for further detailing of the model. Based on this a second version will be developed, possibly using alternative modelling languages.

Another subproject, named User interfaces to construction process models, deals with presentation and communication of process models, and utilizes techniques from research on human-computer interaction. There is a separate paper of this subproject in the ECPPM conference (Berg Von Linde 2000).

3D CAD as a construction tool has been around for several years but has yet not become widely used except for very specialised applications, such as visualisation of architectural designs and model-based structural design. An interesting new evolving use of a 3D-model is to connect it to a time scheduler program in order to help visualise the progress in time of site works, among other things to discover weaknesses in the schedule (this concept is called 4D-CAD). In a MoPo's sub-project, which will be carried out in close co-operation with the he CIFE centre at Stanford University, a case project from the Swedish contractor NCC will be modeled using 4-D methods to study user experiences of this way of working.

The work items for the Phase 1 of the Finnish part of the MoPo-project were as follows:

- Industry requirements capture concerning both construction process support and information requirements.
- Development of a synthesis process modelling methods, called Generic Process Modelling method (GEPM), based on a number of common process modelling methods that have been applied to construction.
- Development of the Process Information Integration Platform (PIIP) as an application of product data technology methods for process information modelling and implementation of the first prototype tool based on the process information model.

These will be described more in detail in the following.

3 CAPTURING INDUSTRY REQUIREMENTS

During the prestudy stages of the MoPo project it has become evident that the usability of a process modelling method is important. Based on interviews with industry experts, who have participated in modelling work, often using the IDEF0 methodology, it is evident that almost as a rule only the experts who have themselves participated actively in the modelling work understand the models. This seems to a large extent to be a result of, on the other hand the iterative development process being an important part of the learning and

consencus making process of the modelling exercise, and on the other hand the rather complex way in which the models are presented. At least in one case (Hoffner 1997), this leads to the company experts choosing a much simpler modelling method despite the evident loss of modelling power.

There thus seems to be a clear dilemma and trade-off between the two criteria of modelling power and usability. In earlier generations of models this dilemma has been difficult to solve, due to the fact that in most modelling methods, the model representation and the model presentation have been one and the same.

The objective of the industry requirements capture work item was on the other hand collect earlier experiences of the partners in process modelling area; and on the other hand to capture industry ideas, needs and requirements for methods and tools for process modelling and support for process design, planning and management. A number of informal meetings or workshops have been held with the industry partners of the MoPo-project for this purpose. Some of the general, main issues identified in the industry workshops include e.g. the following:

- Overall construction project co-operation and co-ordination are a problematic area that would benefit from new developments.
- The partners have quality systems operational, but the construction project specific linking of partners' quality systems would require more support.
- Even in best cases there are still gaps in the integration of enterprise resource planning and control system, construction project and quality management system, and its IT support.
- The management of dependencies between the partners' tasks, and the impact of various kinds of changes and revisions on e.g. designs and plans cause problems and disturbance in scheduling of the process flow.

The industry partners have different background in process modelling using formal methods: on the one hand some of them, like YIT (a general contractor) and Parma Betonila (prefabricated concrete element manufacturer) had already a good amount of experience in IDEF0-modelling e.g. through the Reco-project which concentrated on modelling interactions of various partners (designer, fabricator and constructor) of a construction project. On the other hand some of the MoPo industry partners didn't have experience in process modelling using IDEF0. Because of this various approaches were applied in partner specific work after the general discussions about industry requirements for process support.

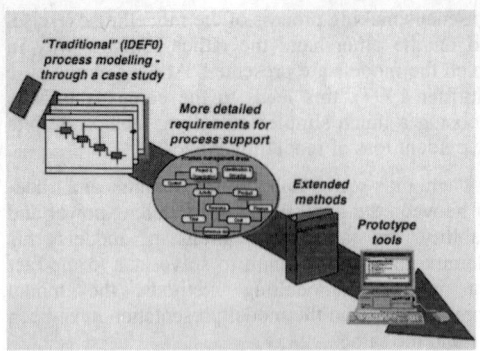

Figure 1. Research and development approach starting from traditional process modelling to capture more detailed requirements, towards extended process modelling methods and tools.

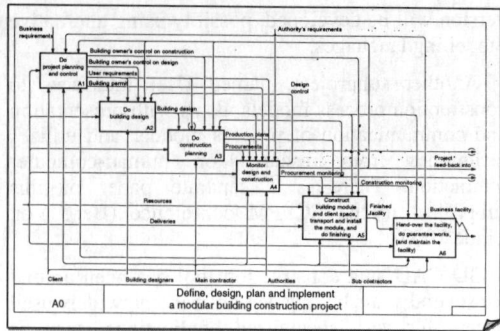

Figure 2. The main IDEF0-diagram of the modular building construction process model developed together with the domain experts of one of the case companies.

With YIT and Parma Betonila the work builds on the top of earlier modelling work of the Reco-project. One of the work items being studied is thus various ways. to present and communicate the process models for "outsiders", for those how didn't participate in the actual modelling effort.

The other approach applied was to start with "traditional" process modelling, i.e. IDEF0, on a subject area of interest for the industry partner. Then use the understanding and experiences from the IDEF0-modelling as a basis for further defining the more detailed requirements for the process support, and extended methods (beyond IDEF0) and model presentations. The requirements then drive also the development of the prototypes, Figure 1. This latter approach was used with JP Talotekniikka (building services and FM consultant) and U Lipsanen (a general contractor).

The process model developed for JP Talotekniikka deal with the process of building condition audit, which is typically done as part of real estate development, facilities management planning, and when selling / buying buildings. JP Talotekniikka provides building condition audit services. The building condition audit model consists of some half dozen diagrams with altogether about thirty activities. The model is complemented with a context model of client's process model related to audits.

For the case of U Lipsanen the process model developed deal with an overall construction project, of a specific project type, from project definition to building hand-over, Figure 2. The process model consists of more than twenty diagrams with altogether about eighty activities. The model is complemented with a simple context model of

clients business process model related to the construction process model.

The workshops and modelling exercises clearly demonstrated that IDEF0 is fairly well accepted by those who have themselves participated in a modelling exercise using IDEF0. One of the difficulties with IDEF0-method is that the diagrams easily get fairly complex. On the other hand if the model are made too simplistic, they also tend to remain trivial, without much explanatory content. Also, the pure hierarchical tree structure of the diagrams forces sometimes to compromise in choosing the model structure. Comprehensive IDEF0-models are often found to be difficult to communicate for others who didn't directly participate into the model development. Hence, it may seem appropriate to use IDEF0 in developing the process model from which various presentations are then derived for communication purposes.

4 FORMAL PROCESS MODELLING METHODS

As part of the GEPM work task, six different methods have been studied. The conceptual model and characteristics of each method were modelled formally using the EXPRESS-G language, which is the graphical counterpart of the EXPRESS data specification language (ISO 1994).

A few authors have earlier proposed conceptual models for construction process information models. The IRMA model (Information reference model for AEC) was developed in the early 90s by a group of leading product model researchers (Luiten et al. 1993). Froese (1995) points out that the central concepts are activity, input, output, control, mechanism or actor, and precedence. STARGEN includes concepts such as contract, resource, flow, and activity point of view (Hannus & Pietiläinen

1995, Pietiläinen & Heinonen 1995). Augenbroe & Amor (1997) have defined process formalisms in construction using an approach similar to Petri Nets.

The methods studied in the GEPM work were project scheduling method, so-called simple flow method, IDEF0, IDEF0v, IDEF3 and Petri Nets. The scheduling method is the name used here to denote the familiar project planning or scheduling technique (critical path, resource levelling, precedence method, PERT, etc.) which has been under development along with software tools since the 1950's. This method is well described in numerous basic textbooks and it is very popular in practice, using software such as MS Project or Primavera. The second method in the list, called here "simple flow method", is an ad-hoc method which has been used e.g. in a general contractor's process modelling effort (Hoffner 1997). It is, however, quite representative of a category of relatively simple flow oriented methods, which can easily be dealt with using simple graphical IT-tools. The four remaining methods are formal methods, which have been described in standards (IDEF0), in proposed standards (Petri Nets) or in the research literature (IDEF0v, IDEF3). It may be noted that yet another method called CYCLONE (Halpin & Riggs 1992) has been used for construction process modelling, but has been left out with purpose from this analysis since it is more aimed for dynamic simulation of process behavior.

The methods above all use a graphical representation that also plays an important role for analysing the developed process models. Features typical for the building construction process, which methods should be able to model, include:

- Process breakdown into activities / tasks, and material and information flows
- Order or dependency of activities, especially on the site
- Iterative activities especially during the design process
- Conditional and probabilistic branching
- Use of feedback for decision-making
- Concurrency, for instance designs of different disciplines are done simultaneously
- Complex organisational schemes, where a project can be subdivided into several layers of subcontracts, and where the organisational scheme may vary from one project to another.

Based on the results from the analysis of the six methods a new generic process modelling method GEPM was defined, Figure 3.

The scope of GEPM is:

- to provide a generic modelling method for construction process modelling
- to enable future software development.

The modelling criteria for the method are:

- activity and task should be separated
- sufficient modelling power should be obtained from a generic conceptual model
- in the implementation internal database storage could be used to generate views for different purposes and methods
- analogy with product data modelling should be observed
- neutral data exchange should be possible.

The graphical presentation of GEPM models was deemed out of scope partly because GEPM-models may be converted into other existing methods, which have already well developed tools, and since graphical user interface development would require more efforts. Also the intention is to store GEPM model in databases, rather than as graphical diagram files, from which different representations can be derived, Figure 4.

The GEPM method has been implemented in a prototype using Lotus Notes / Domino software. Lotus Notes is a groupware tool that uses databases for storing information of various kind. The Domino extension is an application that converts the database into web-environment. It was chosen because it has robust user authentication, various programming language alternatives, and a direct web-interface.

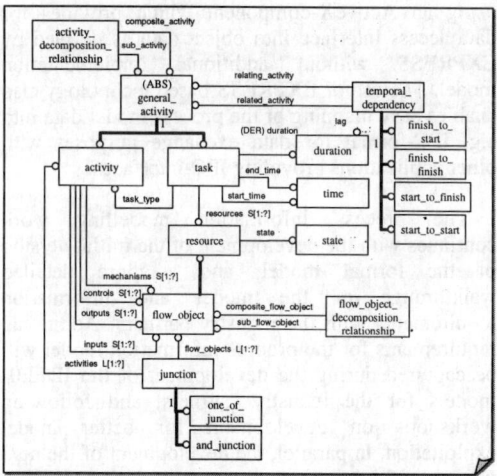

Figure 3. The conceptual model of the proposed generic process modelling method GEPM.

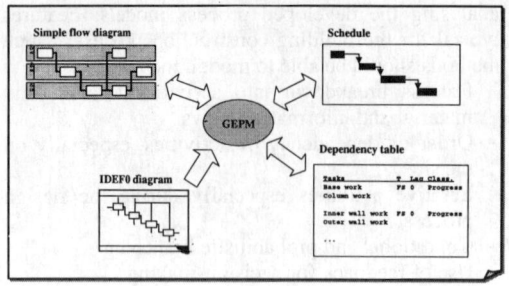

Figure 4. Views for different purposes supported by the GEPM prototype.

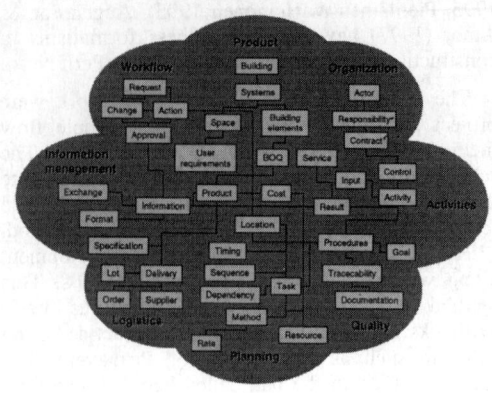

Figure 6. . Views and elements of an informal process information planning model under development.

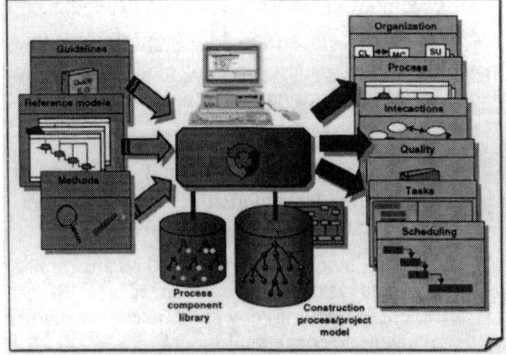

Figure 5. Vision of the PIIP process support prototype tool.

5 PROCESS INFORMATION INTEGRATION PLATFORM

The objective of the Process Information Integration Platform (working title PIIP) work item is to support industry requirements in construction process development by developing a model and prototype tool for construction process design, planning and management. Figure 5 illustrates the overall vision.

PIIP uses product data technology (PDT) approach for construction process modelling. The planned results from the PIIP work item are:

- Captured industry needs and requirements for construction process modelling, and model and tool support.
- A conceptual or information model for construction process information supporting process design, planning and management.
- A prototype implementation of construction process modelling tool to demonstrate support for process design, planning and management. The implementation is based on the construction process information model, together with supporting application functionality.

The work status of the information requirements capture is that based on a prestudy of existing process models and IT tools (STARGEN, IRMA, Froese, Concur, PDM, timescheduling etc.) a rough information planning model with a number of views to be supported has been sketched, Figure 6. A first round of walkthrough of the planning model elements has been done with two industry partners.

The process information model is being developed using EXPRESS data specification language, the standard language in product data modelling efforts like ISO-STEP and IAI/IFC. The prototype implementation will be directly based on the EXPRESS-based process information model by using an ActiveX-component which provides for data access interface into object data as defined by EXPRESS without additional implementation model. The use of EXPRESS-based technology also enables easy mapping of the process model data into e.g. IFC-format for data exchange purposes with other applications providing IFC-interfaces.

The process information modelling work continues with the development of the initial version of the formal model, and a more detailed walkthrough of the model and information requirements with the industry partners. Additional requirements for the process information model will be captured during the development of the IDEF0-models for the industry partners, and follow-up workshops on development for better model exploitation. In parallel, the development of the next version of the process support tool is done.

An initial process model browser tool was developed in the very beginning of the project. The purpose of the early prototype tool was to be able to illustrate some of the basic ideas of the MoPo-project, like object modelling of various construction

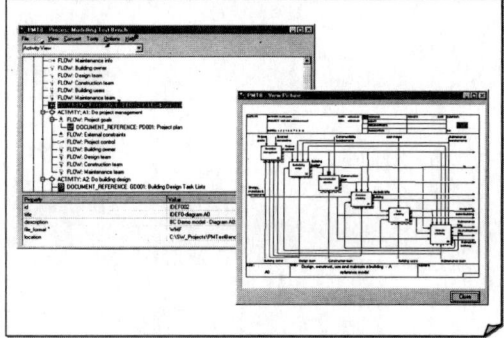

Figure 7. Some early PIIP demonstrator user dialogs.

process aspects, generation of different views for to support various views on the model, and integration of process modelling and model data into other tools like timescheduling. Some user dialogs of the process model browse are shown in Figure 7.

The work status of the further prototype development is that next version of the process modelling browser will be started once the initial process information model has been developed. Both the process information model and the prototype tool will be then developed and fine-tuned in parallel based on the feedback from the workshops with the industry partners.

6 CONCLUSIONS

The experiences from the first stage of the MoPo project have clearly demonstrated the need for this type of research and the industry interest for developing their way of defining their working processes. It has become rather clear that IDEF0 can be a useful for many purposes, and certainly is better than no formal process modelling methodology at all. The presentation format of IDEF0 is, however, difficult for practitioners who have not themselves participated actively in the modelling process.

The GEPM model could offer one way out of this dilemma by offering the means to clearly distinguish between the process definition (which should reside in a database) and its presentation in different views or formats, which could be adapted to the users needs and resemble currently used methods. In the work on the PIIP platform principles used for the development of product data exchange software are being utilized also for the development of process modelling tools, and early results are promising. It should be noted that in line with the overall philosophy of the MoPo project, the two sub-projects described above have proceeded in parallel

but rather independently. This has to do with the fact researchers in the project work towards Ph.D thesis, but are at the same time in most cases part time employed with other work in their respective organizations. Thus we strive to benefit from the synergy and critical mass of a "virtual" project group, at the same time giving each researchers a lot of independence, avoiding the pitfalls of a too tightly planned project, as exemplified by some EU R&D projects.

Acknowledgements:

The other researchers involved in the MoPo-project are Rikard Berg Von Linde, Berndt Lundgren, Mats Knutsson, Robert · Noack and Anders Nilsson, from the Royal Institute of Technology as well as Prof. Ziga Turk from the University of Ljubjana. Colleagues who have provided valuable input include Ghassan Aouad, Matthew Bacon, Andrew Baldwin and Martin Fisher. The project has been financially supported by the TEKES Vera programme and the IT Bygg och Fastighet 2002 program as well as by of the companies YIT, Skanska Oy, JPTalotekniikka, U Lipsanen Oy, Parma Betonila, NCC and Tyréns.

REFERENCES

Aouad, G. et al . 1998. Generic Design and Construction – Process Protocol. University of Salford, Salford, UK.

Augenbroe, G. & Amor, R. 1997. 'Project Control in an Integrated Building Design System'. CIB W78 Workshop on Information Technology Support for Construction Process Re-Engineering, IT-CPR-97, Cairns, Australia, 9-11 July, 1997, pp 53-66.

Björk, Bo-Christer, Lundgren, Berndt, Nilsson, Anders. ProFacil – A model integrating the facilities management process with the building end user's core process. In proceedings of the CIB W55 & W65 - JOINT TRIENNIAL SYMPOSIUM "Customer satisfaction: A focus for research & practice", 5-10 September 1999, Cape Town, South Africa

Froese, T. 1995 'Models of construction process information.' A web document found at url: http://www.civil.ubc.ca/~tfroese/pubs/fro95b_process_mod els/fro95b.html.

Halpin, D. W & Riggs, L. S. 1992. 'Planning and analysis of construction operations.' Wiley & Sons, Inc. New York.

Hannus, M. & Pietiläinen, K. 1995. 'Implementation concerns of process modelling tools.' In: Fischer, M.A., Law, K.H. & Luiten, B. (eds.) Modeling of Buildings Through Their Life Cycle. CIB W78 & TG10 Workshop on Computers and Information in Construction, Stanford 21-23 August 1995. Stanford, CA: Stanford University & CIB. P. 449 - 458. (CIB Proceedings, Publication 180).

Hoffner, L. 1997. 'Projekteringsprocessen vid egen regiprojekt hos JM Byggnads AB. En modellering och analys av processen och dess ingående dokument.' (The design stage of the building process at JM Byggnads AB. A modelling and analysis of the process and the documents involved). Master's thesis no. 318. (In Swedish), Department of

Construction Management and Economics, Royal Institute of Technology, Stockholm, Sweden.

IDEF0. 1993. Integration definition for function modeling (IDEF0). Federal Information Processing Standards publication 183.

ISO. 1994. ISO International Standard 10303 (STEP). Part 11: The EXPRESS Language Reference Manual.

Karhu, V. 1997. Product Model Based Design of Precast Facades. Electronic Journal of Information Technology in Construction. Vol. 2. http://itcon.org/1997/1/

Karhu, V. 2000. Proposed new method for construction process modelling. (accepted to International Journal of Computer Integrated Design and Construction in March 2000).

Karhu, V. & Lahdenperä, P. 1999. A formalised process model of current Finnish design and construction practice. International journal of construction information technology, Vol. 7 No 1, pp. 51-71.

Karstila, K. & Björk, B-C. 1999. Models for the Construction Process – The MoPo-project In Hannus M. and Kazi, A. (edts); Proceedings of the 2nd International Conference on Concurrent Engineering in Construction – CEC99, 25-27 August 1999, Espoo, CIB TG33 and VTT, Finland.

Karstila, K. & Karhu, V. 1999. Status report October 1999 – Finnish Part. MoPo Models for the Construction Process. 15 p.

Kochikar, V.P. & Narendran, T.T. 1994. 'On using abstract models for analysis of flexible manufacturing systems.' Int. J. Prod. Res., Vol. 32, No. 10, pp 2303 - 2322.

Koskela, L. 1995. 'On Foundations of Construction Process Modeling.' Modeling of Buildings through Their Life-Cycle. CIB Workshop on Computers and Information in Construction, August 1995, Stanford University, Stanford. CIB proceedings, Publication 180, 1995, pp 503 - 514.

Luiten, G., Froese, T., Björk, B.-C., Cooper, G., Junge, R., Karstila, K., and Oxman, R. 1993. 'An Information Reference Model for Architecture, Engineering, and Construction.' Proc. of the First Int. Conf. on the Management of Information Technology for Construction, World Scientific, Singapore, pp 391-406.

Merendonk, P. van & Dissel, D. van. (eds.). 1989. Bouw informatie model; versie 3.1. IOP-Bouw. App., Rotterdam, Netherlands.

Pietiläinen, K. & Heinonen, R. 1995. 'STARGEN – Generic modelling of construction processes.' Working paper (unpublished manuscript).

Sanvido, V., Khayyal, S., Guvenis, M., Norton, K., Hetriok, M., Al-Muallem, M., Chung, E., Medeiros, D., Kumara, S. & Ham, I. 1990. An integrated building process model. Technical Report no 1, CIC Research Program. The Pennsylvania State University. University Park, PA, US.

Www-MoPo. MoPo-project's web-site, www.mopo.org

Www-Petri Nets. 2000. A website of Petri Nets at http://www.daimi.aau.dk/PetriNets/.

Zhong, Q., Mathur, K. & Tham, K. 1994. A construction process model by using Design/Idef. CIB W78 workshop on computer integrated construction, Helsinki, Finland, August 22 - 24, 1994. Helsinki: VTT Building Technology & International Council for Building Research Studies and Documentation CIB.

Product and Process Modelling in Building and Construction, Gonçalves, Steiger-Garção & Scherer (eds)
© 2000 Balkema, Rotterdam, ISBN 90 5809 179 1

The development of a process mapping methodology for the Process Protocol Level 2

A. Fleming, A. Lee & R. Cooper
Institute for Social Research, University of Salford, UK

G. Aouad
School of Construction and Property Management, University of Salford, UK

ABSTRACT: The Generic Design and Construction Process Protocol (GDCPP) was created by the University of Salford in 1998. It is a high-level process map that aims to provide a framework to help companies achieve an improved design and construction process. Furthermore, industry interest and acceptance of the framework provided the impetus for further funding to continue the research for an additional three years. The first project concentrated on the high level protocol, and the second aims to develop the sub processes of the original high level map. This paper describes the methodology used to develop the sub processes and explains why a bespoke modeling methodology was developed rather than using the standard process modeling techniques. The methodology enables all of the information relating to the sub processes to be represented as a series of process maps and when viewed holistically, presents an integrated generic decomposition of the processes on the high level map.

1 INTRODUCTION

The British Property Federation Survey (British Property Federation, 1997) identified that one third of major UK clients are dissatisfied with contractor and consultant performance. Similarly, The Egan Report, Rethinking Construction (Egan, 1998), stated that the industry also suffers from low and unreliable profitability, insufficient research & development, and a lack of customer focus. Moreover, these problems typically relate to the industry's adversarial nature, and a profound co-ordination and communication system between the parties is much needed.

The Generic Design and Construction Process Protocol (GDCPP) was created by the University of Salford in 1998 in an attempt to improve the prevailing situation (http://www.processprotocol.com). It is a high-level process map that aims to provide a framework to help companies achieve an improved design and construction process. The map draws from principles developed within the manufacturing industry that include stakeholder involvement, teamwork and feedback, and reconstructs the design and construction team in terms of Activity Zones rather than in disciplines to create a cross-functional team. These Activity Zones are multi-functional and may consist of a network of disciplines to enact specific task of the project, allowing the 'product' to drive the process rather than the function as in a sequential approach. Luck and Newcombe (1996) argue that traditional roles and responsibilities change from project to project, often resulting in ambiguity and confusion; the use of zones potentially reduces this confusion and enhances communication and co-ordination (Cooper et al, 1998). The zones consist of Development Management, Project Management, Resource Management, Design Management, Production Management, Facilities Management, Health and Safety, Statutory and Legal Management, and Process Management, and are located down the left-hand column of the protocol (see Figure 1). The Activity Zones contain high-level processes spanning the duration of a project from inception, through design and construction, and including operation and maintenance. The responsibility for completing the processes may lie with one Activity Zone or be shared.

Furthermore, industry interest and acceptance provided the impetus for further funding to continue the research under the moniker Process Protocol Level 2. One of the primary deliverables of the Process Protocol Level 2 is to create sub process maps of the eight Activity Zones that exist within the original Generic Design and Construction Process Protocol model (Kagioglou et al, 1998).

The maps should provide an increased level of detail and description than the existing GDCPP Map.

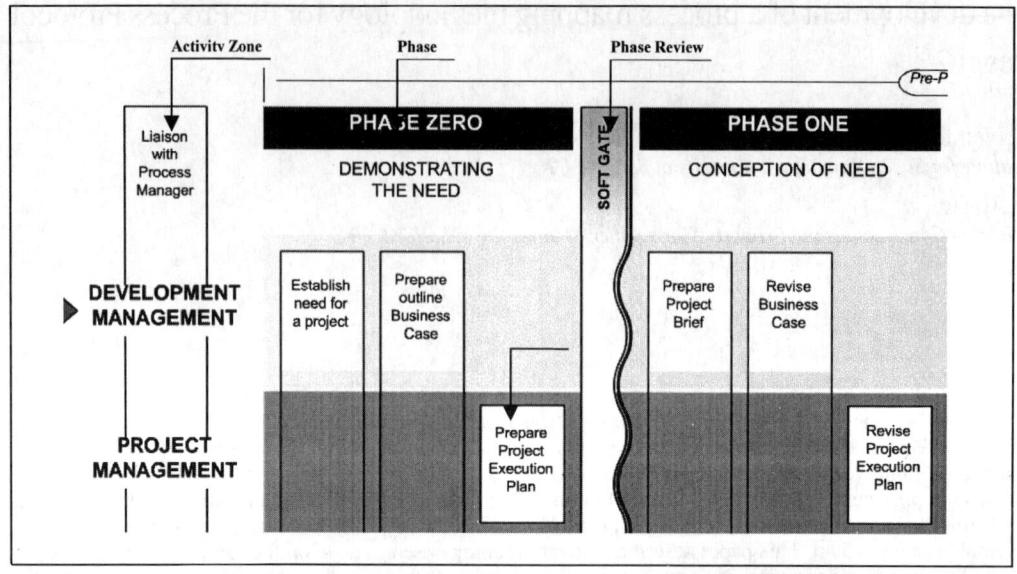

Figure 1 Generic Design and Construction Process Protocol

Moreover, it still should remain generic and adhere to the principles of the GDCPP. The extra level of detail may require the inclusion of sub deliverables as well as sub processes.

2 RESEARCH METHODOLOGY

The methodology for the Level 2 project (Figure 2.) commenced with literature searches of key authors, texts and relevant research projects. A written literature review was prepared and disseminated amongst the project participants. The literature review provided the foundation of knowledge required by the researchers to familiarise themselves with the subject matter and to interview key persons effec- tively.

The methodology stated that ten persons were to be interviewed for each Activity Zone. Key persons were identified and contacted to gain co-operation. The primary purpose of the interviews was to extract the working method/process of the interviewed companies. This information would then be used to

identify innovative and effective practice whilst also monitoring processes and practices that were not effective.

Development of the sub processes for the activity zones was based on the GDCPP guide (Kagioglou, et al, 1998). The GDCPP map was examined horizontally across the activity zones. This was performed process by process in a horizontal fashion. At all times the context of the phase was considered. This formed a skeletal structure that helped extricate the relevant information from the research material that had been collected. Once the information had been collected, analysed, structured and presented in a logical format, the next task would be design a sub process map template that that would effectively present the information.

3 SUB PROCESS MAP DEFINITION

A key requirement of the process maps was that prior understanding of process modelling techniques should not be a pre-requisite to understanding the maps. Process modelling tools such as the IDEF family were considered but felt to be too complex for certain members of the targeted user group. The key was in the representation (Cheung, 1998) of the process and it was felt that none of the tools available met the project's requirements. Therefore it was necessary to develop an original process map template.

After investigation of Visio a diagramming tool it was decided to use it to design and create the sub

Drafting Phase	Issue for Review	Verified Model Phase	Validated Model Review	Final Version
Consulted: Process Protocol Interviews Literature Existing processes	Review internally among project team: Amend Approve	External review: cross activity zone interface check: Amend Approve	Industrial workshops (2): Present Review Amend Approve	Final Review: Submit for Publishing

Figure 2 Sub Process Map Development Methodology

188

process maps. Visio had the ability to attribute information to objects within its diagrams and store this information within its own database. This led to the possibility of exporting data to other applications thus increasing the usefulness of the maps. The number of maps in a series is dependant on the level of detail attributed to a process and the number of processes in the Activity Zone.

A map was created that represented all of the information that the project required and was formally presented to an industrial workshop and conditionally approved. Feedback about the maps format was received. Internal development workshops were held to refine the map. Issues arose from the workshop regarding the modelling of the sub process maps. There was a lack of information relating to the different levels of process modeling. For instance, what criteria is used to distinguish between a level two or three process. It was decided that the distinction is a subtle one that relies on the experience and judgement of the process modeler.

The map produced was discussed and the consensus was that it needed to be simplified. The content of the maps was becoming complex and methods to reduce this complexity were considered whilst still showing all the detail required. A decision was made to only show level 1 deliverables.

Processes that were common to all or some of the activity zones were beginning to be identified and considered as generic process components. An issue that remained unsolved at this stage was how to illustrate the interaction of these intra activity zone components. Process inputs and outputs were indicated at this stage by an arrow headed line and labeled. This was considered to be a temporary measure and would be improved.

3.1 Information structure on the maps

A further workshop was held and the debate focused on the distribution of the content of the maps with respect to the validity of the content and whether it appeared at the correct phase and in the correct sequence. The detail of the level two and three processes was considered as was the level in which the process should belong. Should the process be moved up to level two or down to level three or excluded completely. This allowed the team to familiarise themselves with the content of the maps.

This workshop saw the introduction of phase grouping. Previously it had been noted that activities were often repeated through several phases. Therefore in order to simplify the content of the maps and to avoid repetition activities were grouped together logically. Depending on whether these processes occurred at the beginning of a phase, during a phase or at the end of a phase would determine the standard group in which they would reside.

A further workshop was held which was notable for two main outcomes. The first introduced the principle of process ownership. All of the processes would have an activity zone having overall ownership of a process to ease the co-ordination of a process. The second was the introduction of a new process symbol that illustrated which activity zones were participants to the process/sub process. This solved the problem of how to show intra-activity zone processes. A single glance would now indicate the origin and ownership of a process and what activity zones participated in the process thus validating Activity Zone interfaces. It was agreed to incorporate the new symbol onto the developing process map template together with the activity groups.

An industrial workshop was held to gain industrial validation of the sub process maps. This workshop considered the modeling rules and conventions used to define and model the high level processes into sub processes. A total decomposition modeling technique was adopted to illustrate the decomposition of a process and its sub processes. Logical dependency is represented between processes when they exist.

4 MODELLING RULES

The aim of the modelling work is to provide a visual representation of the sub processes of the activity zones of the Process Protocol map. This is achieved by illustrating 'what' are the sub-processes of the high level processes identified in the Process Protocol map and 'how' these sub-processes interact. As a result it will be possible to provide models for individual phases.

4.1 Modelling conventions

This section describes the main convention types used for the modelling of the sub-processes of the Process Protocol map. The main elements of the template include:

- Phase start up activities
- Map title including phase number, phase title, and Activity Zone name
- The generic top level processes, the two levels of its decomposition and their respective processes
- Ongoing activities
- End of phase activities
- Lexicon
- Map title block including author, version, phase number and title, etc.

The phase start-up, ongoing and end of phase activities are illustrated on separate process maps. They represent activities common to many Activity Zones and represent activities undertaken for most phases.

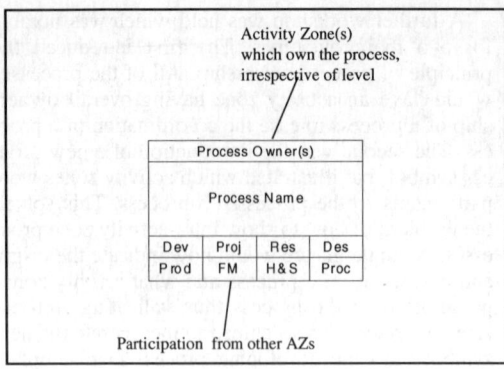

Figure 3 Process Symbol

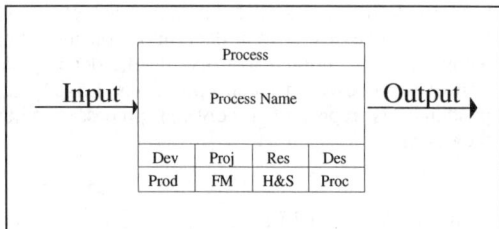

Figure 4 Inputs and Outputs to the Process

4.2 Process representation

The processes and sub-processes are denoted by using the symbol shown in figure 3, which includes:
- Process owner(s)
- Process name (potentially including some description for clarification where required)
- An indication of likely/potential participation from other activity zones in the process

Furthermore, inputs and outputs from a process can be shown as illustrated in figure 4.

4.2.1 Inputs

For clarity, inputs to a process are only shown where they form a logical dependency from another process at that level on the same diagram. All other inputs from different phases or Activity Zones are not shown, but are traceable through the modelling database.

4.2.2 Outputs and deliverables

All processes by definition have an output. Some of these can be called 'deliverables', where the information is in a form (or document) that should be named for easy reference and use in other processes.

The outputs to be named as deliverables may be defined later in the Process Protocol II Project.

The maps only illustrate the level 1 deliverables for simplicity and space purposes. The outputs from all level 2 & 3 processes will be included in the modelling database, which will eventually lead in its inclusion in the process toolkit.

4.3 Process levels

The maps contain three levels that are independent, in that there are no interactions between them. These are defined as follows figure 5 for an illustration:

- Level 1 contains the high level processes and their deliverables as identified in the Process Protocol Map
- Level II contains the sub-processes of the main process at level I (i.e. what the Level I process consists of) and how those sub-processes interact with each other (i.e. how is the Level I process undertaken)
- Level III contains the sub-processes of the processes at level II (what the Level II process consists of) and how those sub-processes interact with each other (how is the Level II process undertaken)

Other attributes related to the process levels include:
- The three levels are separated by a black line
- A single line connects a process at one level with its group of sub-processes at the level below to denote decomposition as shown in figure 5.
- Processes can have a logical dependency within a level and this is shown by an arrow as illustrated in figure 6.
- The participation in a process is shown in the table at the bottom of the process symbol and where such participation does not occur the respective table cell will be knocked back i.e. faded in relation to the other cells.

5 CONCLUSION

As mentioned earlier in this paper, the development of sub process maps for the Process Protocol activity zones was a primary deliverable of the Process Protocol Level 2 project. The aim was to produce sub process maps that would convey detailed design and construction process information to all the participants in the process. Once the research had defined the content of the maps it was then necessary to develop an infrastructure to support the content. Therefore a process map template and set of model-

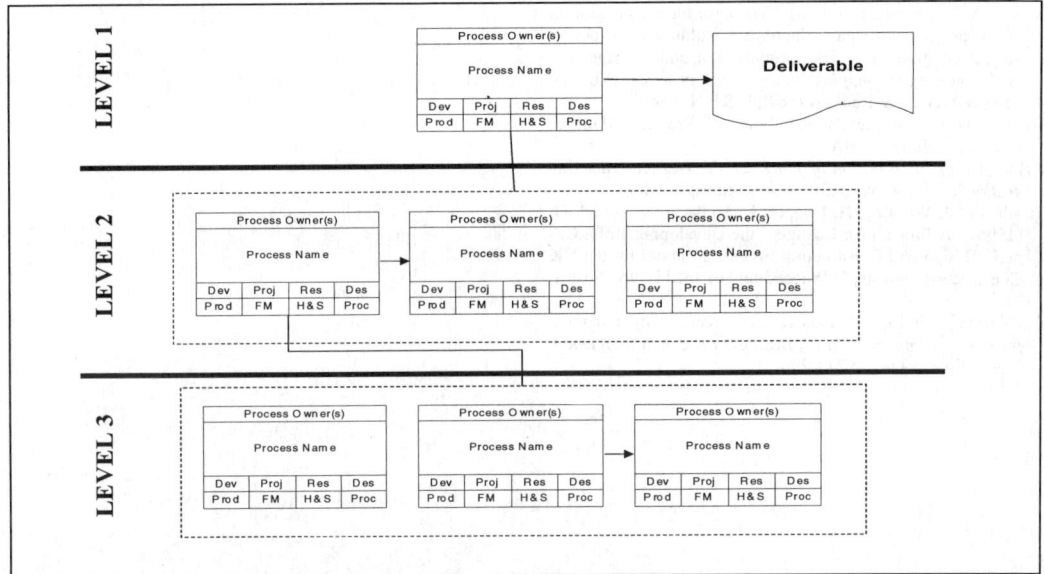

Figure 5 Three Levels of Process Decomposition

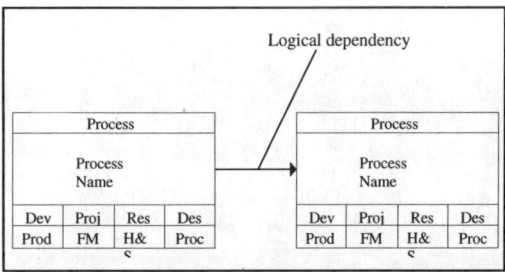

Figure 6 Illustrating Logical Dependency

ling rules were developed to effectively support the presentation of the processes. The eight activity zones thus have an effective medium of representation either holistically or separately.

6 ACKNOWLEDGEMENTS

This research is funded by the EPSRC through the Innovative Manufacturing Initiative. Much of the research underpinned this paper has been conducted by the Process Protocol research team.

REFERENCES

Ball, M. (1988) *"Rebuilding Construction: Economic Change in the British Construction Industry"*. T. J. Press, Padstow.

Banwell, H. (1964) "Report of the Committee on the Placing and Management of Contracts for Building and Civil Engineering Works". H. M. S. O.

British Property Federation. (1983) *"Manual of the BPF System for Building Design and Construction"*. British Property Federation, London.

Champy, J. (1995) "Reengineering Management: The Mandate for New Leadership". Harper Collins, London.

Cooper, R. G. (1990) "Stage-Gate System: A new Tool for Managing New Products". Business Horizons, May-June, pp. 44 – 54.

Cooper, R, Kagioglou, M, ,Aouad, G, Hinks, J, Sheath, D (1998), "The development of a generic design and construction process". European Conference, Product Data Technology (PDT) Days 1998, Building Research Establishment, March 1998, Watford, UK.

Davenport, T. (1993) *"Process Innovation: Re-engineering Work through Information Technology"*. Harvard Business School Press, Boston, MA.

Egan, J. (1998) *"Rethinking Construction"*. Report form the Construction Task Force, Department of the Environment, Transport and Regions, UK.

Hammer, M. & Champy, J. (1997) 5th edition. *"Reengineering the Corporation: A Manifesto for Business Revolution"*. Nicholas Brealey, London.

Kagioglou, M. (1999) *"Adapting Manufacturing Project Processes into Construction: A Methodology"*. Unpublished PhD thesis, University of Salford.

Kagioglou, M., Cooper, R., Aouad, G., Hinks, J., Sexton, M. & Sheath, D. (1998) "Final Report: Generic Design and Construction Process Protocol". The University of Salford, Salford.

Kagioglou, M. Cooper, R. Aouad, G. Hinks, J. Sexton, M. and Sheath D (1998), *"A Generic Guide to the Design and Construction Process Protocol"*, University of Salford

Latham, M. (1994). "Constructing the Team". H.M.S.O.

191

Luck R. & Newcombe R. (1996). "The case for the integration of the project participants' activities within a construction project environment". The Organization and Management of Cosntruction: *Shaping theory and practice* (Vol 2); Langford, D. A. & Retik, A.(eds); E.&F.N. Spon.

RIBA (1991), "Architect's Handbook of Practice Management". 4th edition. RIBA.

RIBA. (1997), *"RIBA Plan of Work for the Design Team Operation"*. (4th edition). RIBA Publications, London.

Sheath, D. M, Woolley, H, Cooper, R, Hinks, J. & Aouad, G. (1996), "A Process for Change – the Development of a Generic Design and Construction Process Protocol for the UK Construction Industry". Proceedings of InCIT '96, Australia.

Yen Cheung (1998), *"Process analysis techniques and tools for business improvements"*, *Business Process Management* Vol. 4 No.4. 1998, pp.274-290

Product and Process Modelling in Building and Construction, Gonçalves, Steiger-Garção & Scherer (eds)
© 2000 Balkema, Rotterdam, ISBN 90 5809 179 1

Process protocol toolkit, an IT solution for process protocol

S.Wu & G.Aouad
School of Construction and Property Management, University of Salford, UK

A.Lee & R.Cooper
Institute for Social Research, University of Salford, UK

ABSTRACT: This paper describes the formation of the Generic Design and Construction Process Protocol (GDCPP), a framework that aims to help construction companies improve their business. Furthermore, industry interest and acceptance has led to further research to develop the generic sub processes and as such, the need for an IT solution emerged to combat the obstacles and problems in implementing the framework. The Process Protocol Toolkit will be composed of a Process Map Creation Tool and a Process Information Management Tool, and their system architecture will be illustrated in this article. The first tool will automate the process map creation process by using the Process Protocol as a framework and allow users to create and customise their specific project process map, and the second aims to manage the process and project information.

1 INTRODUCTION

The business reality of today dictates that if organisations wish to remain successful, then change is endemic. According to Davenport (1993), intense competition and other business pressures on large organisations subsided the quality initiatives, usually in the form of incremental process improvement, that were introduced to improve business performance. Therefore, to achieve substantial business improvement a powerful new tool is required to facilitate the fundamental redesign of work (Davenport, 1993). One such approach is Business Process Re-engineering (BPR; sometimes called process innovation, business process redesign, business engineering or process engineering), "...the fundamental rethinking and radical redesign of business processes to achieve dramatic improvements in critical contemporary measures of performance, such as cost, quality, service and speed" (Hammer & Champy, 1997). Moreover, information technology also plays an increasingly important role in business and subsequently, many software tools have been developed that assist BPR. The purpose of these tools includes project management, co-ordination, modelling, business process analysis, human resource analysis and system development (Klein, 1994).

The UK construction industry is no exception to the need for change as it has been continuously criticised for its less than optimal performance by many government and institutional reports. Most of the reports conclude, time and time again, that the fragmented nature of the industry, lack of co-ordination and communication between the parties, the informal and unstructured learning process, lack of research and development, adversarial contractual relationships and lack of customer focus is what inhibits the industry's performance. In an attempt to improve the situation, Latham (1994) reported that process improvement was one such method and Egan (1998) recommended viewing the construction as a manufacturing process by adopting the techniques developed by the manufacturing industry. Therefore, the Generic Design and Construction Process Protocol project, which is funded by the Engineering and Physical Sciences Research Council (EPSRC) under the Innovative Manufacturing Initiative (IMI), was formed following such recommendations.

The Process Protocol is "a common set of definitions, documentation and procedures that will provide the basics to allow a wide range of organisations involved in a construction project to work together seamlessly", and aims "to map the entire project process from the client's recognition of a need to operations and maintenance" (Kagioglou et al, 1998). The protocol takes the form of a framework detailing the generic design and construction processes within a construction project. The idea was for construction firms to take the map and to use it as a framework to help them to improve their business and through industry interest and acceptance, further funding has been committed to continue the research. It was envisaged that the generic protocol would not be an ad hoc activity, but an ongoing and planned one. Therefore, the framework should not be so prescriptive as to restrict or stifle creativity but be easily adapted and tailored to suit the individual project. This brings the generic protocol down to a secondary-level (Level 2) or product-specific level, which itself can be broken down further to more detailed levels (see Figure 1). The Process Protocol Level II project subsequently aimed to identify such sub processes, however, implementation of the framework also highlighted some significant issues:

- Companies might only adopt part of the Process Protocol model, depending on the nature of the project.
- Some companies have their own working process and are not willing or able to accommodate a new approach.
- The individuals who are responsible for the process modelling and management of a project need detailed knowledge of the Process Protocol.
- Companies need to access any different sources of information and share the information throughout the design and construction process.
- The integration of the process into the information system of the organisation needs to be addressed.
- The opportunities presented by Internet technology for organisations to improve the performance and more effectively reach the parties involved in the project is now being used and the Process Protocol needs to adapt to the technology.

In order to resolve these and other issues, an IT solution, the Process Protocol Toolkit, is also being developed under the Process Protocol Level II project. The Toolkit aims to assist the creation of the process model and to manage the process information of the project based on the Process Protocol framework, and will be discussed in greater detail later in this paper.

2 TOOLS FOR PROCESS MODELLING

As highlighted in the previous sections, increasing competition has forced companies to seek new ways of improving their capabilities, to position themselves for survival in the future. Many companies have adopted a process-oriented view of their business operation, replacing the traditional functional viewpoint to achieve a better integration of operation (Hammer & Champy, 1993). Therefore, software tools to assist such approach have appeared in the software market. These tools can be categorised into two major types, paper based diagramming tools and software enabled analysis tools.

Paper based diagramming tools primarily offer the integration of diagrams and illustrations, together with a wide variety of other features and abilities. Moreover, some of these programs have evolved to become much more than a way to place symbols on a screen for eventual printing or placement in a presentation or report. The most advanced diagramming products are becoming an environment for creating applications (Olsen and Simon, 1997). Most of the tools provide drawing support with templates or shapes, which can be customised to suit individual requirements. The industry standard modelling languages, such as IDEF (**I**ntegrated Computer Aided Manufacturing **Defi**nition), Data Flow Diagram, Entity Relationship Diagram, have been incorporated into these products. The following packages are the major products on the market.

- FlowCharter
- Flow Charting PDQ
- SmartDraw
- Visio

Software enabled analysis tools are more commonly called BPR tools or CASE (Computer Aided Software Engineering) tools and usually encompassed built-in event simulator, static analysis, dynamic modelling and standard database support. These tools are able to produce a descriptive model that attempts to represent the business "as is" or "as to be". Such model can be composed of a number of process definitions including goals, business rules, actions and resource requirements, and expresses the flow of activity between the processes with a combination of diagrams, text and performance measures.

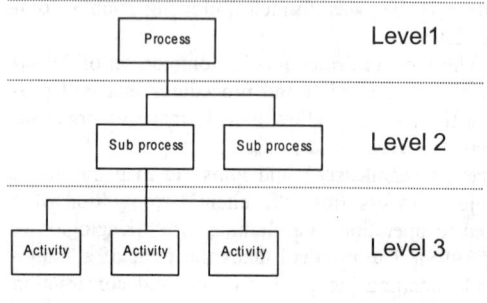

Figure 1 Process Levels

194

Typically, the business model is built using a process modelling (built-in) tool, and they then may simulate the running of the process. However, most tools focus on IDEF methodology and several are based on the Data Flow or Entity Relationship Diagram. Although these process tools provide powerful functions, they can not be effectively used as an IT support for the Process Protocol, because the aim of the toolkit is to help the industry implement the Process Protocol and not to analyse the construction process. In addition, the Process Protocol has its own process modelling methodology which was developed with the industry to facilitate their own simple requirements, though this is not discussed in the extent of this paper. All of the intelligent tools only support standard accepted modelling methodologies, like IDEF, data flow diagram and therefore, the Process Protocol Toolkit needs to be developed to fulfil the role in the project.

3 PROCESS PROTOCOL TOOLKIT

The aim of the Process Protocol Toolkit is to provide an IT solution to address the problems raised throughout the lifecycle of a construction project. As highlighted earlier, it is important and useful to establish the data model of the Process Protocol model before starting any IT development.

3.1 *Process Protocol Map Elements*

The Process Protocol map consists of the following major elements:

1. Process

A set of activities undertaken by multifunctional team is to produce information for other processes or deliverables. For example, 'establish need for project'.

2. Deliverable

Deliverables represent documented project and process information, such as Stakeholder List, Statement of need, project brief, etc.

3. Phase

There are 10 Phases (refer to Table-1) that have been defined in the Process Protocol map to represent the different stage of the whole lifecycle of a construction project.

4. Activity Zone

Nine activity zones in the Process Protocol Map represent the different group of participants involved in a construction project.

5. Phase Review

Phase Reviews are to assure a high quality of work performance by the multifunctional teams at

Table-1 Phases

Phase	Name
Phase 0	Demonstrating the Need
Phase 1	Conception of Need
Phase 2	Outline Feasibility
Phase 3	Substantive Feasibility Study & Outline Financial Authority
Phase 4	Outline Conceptual Design
Phase 5	Full Conceptual Design
Phase 6	Co-ordinated Design, Procurement & Full Financial Authority
Phase 7	Production Management
Phase 8	Construction
Phase 9	Operation & Maintenance

each phase of the project. It will be conducted by a multifunctional senior management group and representatives of the project team. The work, as in the form of deliverables as described in the Process Protocol will be assessed in the Phase Review meeting. The Phase Review report will include key deliverables for the appropriate phase as identified by the project process map.

3.2 *Process Protocol Mapping rules*

Modelling rules for the development of the sub process model are as follows:

1. Process Level

Level I contains the high level processes and their deliverables as identified in the Process Protocol map. Level II contains the sub processes of the original processes at Level I and illustrates any potential relationships between the processes. Level III contains the sub processes of the Level 2 processes and again illustrates any potential relationships between the processes.

2. Deliverable

Each process has an output, and only the output of Level I process output will be shown as 'deliverable'.

3. Logical Relationship

Two types logical relationships were used in the process mapping. The first is process decomposition, in that the process in the high Level protocol can be composed of several processes in a lower Level, as illustrated in the diagram below (Figure 2).

The other type of relationship is logical dependency, which means that one Process is dependent on the completion of another process. (Figure 3)

3.3 *Data model of the Process Protocol map*

The methodology for modelling the data model of Process Protocol map is ERD (Entity Relationship

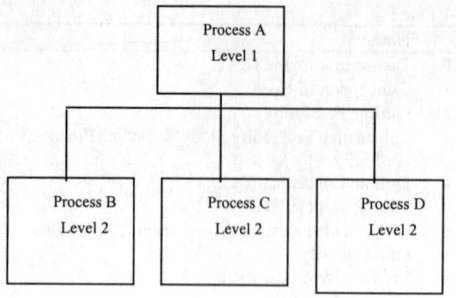

Figure 2. Process A is composed of Process B, Process C and Process D

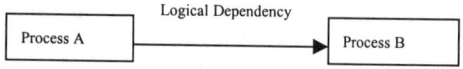

Figure 3. Process B is dependent on Process A

Diagram), which was introduced in the 1970's by Peter Chen to model the design of a relational database from a more abstract perspective. (Chen 1976).

An Entity relationship diagram (ER diagram) uses three major abstractions to describe the data. They are:

- Entities, which are distinct and major elements in the business; i.e. map element 'activity zone'.
- Relationships, which are meaningful interactions between the entities; i.e. entity 'activity zone'. and entity 'process', the relationship between is 'One activity zone has one or more processes.'

- Attributes, which are the properties of the entities and relationships, i.e. name, description of entity 'activity zone'.

The entity relationship diagram in figure 4 represents how the information associated with the Process Protocol map will be stored. This ER model will be finally turned into a relational database for the system implementation by mapping the entities and relationships as database tables.

3.4 *Process Protocol toolkit system architecture*

The aim of the Process Protocol Toolkit is to use information technology to resolve the issues raised during the implementation in the industry. It is suggested that the Process Protocol Toolkit should potentially be composed of two major components, the Process Protocol Map Creation Tool and Process Protocol Management Tool. These will be described in greater detail in the following sections.

3.4.1 *Process Protocol map creation tool*

The Process Protocol map creation tool enables the production of a project process map based on the generic Process Protocol framework. There are three major elements in this tool. They are main creation tool, generic processes data store and project process data store. (see Figure 5)

The main creation tool provides the functions for data retrieval, map creation and map customisation. Users will be able to define their processes, and create the project process map by referring to the generic processes provided by Process Protocol. Furthermore, an export facility will be included in this part of the system. It will enable to present the proc-

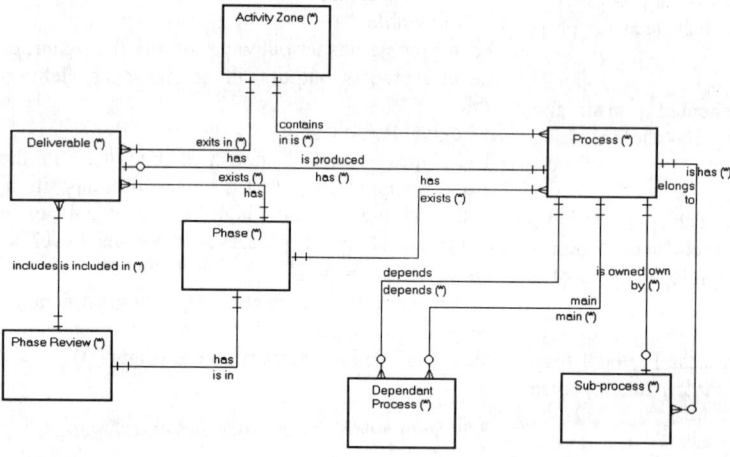

Figure 4 Entity Relationship Diagram

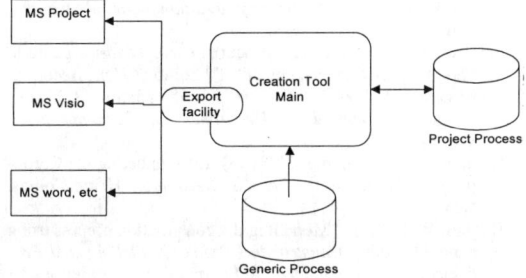

Figure 5 System architecture of Process map creation tool

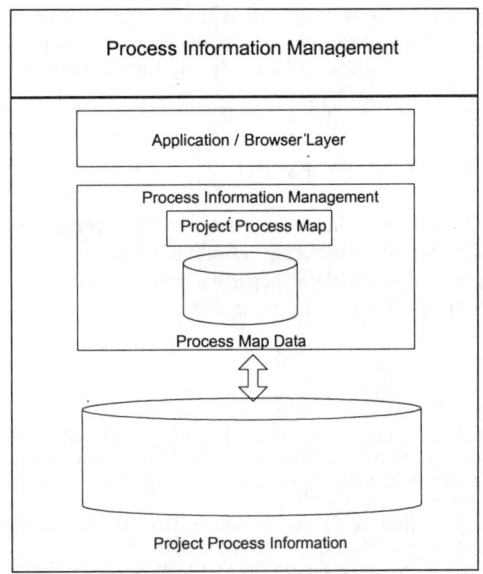

Figure 6 System architecture of process management tool

ess information in different format, such as in MS Word, Excel and Visio.

Generic processes data store is a relational database to store the information associated with generic the Process Protocol map. The Process Protocol map data model established in this paper will also be incorporated into the database structure.

Project process data store uses the same database structure as the generic processes data store, but it only store the process information related to the specific project. The information can be presented in different ways, such as process map in Visio, process report in Word or process matrix in Excel.

3.4.2 *Process management tool*

The Process Management tool aims to control and deliver the project information according to the design and construction process and use the project process map created in the Process Protocol creation tool as a gateway.

The proposed system architecture is presented in Figure 6. It is composed of three major components, they are the Application / Browser, the Process Information Management Layer and the Process Information and Document Data Store.

The Application / Browser component can be built on the web. It will provide an interface for the user to view process maps, documents, drawings and other project information.

The Process Information Management layer includes project process map and map database created in Process Map creation tool. The information in the process map database will be integrated with the Process Information and Document Data Store to enable the user to retrieve the information and document based on the project process map.

The Process Information and Document Data Store is an information repository to hold all the information of the project, including the document, drawings, program information, cost data, etc. A product called 'Information Channel' developed by the Building Information Warehouse, Ltd., which provides very similar functions that are required in

the Process Information and Document Data Store, may be adopted in the implementation.

4 CONCLUSIONS

This paper has highlighted the growing need for change in the UK construction industry, which brought about the formation of the Process Protocol. The Process Protocol aims to be a framework for the industry to adopt in order to improve their business. Construction projects today are usually delivered in a variety of ways, ranging from chaotic to systematic, and there is no industrial standard method of operation. While successful results are the obvious payoffs, it is unwise to constantly rely on luck to salvage the organisational procedure and so such a more consistent process is needed to co-ordinate and control the project team and the project work. This type of generic process approach also should engender more collaboration in a traditionally adversarial industry. However in order to facilitate adoption and implementation it is recognised that an IT tool is needed to facilitate such an adoption enabling the various stakeholders to "buy into" the process.

The Process Protocol Toolkit will be composed of a Process Creation Tool and a Process Management Tool. The Creation Tool will automate the customisation of the original protocol for each individual project, and the management tool will assist in the control and communication of the project information and allow its lessons to inform future projects.

Therefore it is envisaged that the Toolkits, in addition to the protocol, will collectively enable companies to address the problems facing the construction industry today.

5 ACKNOWLEDGEMENT

This research is funded by the EPSRC through the Innovative Manufacturing Initiative. Much of the research underpinned this paper has been conducted by the Process Protocol research team.

REFERENCES

Aouad, G. Kagioglou, M. Cooper, R. Hinks, J and Sextion, M (1999) "Technology management of IT in construction: a driver or an enabler?" *Logistics Information Management,* Vol. 12, No. ½ , pp. 130-137

Aouad,G. Hinks, J. Cooper, R. Sheath, D.M. Kagioglou, M. and Sexton, M. (1998) " An Information Technology (IT) Map for a Generic Design and Construction Process Protocol", *Journal of Construction Procurement* November 1998. Vol 4, No 1, pp. 132-151

Aouad.G, Brandon.P, Brown.F, Child.T, Cooper.G, Ford.S, Kirkham.J, Oxman.R, Young.B. The conceptual modelling of construction management information. *Automation in Construction,* 3, 1995, pp 267-282.

Aouad.G, Kirkham.J, Brandon.P, Brown.F, Cooper.G, Ford.S, Oxman.R, Sarshar.M, Young.B. "Information modelling in the construction industry - The information engineering approach". Published by *Construction Management and Economics,* vol 11, No 5, pp 384-397, 1993.

Beyond the Basics of Reengineering: Survival Tactics for the '90s (1992), Quality Resources / The Kraus Organisation, white Plains, New York, NY

Bradley, P. Browne, J. Jackson, S. and Jagdev, H. (1995), "Business process reengineering (BPR) - a study of the software tools currently available", *Computers in Industry"* Vol. 25, pp.309-330

Brandon, P.S. and Betts, M.(1995), *Integrated Construction Information,* E&FN Spon, London.

British Property Federation. (1983). "Manual of the BPF System for Building Design and Construction," *British Property Federation,* London.

Brown, A (1996a). "Construction Modelling and Methodologies for intelligent information integration" *University of Salford*

Chen, P. (1997) *"The entity – relationship model – Toward a unified view of data",* ACM Transactions on Database Systems 1

Davenport, T H (1994), "Reengineering: business change of mythic proportions?", *MIS Quarterly,* pp.121-127

Davenport, T H. (1993) *"Process Innovation – Reengineering Work through Information Technology"* Harvard Business School Press.

Earl, M J, Sampler, J L, and short, J E (1995), " Stratgeies for business processes reengineering: evidence form field studies", *Journal of Management Information Systems,* Vol.12 No.1, pp.31-56

Edwards, C. and Peppard, J W (1994), "Business process redesign: hype, hope or hypocrisy?", Journal of Information Technology, Vol. 9, pp.251-266

Egan, J (1998) . "Rethinking Construction". DETR

Fortier, P J (1997). *"Database Systems handbook",* McGraw – Hill

Franks, J. (1990). "Building Procurement Systems: a guide to building project management". 2nd Edition CIOB, Ascot.

Galhenage, G.P (1996). " Comparison of traditional engineering and CE approach". *Department of Computer Engineering.*

Hammer, M. and Champy, J.(1993), Reengineering the Corporation: *A Manifesto for Business Revolution,* HarperCollins, New York, NY.

Hughes, W. (1991). " Modelling the construction process using plans of work". *Construction Project Modelling and Productivity* - Proceedings of an International Conference CIB W65, Dubrovnick, 1991

Kagioglou, M. Cooper, R. Aouad, G. Hinks, J. Sexton, M. and Sheath D (1998*) "Final Report: Generic Design and Construction Process Protocol",* University of Salford

Kagioglou, M. Cooper, R. Aouad, G. Hinks, J. Sexton, M. and Sheath D (1998), *"A Generic Guide to the Design and Construction Process Protocol",* University of Salford

Klein, M M (1994), "Reengineering methodologies and tools", *Information System Management,* Spring, pp.30-35

KPMG & CICA (1993), *Building on IT for Quality,* London

Latham, M. (1994). "Constructing the Team". H.M.S.O.

Lockey, S R. et al. (1994). "The Combine data exchange system". *Proceedings form the first ECPPM Conference,* Dresden.

Luck R. & Newcombe R. (1996). "The case for the integration of the project participants' activities within a construction project environment". The Organization and Management of Cosntruction: *Shaping theory and practice* (Vol 2); Langford, D. A. & Retik, A.(eds); E.&F.N. Spon.

Luck, R. McGeorge, D. and Betts. M. (1997). *"Reaserach Futures: Academic Responses to Industry Challenges",* Construct IT Centre of Excellence

Makey, P. (1995). "Business Process reengineering strategies, Methods and tools". *Buttler group*

Masterman, J W E. (1992). "An Introduction to Building Procurement Systems" *E& F.N. Spon;* London

Ould, M A (1995). *Business Processes – Modelling and Analysis for Reengineering and Improvement,* Wiley & Sons, New York, NY

Ritchie, B. Marshall, D and Eardley A. (1998), *"Information System in Business",* International Thomson Business Press

Sheath, D M, Wolley, H, Cooper, R, Hinks, J & Aouad, G (1996). "A process for change – the development of a generic design and construction process protocol for the UK construction industry". *Proceedings of inCIT'96,* Sydney, Australia.

Wright, D T. and Burns, N D (1996), "Guide to using the WWW to survey BPR research, practitioners and tools", *IEE Engineering Management Journal,* Vol.6 No.5, October, pp.211-216

Yen Cheung (1998), "Process analysis techniques and tools for business improvements", *Business Process Management* Vol. 4 No.4. 1998, pp.274-290

Yu, B and Wright, D T, (1997), " Software tools supporting business process analysis and modelling" *Business Process Management* Vol. 3, No.2, 1997, pp.133-150

Towards a unified specification of the construction process information: The PSL approach

A. F. Cutting-Decelle
University of Savoie/ESIGEC/LGCH, Chambéry, France

C. J. Anumba, A. N. Baldwin & N. M. Bouchlaghem
University of Loughborough, UK

M. Gruninger
University of Toronto, Ont., Canada

ABSTRACT: As the use of IT in manufacturing or construction has matured, the capability of software applications to interoperate has become increasingly important. Standards-based translation mechanisms have simplified integration by requiring only a single translator. This challenge is especially apparent for process information, used by many software applications, each in a different way. The primary difficulty is that they sometimes associate different meanings with the terms : both their semantics and their syntax need to be considered when translating to a neutral standard.
The Process Specification Language (PSL) creates a neutral, standard language for process specification to serve as an interlingua to integrate multiple process-related applications. This interchange language is unique due to the formal semantic definitions (the ontology) that underlie the language.
The aim of this paper is to analyse the « *applicability* » of PSL to the construction sector through the example of a generic construction process.

1 INTRODUCTION

In all types of communication, the ability to share information is often hindered because the meaning of the information can be drastically affected by the context in which it is viewed and interpreted (Cutting1 2000). This is particularly true in construction, since, in addition to the complexity of information and problems of information exchange, there are additional problems due to the number of the actors involved in the construction process and the diversity of the information handled during the building's life cycle (design, construction, operation of the building). This results, at the engineering design level, in a poor understanding of the contents of the messages between the actors and with the site.

At the same time, there is an increasing need for improvement of the conventional design and construction process in the construction industry, mainly related to the poor performance commonly associated with building projects (Cooper 1998).

Recently, there has been an increased need for the development of process modelling concepts, particularly in terms of how *Information Technologies* (IT) are used and could be used to support the overall life-cycle process (Karhu 1999). A generic process modelling method would be suitable for describing the building construction process from the different points of view whose integration defines the whole project.

A first stage in the integration of the construction information lies in the development of a common language enabling all the actors of the construction process to share the same semantic concepts intrinsic to the capture and exchange of information used in the process.

Motivated by this growing need to share information process in the manufacturing environment, the PSL project is aimed at providing a generic *Process Specification Language,* as a language focused on the description of process, building on existing modelling methods. Originally developed for a manufacturing environment, and given the similarities between this sector and the construction sector, there is a need for testing the applicability of the PSL concepts to the representation of the construction process.

2 TOWARDS A GENERIC MODEL OF THE CONSTRUCTION PROCESS

2.1 *The concept of "process"*

The concept of process lacks a commonly agreed definition. A typical definition is : « *a set of partially ordered steps intended to reach a goal* » (Humphrey 1992, referred in : Koskela 1995).

There are four common perspectives to processes (Curtis 1992) :
- Functional : representing what process elements are being performed, and what flows connect these elements ;
- Behavioral : representing when process elements are performed, and how they are performed through feedback loops, iteration, decision making conditions, etc. ;
- Organisational : where and by whom process elements are performed ;
- Informational : a perspective of the informational entities produced or manipulated by the process.

In the functional view, processes consist of activities, that together achieve the purported goal. In addition, auxiliary concepts such as artifacts (products of activities) can be used for process representation.
In a behavioral perspective, processes may consist of precedence relations or information and material/ information flows, with the time explicitly represented. Flow process concepts focus on what happens to material and information in timeline.
In an organisational perspective, processes may consist of agents (performing activities) and roles (set of activities assigned to an agent). Also, the process may be viewed as composed of a supplier-customer partnership.
In an informational perspective, processes consist of data, objects, documents, etc.
In principle, these perspectives, when combined, produce a complete model of a process. However, in current practice of process modelling, the functional perspective (as provided by SADT method) often dominates : activity is seen as the basic construct, and this process concept only achieves one goal : « how to obtain the result ».
Of course the answer to this question is sufficient for achieving the process ; however, it does not exhaust all improvement potential. There are two other relevant goals, that should generally be tackled : how not to consume unnecessary resources (Koskela 1995) and How to ensure that the result corresponds to requirements. In order to achieve these goals, contributions from behavioral and organisational perspectives are needed. We will see in this paper that, as a language, PSL is capable of expressing these different points of view, once they have been represented using modelling methods dealing with these multiple views (such as IDEF3, for example).

2.2 *An approach of the construction process*

Information handled during the construction process can be divided into several categories (Björk 1992) :
- First, information must state facts : such as design documents, which are the results of design decisions. Information to be transferred between computing systems in the construction process is mostly of this type. This information has also to define goals and requirements which a particular project must fulfil. The third category of information states rules which restrict facts, but which apply in general and are not tied to a particular project. These three categories of information can be called « facts », « constraints » and « knowledge ». From a programming language point of view, facts can be constructed using assignment statements, requirements are mainly represented by inequality operators (or algorithms) and knowledge through knowledge based systems ;
- The second point provides a semantic approach dividing information into project-specific and more general information. Facts can be both project specific and general. Constraints are mainly project-specific and knowledge is usually general in nature ;
- The third point of view concerns the presentation and categorises the types of documents used to present the information for human interpretation. Some typical presentation formats used in construction are : drawings, schemas, realistic visualisations, written specifications, calculation results, bills of materials, contracts, orders and various tendering documents.

We will here limit our study to project-specific information, focusing on the semantics of the information. The reason of this choice comes from our primary concern to study information management within construction projects. The information to be communicated to other parties in the construction process mostly consists of factual information. Clearly constraints are very important in the early briefing stages of projects and in quality assurance applications. Knowledge mainly resides in application programs and its effect on the actual transfer of data between project participants will need to be examined further.

2.3 *Modelling of the construction process*

Several process models have been developed in the domain of construction, among which the MoPo model (Cooper 1998), covers the whole construction life cycle. Other models mainly focus on the design stage such as the ADePT model described in (Austin 1996). Some process models introduce concurrent engineering features, such as the model presented in (Anumba 1996), or client requirements (Kamara 2000).

It is interesting to make a synthesis of the common features of these models, thus leading to a generic process representation. One of these generic representations of the construction process, as provided by (Björk 1992), consists in three main categories, which are : *activities*, *results*, *resources*. An *activity* uses *resources* to produce *results*. Traditional construction classification systems often tend to equate results to buildings and their parts. This is due to a desire to distribute total construction costs over building parts, which is useful for cost analysis purposes. It is, however, evident that information (mostly delivered as documents) and services are other important sub-types of results.

The schema of the Fig. 1 (adapted from Björk 1992) gives an EXPRESS-G (ISO10303-11 1994) representation of some objects of a generic construction process. These include :

- **Activity** : the kernel of the model, with relationships with most of the other objects of the model : an activity may have relationships with the result it produces, the resources it uses and the agents performing it.
- **Result** : an example of an entity type needed for classification purposes, but which is intermediate in nature since most of the relevant information about results will be defined in the class descriptions of the sub-types of results.
- **Physical _object** : any physical object with shape and location. Both characteristics may be dynamic.

- **Service** : results of activities which are not physical objects or documents (information), such as « guarding the site ».
- **Agent** : any organisation, person, machine, or facility which participates in the activities of the project. An agent performs some activity in the construction process. A fundamental aspect of an agent (distinguishing him from a product) is that he/it has an existence outside the project and usually participates in several projects.
- **Resource use and cost** : each activity in the project demands a number of inputs in the form of resources which are « consumed » or used. The actual use may be measured in manhours, tons, squaremeters, etc. A clear distinction is made between resource use and the resources entity in itself, which may be documents, materials, machines or persons. The use of any resource involves a cost, which in most cases can be measured by the amount of the resource consumed, but in some cases by the opportunity cost of that resource, that is the cost the customer accepts to pay for the resource in question.
- **Management_activity** : a super-type of the different types of management related to a construction project, such as technical management (planning, logistics, QS activity), document management (drawings, bills of quantities, calculation notes) and financial management.

This model of the construction process will provide the basic example on which the PSL concepts will be mapped in this paper.

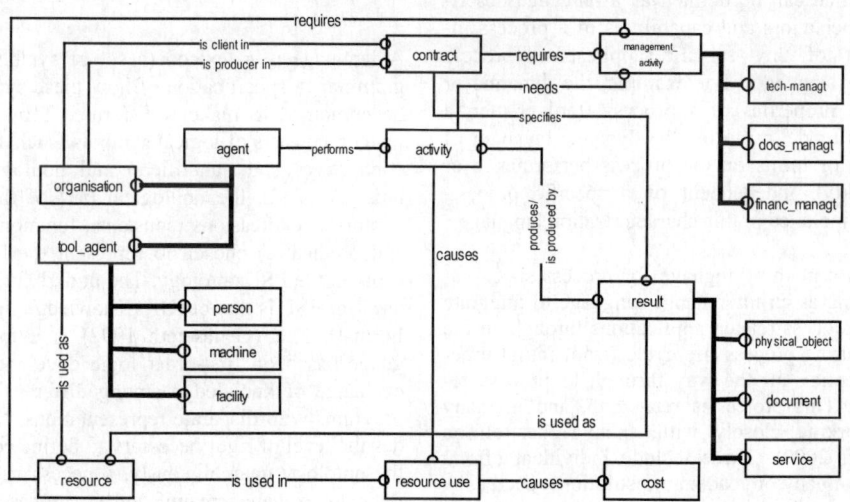

Figure 1 : EXPRESS-G diagram of a generic representation of the construction process (Björk 1992)

3 THE PSL (PROCESS SPECIFICATION) LANGUAGE

The objective of PSL (initially developed by the National Institute of Standards and Technology, NIST, US) is to create a process interchange language that is common to all *manufacturing* applications, generic enough to be decoupled from any given application, and robust enough to be able to represent the necessary process information for any given application. This representation would facilitate communication among the various applications because they would all have a common understanding of concepts to be shared. This language is currently being standardised at the international level, by the ISO TC184/SC4 committee (« *Industrial Data* »), now at the level of Provisional Work Item ISO 18629 (ISO 18629-1 2000).

PSL specifies a language for the representation of process information, limited to the realm of discrete processes related to manufacturing, including all processes in the design/manufacturing life cycle. Business processes and manufacturing engineering processes are included in this work both to ascertain common aspects for process specification and to acknowledge the current and future integration of business and engineering functions.

The goal of the project is to create a process *specification* language, not a process *characterization* language. A process specification language is a language needed to specify a process or a flow of processes, including supporting parameters and settings. This may be done for prescriptive or descriptive purposes. The language is composed of an ontology and one or more presentations.

This is different from a process characterization language, which can be defined as a language describing the behaviors and capabilities of a process independent of any specific application (process *modelling* language). For example, the dynamic or kinematic properties of a process (tool chatter, a numerical model capturing the dynamic behavior of a process or limits on the process performance or applicability), independent of a specific process, would be included in this characterization language.

PSL is a neutral language for process specification serving as an interchange language to integrate multiple process-related applications throughout the manufacturing process life cycle (from initial process conception all the way through to process retirement). This project is related to, and in many cases working closely with, many other efforts (Schlenoff 2000). These include individual efforts (single company or academic institutions) such as A Language for Process Specification (ALPS) Project, the Toronto Virtual Enterprise (TOVE) Project, the Enterprise Ontology Project, and the Core Plan Representation (CPR) Project. In addition, the PSL project is in close collaboration with various projects (involving numerous companies or academic institutions) such as Shared Planning and Activity Representation (SPAR) Project, the Process Interchange Format (PIF) Project, and the WorkFlow Management Coalition (WfMC). Most of these efforts have been taken into consideration in the development of PSL, the language benefitting from the experience gained through the different projects analysed in the study (Knutilla 1998).

The primary component of PSL is an ontology designed to represent the primitive concepts that, according to PSL, are adequate for describing the basic manufacturing, engineering, and business processes. An *ontology* is lexicon of specialized terminology along with some specification of the meaning of terms in the lexicon. Note that the focus of an ontology is not only on terms, but also on their meaning. We can include an arbitrary set of terms in our ontology, but they can only be shared if we agree on their meaning. It is the intended *semantics* of the terms that is being shared, *not simply* the terms.

The challenge is that a framework is needed for making the meaning of the terminology for ontologies explicit. Any intuitions that are implicit are a possible source of ambiguity and confusion. For the PSL ontology, we must provide a rigorous mathematical characterization of process information as well as precise expression of the basic logical properties of that information in the PSL language. In providing the ontology, we therefore specify three notions: language, model theory, proof theory.

3.1 *The language*

A language is a lexicon (a set of symbols) and a grammar (a specification of how these symbols can be combined to make well-formed formulas). The lexicon consists of logical symbols (such as boolean connectives and quantifiers) and nonlogical symbols. For PSL, the nonlogical part of the lexicon comprises expressions (constants, function symbols, and predicates) chosen to represent the basic concepts in the PSL ontology. The underlying grammar used for PSL is that of KIF (Knowledge Interchange Format). KIF (Genesereth 1992) is a formal language based on first-order logic developed for the exchange of knowledge among different computer programs with disparate representations. KIF provides the level of rigor necessary to define concepts in the ontology unambiguously, a necessary characteristic to exchange manufacturing process information using the PSL Ontology.

3.2 Model theory

The model theory of PSL provides a rigorous, abstract mathematical characterization of the semantics, or meaning, of the language of PSL – an abstract representation of the primitive concepts of PSL. This representation is typically a set with some additional structure (e.g., a partial ordering, lattice, or vector space). The model theory then defines meanings for the terminology and a notion of truth for sentences of the language in terms of this model. Given a model theory, the underlying theory of the mathematical structures used in the theory then becomes available as a basis for reasoning about the concepts intended by the terms of the PSL language and their logical relationships, so that the set of models constitutes the formal semantics of the ontology.

3.3 Proof theory

The proof theory of PSL is perhaps its most important component. It consists of three components : PSL Core, one or more foundational theories, and PSL extensions :

- **PSL Core :** PSL Core is based upon a precise, mathematical first-order theory, a formal language, a precise mathematical semantics for the language and a set of axioms that express the semantics in the language. There are four primitive classes, two primitive functions, and three primitive relations in the ontology of PSL Core. The classes are OBJECT, ACTIVITY, ACTIVITY_OCCURRENCE and TIMEPOINT. The four relations are PARTICIPATES-IN, BEFORE, and OCCURRENCE-OF. The two functions are BEGINOF, and ENDOF. ACTIVITIES, ACTIVITY_ OCCURRENCES, TIMEPOINTs (or POINTs for short), and OBJECTs are known collectively as entities, or things. These classes are all pairwise disjointed.
- **Core Theories :** The purpose of PSL Core is to axiomatize a set of intuitive semantic primitives that is adequate for describing basic processes. Consequently, its characterization of them does not make many assumptions about their nature beyond what is needed for describing those processes. The advantage of this is that the account of processes implicit in PSL core is relatively straightforward and uncontroversial. However, a corresponding liability is that the Core is rather weak in terms of pure logical strength. In particular, the theory is not strong enough to provide definitions of the many auxiliary notions that become needed to describe an increasingly broader range of processes in increasingly finer detail. For this reason, PSL includes one or more *core theories*. A core theory is a theory that axiomatizes new primitive concepts not found in PSL-Core, but which are needed to provide rigorous semantics for other terms in PSL.

- **Extensions :** The final component of PSL consists of PSL *extensions*. Roughly speaking, a PSL extension gives one the resources to express information involving concepts that are not part of PSL core. Extensions give PSL a clean, modular character. PSL core is a relatively simple theory that is adequate for expressing a wide range of basic processes. However, more complex processes require expressive resources that exceed those of PSL core. Rather than clutter the PSL core itself with every conceivable concept that might prove useful in describing one process or another, a variety of separate, modular extensions have been (and continue to be) developed that can be added to PSL core as needed. In this way a user can tailor PSL precisely to suit his or her expressive needs. To define an extension, new constants and/or predicates are added to the basic PSL language, and, for each new linguistic item, one or more axioms are given that constrain its interpretation. In this way one provides a « semantics » for the new linguistic items. A good example of such an extension is the theory of timedurations below. The PSL core itself does not provide the resources to express information about timedurations. However, in many contexts, such a notion might be useful or even essential. Consequently, a theory of timedurations has been developed which can be added as to PSL core, thus providing the user with the desired expressive power.

3.4 Current structure of the language

To date, the language is built on the following two categories of extensions :
- PSL core and outer-core (small set of extensions that are so generic and pervasive that they have been put apart), introducing primitive concepts of the language : core, activity occurrences, atomic activities, complex activities, occurrence trees, activity performing, subactivity occurrence ordering, integer and duration, resource requirements theory, resource sets ;
- PSL extensions, introducing new definitions : ordering relations (complex sequences), nondeterministic activities, ordering relations over activities, junctions, duration, reasoning about states, interval activities, states, temporal ordering relations, reasoning about resource divisibility, resource roles, reasoning about resource usage, capacity-based concurrency, substitutable resources, fixed resource sets, homogeneous resource set, inventory resource sets, resource pools, resource set-based actions, resource paths, processor actions.

All the extensions are written using the KIF syntax, under the form of basic axioms and related definitions, in the following way: (excerpt from Occurrence trees extension : occ_tree.th)

Occurrences form a tree.

```
(forall (?a1 ?a2 ?occ1 ?occ2)
    (=> (=    (successor ?a1 ?occ1)
              (successor ?a2 ?occ2))
        (and    (= ?a1 ?a2)
                (= ?occ1 ?occ2))))
```

4 APPLICATION OF PSL IN CONSTRUCTION

To date, PSL is mainly used in the manufacturing environment. One of the aims of our research work is to provide elements of convergence between the two sectors, manufacturing and construction, through a common representation of process information.

4.1 Method of work

The application of PSL to the construction sector requires several stages described here on the basis of the example in the Fig. 1. These stages are not necessarily sequential (Cutting2 2000).

- *Stage 1 : Identification of extensions relevant to the processes to be represented in PSL* : since the beginning, the development of extensions·has proceeded on an « as-needed » basis ; the initial PSL ontology was developed using a single scenario, the EDAPS (Electromechanical Design and Planning System) scenario developed by Steve Smith at the University of Maryland (CIM). The concepts introduced were defined and modelled within PSL and later extended as other scenarios were explored, such as the pilot implementation in the domain of scheduling between ProCAPP (KBSI) and the Scheduler 4.3 software (ILOG). If there is not any extension corresponding to the concepts to be expressed with PSL, it may be necessary, either to adapt the process model and/or its representation, or else to develop new extension(s).

- *Stage 2 : Elaboration of a synthesis model of the construction process* : generic enough to encompass the different stages of the construction and the diversity of the actors involved, but precise enough to provide elements of information suited to the interoperability among categories of software corresponding to different stages of the building life cycle and usually considered as incompatible in terms of data representation.

- *Stage 3 : Expression of the synthesis model using a formalism already identified as such within PSL* : this stage is not mandatory, however the work of translating a process model into PSL is largely simplified. This is the case when using the IDEF3 process representation (Mayer 1995), since several PSL extensions make use of elements coming from this method. However, a research work conducted by (Cioccoiu 1998) provides the possibility of a nearly direct translation from IDEF3 to PSL.

- *Stage 4 : Translation of the concepts expressed in the process model into the PSL process language.*

Applied to the example in the Fig. 1, considered as a first approach of a generic process model, the IDEF3 representation of the process model shows the following features, depicted in the Fig. 2 below. This paper presents only the early stages of the work.

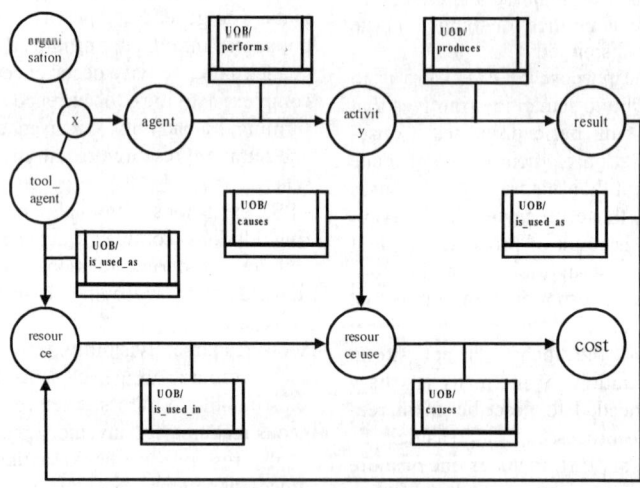

Figure 2 : Enhanced transition schematic representation of a part of the Fig. 1 using IDEF3

Note : in order not to get a too complex schema, all the elements in the Fig. 1 have not been represented in the Fig. 2.

Then, the following stage will consist of expressing the concepts of the schema into the IDEF3 language, corresponding to the graphical representation : each item of the graphics can be expressed in the IDEF3 language, such as : UOB (Unit of behaviour) and UOB-use declarations, processes, links, junctions.

The PSL representation of an IDEF3 schematic is a set of KIF sentences that define a PSL theory. The translation process can be described by a set of meta-theoretic *compilation rules* that associate KIF sentences with the IDEF3 constructs. Writing such compilation rules can also be seen as providing a formal, declarative semantics into PSL to IDEF3 constructs.

Once the compilation rules are written, implementing the translator becomes possible, using for example the lisp macros provided by the compilation. It can also be possible to provide the compiler with a KIF expression simplifier, leading to a simple syntax using typeless quantifiers and standard operators.

4.2 *Benefits of the PSL approach in construction*

- An important feature of this approach is related to *the role of PSL as « interlingua »* : through the ontology on which the language is built, it becomes possible to find, for as many concepts as mentioned in the extensions, a common generic representation, thus enabling exchanges of information among software applications traditionally non-interoperable.

The PSL approach thus contributes to the definition of interoperable process models, providing rules enabling this interoperability. Then, the integration of these rules may lead to a general methodology of development of synthesis process models.

This point is very important, since many process models have been developed, some of them are already implemented in software tools, but none of them are, to date, able to directly exchange process-related information.

- It is also interesting *to compare the PSL approach with other synthesis approaches* : here we will base our analysis on the example of the GEPM (generic process model) provided by the MoPo project, since it is one of the more developed to date.

First of all, the final objective is not the same : « the MoPo project aims at providing new methods, reference process models and (IT) tools to support construction process improvement through system-

atic process analysis and design/planning using construction modelling approach » (Karstila 1999). The expected results of the project are, among others, to develop a « generic construction process information model capturing information requirements for construction process modelling, reference construction process models to be re-used in e.g. company specific modelling efforts for process descriptions ».

In this kind of project, the main efforts concern the development of (synthesis) process models.

However, while MoPo is aimed at providing process *modelling tools*, an initiative such as PSL targets the *litteral expression* to be given to these models (with the final objective of computer-based exchanges among different categories of software, with different and most of the time incompatible data representations).

In other words, it is also possible to say that there is no concurrency in their use : PSL, as other process *specification languages* will act downstream with respect to modelling methods/tools such as the MoPo initiative.

Besides, the use of generic process modelling tools can be considered as a mandatory step of a process representation approach using PSL, since the genericity of these tools contributes to guarantee the validity of the results provided by PSL !

The added value of the PSL approach comes from its « *interlingua* » feature, however, it also comes from the synthesis models on top of which the language is built !

5 PERSPECTIVES OF THE WORK

One of the stages of the on-going work is to make a synthesis of the different approaches in terms of process modelling currently available, these include : the IRMA (Information Reference model for AEC) initiative (Froese 1993), the Unified approach model (Björk 1992), Process Protocol II (PPII 1999), ADePT methodology and tools (Austin 2000), MoPo project (Karstila 1999), DFD model of a construction company operation (Fisher 1992), COSMOS and STAR project (Hannus 1999), Model of construction process information (Froese 1994), various papers dealing with process models, *Organisation of information about construction works, Part 2 : Framework for classification of information, ISO DIS 12006-2, 1999.*

The following stage consists of making a synthesis of these models, in order to create a generic process model applicable to several stages of the life cycle of a building (design, construction, operation, maintenance, decommissioning, facility management), identify the milestones of process models for construction (mandatory features, common elements, etc.) and identify the commonalities/differences with manufacturing industry : comparison of requirements of manufacturing/construction industries.

Then, we plan to propose an IDEF3 representation of the model coming from the synthesis work, in terms of process schematics and object schematics.

We shall then analyse the results of the mapping to IDEF3 representation, in terms of loss of information, incompleteness of the model, impossibilities, etc. in order to propose a translation of the IDEF3 representation into PSL, with new extensions if they are necessary.

The validation of the work will be conducted using an example of a real construction project, in order to test the concepts of the example of a real test case.

6 SUMMARY AND CONCLUSIONS

This paper has discussed the applicability of the Process Specification Language (PSL) to the construction industry and used an example to illustrate the representation schema. There is much scope for the use of PSL to facilitate the specification and exchange of process information in the construction industry. The research project on which this paper is based is exploring this and intends to deliver appropriate PSL extensions that will facilitate its deployment in the construction sector.

ACKNOWLEDGMENTS

The work described in this paper is the subject of a research contract funded by the Engineering and Physical Sciences Research Council (UK) and being conducted in collaboration with Loughborough University (UK).

REFERENCES

Anumba C.J., Evbuomwan N.F.O., 1996. A concurrent engineering process model for computer-integrated design and construction, in *Information Processing in civil and structural engineering design*

Austin S. Baldwin A.N., Newton A., 1996. A data flow model to plan and manage the building design process, *Journal of Engineering Design*, Vol 7, N° 1

Austin S. Baldwin A.N., Baizhan L., Waskett P. 2000. Analytical design planning technique (ADePT) : a dependency structure matrix tool to schedule the building design process, *Construction management and economics* (2000) Vol. 18

Björk B.C. 1992. A unified approach for modelling construction information, *Building and Environment*, Vol. 27, N° 2

CIM http://www.isr.umd.edu/Labs/ CIM/cimcontent.html

Cioccoiu M. 1998. Translating IDEF3 to PSL, *Technical Report*, University of Maryland, 98-63

Cooper R., Kagioglou M., Aouad G., Hinks J., Sexton M., Sheath D., 1998. The development of a generic design and construction process, *European PDT Days*

Curtis B., Krasner H., Iscoe N. 1992. Process modeling, *communications of the ACM*, 35(9)

Cutting-Decelle A.F., Michel J.J. 2000. Representation of industrial information through the joint use of ISO 15531 MANDATE and PWI ISO 18629 PSL : a contribution to the factory of the future, *PDT Europe Conference*, Noordwijk

Cutting-Decelle A.F., Michel J.J., Schlenoff C. 2000. From manufacturing to construction : towards a common representation of process information : the PSL approach, *Concurrent Engineering Conference*, Lyon

Fisher N. & Yin S.L. 1992. Information management in a contractor, *Thomas Telford Ed.*

Froese T. 1994. Developments to the IRMA model of AEC projects, *Computing in Civil Engineering : proceedings of the First congress*, Khalil Khozeimeh Ed. ASCE, Vol.1 June 20-22

Genesereth M., Fikes R., 1993. Knowledge interchange format (Version 3.0) – Reference manual, *Computer science department, Stanford University*, Stanford

Hannus M. 1999. http://cic.vtt.fi/projects/star/star1/brochure.html

Humphrey W.S. & Feiler, P.H. 1992. Software process development and enactment : concepts and definitions. *Tech.Rep. SEI-92-TR-4, Pittsburgh : Software Engineering Institute*. Carnegie Mellon University

ISO 10303-11 1994. Industrial automation systems and integration - Product data representation and exchange, ISO IS 10303-11, Part 11 : Description methods : EXPRESS language reference manual, 1994

ISO 18629-1 2000. Industrial automation systems and integration - Process specification language, ISO WD 18629-1, Part 1 : General overview

Kamara J. M., Anumba C. J. & Evbuomwan N. F. O. 2000. A Process Model for Client Requirements Processing in Construction, Business Process Management Journal (in press).

Karhu V. 1999. Formal languages for construction process modelling, *CEC 99 (Concurrent Engineering in Construction) Conference*, Helsinki

Karstila K. 1999. Models for the construction process - The MoPo-project, *CEC 99 (Concurrent Engineering in Construction) Conference*, Helsinki

Knutilla A., Schlenoff C., Ray S., Polyak S.T., Tate A., Cheah S.C., Anderson R.C. 1998. Process specification language : an analysis of existing representations, *NISTIR report n° 6160*, NIST

Koskela L. 1995. On foundations of construction process modeling, *CIB W78*, Stanford

MayerR.J., Menzel C.P., Painter M.K., de Witte P.S., Blinn T., Perakath B., 1995. Information integration for concurrent engineering (IICE) IDEF3 process description capture method report, *KBSI Inc, AL-TR-1995*

Process Protocol II 1999. http://PP2.dct.salford.ac.uk

Schlenoff C., Gruninger M., Tissot F., Valois J., Lubell J., Lee J., 2000. The process specification language (PSL) Overview and version 1.0 specification, *NISTIR 6459,* NIST

A usability study of a construction related IDEF0 model

R. Berg von Linde
Royal Institute of Technology, Stockholm, Sweden

ABSTRACT: Process models have been used for several years to describe and manage enterprises. In some cases have simple modelling methodologies been used, in other cases more advanced, e.g. IDEF0. Imprecise critique has been levelled against IDEF0 and the models that this methodology produces. In order to make powerful and useful process models is more knowledge about usability concerns and process modelling needed. This paper presents the results of a usability study that examines a particular process model visualised according to the IDEF0 conventions. Project manager professionals have participated in the tests. The most severe usability problems found were lack of overview, memory overload, and lack of measures to control how details are displayed. Suggestions are made to improve the usability, e.g. to make use of miniature diagrams to display context and develop tools for control of how details are displayed.

1 INTRODUCTION

Process models have many applications. A simple procedure, e.g. internal mail delivery at an office, can be effectively described with a process model. The same approach can be used to get the input to development of advanced IT support systems e.g. enterprise resource planning (ERP) systems. If the discussion is limited to business process models, the following arguments for producing process models are relevant according to Eriksson & Penker (2000):

- The process model gives a better understanding of an existing business
- The model can act as the basis for creating suitable information systems that support the business
- The model facilitates improvement of the current business structure and operation
- The structure of an innovated business can be presented by means of a process model
- Benchmarking on the model level can be performed, that is experiment with new business concepts or study a concept used by a competitive company
- Outsourcing opportunities can be identified in the business process model.

The process model can give a better understanding of the existing business as well as present the structure of an innovated business. If this is held true, the model can be considered as a pedagogical tool. There are companies that have used business process models as basis for graphical user interfaces to information systems describing working procedures. A model created to act as a basis for developing information systems, does not need to have the same visual abilities as a model that is used as a navigator to an information system. There is a difference between using process modelling as a tool for analysing businesses or demonstrating businesses to employees.

The process modelling methodology IDEF0 have been used by researchers affiliated to the construction industry for several years (Sanvido et al. 1990). The practitioners of the industry have not used this methodology in any larger extent. What are the problems that make IDEF0 an unattractive methodology for practitioners? The criticism of the methodology is not expressed precisely enough to make it possible to develop more usable methodologies based on the experiences from using IDEF0.

This report works with the hypothesis that the information content of the IDEF0 methodology is sufficient to serve its purpose, but the graphical user interface is an obstacle that must be handled if the process models ever will be useful for practitioners of the industry.

1.1 *Usability and information visualisation*

User interface professionals use the term usability. Usability is according to Nielsen (1993) a narrow concern compared to the larger issue of system acceptability. Figure 1 presents a model of the at-

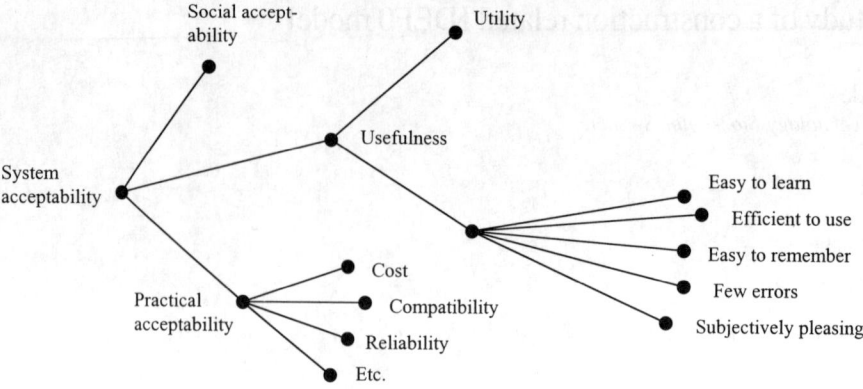

Figure 1. The figure shows a model of the attributes of system acceptability. Revised from Nielsen (1993).

tributes of system acceptability. Usefulness (whether a system can be used to achieve some desired goal) can be broken down to utility and usability. Utility answers the question of whether the system has the needed functionality and usability is the question of how well users can use that functionality. Usability is a part of a field known under the name HCI (human-computer interaction). Information Visualisation represents another branch of HCI and addresses the problem of information retrieval and visualisation. A typical application that has been explored is interactive graphics and animation technology to visualise and making sense of larger information sets. One example of that is the hyperbolic browser for visualising large hierarchies by Lamping et al. (1997).

2 METHOD

Usability tests were conducted on a particular IDEF0 model. The model illustrated the interface between a supplier of prefabricated concrete elements and the customer. The test methodologies used are well documented by several researchers working with human-computer interaction, e.g. Nielsen & Mack (1994) and Redmond-Pyle & Moore (1995).

Usability testing is a methodology that is carried out with real users. The method of thinking aloud protocol was used and interviews were conducted to collect data when the usability tests were performed. The thinking-aloud method has traditionally been used as a psychological research method (Ericsson & Simon 1984), but it is also an accepted method for practical evaluation of human-computer interfaces. One of its strengths is the wealth of qualitative data it can collect from a small number of test persons (Nielsen 1993).

3 THE USABILITY TEST

A process model with a well-known content was presented to a group of test users. The process model was set up according to the rules given by the IDEF0 methodology (NIST 1993). The test users were asked to do a number of tasks, thinking out loud meanwhile and being recorded and observed by the test administrator.

3.1 Participants

Six persons were chosen as participants in the usability test. Three of the participants were senior project managers and three of them relatively new on their positions as project managers. All participants were assessed to have good knowledge about project management (the content of the process model) and little or no knowledge about process modelling. Their knowledge about process modelling was found out by a number of questions asked before the actual test session.

3.2 Examined process model

The IDEF0 process model that was used in the study described manufacturing of prefabricated concrete elements. This choice of model was determined by a number of facts. It was desired that the model should describe a business that was familiar to the test persons since problems understanding terms used in the model would be disturbing. Knowledge about the modelled business makes it also possible to discuss however the process model presents a sound picture of the business.

Viewpoint of the process model was also chosen so that it would be familiar to the test persons. The model was made from the customer's point of view,

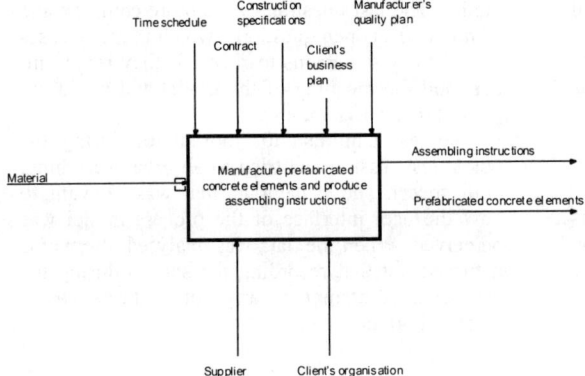

Figure 2. The topmost box of the IDEF0 test model, usually named the A-0 box.

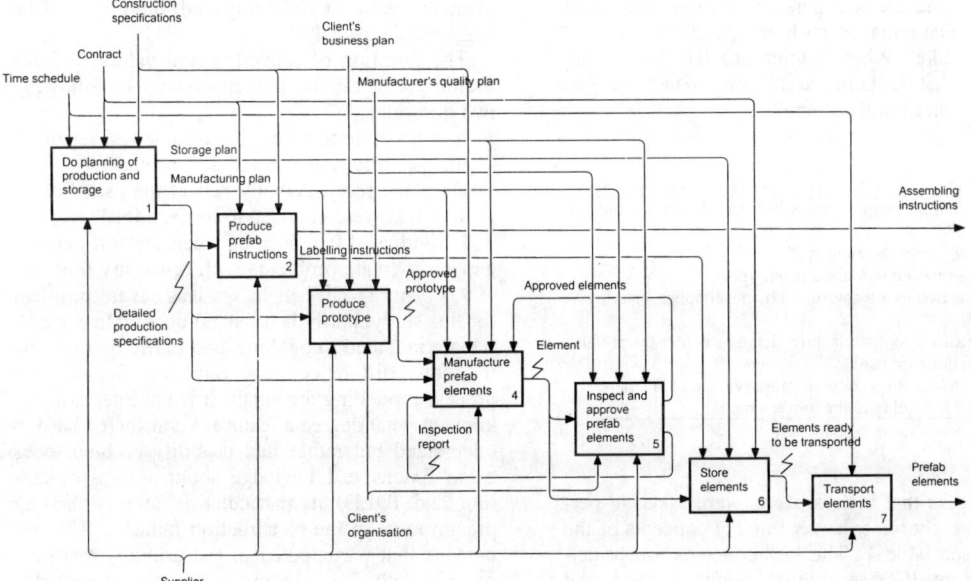

Figure 3. The A0 box, that illustrates manufacturing of prefabricated concrete elements and production of assembling instructions.

which is how project managers normally see manufacturing of prefabricated concrete elements.

The process model consisted of a top activity that was broken down in two levels. The number of diagrams were consequently few. The two topmost levels of the model are depicted in figure 2 and 3. All labels were written in the test persons' native language, which is Swedish, and the figures above are thus translated. The total number of diagram pages was nine. Each diagram of the model was printed on a single page and all pages were kept together by means of a binder when the model was handed over to the test persons.

3.3 Procedures

The usability test was composed of four sections and lasted no more than 45 minutes. The first part was an introduction given by the test administrator. This introduction explained the test procedures and estimated time consumption. It was emphasised that the test person was not the subject to the test but the process model itself. The next part of the test was a written questionnaire, which surveyed the participants' knowledge about process modelling in general. Knowledge about the IDEF0 methodology was particularly investigated.

This introduction was followed by the actual test activities. A written instruction, which explained how IDEF0 models are intended to be interpreted, was handed over. The test person was given as much time as needed to feel comfortable with the methodology. Questions were allowed and in most cases were elucidations necessary. A short introduction to the test model was given, especially the context and the viewpoint of the model. At this stage the written exercises were handed over to the test person. The participants were told that they could use as much time as needed to carry out the tasks. A thinking out loud procedure was used during this part of the test and everything that was said was recorded and could be analysed afterwards. The test administrator played an important role during this part of the usability test. It is easy e.g. to get impatient and don't let the test persons try as hard as they can, which may result in loss of valuable data and the test persons become dependant on the test administrator's help. Questions must not be asked like "What is your problem here?" but rather "What is your goal?" or "What are you thinking you should do here?"

Table 1. Tasks that the test persons were asked to perform.
1. How does the supplier know how the elements should be packed?
2. Who approves the elements?
3. What is needed to make a prototype?
4. What activities are controlled by the supplier's quality plan?
5. Is it possible to manufacture elements before the prefab instructions are made?
6. Explain how inspection and approval should be performed according to the process model!

The tasks that the test users were asked to perform were chosen to cover the key concepts of the model, see table 1. The key concepts are principally about ICOMs (input, output, control and mechanism are usually referred to as ICOM in an IDEF0 model) and activities and their relation to each other and their execution over time. To examine the understanding of mechanisms the test persons were e.g. asked to explain who is, according to the process mode, responsible for some given tasks that was covered by the model. The understanding of how activities relate to each other and the fact that time and sequence is not explicitly given by the model, needed more open questions. One question was e.g. "Explain how inspection and approval should be performed according to the process model". The order of the tasks was carefully chosen with regard to increase the test person's confidence. The first tasks were fairly simple and helped to test persons to get an overview of the

model. The last questions were more complex and formulated as open questions, which in many cases forced the test persons to reveal if they really understood the meaning of the model and could explain it in a convincing way.

The test administrator took notes during the whole test session and tried to observe everything, both speech and behaviour, that was relevant to how the user interface of the process model was perceived. When the data was analysed afterwards it turned out that recording the speech during the test person's attempt to carry out the tasks was of great importance.

3.4 Reliability and validity

The usability test was carried out to get an indication of possible usability problems, not to make a complete investigation of all existing ones. However, the issue of reliability and validity must be considered seriously.

The question of reliability and validity of this usability test can be broken down into following two questions:
1 Is it possible to recreate the same test results if the usability tests were to be repeated?
2 Does the result actually reflect the usability issues that were intended to be examined?

Individual differences between the test persons generally create problems with reliability (Nielsen 1993). This is nevertheless a less severe problem for this study, partly because no quantitative analyses are intended to be done, and partly because the individual differences concerning knowledge about process modelling are small. It is not interesting to know at what degree a feature is considered hard to understand but rather that usability problem does exist. Extensive knowledge about process modelling and IDEF0 in particular is unusual among practitioners in the construction industry. The test persons that participated in the usability test possessed, with few exceptions, similar knowledge about process models.

Assessing validity of the test involves understanding of methodological issues (Nielsen 1993). The first concern is whether the given tasks are relevant and addresses the concepts that are significant for understanding the process model. This is managed by compilation of tasks that both concerns detailed facts housed by the process model and tasks that are more comprehensive and calls for a deeper understanding of the model. The second concern is confounding effects that may lower the validity of the usability test. One confounding effect is e.g. if the content of the model or expressions used in the model is difficult to understand (distinguish this from the problems of understanding the design of the model may cause). By using a test model that illustrates a business familiar to the test persons, this kind of confounding effects can be managed. Another confounding effect is the quality of the instructions that were given the test persons before they performed the tasks. It is a risk that the usability test only reflects how good the instructions were written, not how good the test model was at communicate its content. To eliminate this, the same written instruction was given to all test persons.

On purpose, the model was presented as a number of pages kept together in a binder, not as a computer application. This was to eliminate the problem of separating results that were caused by difficulties managing the computer application from trouble interpreting the model.

4 EVALUATION OF THE RESULTS

The recordings and the notes taken during the test sessions were analysed and the findings are given below. A general finding from the tests is that the participants in most cases acted very similar and asked nearly the same questions about how to interpret properties of the model.

4.1 Findings

Most of the basic concepts of the IDEF0 methodology were understood rather rapidly. There was no problem understanding the hierarchical nature of the model. Breaking down an activity into subtasks is the normal procedure when time schedules are made and all the participants had experience from making or reading time schedules. The meaning of input and output was also easy to understand. Controls and mechanism were however considered confusing at the beginning but after some practice with the model were also those elements understandable. Some of the participants even started to question how input and control had been used in the model, which showed that they had fully understood the meaning of ICOMs.

Understanding in what sequence a process can be performed turned out to be difficult. Some participants interpreted the model as a workflow and strictly read the model from left to right. Few of the participants managed to decide the order of the activities by analysing the dependencies between the activities.

With few exceptions the participants tried to solve the tasks by searching the answer on the lowest level of the model. The effect of this behaviour was that the context was lost and the answers that were given were often incomplete. Some of the participants explained that they choose to look on a lower level since it looked less complicated

with fewer arrows and had content with a less wide scope.

All participants complained about the amount of arrows and that it was difficult to separate them from each other and decide which activity they were connected to.

It was also said that it was difficult to keep in mind how the diagram on a connected higher level looked like, which is sometimes needed to understand where input or control come from. This need of information from several diagrams at the same time causes a great deal of browsing through the model.

Those who had some kind of previous experience from working with process models had in general a more positive attitude towards the IDEF0 methodology compared to those with no previous experience. Some of the participants claimed that they definitely could use the IDEF0 methodology in their everyday work. A project manager claimed that he had enough problems making the project members understand the time schedule of the project and IDEF0 models would certainly not be easier.

Some of the participants said that they changed their opinion about the model while they worked with it and found it less difficult to understand after a while. The first confrontation with the model was considered as stressful but accustoming to the model took less time than expected by the participants. "It's a matter of habit" as one of the participants expressed it.

The last task of the usability test was more open than the previous tasks. This was to check if the participants understood the full meaning of the model and could make a verbal translation of the model. Those who worked with a top-down approach could deliver a more complete and elaborate answer than those who tried to find the answers only by looking at the lowest levels. It was clear that the lack of overview caused a lot of false conclusions.

5 DISCUSSION

The findings from the usability test can be used as a basis from which suggestions for a better user interface can be developed. Usability problems that should be taken care of are:
- Lack of overview
- Memory overload
- Too many details when not needed.

The test model was printed on paper, which limited the users' interaction with the model considerably. If a computer based user interface is used it is possible to do something about usability problems that a paper based model never can get rid of.

The context of a diagram page should always be clear. One way of achieving this is to display a miniature of the mother diagram on the same page as the currently observed diagram. It should also be possible to show child diagrams without leaving the current diagram. Making a window containing the child diagram pop up temporarily as long as the pointer is held over an activity box could do this. These measurements give a better overview and decrease the memory load of the users. Most IDEF0 modelling tools (e.g. Bpwin which is made by Logic Works) shows the context by means of a tree view similar to Windows' file browser. Research (Zaphiris et al. 1999) has however shown that sequential menus are more effective than hierarchical. A good solution may be to combine the tree view and miniatures of above and under laying diagrams on the same screen.

Details that are not needed must be possible to hide. One way of doing this is to make it possible to only show ICOMs connected to chosen activities or make it possible to only show one kind of ICOM at a time.

The participants had problems separating the arrows and it was especially disturbing that it was difficult to decide who was responsible for what activity. Colour encoding of the arrows or the activities is one way of addressing this problem.

The target users of today's modelling tools are in most cases process modellers or process analysts. This is why the software features supporting drawing and analysing processes are well developed and little effort has been made to develop effective ways of displaying process models. Some tools make it possible to convert the models to web enabled formats. This facilitates distribution, but the web models are usually simple images without any navigation aids or interactivity means. Process management involves process mapping and discussions about the processes and this is why comprehensible and usable process models are needed.

The next step in this research is to build prototypes and implement these suggestions for usability improvement and repeat the usability test on these prototypes.

6 CONCLUSIONS

Usability testing is a worthwhile way of finding weaknesses in process models and their ability to communicate its content.

The most severe usability problems found were lack of overview, memory overload and lack of measures to control how details are displayed. A computer aided user interface makes it possible to address the majority of the usability problems found in this study.

REFERENCES

Ericsson, K.A. & Simon, H.A. 1984. *Protocol Analysis: Verbal Reports as Data.* Cambridge, MA: The MIT Press.

Eriksson, H-E. & Penker, M. 2000. *Business modelling with UML: business pattern at work.* New York: John Wiley & Sons, Inc.

Feldman, C.G. 1998. The practical guide to business process reengineering using IDEF0.New York: Dorset House Publishing.

Lamping, J., Rao, R. & Pirolli, P. 1995. A Focus+Context Technique Based on Hyperbolic Geometry for Visualization Large Hierarchies. *Proc. ACM CHI '95 Conf. Denver, USA, May 1995.*

Nielsen, J. 1993. *Usability Engineering.* San Francisco: Morgan Kaufmann Publishers, Inc.

Nielsen, J. & Mack, R. 1994. *Usability inspection methods.* New York: John Wiley & Sons, Inc.

NIST (1993). *Integration Definition for Function Modeling (IDEF0).* Federal Information Processing Standards Publication 183 (FIPSPUB 183). Springfield: the National Technical Information Service, U.S. Department of Commerce.

Redmond-Pyle, D. & Moore, A. 1995. *Graphical user interface design and evaluation: a practical process.* Hertfordshire: Prentice Hall.

Sanvido, V., Khayyal, S., Guvenis, M., Norton, K., Hetrick, M., Al-Muallem, M., Chung, E., Medeiros, D., Kumara, S. & Ham, I. (1990). *An integrated building process model.* Technical Report no 1, CIC Research Program. University Park, PA, US: The Pennsylvania State University.

Shneiderman, B. 1998. *Designing the user interface: strategies for effective human-computer interaction.* New York: Addison Wesley Longman, Inc.

Zaphiris, P., Shneiderman, B. & Norman, K. 1999. *Expandable Indexes Versus Sequential Menus for Searching Hierarchies on the World Wide Web.* Available at ftp://ftp.cs.umd.edu/pub/hcil/Reports-Abstracts-Bibliography/99-15html/99-15.html

APPENDIX – INTRODUCTION TO IDEF0

The IDEF0 (Integration Definition for Function Modelling) is a subset of SADT (Structured Analysis Design Technique). Applying IDEF0 to the analysis of an enterprise results in a two-dimensional graphical model of the enterprise. IDEF0 models are made up of boxes and arrows, where the boxes represent happenings (activities and the arrows represent the interface between those happenings – the things (Feldman 1998). The model is hierarchical in nature, see figure 4. The activities are shown as rectangular boxes. One activity can be broken down into a set of activities, which shows a more detailed view.

A box is named by an active verb or verb phrase and it has a number, which appears in its lower-right corner. Input is converted by the activity to produce the output, under the direction of the control information (Feldman 1998). Each of the four sides of the box may have one or more arrows, which meaning is decided by which side it enters

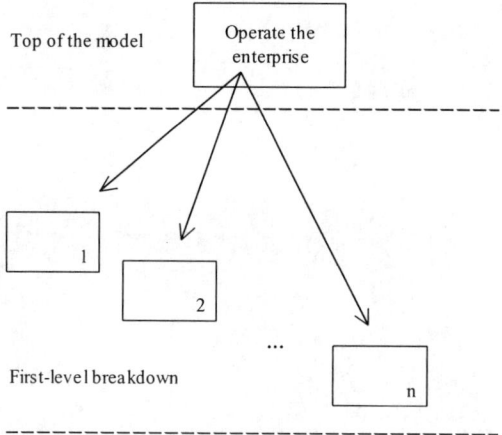

Top of the model

First-level breakdown

Figure 4. Breakdown of activities. Revised from Feldman (1998).

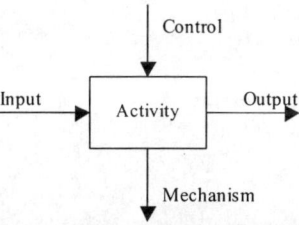

Figure 5. Syntax of the activity box.

or leaves, see figure 5. Arrows entering the left side of the box indicate input, arrows leaving the right side of the box represent output, arrows entering the top of the box show control, and arrows entering the bottom of the box indicate mechanism. Input, output, control and mechanism are often referred to as ICOM in an IDEF0 model. The arrows are labelled with a noun or a noun phrase, which describes the content of the arrow.

Significant for an IDEF0 model is that the model is presented from a chosen point of view and it is independent of time. The placement of the boxes in the diagram pages illustrates what activities dominate other activities. Boxes placed to the left and above other boxes dominate the other boxes. It is in other words not possible to read a model from left to right and assume that the activities will be performed in that order.

The PROMOTE project: Process oriented knowledge management

D. Karagiannis
BOC GmbH, Department of Knowledge Engineering, University of Vienna, Austria

R. Telesko
Department of Knowledge Engineering, University of Vienna, Austria

ABSTRACT: PROMOTE is a EU project dealing with knowledge management and running in the IST programme, Nr.: IST-1999-11658 (IST programme 1999). The overall goal of the PROMOTE project is to develop an integrated framework for process oriented knowledge management based on the existing business process management toolkit ADONIS®, to validate it by developing a product named PROMOTE® and to test it with end-user companies from the financial and insurance sector. Information about the project can also be found under the website http://www.boc-eu.com/promote.

1 INTRODUCTION

The overall goal of PROMOTE is to adapt the existing "Business process management systems" methodology for realising "Process oriented knowledge management", to validate it by developing a product named PROMOTE®, and to test it in end-user companies from the financial services sector. PROMOTE guides the accumulation, retrieval and distribution of product-process related knowledge and employees' know-how, and serves as an on-line support tool for knowledge managers as well as for the employees who generate and use knowledge.

The product will be based on a meta-model to be developed within PROMOTE that can be customised to suit a whole range of knowledge management approaches as long as these are expressed in terms of activities. The business process view provides an integrating framework, which enables a focused analysis of knowledge assets and requirements in a company. By relating knowledge management to business process modelling, one gains a basis for knowledge representation and it is guaranteed that knowledge is represented in the context of its generation and use.

The process based knowledge management approach will be applied in three investigation areas:
- the business processes supporting the delivery of products/services
- the product/service development process
- the human capital of the company

We adopt this division because it covers all the important parts of the strategic, operational and human resources aspects of knowledge management. Therefore it provides a clear framework of where to start, how to start and what issues to address.

2 STRATEGIC GOALS AND POSITIONING OF PROMOTE

2.1 *Positioning PROMOTE in knowledge managment*

There is no turnkey knowledge management solution available in the market today. Several methods and tools exist that employ intelligent techniques developed by research in various disciplines such as artificial intelligence, psychology, databases, Computer Supported Cooperative Work (CSCW) etc. These tools however lack functionality that
- enables them to perform as complete solutions
- support thousands of end-users (Existing applications are only suitable for departmental solutions.)
- guarantees their proper integration into the enterprise's value chain and the successful application of them to products and services

It is simply not sufficient if a tool performs navigation and retrieval as long as these functions are not aligned with business processes and corporate goals. In PROMOTE this integration will be realised in a process based approach. PROMOTE® will provide the opportunity for breakthrough research in linking organisational knowledge with business processes.

PROMOTE® does not aim to compete with other tools for knowledge management, such as GroupWare tools (e.g. Lotus Notes®) or data mining software. The tool to be developed in PROMOTE fits synergistically into this tool market. PROMOTE® is a tool to configure such specific tools and to integrate them into a coherent knowledge management approach in line with a company's business processes. On the technological side it characteristically contributes to the European technological progress.

The transformation of the nature of competition in the late 20th century poses some significant challenges for the economies of the European countries. Competition is increasing especially in the services sector not only because regional markets are liberalised but also because the nature of services and products is changing rapidly. New products are introduced more frequently and the era of mass marketing is almost over. Products demonstrate a higher degree of customisation to specific customer groups. Product and service innovations and the availability of communication means lead to an extremely competitive environment. The knowledge content of services and in our case of financial services is augmented. Companies will need tools like PROMOTE® to be able to manage their knowledge and transform it to innovative and valuable solutions for their customers.

Companies in all sectors realise that full mobilisation of their knowledge assets and reconfiguration into new productive forms that foster continuous innovation are the most critical success factors in competition. Therefore European corporations must be highly innovative in order to stay competitive in today's global markets.

PROMOTE® aims to become an enabler, a support tool for knowledge-intensive work in such environments. Which means that PROMOTE® will help companies in fully exploiting their knowledge pool. Building knowledge contributes to an individual's wealth, to an organisation's wealth and to a society's wealth. The next step is leveraging knowledge and this is where the greatest potential for advancement lies. There is a great need today for putting knowledge into work. PROMOTE aims in accomplishing this target in the corporate environments.

Knowledge intensive work has different characteristics from other ordinary activities. Employees while using PROMOTE® and giving their feedback will gain the experience (e.g. corrections, update) of being the company's main capital. Employee empowerment is a natural consequence, increases in specific qualifications but also in general ones that as competencies in self-guided information search and learning will raise employability.

The demand for highly knowledgeable employees is ever increasing in today's industries. In addition, employees' qualifications change from serving a strategy-fulfilling need to a strategy-forming one: Employees' knowledge is increasingly often seen as a means to enable a company to set the right strategic goals for itself. Instead of training employees solely to enable them to fulfill goals others have set for them, in many sectors companies have to increasingly rely on the creativity and competence of their employees to set their own goals and to contribute to the company's strategy. This is a core idea behind the vision of a learning organisation. We see knowledge management methods as an essential step to foster the learning organisation. If a company does not learn very rapidly, it will not be able to compete in today's global markets.

PROMOTE will improve current working conditions. When using the PROMOTE® tool, an easier retrieval, maintenance, delivery and distribution of organisational knowledge in a user-friendly environment is enabled. Employees do not any longer loose time or feel frustrated when they are looking for another persons' expertise in the company.

By handling properly issues such as who, when and what can contribute to the organisation's knowledge base, a new team-oriented working environment is built. Issues like power distance, hierarchy and authority reach differing meanings and question a lot of assumptions we currently have about working life and culture.

PROMOTE is based on multidisciplinary research: computer science, business administration, psychology and organisational research are combined. Today, multidisciplinary work is seen as a necessary approach to handle modern complex socio-economic and technological problems.

2.2 The PROMOTE consortium

The coordinator of PROMOTE is BOC (Austria), developer of ADONIS®, the most powerful business process management toolkit available today. ADONIS® (Karagiannis 1995, Karagiannis et al. 1996) supports the whole cycle of business process optimisation from knowledge acquisition to performance evaluation.

The business areas of BOC are further development and marketing of ADONIS®, consulting and project management in business process optimisation projects in the financial services sector and consulting in workflow technology applications.

BOC assists it's customers in identifying their Information technology (IT) potential, optimising their business processes, better utilise their knowledge assets and optimally deploy their human and IT resources.

BOC has vast experience in Business process management (BPM) projects in the financial services sector. Since BPM projects are focused on practical application of knowledge, BOC can leverage this expertise in PROMOTE. Furthermore the accumulated experience in change management techniques that BOC has used, can be successfully transferred, when necessary, in PROMOTE. BOC has conducted two EU projects in the BPM area, namely REFINE (REFINE 1998) where ADONIS® was tailored for the insurance sector and ADVISOR (ADVISOR 1999) where the ADONIS® toolkit is currently extended to cover business process-oriented learning and training activities.

The user organisations in PROMOTE are FIDUCIA (Germany) and INTERAMERICAN (Greece).

FIDUCIA is the largest computing centre service of the co-operatively organised bankers' syndicate in Germany. Together with its subsidiaries it supports a large clientele form various industries. The performance reaches from the conversion of complex data processing tasks up to the outsourcing of complete applications.

INTERAMERICAN, an insurance company in Greece, has a strong position in the domestic market and also significant operations in other European countries.

Both end-users have vital interest in exploiting the know-how and the tools that the PROMOTE project will produce.

3 PROMOTE: METHODOLOGY AND PRODUCT

3.1 Building and testing a knowledge management toolset in PROMOTE

The basic assumption in PROMOTE is that many central aspects of a company's knowledge are connected to the business processes realised in that company (also the ones that span the company's boundaries), some in explicit form (e.g. documents), many in implicit form. We speak of implicit knowledge with two connotations: Implicit because it is not documented/externalised but in the "head" of (single) employees, and implicit because is not mentally represented but only tied to practice.

An important goal of the PROMOTE project is to make both kinds of implicit knowledge around business processes explicit, not only in form of text documents, but also in form of (multimedia) case descriptions, models, and enriched multimedia documents. To accomplish this, knowledge activities in a company must be thoroughly integrated with a company's strategic goals as well as its business processes and with the development of its human resources. The notion of KMP is one of the central concepts in PROMOTE for a better understanding of knowledge activities in a company.

3.2 The concept of knowledge management process (KMP)

Today's knowledge management approaches are often formulated in rather general terms, lacking a tool set to support their implementation in organisations. The PROMOTE consortium will develop the PROMOTE® product by adapting the model-editor of the ADONIS toolkit. For capturing the various KMPs with different degrees of formalisation in the different investigation areas, a graphical model-editor that incorporates a powerful modelling language will be developed.

KMP modelling deals with the representation of acquisition, search & retrieval and maintenance of domain knowledge.

The tool set to be developed and tested in PROMOTE will allow to model KMPs in a form similar to the modelling of business processes. Note that with this tool knowledge processes are modelled independently of specific IT tools and platforms.

KMP modelling deals basically with the kind of entities and activities PROMOTE will help to manage and realise. In PROMOTE by introducing KMPs, it is not aimed to simulate cognitive abilities of human beings like in artificial intelligence, but to provide efficient means for amplifying the intelligence of employees.

In PROMOTE generic and specific KMPs will be developed. These processes will be initially designed for the end-users. By means of the meta-architecture of PROMOTE®, KMPs will then be transferred to the financial services sector and later on to other sectors. Companies will be able to take PROMOTE® and a large library of KMPs, customise them according to their specific needs and apply them in parallel with their business processes. In addition, users will have the possibility to define their own KMPs depending on the objectives they have (a new product development/introduction, market monitoring, competitor analysis etc.).

The consortium believes that in different industry sectors there exist quite different knowledge needs and processes. Even within one sector or company, different knowledge objectives may exist that determine which knowledge is valuable and also influence how it is acquired, stored and used. Generic knowledge management approaches need to be adapted to these different objectives and tools need to be customised accordingly.

The architectural basis for PROMOTE® will assure that knowledge managers can design knowledge management strategies and measures in their organisations relying on different knowledge management approaches, on different departmental and company-specific needs, and independent of the specificity of the underlying IT. An immediate advantage is that the PROMOTE approach - by abstracting away from the information level - allows for a fully integrated approach towards knowledge management. A knowledge manager will even be able to plan and monitor knowledge activities which are not (or at least not at planning time) realisable with IT. Since the knowledge level is not restricted by the information level, one can for instance plan for brainstorming sessions or conferences as a knowledge acquisition method without having to commit oneself to specific IT or to IT at all: "Human-only" knowledge management measures can be integrated with those that involve IT. The knowledge management strategy of a company can be planned and monitored without being constraint by today's IT solutions. Rapid implementation is guaranteed by supporting the mapping from the meta-level (KMPs) to whatever IT exists or is under development.

The PROMOTE® product will be integrated into the end-user IT environment and be linked with existing tools for process modelling, collaboration, document management etc. Substantial end-user involvement will allow for testing and evaluating the tool under realistic conditions.

3.3 The PROMOTE® architecture

Figure 1 shows the different levels of the PROMOTE approach for knowledge management.

The PROMOTE approach is also concerned with the implementation of the following management activities in PROMOTE® which can be seen as a meta-tool for different purposes:

- allowing for different knowledge management approaches by using PROMOTE®
- modelling and realisation of the KMPs
- supervision, assessment and evaluation of PROMOTE® by a knowledge officer

The method to be developed will allow knowledge officers and employees in organisations to plan, configure, monitor and evaluate the core activities of PROMOTE in order to reach their knowledge objectives.

The PROMOTE methodology consists of the following steps (Fig. 2) which are described now in more detail.

3.4 The PROMOTE® methodology

3.4.1 Strategic decisions - The awareness phase

In PROMOTE each end-user focuses on a particular investigation area. In the first step, users consider knowledge management as a task that will only be successful if accompanied by appropriate management strategies and policies. Strategic decisions concerning knowledge management deal with points like:

- linking the knowledge management initiative with corporate objectives, in other words, deriving the extent and the content of the knowledge management initiative from predefined business goals. Company-specific knowledge management procedures and objectives are to be formulated here.
- identifying activities in the investigation areas that have a strong knowledge leverage potential.

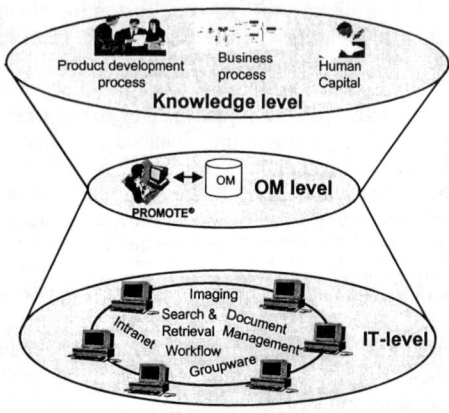

Figure 1: The PROMOTE architecture

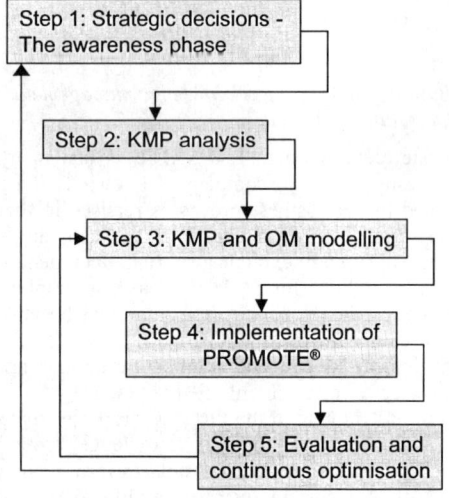

Figure 2: Steps in PROMOTE

- identifying the so-called "soft factors". Experience has shown that the impact of cultural and psychological factors on knowledge management projects is tremendous and companies should readjust and take complementary actions to tackle these issues effectively.
- Defining criteria for evaluation. Based on the PROMOTE approach, knowledge activities are planned and modelled. This does not only allow for a clear implementation of the activities in the operational field, but also for a thorough evaluation. Expectations about the effects of knowledge activities can be formulated based on the models and empirical observations can be matched against these expectations.

3.4.2 KMP analysis

In a business process, knowledge exists in the form of data and information in combination with experience, communication, reflection, expertise, techniques and cognitive abilities. PROMOTE starts from a careful modelling analysis of the existing business processes (i.e. analysis of work practice) and identifies what kind of knowledge is created/used and by whom in the context of the respective business processes. Even if a company has yet not taken explicit measures to measure its knowledge and assess its knowledge processes, they exist in an implicit form, connected to the activities of employees in the context of their daily work.

Starting from the modelled business processes of selected trial cases, this step aims to identify knowledge contents, knowledge flows, and knowledge sources. The complete cycle of knowledge acquisition, storage, distribution and usage is to be covered in the course of this analysis. In the course of identifying processes also the human resources required to realise these processes will be identified and described in terms of the functions they have for the knowledge processes.

Identification of these activities can be done in a top-down as well as bottom-up manner, building on an analysis of the respective processes in the user trials. These activities are partly organisational (such as providing for specific communication routes and forms), partly IT support, partly training and human resource related.

Knowledge acquisition refers to the identification of the relevant knowledge to be represented in an organisational memory (OM) (Hinkelmann et al. 1998). Our focus is on assessing how knowledge is related to processes and their execution. Empirical research will be done at the user sites beginning with the identification of knowledge contents. Some of the tasks involved in this step concern:
- knowledge acquisition methods
- identifying know-how of employees
- capturing practice and experience
- capturing brain storming session outcomes
- capturing knowledge distributed across a number of employees who perform a process
- integrating "hard" information, for instance from benchmarking

In PROMOTE FIDUCIA focuses on the aspects and influences of the companies human capital in the process based knowledge management system. For FIDUCIA human capital is the accumulated practical knowledge, skills and experience of the employees that make them productive in the sense of the companies targets. Human capital is one of the most important asset FIDUCIA has access to. It is the source of business success in the existing market and is the fundamental basis for the continuous competitive advantage in the future.

The selected investigation area for INTERAMERICAN is the legal protection insurance claims management. Legal protection insurance protects against eventualities, which either increase the liabilities of persons/companies or decrease the level of their assets. The intention is to focus on reducing the claims process execution time with the view to increasing customer satisfaction by providing user friendly knowledge objects linked to business processes.

3.4.3 KMP modelling and optimisation, OM modelling

In step 3 the KMPs as well the OM using the results of analysis of step 2 are modelled.

Both processes and OM are at this point specified conceptually, using an adequate ontology and corresponding modelling tools, not in terms of specific IT.

3.4.3.1 KMP modelling

One of the big issues in PROMOTE will be the development of an adequate and complete KMP language capturing all different types of KMPs. In the following a list of possible KMPs is given.
- Acquisition: Integrating new knowledge in the OM (Direct acquisition, guided acquisition, automatic acquisition, interaction knowledge producer – knowledge consumer)
- Search and retrieval: Retrieving knowledge from the OM (Querying, full-text search, entity-based search, knowledge navigation assistant, knowledge selection, information agent)
- Maintenance: Management functions of the OM (Archive, validation, feedback, knowledge format transformation etc.)

There exists a lot of approaches for process oriented knowledge management (Abecker et al. 1999, Nissen et al. 2000, Reimer et al. 2000, Remus et al. 2000), yet all these approaches do not specifically aim at developing a language for capturing different types of knowledge activities.

For modelling simple types of KMPs elements of

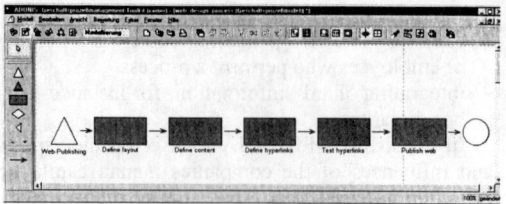

Figure 3: KMP Web publishing

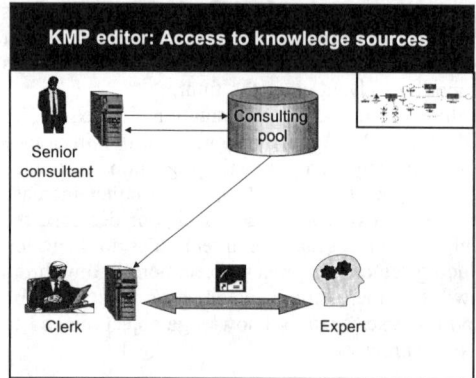

Figure 4: KMP Access to knowledge sources

event-based process chains (EPCs) can be used. This is shown for the KMP „Web publishing" in Figure 3, where some basic activities to set up a webpage are modelled using the model-editor of the business process management toolkit ADONIS®.

A more complicated case which requires additional modelling constructs is shown in Figure 4, where access to some knowledge sources is designed.

Only two roles of employees are distinguished here, a normal clerk and a director. Both have access to the consulting pool, yet the director is also allowed to update the consulting database. This is indicated through the direction of the arrows. For the clerk it is also possible to contact an expert via e-mail when solving a specific problem. The context of this KMP is given by a business process depicted in a corner of the model editor.

KMP models will also be enriched with the notion of "user groups" in order to adapt knowledge acquisition and access to specific user-group needs and capabilities.

Like in the business process management toolkit ADONIS® there will exist three levels for KMP modelling:

- Level 1: KMP meta model: On this level the generic elements of a KMP are defined, e.g. KMP start, condition, action, parallel action, KMP end.

- Level 2: Customisation of KMP elements for the end-user: In order to guarantee flexibility of the KMP modelling for different end-users in PROMOTE, a customisation of the generic KMP elements for specific requirements is appropriate.
- Level 3: Specific KMP: This might be for example the specific KMP for web publishing shown in Figure 3.

3.4.3.2 OM Modelling

A common approach in knowledge management to designing an OM is to take a document-centric, technical perspective: Which of the documents and database records produced in a company contain "knowledge" and how can that "knowledge" be distributed to employees?

In PROMOTE, we see the OM from a learning perspective - both individual and organisational learning - and ask: How can we represent information in an OM so that people can rapidly turn it into knowledge? (On the side: We believe that only people have "knowledge" because they, but not machines, are able to use information to solve new problems. IT provides data that need to be turned into knowledge by people in the context of problem solving. The art of designing an OM is to provide the right data at the right time in the right format.)

Our innovative approach consists in providing end-users with knowledge objects instead of documents or records. A knowledge object can be one of many things: Case descriptions capturing an experience or an instance of best practice, rules and procedures, a model of a business process, comments on a business process, etc. Knowledge objects are not so much defined in terms of what they contain as by how they can be generated, accessed and modified. Knowledge objects are information structures which are designed in manner so that they can be quickly turned into knowledge (be used for solving problems).

Building on a generic model of an OM, which is to be developed in the project, it will be formally described how KMPs and their outcomes are stored. Note that this OM will not only allow to store process outcomes (such as documents), but also the knowledge generation and use processes. This is important to have this contextual information represented in the OM.

The PROMOTE OM will store knowledge representing artefacts (documents, multimedia documents, process models, formal knowledge representation models such as conceptual graph, databases, informal as well as formal case representations etc.) together with a description (in form of KMPs) of the processes and human resources which produce and use these artefacts. In addition, it will structure its content (the knowledge objects) in a manner that is optimised for information navigation and learning. A first set of requirements to operationalise this goal are that information should be

- presented in the problem solving context it is most useful
- associatively linked to other pieces of information
- represented in multiple forms covering multiple perspectives
- take a narrative, story-like, application-oriented form
- make the information source explicit to that its credibility can be judged
- shareable with other people

It is important to understand that the OM at this point is mainly a model of an OM, although a highly specified model, at this stage customised to a company's specific knowledge objectives and approach. As the functionality of the OM must eventually be realised by means of the specific IT used in a company, it is not desirable to commit oneself to specific IT (e.g. specific database and intranet tools) at this stage. In case the available functionality of the existing tools at the end-user sites is not sufficient to cover the necessary knowledge management activities, necessary extensions of the functionality will need to be implemented.

3.4.3.3 Knowledge Management Optimisation

Having a formal model of a company's KMPs and its OM requirements, improvements („knowledge process re-engineering") upon the current state can be suggested, modelled and analysed. For instance, the knowledge officer can plan for additional and/or more systematic knowledge acquisition activities, for more user-friendly access mechanisms, design different tools and access mechanisms for different user groups in the company, and so on. PROMOTE will support the optimisation phase by means of a tool set with generic knowledge management activities (e.g. various acquisition methods), a set of templates covering prototypical knowledge management tasks (to some extent sector-specific and already tailored towards specific knowledge objectives), means to rapidly model knowledge management scenarios using a graphical model-editor, and where appropriate the means to simulate a KMP.

3.4.4 Specification and implementation of PROMOTE®

Once the knowledge-related processes are identified (step 2) and conceptual models of KMPs and OM are developed (step 3), the next step is to implement the PROMOTE® product and to map the models onto company-specific IT platforms which serve than as the infrastructure which supports employees in knowledge-intensive activities.

The main issues when implementing PROMOTE® are:

3.4.4.1 Interface to ADONIS®

The interface to ADONIS® enables access to business process model information for the PROMOTE® product (e.g. description of activities with ADONIS® notebooks).

3.4.4.2 PROMOTE OM

Based on the model-editor of the ADONIS® toolkit for a graphical description of business processes a model-editor for KMPs will be developed aiming at an explicit representation of knowledge acquisition, retrieval and use processes. A KMP describes how specific knowledge in a company is generated, retrieved and updated. In this context also the structure and content of the end-user OM and different accesses to it will be modelled.

Note that the PROMOTE OM consists of the relevant business processes, the KMPs and the knowledge objects in the OM as shown in Figure 5.

3.4.4.3 Integration into the end-user IT infrastructure

The PROMOTE® product is not a stand-alone system and therefore needs to be integrated into the existing end-user IT infrastructure (collaboration tools, document management systems etc). The interface of PROMOTE® will be constructed by using current web programming languages and tools (e.g. Java). PROMOTE® will support customising functionality on three different levels:

- End-users representation: Different OM representation for different user groups
- Customising to different knowledge management approaches
- Easy configuration by the user: Customising the PROMOTE® interface according to end-user needs

3.4.5 Evaluation and continuous optimisation

The use of PROMOTE® will imply several advantages for the organisation: Employees will be

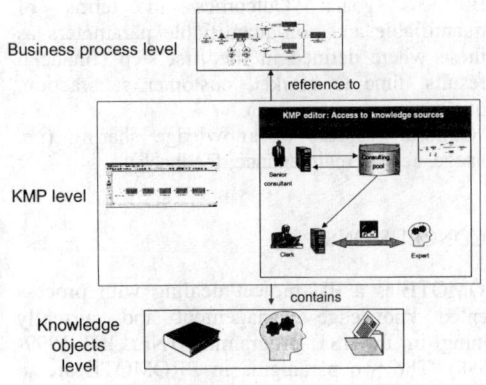

Figure 5: Structure of the PROMOTE® OM

supported with knowledge that goes deeper (and is more case-specific) than captured in standard organisational handbooks and process models (extending depth); this should allow for more flexible case processing.

Employees will be supported in getting knowledge about a whole case, not only isolated information about single activities (extending scope); this should move employees closer to the customer and again make them more flexible in their case processing.

Employees will also be enabled to formulate suggestions for process improvement not only in form of documents and annotations to documents, but in form of process models; this makes it possible to rapidly evaluate the quality of employees suggestions (for instance, by simulating the effects of the proposed changes and comparing them with the status quo) and to rapidly integrate them – once validated – into the organisational practice.

Employees will be motivated to contribute their knowledge to an OM because of the various motivational factors (avoiding extrinsic incentives as much as possible) built into the PROMOTE® approach (rapid input, graphical and multimedia-based input, knowledge sharing, rapid feedback, rapid implementation of high-quality suggestions, assuring intellectual ownership etc.).

The high degree of customisation, made possible by the clear meta-architecture of PROMOTE®, enables fast integration into the end-user companies IT and knowledge management strategy. Evaluation will take place in various dimensions:

- Technical: Evaluation of the PROMOTE approach in terms of technical feasibility.
- Organisational: How does knowledge management affect organisational structure and process execution?
- Socio-psychological: Issues such as barriers, participation, acceptance, work satisfaction, usability, learnability and sustainability, reward mechanisms.
- Business goals: Outcomes in terms of quantifiable and non-quantifiable parameters as these where defined in the first step (financial results, time to market, customer satisfaction, quality improvement etc.).
- "Cultural" aspects of knowledge sharing (i.e. usage, access, maintenance, feedback).

4 CONCLUSIONS

PROMOTE is a EU project dealing with process oriented knowledge management and currently running in the IST programme (Nr.: IST-1999-11658). The two paradigms in PROMOTE are to link knowledge management with the business processes in a company and the intention to model knowledge management processes and knowledge objects in a way similar to business processes.

In PROMOTE a methodology for process-oriented knowledge management and a product is developed which will be tested with end-user companies from the financial and insurance sector.

REFERENCES

Abecker, A. & Bernardi, A. & Sintek, M. 1999. Enterprise Information Infrastructures for Active, Context-Sensitive Knowledge Delivery. *Proceedings of ECIS'99 – The 7th European Conference in Information Systems,* Copenhagen, Denmark. June 1999.

ADVISOR 1999. http://www.boc-eu.com/advisor Access 19/06/2000.

Hinkelmann, K. & Abecker, A. & Bernardi, A. & Kühn, O. & Sintek, M. 1998. Towards a Technology for Organisational Memories. *IEEE Intelligent Systems & Their Applications* 13(3): 40-48.

IST programme 1999. Overview of the programme. http://www.cordis.lu/ist/overview.html Access 19/06/2000.

Karagiannis, D. 1995. BPMS: Business Process Management Systems, *ACM SIGOIS Bulletin.* 16: 10-13.

Karagiannis, D. &. Junginger, S. & Strobl, R. 1996. Introduction to Business Process Management System Concepts. In: B. Scholz-Reiter, E. Stickel (eds.), *Business Process Modelling,* 81-106. Berlin: Springer.

Nissen, M.E. & Kamel, M.N. & Sengupta, C. 2000. Torward Integrating Knowledge Management, processes and Systems: A Position Paper. In Steffen Staab et al. (eds.), *Proceedings of the AAAI Spring Symposium Series 2000 Bringing knowledge to business processes,* Stanford, CA, March 2000.

REFINE 1998. http://www.boc-eu.com/refine Access 19/06/2000.

Reimer, U. & Margelisch, A. & Staudt, M. 2000. A Knowledge-Based Approach to Support Business Processes. In Steffen Staab et al. (eds.), *Proceedings of the AAAI Spring Symposium Series 2000 Bringing knowledge to business processes,* Stanford, CA, March 2000.

Remus, U. & Lehner, F. 2000. The Role of Process-oriented Enterprise Modeling in Designing Process-oriented Knowledge Management Systems. In Steffen Staab et al. (eds.), *Proceedings of the AAAI Spring Symposium Series 2000 Bringing knowledge to business processes,* Stanford, CA, March 2000.

Product modelling

Will the use of IFC change the CAD landscape

Peter Muigg
Computer Anwendungen Muigg, International Technical Coordinator of the German Speaking Chapter of the IAI, Germany

ABSTRACT: The purpose of this paper is to give an indication to what extent the use of IFC files may change the way building projects are being done in the future. The author is a both a vendor for Architectural CAD Software as well as a high level member of the International Alliance for Interoperability (IAI). So, the views given reflect both the assumptions from a the perspective of a software vendor as well as the assumptions of the IAI.

1 THE NEED FOR A NEW FORMAT

The need for a new data exchange formats is mainly driven by the fact that CAD vendors are all shifting towards object oriented CAD systems. As long as CAD was mainly used for drafting, elementary geometry (lines, arcs and circles) was used to draw floor plans. Therefore, exchange formats like DXF were sufficient to exchange drawing, because elementary geometry can be understood by all CAD systems.

The main difference between conventional drafting systems and object oriented CAD systems (OO-CAD) is, that OO-CAD uses objects for things like walls, doors, beams etc. In order to exchange these objects between different CAD vendors (without breaking them down to elementary geometry) it is necessary to define standards for objects so that every CAD system is able to work with them. To define an international standard how to describe the objects used in the building industry that can be used by both CAD vendors as well as other software packages and end users was one of the driving forces behind the IAI.

2 THE PROMISE OF INTEROPERABILITY

The other driving force is the need for more "Interoperability", i.e. to reduce the loss of information between, say, architects and building engineers. IAI stands for "International Alliance for Interoperability". So, it is not only the geometry that is being taken care of, it is also the "background information" about the objects involved. This is where the term "Building Model" comes into play. Loss of data does occur when, for example, an architect who has already defined spaces for the building under design sends a "stupid" CAD drawing to a building engineer who needs to calculate the heat loss. Even though that the architect has already defined spaces, the building engineer will most likely not be able to use this information but will have to "rebuild" the spaces from the lines that form the boundary of spaces. In IFC, the information about spaces will be preserved, together with information about which walls form the boundary of the spaces and what openings are in the walls forming the space etc. So, the building engineer will be able to perform heat loss calculation without having to enter information that was already there. This is not only faster, it is also much less error prone. All he will have to do is to supply additional information needed for his calculation. This information again will be kept in the model, so that, for example, a facility management system can take advantage of it.

3 THE STATE OF IFC DEVELOPMENT

Before looking into the future and trying to predict what the impact of IFC might be for the use of CAD, a short overview about the current state of IFC development will be given. What are the methodologies used, what is being covered in IFC today (which domains), what will be added in the future (the "IFC Roadmap" of the IAI). The structure of the IAI as an international organisation and the working procedures of the domain groups in the various chapters as well as the mechanisms of the so called International Technical Management (ITM) will be described. Most important in the context of this paper is the current state of IFC implementation in shipping software products – both CAD and non-CAD. An overview about the current status will be given. By the time when the conference actually takes place, results from projects which are now being started where the exchange of IFC will be tested in practice should be available and will be presented.

4 WILL USERS ADOPT OO TECHNOLOGY ?

The question if and how the use of IFC will change the CAD landscape also depends on a number of factors which are not directly related to the quality and usability of IFC data exchange.

The key question is: To what extent will users shift towards the use of OO – CAD ?

Today, the reality in architectural offices around the world is that during early design stages, architects do use 3D capabilities of their systems for presentations and studies. In later stages, most of the drawings are still created in 2D – independent of the CAD system they actually use (i.e. even if a CAD system does have OO features, drawings are still created using lines instead of wall objects). The reasons why this is still the case are manifold. One reason may be that that the ease of use of OO CAD is still not quite as good as it should be. Another reason may be that still many architects prefer to use CAD as an electronic drafting board because this suits their traditional working style better. The main reason, however, might be that it is still considerably faster to create a 2D drawing using conventional geometry than using a model based system with objects.

This, of course, also depends on what the customers of an architect do expect as result of the architect's work. If a customer is happy with a pack of paper drawings, architects will be inclined to deliver this output with minimal effort and cost – and if 2D CAD is the fastest way to produce this result, they will continue to use it. However, if developers and building owners ask for more – like the delivery of a ready to use FM system reflecting the "as built" state of the building - things will be different because then the use of model based cad will show it's additional benefits – and the ability to exchange information using IFC is one of these benefits.

The most important goal of the IAI is to improve the quality of buildings and to reduce cost during all cycles in the life of a building. From design through construction and the actual use of the building until it is being demolished. The very strong support the IAI is getting from both vendors as well as software users and building owners is a strong indication that there is a very good chance that the IAI will be able to deliver on this promise. As soon as building owners begin to realise the true power of IFC, they will demand that designers, building engineers, structural engineers and facility managers be able to use and deliver IFC data.

The quality of IFC support will become a key factor in evaluating a piece of software to be used in the building industry. One of the latest developments in the IAI is a very close collaboration with activities regarding XML. In the US, a group has formed called aecXML and has now become active member of the IAI in North America. Similar activities in the European Union are under way. Making use of XML is particularly important for the bidding and purchase side of the building process.

5 CONCLUSION

If IFC will change the CAD landscape in terms of vendors on the market is something that depends on the vendors strategy regarding IFC support. But what will change is the way professionals in the building industry work in the future – and I strongly believe that IFC will be a key factor – both for CAD software as well as other software used in the building industry.

Product and Process Modelling in Building and Construction, Gonçalves, Steiger-Garção & Scherer (eds)
© 2000 Balkema, Rotterdam, ISBN 90 5809 179 1

A proposed extension of the IFC project model for structural systems

M. Weise, P. Katranuschkov & R. J. Scherer
Technische Universität Dresden, Germany

ABSTRACT: Using the ability to extend the Industry Foundation Classes (IFC) the article introduces to a concept of a possible extension of the IFC by the structural engineering domain, which is not supported in the current release. To fulfil the concept of the IFC a structural extension is proposed, which provides a two step process for defining structural assumptions. In the first step the building structure is described in a more general way as needed for planners from other domains. The second step extends those information about more detailed structural assumptions used only by a structural engineer. This is a starting point for the usage in proprietary structural applications. The highest benefit of the proposed extension is the realized bilateral connection between architectural information, the general description of the building structure and the structural analysis models used for the calculation. This is fulfilled by a minimal extension of new IFC classes and the usage of property set objects offered by the IFC model.

1 INTRODUCTION

A necessary basis for a common computer aided planning, construction, and maintenance of a building is an efficient exchange of information between all planners. Therefore, a common data model such as proposed by the International Alliance for Interoperability (IAI) is needed. To fulfil this necessary basis the IAI, an organization founded by leading software developers for the building industry and building companies from important industrial countries all over the world, is developing the Industry Foundation Classes (IFC).

Modelling the whole field of the building industry is very complex indeed. Therefore the IAI preferred a way of an incremental development of the anticipated project model. To support this intention an extensible architecture for the different domain models is provided. This provided architecture and the lack of a possibility to describe structural domain information – currently supported are the domains of architecture, HVAC engineering and facility management – was the starting point for the investigation of the extensibility of the IFC and for the elaborated proposal for a structural domain extension.

The article gives a general overview of how to extend IFC, a more detailed description of the main ideas and partial information about the technical specification of the proposed extension. For conclusion the expected benefit is outlined.

2 EXTENSIBILITY OF THE IFC

2.1 *Overview of relevant characteristics*

As outlined before, the extension of the IFC by adding different domain models is a major feature and most interesting for this work. The expected profit should be clear. First concentrating on domains with the most attractive benefit of seamless information sharing and then adding other domains to enable step by step a wholly integrated computer aided work. A really pragmatic approach. Reasoned by this situation questions are arising about the extensibility and the main goals of the IFC project model.

2.2 *Main goals of the proposed extension*

Information sharing is a general goal of the IFC project model. In this context an extension has to provide and to deal with sharable information first, other desirable features have to make compromises. The mentioned problem that occurs here is the discrepancy between domain-independent information sharing and the wish to assimilate the whole domain-specific information.

Furthermore the described extension has to deal with a minimum definition of new classes. This takes into account the wish to avoid an unnecessary growing of the IFC project model and therefore it helps achieve an easier implementation and maintenance of the model. Under this conditions it is understandable that it is not achievable to cover the whole structural engineering domain. This proposed extension is an attempt to cover the information needed most urgently with the highest benefit for the structural engineer as well as for the other planners.

2.3 Architecture of the IFC

A necessary requirement to understand the proposed extension is the knowledge about the main ideas of the architecture of the IFC. Therefore it is useful to outline a short overview here. For a more detailed description the IAI (1999a) is recommended.

The differentiation of the IFC project model into several layers and the so called ladder principle, which is using those layers to describe a rule for possible references between classes, is the most interesting aspect. Therefore, depending on their "specialization" each class belongs to exactly one layer. If a class keeps more common functionality, it belongs to a more common layer. In general, each class can have references to classes which belong to the same or a more common layer.

At this point the different layers become interesting. Especially the Domain layer and the Interoperability layer have to be mentioned here. Each class belonging to the Interoperability or a more common layer is used to share domain-independent information. In accordance with that fact the domain layer keeps all domain-specific information. It must be noticed that each domain, and therefore the classes belonging to

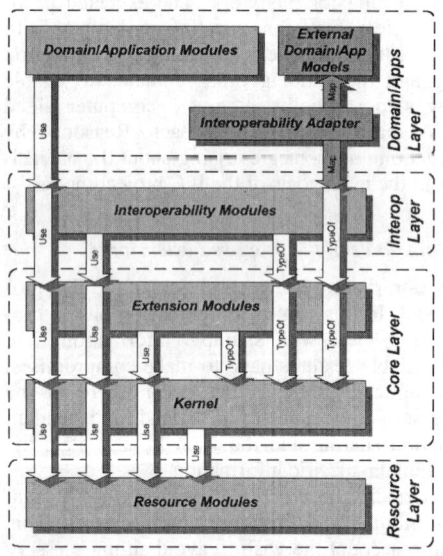

Figure 1. Layering concept of the IFC architecture (IAI)

this domain, is independent from each other domain. Figure 1 illustrates the valid references between those layers.

2.4 Additional rules for extending the IFC

In addition to the presented architecture of the IFC there are other rules which have to be considered for an extension. This includes the used object-oriented concept and rules for classifying and modelling.

The mentioned object-oriented concept and the rules for classifying have a major moderate affect on the hierarchy chart of the IFC classes. For the creation of new classes only the kind of the elements is relevant and neither their function nor membership to a system. From the technical point of view the single inheritance of subtypes, the substitution principle of Liskow and the exclusion constraint in supertypes shall be named as a selection which differs from the most known object-oriented concepts like STEP/EXPRESS [ISO 10303-11 IS. (1994)] or UML [Rumbaugh J., Jakobson I., Booch G. (1998)].

The rules for modelling, for example the rules for naming or those of how to handle m to m relationships, are not less important. For a more detailed description the Annex B of [IAI. (1999b)] is recommended.

3 OVERVIEW OF THE PROPOSED EXTENSION

3.1 The scope of the extension

The purpose of this extension is not to cover the whole domain, but to keep the main decisions from the structural engineer and to provide them for other planners. The full range of those structural information holds for other structural engineers and part of it to planners from other domains.

For a structural extension of the IFC the description of the general building structure, the used structural analysis models with their main mechanical assumptions, the loads, and the results are considered useful. The support of detailed analysis models, for instance as needed for FE analysis, dynamic investigations or stability problems, the structural design and the definition of some special loads, like prestressed loads, temperature or eccentric loads is out of scope. But this does not mean that there is no possibility to deal with this information in future extensions if needed. It must be noticed that above described "lack" of information, which causes a less voluminous model, is an advantage for handling IFC project data and receiving an early first version, which is an important demand of industry.

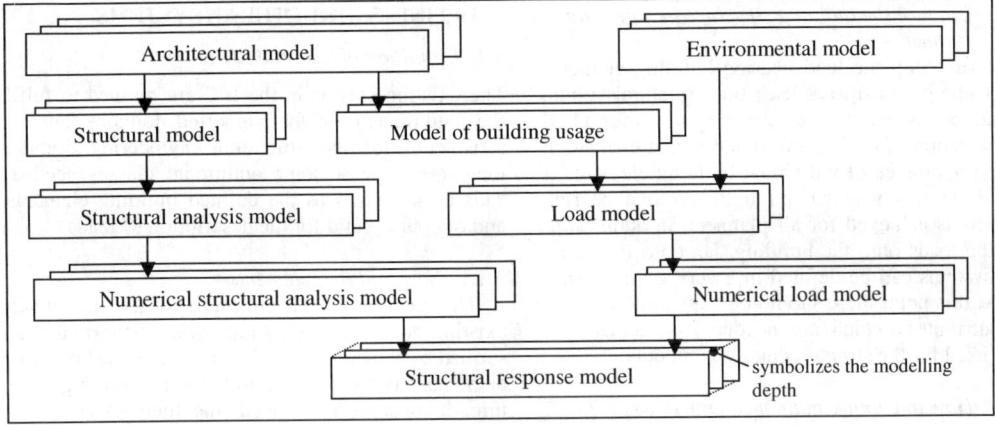

Figure 2. The typical abstraction process within the structural engineering domain.

3.2 Additional considerations

Beside the covered information there are some other considerations that have to be taken into account to develop a well suitable structural domain extension. At first there is the typical process for finding and defining the building structure that has to be supported. A general overview is given in figure 2. In detail, it offers the possibility to capture structural information in both a more general or a more detailed way. Furthermore, for the definition of the building structure it is desirable to enable the structural engineer to choose his own preferred way and not to force him into predefined ways of working. For instance, the structural engineer should have the chance to deal with both planar and spatial mechanical models. Furthermore the capture of the connection between architectural elements and the mechanical representations is also very important. How to fulfil those rarely described intentions is outlined in the following.

3.3 The concept of the proposed extension

The IFC architecture promotes a solution with two separate parts, which also matches the described considerations.

1.) A general, i.e. a "semantic" description of the building structure with the aim to provide needed structural information for planners mainly from other domains. Within the IFC architecture the Interoperability layer or a more common layer has to be used to introduce new classes.

2.) The description of the building structure as needed only for the structural engineer. From the IFC point of view a structural engineering domain encapsulated in the area of the Domain layer is introduced. It

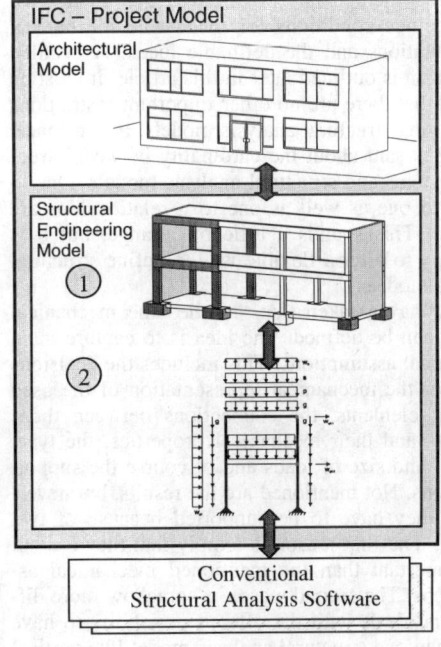

Figure 3. Shows the supported process and the two separated parts of the proposed extension.

should be used to store the main mechanical assumptions and as a starting point for the needed calculations. Therefore a mapping to the preferred proprietary analysis application is necessary.

This separation breaks down the definition of the building structure into two steps [Fig. 3].

3.3.1 *Capture the building structure in a general manner*

In the first step the load bearing building elements have to be grouped to at least one structural system that can be divided into several subsystems for a better distinction of the bearing function and evaluation of the importance of individual building elements if needed. In this way the publicity of load bearing elements is achieved for all planners. In addition to that, the loads onto the building, i.e. onto the structural systems can be defined in a very general manner. At this point there should be kept some elementary attributes, which are needed to calculate the loads used for the structural analysis models.

3.3.2 *How to capture main mechanical assumptions*

The second step is to derive structural analysis models from these previously defined structural systems. For the connection only a weak binding is used, but of course there exist some rules. The mentioned rules include conditions for the usable mechanical representations and the definable loads. A detailed description is outlined later in this article. It must be noticed that there are no other important restrictions for deriving structural analysis models. For instance, nothing is said about the cardinality between structural systems and structural analysis models. One to one, n to one as well as one to n relationships are possible. This sounds a little bit strange, but it is necessary to offer a flexible way to define structural analysis models.

After haven taken that "hurdle" the mechanical models can be defined. The idea is to capture main mechanical assumptions. This includes the decisions made for the mechanical representation of the used building elements, the connections between those elements and their mechanical properties, the type, location and size of loads and of course the support conditions. Not mentioned are the results, but nevertheless they have to be supported because of two reasons. The first reason is simply that they are not less important than the mentioned mechanical assumptions. The second reason is somehow more difficult and deals with the offered possibility to have more than one structural analysis model like vertical and horizontal load bearing systems. In such a case there exist dependencies between these models, of course. The intention is to capture those dependencies. Dependencies are understood in such a way that a reaction for one model can be an action for the other model. Therefore in order to keep those dependencies properly it is a necessary to capture results.

A more detailed description of handling those situations and related aspects is given below.

4 DEFINING STRUCTURAL SYSTEMS

4.1 *Definition of new classes*

The existing classes in the IFC are defined to fulfil the requirements of the supported domains, but for satisfaction of the structural engineering domain there are at least some additional classes needed. This is addressed to the defined building elements and of course valid for the description of loads.

4.1.1 *New building elements*

The intention of this concept is to use the already existing building information, which is normally described by the architect. Therefore, the used building elements have to be taken to form the building structure. It must be mentioned that there is especially one category of building elements that do not bother the design aspects from an architect, but are needed to carry loads. Those building elements, which are currently not part of the IFC project model, are the foundations. To avoid a "misuse" of other building elements, like the usage of IfcWall to describe a strip foundation, an extension of the IFC for foundation elements is proposed. Therefore two classes, IfcDeepFoundation and IfcShallowFoundation, and a few property set objects for further differentiation were introduced. It must be noticed that the introduction of those classes is not cogent, the existing building elements are rather independent from the proposed description of the building structure and the structural analysis models, they have no influence on the concept. The new building elements are mainly introduced to provide a possibility to describe foundations in a semantically correct way.

4.1.2 *Description of actions*

Actions are to be defined to deal with the physical behaviour of buildings. However, they are most associated with the structural engineering domain. But this is thought in a too short range. For instance temperature is needed in the structural engineering domain as well as in the HVAC domain.

Taking this into consideration a new general class – IfcAction – is introduced to represent actions in a very general manner. The idea is to differentiate the type of actions according to a special type attribute and to keep their individual features in type specific property set objects.

4.2 *Definition of structural systems*

As outlined before, all bearing building elements have to be grouped to at least on structural system. This is the way just to get the information whether a building element has a bearing function. This is knowledge that other planners need, for instance architects.

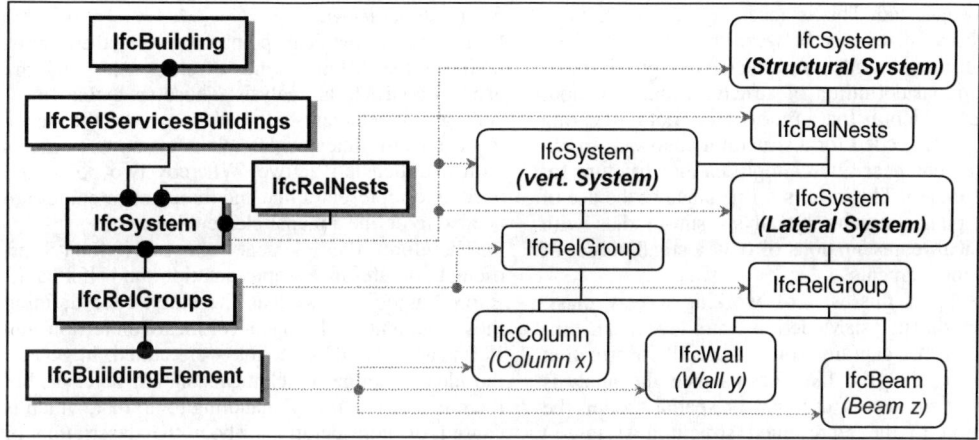

Figure 4. The used IFC entities and an example for the usage of a hierarchical description of structural systems.

4.2.1 *Grouping building elements*

To group building elements the functions provided by the IFC project model are used. A structural system is described by an instance from the class IfcSystem with a special setting for one attribute to be able to distinguish it from other systems. For grouping the bearing elements the relationship IfcRelGroups is used, which is explained in detail in [IAI (1999d)].

Often it may be useful to distinguish between different bearing functions and to have the possibility to evaluate the importance of individual building elements separately. Therefore a way to define more than one structural system and to arrange those structural systems hierarchically is foreseen.

4.3 *Hierarchically description of structural systems*

First of all it must be mentioned that one building element can belong to more than one structural system. In this case the building element has most probably more than one bearing function. For instance a wall usually carries vertical loads and may have also a function in the lateral system of the building. In this way it is possible to separate building elements which differ in their structural function. Using different structural systems is an opportunity to evaluate the importance of individual building elements. This is realized in two ways.

First, it is possible to arrange structural systems hierarchically by means of the IFC relationship IfcRelNests. It is used to subdivide into structural systems in smaller "pieces" with a more specialized structural function. An example is given in figure 4. In this figure the mainly used classes as well as a possible usage of those classes is shown. In this case the real general building structure is subdivided into

a lateral system and a bearing structure for vertical loads. Using this option it must be known that each building element of any specialized structure is also part of the structure, of course.

Second, it is possible to define a load flow between those different structural systems. The idea is to describe which structural system supports which of the other ones. If a foundation system is additionally subdivided, it is obvious that the foundation system supports the lateral and vertical structural bearing system and it can be defined. This is the way to define functional relationships between structural systems and it is realized by the introduction of property set objects.

4.4 *Handling actions*

A next step is to define actions and to assign them to the above outlined structural systems which deal with them. Those actions can be seen as a starting point for the load flow to the ground. The assignment to structural systems is also realized by property set objects.

5 DEFINING STRUCTURAL ANALYSIS MODELS

5.1 *Defining structural analysis models*

The above outlined definition of structural systems is used to capture more general structural information. It supports the walkthrough process of finding an appropriate building structure. This fulfils the requirements of planners from other domains, but doubtlessly additional information are needed to support analysis applications. Therefore deriving structural analysis models from those structural systems is suggested. This is necessary because there is an important difference between the intention of defining structural models and the desired providing of an individual definition of structural analysis models. The problem is that in most cases not all bearing elements are needed for a structural analysis. An experienced engineer often simplifies the structure for the calculation. This results for instance in the use of several planar structural analysis sub-models with mechanical representations of only a small subset of the bearing elements.

To support this way of working a new class, IfcStructuralAnalysisModel, is introduced. The intention is to capture some special information needed for analysis, like for instance the analysis method or the use of a planar or spatial model, the connection to the represented structural system(s) and of course the used mechanical representations belonging to this model. It must be noticed that there are some constraints for the disposable mechanical representations and the definable loads. In more detail, it is only possible to use those mechanical representations of building elements, which are even part of the represented structural system(s). Keeping this in mind, it is understandable to recommend to be more careful in defining structural systems, not for avoiding handicaps in the later work, more over it can be useful in defining structural analysis models.

5.2 The used topological representation

As above marginally outlined, the structural analysis models deal with mechanical representations and not with the "real" building elements. Therefore and in order to cover the representation of mechanical connections, the support conditions and the load ports a topological representation is proposed. This is the most sophisticated part of the proposed extension. In the following only a compact explanation is given.

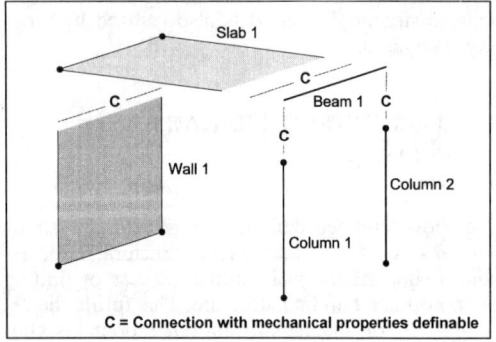

Figure 5. The principle of the used topological description of structural analysis models.

5.2.1 Used elements
From the mechanical point of view, there exist mainly three different kinds of elements, which are used in a structural analysis. Those are point, linear and planar elements. In other words the range reaches from dimensionless elements up to elements with a dimension of two. Whereby two points are needed to represent a line and three or more lines are needed to define a planar element [Fig. 5].

Therefore the element classes IfcPointRepresentation, IfcLinearRepresentation and IfcPlanarRepresentation as well as the classes for defining those elements – IfcPlanarToLinearConnection and IfcLinearToPointConnection – are added. In general the idea to define the elements should be clear, but there is a little "tricky" handling to do that, which is caused by considerations about the description of mechanical connections. The most needed features for that "tricky" handling are captured in the abstract super classes IfcStructuralConnection and IfcStructuralRepresentation.

5.2.2 The class IfcStructuralConnection
This class is used to define common properties for the subclasses IfcPlanarToLinearConnection and IfcLinearToPointConnection, which are used to define planar and linear representations respectively with the mechanical behaviour of their connections. Taking into consideration that some frequently needed connection types can be described semantically without using complicated mechanical properties the attribute PredefinedType is introduced to distinguish between the easily describable rigid and pinned connection and more complicated connection types. For such other types of connection special property set objects are used to describe the mechanical behaviour. At this point it must be mentioned that an easy description of a rigid connection is of special interest in the proposed concept.

5.2.3 The class IfcStructuralRepresentation
A structural representation can be an element, a "node", a support, a load port – load port stands for a place where a load acts on the structural system – or any combination of them. This differentiation is defined by a special attribute and is used to assign the right property set objects to the structural representation. Beside that there exists an inverse relationship to the structural analysis models in which the structural representation is used, an optional relationship to material properties needed for defining element behaviour and the most interesting BelongsTo relationship.

The name BelongsTo points to the purpose of that relationship. Any instance of IfcStructuralRepresentation belongs to exactly one IfcBuildingElement or another instance of IfcStructuralRepresentation. To which one it belongs depends on its type. If a struc-

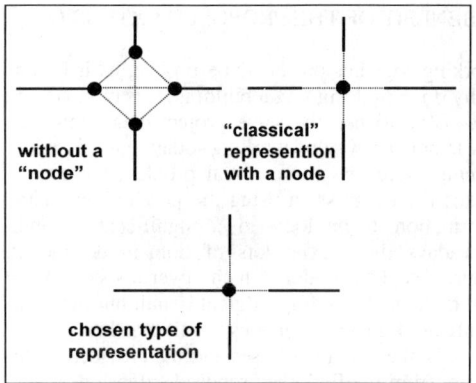

Figure 6. Different types for the definition of connections and their mechanical properties. The dotted lines symbolize definable mechanical properties for the connection.

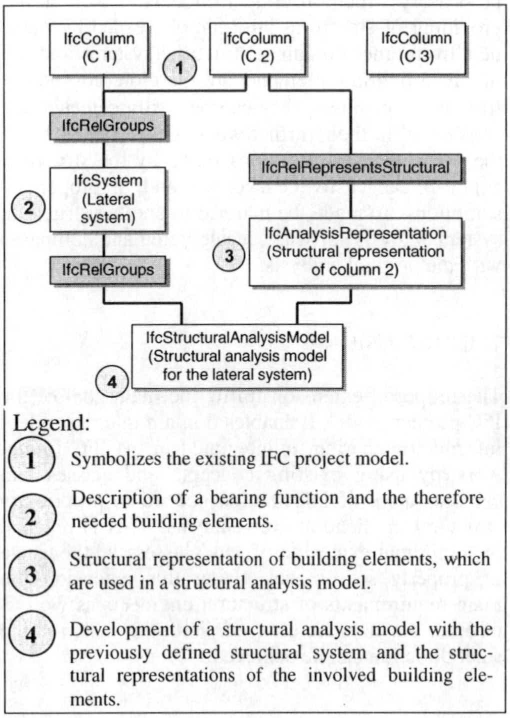

outlined before, for defining linear or planar structural representations other instances of IfcStructuralRepresentation are needed. In the case of a linear representation two "node" points are used. Normally, such "nodes" are used by more than one linear representation and therefore the usage of the BelongsTo relationship has to be explained in more detail.

Such a "node" belongs to exactly one other structural representation, here to a linear element. With this relationship a constraint of a rigid connection between these two structural representations is defined. This deals with the fact that for a mechanical connection of n elements the definition of only n-1 mechanical connection properties is needed. Figure 6 illustrates these considerations.

5.3 Connection to building elements

The connection between building elements and their mechanical representations is realized by the relationship IfcRelRepresentsStructural. [Fig. 7]

This connection has an effect on the selectable material properties and the used coordination system. In more detail, the mechanical representation can use only one of the material defined by the connected building element.

A coordination system is only needed for point representations, because within the used topological representation they are defining the location of each other element. To find the right coordination system of a point the BelongsTo relationship has to be traced until a building element is reached.

5.4 Handling of loads and results

First of all, there exists the possibility to define point loads, different kinds of linear loads and planar loads, which can be organized to load groups, load cases or load combinations. This corresponds to the typically usage within the structural engineering domain and supports the traditional and most practical way of handling loads. Therefore the classes IfcPointAction, IfcLinearAction, IfcPlanarAction, their abstract superclass IfcPhysicalAction and for grouping loads the class IfcLoadGroup are defined. This classes are used to describe both actions and reactions, because an action is in principle a reaction with an inverse signed value. Therefore only one new class, IfcStructuralResultGroup, is additionally needed. To fulfil that the classes IfcPhysicalAction and IfcStructuralResultGroup are of interest.

5.4.1 Definition of actions

The most interesting point is the connection between an action and the structural representation on which it acts. Therefore a structural representation or more precisely a load port within the same dimension as

Figure 7. The connection between building elements, analysis representations, structural systems and structural analysis models by using an example.

tural representation is used to represent a building element – for instance a linear representation is used to represent a column –, it is of the type "element" and must be connected to this building element. As

235

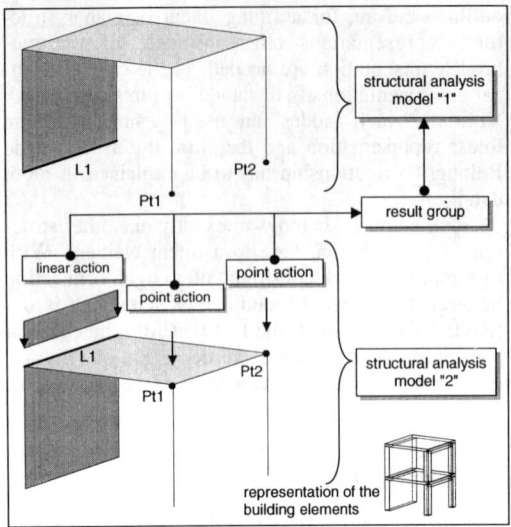

Figure 8. The connection of these two structural analysis models by using the same topological objects and the handling actions to define dependencies.

the load must be defined. With this load port the location and the addressed structural representation are defined. The amount and the direction of an action is then described through special property set objects.

5.4.2 *IfcPhysicalAction*

Major features for describing loads respectively results are realized through this IfcPhysicalAction. This includes information about the used coordination system, the type (action or reaction) and the membership to a structural analysis model.

It must be noticed that the type attribute of IfcPhysicalAction can be derived from the different types of its membership to structural analysis models. Therefore this attribute is not absolutely necessary, but it covers some performance considerations.

5.4.3 *Action versus Reaction*

Using the possibility to describe several interacting structural analysis models this concept is used to capture those connections. Knowing that a reaction for one structural analysis model can be an inverse signed action for another the connection is captured through the usage of the same physical action object for both structural analysis models [Fig. 8].

To do so there are some additional considerations that have to be taken into account, for instance for the used load combination factors or to distinguish actions from reactions. These facts are only mentioned here. For further details see Weise, M. (1999).

6 BENEFIT OF THE PROPOSED EXTENSION

Working together on the same project, as it is normally done in designing a building, it is obvious and reasonable to use the same project data. However, this is not the way of working today, but this is becoming more and more a real problem, because it brakes the progress in using the possibilities of the information technology in a significant manner. Nowadays there exist lots of domain dependent, proprietary data models which cover a special view of the planned building in great detail, but the commonly used information does not fit together.

Using the existing classes and the provided architecture from the IFC, the proposed extension fills an information gap between the structural engineering domain and other domains. The structural engineer is able to use already captured building information and has the possibility to associate structural assumptions with them. This is the most valuable achievement and enables a more easier mapping to proprietary structural domain models as well as the providing of structural information needed by planners from other domains. Structural systems and the involved building elements are therefore not hidden from other planners, they can be easily detected and considered in their further work. On the other hand, the mechanical assumptions made by the structural engineer can easily be checked and, if proper assumptions are made, be mapped to another structural system analysis software, recalculated and compared with the original analysis.

7 CONCLUSION

The proposed extension fulfils the main goal of the IFC project model. It enables domain interdependent information sharing, is integrated in the IFC framework by using existing concepts and classes and captures a lot of information needed in proprietary structural applications for analysis. This is reached by a minimal extension of new classes and the usage of property set objects. Using this extension the main requirements of structural engineers as well as of other planners are covered. With this a step in the right direction can be achieved.

8 REFERENCES

IAI ST Domain Japan chapter. Draft 2, 1998. *IFC R3.0 Domain Project Documentation, [ST-2] Reinforced concrete structure and foundation structure*. IAI.
International Alliance for Interoperability (IAI). 1999a. *IFC Object Model Architecture Guide*. Oakton/ Virginia: IAI.

International Alliance for Interoperability (IAI). 1999b. *IFC Specifications Development Guide*. Oakton/ Virginia: IAI.

International Alliance for Interoperability (IAI). 1999c. *IFC Object Model Guide - Specifications*. Oakton/ Virginia: IAI.

International Alliance for Interoperability (IAI). 1999d. *IFC Object Model Reference - Specifications*. Oakton/ Virginia: IAI.

ISO 10303-11 IS. 1994. *The EXPRESS Language Reference Manual*. ISO TC 184/SC4. Geneva: ISO.

Rumbaugh J., Jakobson I., Booch G. 1998. *The Unified Modeling Language Reference Manual*. Addison-Wesley: N.Y.

Ward, Michael A.; Watson, Alastair S. Working Draft, 1996.: *ISO 10303 Part 230 Building Structural Frame: Steelwork*. University of Leeds.

Weise, M. 1999. *Konzeption und Validierung eines objektorientierten Tragwerksmodells für den Bereich Hochbau auf der Basis des IFC-Projektmodells 2.0*. Dresden: Diploma Thesis.

Integration of resources and scheduling information for building and construction, linking IAI/IFC and ISO 13584

R.Jardim-Gonçalves, R.Tavares & A.Steiger-Garção
Università Nova de Lisboa, Facultat CiêncIas e Tecnología, Departamento Engenharia Electrotécnica, UNINOVA, Instituto de Desenvolvimento de Novas Tecnologias, Portugal

A.Grilo
Fordesi – Formação, Desenvolvimento e Investigação, Lisboa, Portugal

ABSTRACT: Integration of resources and scheduling information in building and construction environments is essential to support and automate its management tasks. Nowadays there are not many proposals to represent catalogues of resources ready to be adopted by that industry. The ESPRIT 25559 Project, Supply Chain Management in Construction Industry – SUMMIT, developed an IAI/IFC-based integrated platform for distributed information flow. This paper presents the results of the research done to support the integration of applications in such platform, and gives a proposal to extend such platform to integrate catalogues of resources using standard ISO 13584 – PLib.

1 INTRODUCTION

The ESPRIT 25559 Project, Supply Chain Management in Construction Industry – SUMMIT developed an Electronic Data Interchange (EDI)-based communication infrastructure between the various partners involved in the manufacturing and construction of prefabricated houses, which automates the tendering, ordering, delivery, invoicing and payment processes, in a context of heterogeneous IT systems [3].

The adopted methodology focused on: *i*) the analysis of the current business processes information infrastructure and grounded on the possibility of integration to support derive of new business processes. *ii*) seamless integration of the most advanced technologies, like e.g. Internet, Workflow, STEP/EDI Standards, Procurement Systems both in terms of current R&D developments and commercial applications.

SUMMIT is dedicated to the exploitation of potentials of both EDIFACT and IFC approaches to product and process data communication, by a concept of combining the complementary nature (business versus technical data) of both approaches. This, together with an EDI system based on an inter-organizational workflow system, enables a better coordination between the companies on the supply chain, from the client's representative (project manager), to the contractor, and suppliers. Major results by SUMMIT are:

- A procurement process model exploiting the potential of pure digital data exchange and system interoperability
- EDIFACT messages sub-sets for the procurement of house systems and equipment
- IFC product/process model enabling the representation of project management data for prefab houses
- An EDI communication infrastructure featuring secure and compatible information exchange for the entire production process
- A distributed information flow component based on EDI and inter-organizational workflow infrastructure, supporting process definition and process execution in the boundaries of autonomous organizations
- Use of commercial application for fast-track implementation of the ICT infrastructure, and linkage to widely used commercial and technical applications.

1.1 *SUMMIT Developments in Project management data*

EDIFACT does not provide the solution to the whole spectrum of information exchange between heterogeneous IT applications within the supply chain (see description of SUMMIT results regarding EDI and EDIFACT in [1,2,4]). In SUMMIT it was decided that the exchange of commercial and administrative data using EDIFACT standards, would be complemented with the exchange of project management data but using a more adequate standard for the exchange of product and process data.

Failed an initial attempt to use developments in the Building Construction Core Model (BCCM) by the lack of in-depth work, SUMMIT focused on the developments on the Industry Foundation Classes (IFC) within the work of the International Alliance for Interoperability (IAI) [5]. SUMMIT worked in the Project Management Domain, in particularly in the Domain project PM-1 Scheduling, in co-operation with current IFC 3.0 Release working groups on that area. However, for the implementation, it were used the IFC 2.0 Release Specifications, issued in April 1999, which incorporated much of the work being developed for the IFC 3.0 Release.

In the IAI/IFC approach, the construction scheduling process creates a construction schedule using the objects across the IFC Model. In general, construction schedules will be developed through analyzing the Task objects and Resource Use objects created when developing the Cost Estimating, and aggregating them into Construction tasks at the appropriate level for scheduling. This process must consider the size and complexity of the model. In situations when there was no cost estimating, construction scheduling process will be different, implying a querying to the shared project model, and the objects found will be used as the basis for developing construction tasks. In both situations, time duration of the tasks will be estimated and construction sequences will be identified. The construction schedule is then created and analyzed, and after completion of the schedule, dates will be embedded in the IFC Model.

Despite the IFC/IAI defines a clear model for the development and integration of scheduling and project management data with other components of the building process and product, there was an underly-

ing philosophy of SUMMIT for systems integration of existing commercial applications and R&D developments, rather than an approach of major software developments or programming. Thus, the adopted approach (depicted in Figure 1) is grounded on the concept that scheduling applications are linked through ODBC (or other dynamic link) to an IFC scheduling database. The format of the database is based on current specifications of IFC 2.0 Release. As a result no IFC compliant scheduling applications, on top of the database an IFC interface (combined with an editor/browser) simulates a future IFC compliant project management application.

The technological complexity of the SUMMIT project, was to make simple (and using mostly commercial applications and state-of-the art R&D developments) the implementation of an integrated system combining IFC and EDIFACT information in a common communication infrastructure defined by an Internet-based inter-organizational workflow system, that co-ordinates the business and management information flows between project manager, contractors and suppliers in the various stages of building prefab wood houses.

To have a library of resources described in a standard format could help in scheduling tasks, since the process of search, selection and assignment of resources to tasks could be done in a normalized way, independently of the source supplier.

Although out of the scope of SUMMIT project, after project conclusion this approach was considered and implemented at UNINOVA, based on ISO13584 – PLib for representation and exchange of libraries of resources, and with a direct link with the developed SUMMIT model.

2 THE TRANSLATOR MANAGER

The Translator Manager (TM) is the application developed for SUMMIT to support the integration of applications in the domain of workflow management for Building and Construction environmemnts, when using the SUMMIT model as the support for the data exchange.

The main role of TM is to put available the data to the applications joining the integrated platform, in the place and format compatible with the several those applications. TM is able to convert:

- STEP [6] data into Microsoft Project data.

- Microsoft Project data into STEP data.

- STEP data into Microsoft Excel data

- Microsoft Excel data into STEP data

- STEP data into Microsoft Access data

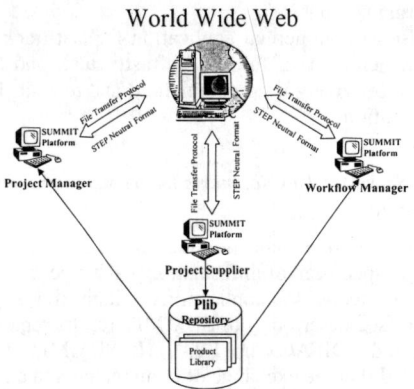

Figure 1. The SUMMIT integrated environment

- Microsoft Access data into STEP data

  The TM architecture is based on:

- SDAI repository, where STEP data is, or will be stored in the STEP platform.

- SDAI model, where STEP data is, or will be stored in the STEP platform.

- input_file/output_files for data import/export. If the data conversion is from STEP data, then the input file name is the configuration file of the STEP platform . Otherwise the input file, is a CSV format file or an Access database file.

2.1 TM's Interfaces for data translation and exchange

The interface of the Translator Manager to the exterior is established using a set of drivers. TM uses a CSV driver, the code generated by Genesis and DAO objects.

2.1.1 CSV Driver

The manipulation of CSV data is made through a text *driver* developed to handler the data separated by a character separator. It is implemented in C++ and based on the following classes: class *TDB*; class *TDB_Line*; class *TDB_SlimField*; and class *TDB_FatField*.

- The *TDB* class handles the complete file. It has methods to read from and write to a file. The output and input parameters are objects of class *TDB_Line*. Below is presented the structure of this class:

```
// -- TDB ~ Text DataBase class ----------------------------
class TDB {
    FILE*   fileDB;
    char    fieldSep, aggrDel, subSep;
    int     maxLineLen, maxFieldLen;
public:
    TDB(char* fileDBName, char inFieldSep, char inAggrDel, char inSubSep, int
inMaxLineLen, int inMaxFieldLen);
    ~TDB(void);
    void        writeLine(TDB_Line* inLine);
    TDB_Line* readLine(void);
    void        readDB(UNI_List& excelLine);
    void        showContents(void);
    char*       nameOf(void);
};
```

- *TDB_Line* class handles the lines of text. It retrieves and stores a field (simple or complex) from and to a line. The parameters are objects of class *TDB_Field*, which is an abstraction of the two classes that implement the CSV fields: *TDB_SlimField* and *TDB_FatField*.

- *TDB_SlimField* class handles simple CSV fields. The input and output data of this class are simple strings, which represent the values.

- *TDB_FatField* class handles complex CSV fields. The input and output data of this class are objects of the class *TDB_SlimField*. This class separates the values of the different simple field by the character separator for the aggregate fields. It also can retrieve the complete complex field as a simple string, which is the value of the aggregation of fields with simple values.

The constructor of the *driver* is the association of a CSV file associated with an object of the class *TDB*. The manipulation of the several CSV fields is made through the *TDB_Field* (*TDB_SlimField* and *TDB_FatField*) objects in TDB_Line objects, which are part of the CSV database.

The reading process reads lines of text. The identified fields in there can be simple or complex.

The class *TDB_Line* provides a way to determine if the field is simple or complex. It also decides to get the value of the field or, to retrieve the simple fields inside a complex CSV field. The fields in a complex field are handled like simple fields, before decomposition.

The writing process is in the opposite way. The fields are written to a line, and the object line provides the mechanism to separate the fields with separator character. The same procedure is done for complex fields. The simple fields that compose the complexes are written into the *TDB_FatField* object that provides the mechanism to separate them with the separator character in the string value.

The *TDB* object representing the database, receives *TDB_Line* objects and saves them on the CSV file associated with it.

2.1.2 STEP – Genesis Code

The code generated by Genesis is a C++ high-level library that helps the access to the STEP data via the Standard Data Access Interface (SDAI) of the STEP standard [7,8,9,10].

This layer of code abstracts the user of the complexity of SDAI, virtualizing most of its functionalities, avoiding the need of having deep knowledge about SDAI functions to implement STEP, and since the code is generated automaticaly based on a EXPRESS defined model, avoids the production of implementation errors when using it.

2.1.3 DAO objects

The DAO objects are used to handle the Microsoft Access data, and establish the programatic interface between TM and the database.

The process to access the Microsoft Access database is seamless to the one described for the access to the CSV database. There is an object that represents the database and other to gather the data rows.

First, it opens the database and associates the object *CDaoDatabase* with the database file, and the rows are collected in the object *CdaoRecordset*. Therefore, the fields in each row are retrieved by name of the field.

The complex fields are retrieved as simple fields, and separated afterwards using the separator character.

2.2 *The sequence of actions of TM*

The behavior of TM is conducted by the sequence of actions as depicted in the diagram of Figure 2.

After decision by the user of what kind of translation wants (i.e., CSV/Access -> STEP or vice-versa), it is selected the local data file and the STEP platform.

In case of a local translation to STEP, after translation done the resultant file with the data in STEP Neutral Format is written in the file system. For a remote choose, the server is contacted and the STEP file uploaded there.

For the local translations from STEP, the procedure is to select and read the file with data in STEP neutral format, and proceed with the translation. In the remote case, after contact with the server it uploads the file, reads it and then proceeds with the translation.

3 ISO 13584 (PLIB)

The ISO 13584, Parts Library (PLib), is an International Standard for the computer-interpretable representation and exchange of part library data (i.e., Dictionary, Catalogue and Functionalities) [6].

PLib provides a neutral mechanism capable of transferring parts library data, independent of any application that is using a parts library data system. It does not specify the content of a supplier library.

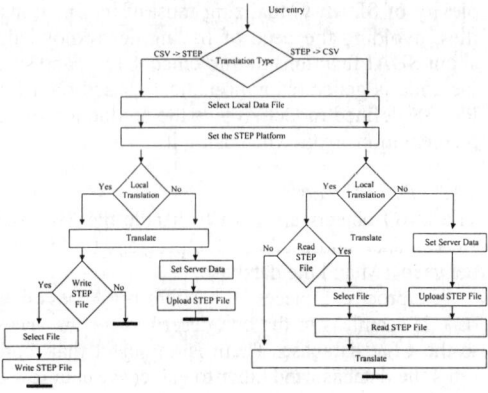

Figure 2. TM: Sequence of actions

The content of a supplier library is of responsibility of the library data supplier.

The library management system used in the implementation of the structure defined in ISO 13584, and any interface between this system and a user of the system, is the responsibility of the library management system vendor and is not specified in this standard.

PLib is organised in a set of Partsand includes:

- the representation of parts library information, including libraries of components and libraries of assembled parts;
- the exchange of parts library information, including storing, transfer, accessing, updating and archiving.
- the ability of a part supplier to describe the part the supplier provides without any reference to any other libraries or external dictionaries.
- the definitions of mechanisms that permit reference to standardised dictionaries from a supplier library (when they are available).

3.1 *PLib capabilities*

PLib has two major capabilities:
1. Definition of dictionaries of technical properties / components
    - Defining / exchanging / referencing
    - Already used by standard committees: IEC 61360, ISO TC 29, etc.
2. Full electronic exchange between suppliers and users
    - Complete catalogues
    - Support for selection / behaviour / representation / corporate integration
    - May be identified / referenced in STEP product data

The use of PLib implies two main structures: Dictionary and Catalogue.

3.1.1 *Levels of use of library parts in product data*
PLIb identifies three levels of use of library parts in product data:
- Level 1. At this level, all information about a part in a System A will be transferred to a System B exclusively by means of ISO 10303 (STEP). This means that System A, with direct interaction with its Supplier Library described in Plib format, will get all required information about a part and, after collect the information about the part (e.g., geometry) will translate and transfer it to System B using STEP representation.
- Level 2. At level 2, only that information that is necessary to describe the same part from a Library from System A to System B is trans-

fered. This implies that Library of System A and Library of System B both must contain the same and complete information about the part. Therefore, this level allows at setup stage an exchange of Library's information between systems using Plib representation, impling that for product data exchange between systems A and B only references to those parts described in their library need to transfered described in STEP.

- Level 3. Information transfer from System A to System B at Level 3 is done without any assumption about the content of Library at System B. This means the information transfered from System A to System B contains a subset of Library of System A. In this level Plib (Supplier Library) and STEP (Product Data) representation are used together in the data exchange channel.

3.2 *Conceptual model of a supplier library*

The conceptual model of a supplier library is described in Part 10 of ISO 13584 – PLib. It uses the OO paradigm for representation of class, and it is structured hierarchically following the simple inheritance relationship. It considers:

- a family of parts, i.e., a general model class
- the representations of the different parts of a parts family that belong to the same representation category; i.e., functional model class
- the definitions of the different representation categories capable of being provided for any family of parts, i.e., a functional view class

3.3 *Description of a supplier library*

Each parts family that constitutes a supplier library has two levels of description:

1. Describes the concepts about the parts family. Each parts family is formally defined by a set of attributes, that constitutes the dictionary definition of the corresponding parts family. It is represented as a dictionary_element, describing only the abstract concept the parts family corresponds to. It does not define the set of parts that constitute the parts family. Therefore, a dictionary_element may be exchanged and stored in the semantic dictionary of the receiving system.
2. Describes the set of parts that belongs to the family. Constitutes the library specification of the corresponding parts family, and it is represented as a content_item intended to be stored in a user library. The content_item specifies, implicitly, all the parts that belong to the parts family.

3.4 *The BSU mechanism*

When using PLib, the Basic Semantic Unit (BSU) mechanism provides the independence between dictionary_elements and content_items.

Each piece of information intended to constitute a dictionary entry is therefore represented through three entities:

1. the basic_semantic_unit (BSU) entity, that carries the universal identification of this piece of information
2. the dictionary_element entity, that contains the set of attributes that constitutes the dictionary description of this piece of information. e.g., name, definition and type of value
1. the content_item entity, that represents the possible values of this piece of information. e.g.; the content_item of a class, represented by the class_extension entity, specifies the possible instance values of this class.

3.4.1 *Reference between several EXPRESS schema populations via the BSU mechanism*

When a dictionary entry is intended to be referenced in some population, only its BSU entity instance is requested to belong to this population.

So, reference to a BSU stands for a reference to the complete piece of information that constitute a dictionary entry, whether its dictionary_element and content_item entities belong to the same population or not.

This will ensure independence between:

- the dictionary_element entity instance
- the content_item entity instance
- the BSU entity instance for any dictionary entry.

This mechanism also provides a means to implement references between dictionary entries that belong to different EXPRESS schema populations, possibly stored in different exchange files or data repositories.

4 LINKING THE SUMMIT MODEL WITH PLIB

One of the sensitive aspects to support the optimization of the scheduling tasks is the selection and assignment of resources.

These resources are very often provided from different suppliers and the information referent to the available resources is most of the times not made available electronically.

Even when it is, the data formats are not compatible between the different sources, meaning that applications managing with this information need to have translators to all of them they intend to support, to be able to use electronically such information.

To have available catalogues of resources described in a standard format will open the way to go towards a complete integrated scheduling system.

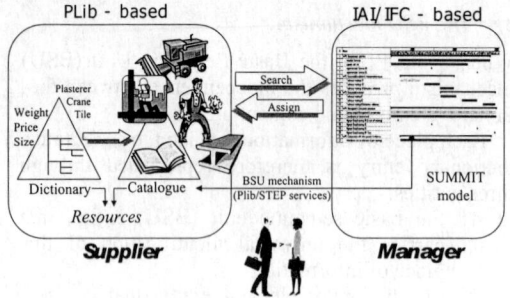

Figure 3. Integration of SUMMIT model with PLib.

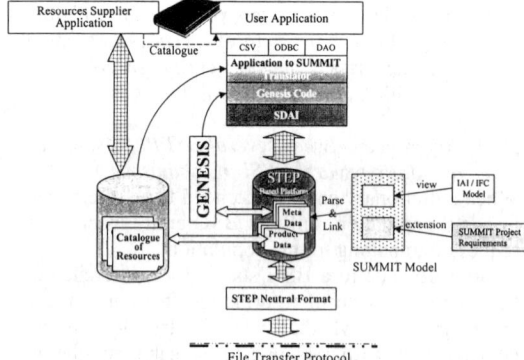

Figure 4. Extended architecture of SUMMIT to support PLib

Based on the platform resulting from the SUM-MIT project, it was done one extension to support the use of such catalogues modeled using PLib.

Figure 3 depicts the general architecture of this approach.

The supplier describes its catalogue of resources using the PLib methodology, based on a dictionary previously agreed by those operating in this field.

Therefore, this catalogue is ready to be sent to those parties supporting this standard and interested in such data.

On the Manager side, he can immediately consult and search data on this catalogue, or on any other also adopting this standard.

For the assignment of resources to the tasks, the PLib's BSU mechanism should be used as a reference from one task to one resource. This avoids the obligation to import of all data referent to the resource for the schedule data representation, as it happens nowadays when using pure STEP.

Using this reference mechanisms, and having the Manager access to the catalogues describe in PLib (from different suppliers if he wants), he is able to consult the catalogues, search for those resources

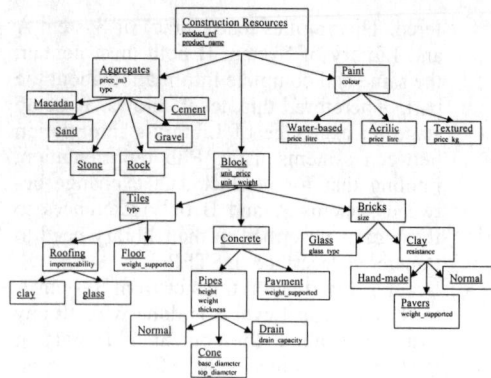

Figure 5. Example of a dictionary for a catalogue of resources

wanted, and select and assign to the tasks those he is interested to have.

Figure 4 presents the extended architecture of SUMMIT to support the supplier's libraries using ISO13584 – PLib.

Figure 5 depicts the example of the dictionary used to represent the structure of the catalogues used on this platform.

Below there is an extract of the description of the supplier library in STEP Neutral Format, following the PLib representation based on the dictionary presented in figure 5.

```
ISO-10303-21;
HEADER;
FILE_DESCRIPTION(('SUMMIT Material Resources'),'2;1');

...
FILE_SCHEMA(('ISO13584_G_M_IIM_SCHEMA'));
ENDSEC;
DATA;

...
#1369=ITEM_NAMES(LABEL('Construction
Resources'),(),LABEL('ConsRes'),$,$);
#1183=PROPERTY_DCU('0001U','001',#1370),
#1184=NON_DEPENDENT_P_DET(#1183,#1374,'001',#1182,TEXT("),$,$,$,$,(),$
,$,#1017,$);
#1182=ITEM_NAMES(LABEL('Product name'),(),LABEL('prod_name'),$,$);
#1190=PROPERTY_BSU('0001a','001',#1370);
#1191=NON_DEPENDENT_P_DET(#1190,#1374,'001',#1189,TEXT("),$,$,$,$,(),$
,$,#1017,$);
#1189=ITEM_NAMES(LABEL('Product reference'),(),LABEL('prod_ref'),$,$);
#1364=CLASS_BSU('0011','001',#1375);

...
#1175=ITEM_NAMES(LABEL('Price per m3'),(),LABEL('price_m3'),$,$);
#1346=CLASS_BSU('0111','001',#1375);
#1347=MATERIAL_CLASS(#1346,#1374,'001',#1345,TEXT("),$,$,$,#1364,(),(),$,(
),(),$);
#1345=ITEM_NAMES(LABEL('Macadam'),(),LABEL("),$,$);
#1340=CLASS_BSU('0211','001',#1375);
#1341=MATERIAL_CLASS(#1340,#1374,'001',#1339,TEXT("),$,$,$,#1364,(),(),$,(
),(),$);
#1339=ITEM_NAMES(LABEL('Sand'),(),LABEL("),$,$);
#1281=MATERIAL_CLASS(#1280,#1374,'001',#1279,TEXT("),$,$,$,#1352,(),(),$,(
),(),$);
#1279=ITEM_NAMES(LABEL('Bricks'),(),LABEL("),$,$);
#1032=PROPERTY_BSU('0321','001',#1280);

ENDSEC;
END-ISO-10303-21;.
```

REFERENCES

[1] Grilo, A., "The development of electronic trading between construction firms". Published PhD Thesis in the University of Salford, 1998.

[2] Almeida, L., Grilo, A., Rabe, L., Duin, H. (1998), "Implementing EDI and STEP in the Construction Industry", *2nd European Conference of Product and Process Model in the Building Industry – EC-PPM'98*, Watford, UK, Proceedings.

[3] SUMMIT, (1998). "Project summary". http://www.exnorm.com/r&d/summit/pages/summary.htm

[4] Rabe, L., Hagen, N., Duin, H., Petersen, H., Haubler, G. (1998), "Report on Harminsed Message Contents at ExNorm". SUMMIT Internal Report.

[5] IAI/IFC, http://www.interoperability.com/

[6] ISO TC184/SC4, http://www.nist.gov/sc4/

[7] Jardim-Gonçalves, R., Silva, H., Vital, M. Sousa, P., Steiger-Garção, A. and Pamiés-Teixeira, "Implementation of Computer-Integrated Manufacturing Systems using SIP: CIM case studies using a STEP approach", Special issue of International Journal of Computer Integrated Manufacturing on Design and Implementation of Computer Integrated Manufacturing Systems. Vol.10, Jan-Aug. 97.

[8] Jardim-Gonçalves,R., Pimentão, J.P., Sousa, P.; Vital, M., Silva, H. and Steiger-Garção,A. "Integrating applications for the construction industry using a STEP-based integration platform (SIP) - Seeking a life-cycle-oriented modelling", Procedings of Workshop in Modelling of buildings through their life-cycles, pp. 211-222, University of Stanford, Aug 95

[9] Jardim-Gonçalves, R., Pimentão, J., Vital, M., Sousa, P., Silva, H. and Steiger-Garção A., "RoadRobot project - An experience using STEP in construction", 13th ISARC, Tokyo, June'96.

[10] Sousa, P., Pimentão, J., Jardim-Gonçalves, R., Steiger-Garção, A., "A STEP (ISO-10303) based architecture for customer-manufacturer integration - approach and results of EU-ESPRIT project funStep" The Third International Conference on Technical Informatics - CONTI98, October 29-30, 1998, Timisoara, Romania, as a volume of Transactions on Automatic Control and Computer Science, 98.

Product and Process Modelling in Building and Construction, Gonçalves, Steiger-Garção & Scherer (eds)
© 2000 Balkema, Rotterdam, ISBN 90 5809 179 1

STEP based application protocol under UML for a specific building code

I.A. Santos
Algarve University, Faro, Portugal

F. Hernández-Rodríguez
Sevilla University, Spain

ABSTRACT: This paper proposes the use of a single modeling method for the development of application protocols within the ISO-STEP standard. In particular, the use of UML (Unified Modeling Language) is proposed. To demonstrate feasibility, the development of an AP in the AEC field is described; specifically an AP for the Portuguese code for urban building (RGEU, "Regulamento Geral das Edificações Urbanas"). The ultimate goal of on going work is to establish a conceptual framework for the later support of the building design conformance checking to the mentioned code.

1 INTRODUCTION

The ISO STEP standard introduced a technology based in the concept of Product Modelling (Bloor et al. 1995), which represents an evolution from traditional approaches regarding some important aspects (Hernandez-Rodríguez 1995): (1) it includes the definition of the formal language EXPRESS to specify all the information requirements within the standard; (2) its architecture defines specific functions for different series of components, like Integrated Generic and Application resources, or Conformance Testing; (3) the development strategy is based upon the concept of Application Protocol, which guaranties the independence of the information requirements specification from any particular implementation within an application domain.

An Application Protocol (AP) includes the specification of the scope, the context and the information requirements of an application domain. The process of AP's development consists of the following phases: first, an Application Activity Model (AAM) is built, which denotes the threads, the data flows and the functional requirements of the domain, making use of an appropriated description method like IDEF0 language (Softech 1981); second, an Application Reference Model (ARM) is developed, which describes the information requirements of the application domain under a formal modeling language, such as IDEF1X (D. Appleton C. Inc. 1985), NIAM (Nijssen et al. 1989) or EXPRESS-G (ISO 10303-11:1994); finally, the correspondence between the application information requirements and the STEP Integrated Resources is established by means of a set of Application Interpreted Constructs, resulting in an Application Interpreted Model (AIM) specified under the EXPRESS language.

Although this methodology continues to be effective in its basic principles, the above enumeration shows some important limitations, mainly due to the need of using different (and partially incompatible) methods for describing each of the three models. Among these limitations are: (1) the models obtained are clearly disconnected, specially the AAM and ARM models; (2) to correctly interpret the different models one must have a deep knowledge of several modeling languages; (3) some methods are only oriented to human interpretation, not to computer representation and processing. The negative effect of these limitations concerns not only the AP's development but also the support to its implementation and the general acceptance of STEP.

In order to avoid such difficulties, this paper proposes the use of a single modeling method for AP's development, based upon the existence of more recent and powerful modeling languages. In particular, the use of UML (Unified Modeling Language) is proposed.

To demonstrate the feasibility of this proposal, the development of a STEP AP for the AEC sector is described, which relates to the Portuguese code for urban building (RGEU, "Regulamento Geral das Edificações Urbanas" (RP 1997)). Among other resources, STEP dedicates part 106 (ISO 10103-106:1997), included as Integrated Application Resources, to the so-called Building Construction Core Model (BCCM). The more recent IAI group (International Alliance for Interoperability) is also dedicated to the AEC sector and it has been developing the IFC standard (Industry Foundation Classes (IAI 1997)) from a former close collaboration with the ISO technical committee responsible for BCCM. For the reasons exposed later, the STEP-BCCM is choose as the basis for the proposed AP.

The content of this paper derives from an on going work for a PhD thesis at the Departamento de Ingeniería del Diseño of the Universidad de Sevilla, in which the ultimate goal is to establish a conceptual framework for the automation of building design conformance checking to the mentioned Portuguese code.

Many results that relate to the present subject could not be shown here because of their extension, although we believe there are enough to emphasize the most important aspects.

Since we intend to use UML as the main description method, it will be necessary not only to describe the development of our particular application model but also to translate the actual EXPRESS specification of the STEP model and still relate the final interpretation constructs back to EXPRESS. For this reason, this paper includes the proposal of an UML-EXPRESS mapping, under the concept of a UML Profile, which is then used for the description of the STEP resources.

2 UML PROFILE FOR EXPRESS

2.1 *EXPRESS versus UML*

The EXPRESS language (ISO 10303-11:1994) is defined by STEP for the formal specification of the information requirements of other parts of the standard in a way understandable by both humans and computers. It is partially based on some programming languages (such as Ada, C, C++, Modula-2, PL/I, Pascal and SQL) but it is not a programming language *per se*, because the main goal is the description of objects in terms of entities whit their attributes, instead of control. EXPRESS is a textual language but some constructs have an equivalent graphic symbol. Together, these constructs embody the subset called EXPRESS-G, which is much more suited for the communication and readability of the information.

However, we believe that EXPRESS has some important limitations: (1) using EXPRESS-G to increase readability, greatly diminishes the expressive capacity of the language; (2) EXPRESS is not entirely object-oriented (for example, an entity is not the same as a class, because algorithmic information may have an independent scope), which means probably less capacity to address complex problems and integration with recent modelling technologies; (3) EXPRESS basic constructs (schema, entity, type,

algorithm and rule) are highly limited, from the semantics point of view.

On the other hand, UML (OMG 1999) is basically a visual modeling language that has inherited from the prior main object-oriented methods (like OMT, Booch and OOSE). As a consequence, it uses a standard graphic notation and it is more expressive, clean and uniform than its predecessors. In the UML's architecture, a language is considered a metamodel of reality and, at the higher level of abstraction (so called the meta-metamodel), we find a set of concepts that describe the main language itself, together with some mechanisms that one may use to extend the former semantic constructs. Its objectives (Booch et al. 1998) extend the representation of a system's information requirements to a life cycle perspective, supporting analysis, design and specification, both in static and dynamic views. Being an OO language, the class and the interrelation become the most important elements

UML has been standardized by OMG (Object Management Group) and already accepted as a specification language by other ISO standards (like ISO 19100 series). Major software builders support UML and tools have been developed for models diagramming, and also facilitating the implementation of such models under programming languages such as Java, C++ and Visual Basic.

Generally compared with EXPRESS, the UML has the advantage of satisfying a larger and effective set of requirements, as are: (1) it supports an immediate integration between a product's model and the model of a system that operates the product; (2) it is more flexible; (3) it is more recent (thus being natural to the modelling concepts and methodology of actual paradigms); (4) it has a much more expressive graphic notation; (5) it is more generally accepted (under OMG control).

2.2 *UML profile for the EXPRESS language*

In the development of the UML profile for the EXPRESS language, the UML mechanisms stereotype and restriction become largely adopted as a solution for the distinctive integration of EXPRESS constructs under the general unifying concepts of class and package. Basically, a stereotype represents some semantic specialization of a former standard element, as defined by a set of restrictions.

In short terms, each UML element included in the profile consists in a package, a class derived from an existing stereotype, or a new stereotype for a package or class.

As Figure 1 shows, there are two main packages: one for new defined stereotypes of packages and classes; and another for classes representing EXPRESS data types (defined under the pre-existing

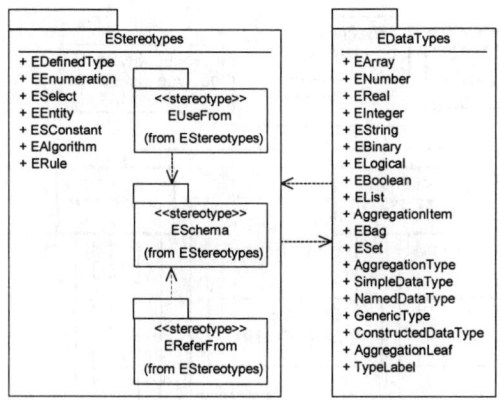

Figure 1. UML profile for EXPRESS – packet diagram

UML stereotypes "enumeration" and "type"). In addition, there are three other nested packages that really represent new defined package stereotypes. All names begin with the literal "E" for a better recognition.

Without getting into detail regarding the definitions, restrictions and relationships among each profile element, some other aspects must be clarified, which relate to generalization, association and multiplicity:

(1) The concept of generalization/ specialization relationship is similar in both languages concerning inheritance, polymorphism, "one of"/ "and or" restrictions and the possibility of indirect instantiation of a superclass. The major problem is that EXPRESS reserves generalization for entity elements, so the profile must reflect this condition by a specific set of rules: (a) classes of stereotype "EEntity" are the only ones that may participate in generalization relationships, with the exception of the following cases; (b) a class of stereotype "ERule" is always a supertype of those classes affected in the EXPRESS corresponding definition; (c) a data type may be a subtype of an abstract data type.

(2) The OO/UML concept of association is implicit in EXPRESS when an entity attribute identifies another entity as its domain. In these cases, the EXPRESS identifier of the attribute becomes the UML identifier of the role played by the "second" class, which then is enough to understand the meaning of the association. If an EXPRESS inverse attribute is also specified it becomes the role of the "first" class. Those EXPRESS attributes like "is part of", "has" or "belongs to" can be considered special cases, corresponding to a UML aggregation association, because of the strong sense of ownership that is implied.

(3) The concept of multiplicity has a slightly different meaning in the two languages. EXPRESS uses simple, defined and named data types for defin-

ing a domain of values, while aggregation data types are used to specify the number of times that a certain domain of values may participate in a relationship for a single domain on the opposite side (which EXPRESS calls cardinality). To achieve complete identification of the EXPRESS constructs for multiplicity under UML, the EXPRESS aggregation type expression is simply converted into a similar expression (an instance of a special class named "TypeLabel") and is shown as an adornment that describes a restriction. As an alternative, the UML model for multiplicity specification can still be used when EXPRESS cardinality is one-dimensional and exclusive, with the occasional need of the UML standard restrictions "ordered" or "unique".

For a more detailed understanding of this profile, the complete set of UML diagrams is documented in Santos et al. 1999a.

3 UML SPECIFICATION OF THE BCCM

3.1 Why BCCM?

The Building Construction Core Model (ISO 10303-106:1997) is the part of STEP that must specify the integrated application resources for the building industry to support the development of Application Reference Models related to specific fields, the interchange of information and the interoperability between applications.

The development of the BCCM was intense from 1995 to 1997 (Wix 1997) but it stopped after that, possibly because a disagreement about the balance between the necessity of abstraction and the convenience of high-level semantics. Since then the former parallel standard IFC has been the one choose by some software implementations as it represents the last preferences of the software industry view (it includes objects like "door" and "beam" and it remains more attached to geometry) while still making use of many BCCM and STEP resources (including the description and the implementation methods).

Even so, we consider the BCCM as representing a major reference because it benefits entirely from the STEP technology. Besides, ISO standards are published while the access to IFC specifications has been restricted to the supporting IAI members.

3.2 The BCCM

The BCCM resources are considered under four different natures: Products – the results of BC activities; Resources – the equipment and tools necessary for BC activities; Processes – the BC activities; and Controls – the conditions and ruling that affect BC activities. The explicit graphic representation of building elements is considered out of scope of the BCCM but there is a natural connection with STEP part 225 to this purpose.

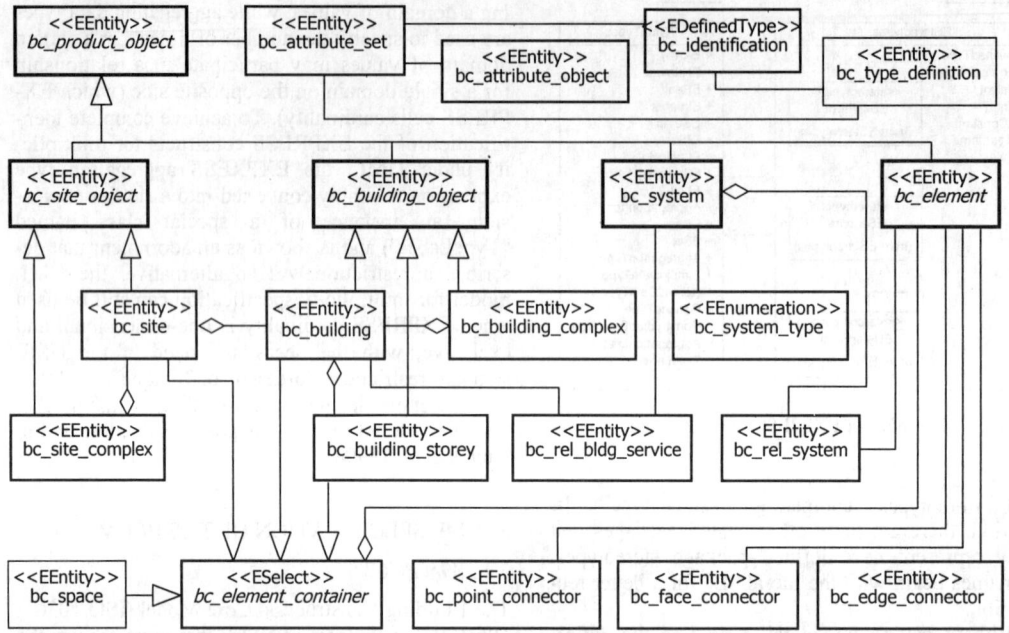

Figure 2. UML specification of BCCM – partial and simplified class diagram

This study uses the version T200 of the BCCM (a working draft approved by ISO TC184/SC4 meeting in Chester, UK, March 1997) as described by a set of 27 EXPRESS-G diagrams which have been converted to UML class diagrams under the referred mapping profile. For the purpose of our application, we then eliminated clearly unnecessary objects (those provided as Controls, Resources and Processes) and reorganized diagrams to get a more adequate view of the model.

As an example, Figure 2 shows a simplified view that corresponds partially to the EXPRESS schema "BC core". The class attributes, the association adornments and the external references were elided because of space limitations. The thick line icons identify the major classes and the thin line those that are better presented elsewhere (the use of icon differentiation by line or fill colors, or even by changing the graphical layout, is suggested for when it clearly improves readability).

The concept of Product has deep propagation in the model, starting with the subtypes site, building, system and element. By inheritance, the majority of the model objects are associated with a set of zero or more extended attribute sets, which represents an important feature for supporting specialization. Besides, the concept of Type Definition was lately introduced in the model to make possible the attachment of additional unstructured information to many of the leaf objects with higher semantic sense

(namely filling, opening, covering, layered and profiled elements, together with space, zone, equipment, fixture and system).

The kind of simplification presented in these diagrams, yet preserving the entire model specifications, was not possible to achieve with EXPRESS or EXPRESS-G methods, although it considerably helps analysis and communication.

For a more detailed understanding of the UML specification of the BCCM, the complete set of diagrams is documented in Santos et al. 1999b.

4 INFORMATION REQUIREMENTS OF RGEU

4.1 RGEU description

What really counts in a building is its interior empty space, which is merely a side effect of construction. So, it's not strange that the primary phases of building architectural design are based on a largely common building concept and are almost exclusively oriented to satisfy functional requirements derived from space layout, not from technical questions.

Usually, these functional requirements are partially established at a general level by some state or local regulations and are complemented by each concrete building program. In the Portuguese case the set of state regulations that best fits to this purpose was published back in 1951 but it still remains as a major reference for the architects work (RP 1997).

250

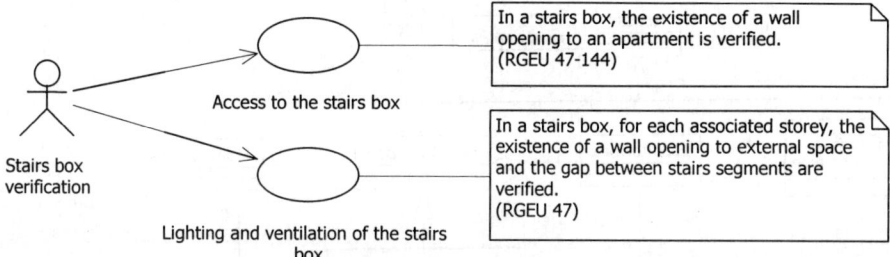

In a stairs box, the existence of a wall opening to an apartment is verified. (RGEU 47-144)

In a stairs box, for each associated storey, the existence of a wall opening to external space and the gap between stairs segments are verified. (RGEU 47)

Stairs box verification

Access to the stairs box

Lighting and ventilation of the stairs box

Figure 3. Example of information requirements description

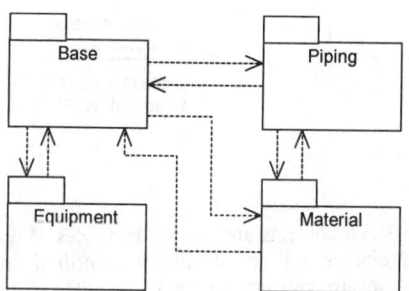

Figure 4. RGEU reference packages

According to our objectives, the application domain was first circumscribed within RGEU, by which some rules were filtered out because the entire content is poorly structured, frequently subjective and sometimes simply inadequate.

4.2 *Information requirements*

The RGEU document is actually written in textual Portuguese language. Its included rules are based on a certain view of a building upon which they specify how exactly it must be regarding some characteristics. They are referenced in numbered "articles" (small groups of one or more rules) and those selected to the effective application domain are divided into eight named major groups.

Our issue is to define the information requirements for the representation of those rules and to organize them into a consistent information reference model. This does not mean the simple conversion of actual building specifications into some other language, neither the development of procedures for the verification of a design conformance to the rules (which can be a next step but is beyond our present goal).

The methodology used for analysis considered two important stages: first, an OO perspective of the relevant information is traced out from the common language specifications, based in the recommenda-

tions of the Object Modelling Technique (Rumbaugh et al. 1991); second, the information requirements of each identified rule were reviewed under the referencing vocabulary of potential classes and respective attributes, using the representation resources of UML.

As a result of the second stage analysis, the UML use case diagrams with annotation descriptions have been used to reference all the information requirements. Figure 3 presents an example of this. The UML actor identifies the class to which the verification most closely relates.

The entire collection of use case diagrams is documented in Santos et al. 1999c. They correspond to the UML modelling layer Use Case View.

4.3 *Reference model*

After having traced the potential classes and identified all information requirements of the application domain, the emphasis passes from the domain analysis to a model design, by which integration and consistency must be guaranteed. Again, the UML resources used in this phase gave a good support to design development and validation, mainly due to the possibilities of good diagram integration, simplification and easy communication.

An early design decision was made regarding the organization of the potential classes into four UML packages, as shown in Figure 4. Together, these packages contain 58 classes. The Base package is the most important one. It includes 46 classes that represent the space and the prior physical elements of a building.

At the end of the design process the entire model is specified by a set of 10 class diagrams, referring the following aspects: physical base elements, space, compartments, access, external space, spatial connection, enumeration types, equipment, piping and material.

As an example, Figure 5 presents the complete class diagram referring Access information. As an example of enumeration items specification, Figure 6 presents the class *wall*, together with an enumera-

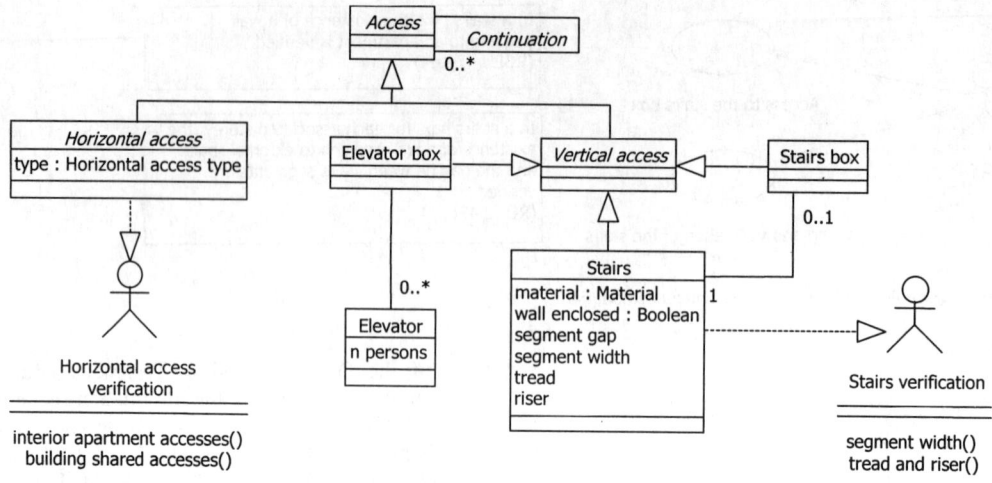

Figure 5. RGEU reference model – complete class diagram of accesses

Wall	
type : Wall type	
structure : Material	
width	
height	
orientation	
aerial : Boolean	

<<enumeration>>
Wall type
outer wall
outer separation wall
interior wall
interior separation wall

Figure 6. Example of enumeration items definition

tion that represents the domain of the values for its "type" attribute.

The UML concept of actor used in the diagrams represents some kind of action that is performed from outside the focused system. It has also been used at this stage to maintain a good reference to the previous use case diagrams, identifying those classes from which verifications must be started.

4.4 Validation

The final phase of the modeling design process is dedicated to the model validation and last refinement. It must be assured that the model specifications respond to all the application information requirements as they were traced at the beginning, which means having the right classes with the right attributes and collaboration relationships.

There are several UML resources that can be used for this purpose. Considering the static nature of the information involved, the UML object diagrams have been choose. For each use case diagram, a corresponding object diagram was built, which validates the model for the specific information requirements. Figure 7 shows the example of an object diagram that relates to a use case situation presented

in Figure 3. The objects are generic instances of the identified classes and the attributes identified are those that inform the navigation or the data to be checked. The navigation starts from one instance of the class associated with the actor that represents that same verification process. The vertical path means a generalization relationship and the triple icon corresponds to a "one to many" association.

The entire collection of class diagrams and object diagrams is documented in Santos et al. 1999c. They correspond to the UML modelling layer Logical View.

The possibility of describing concrete situations by such diagrams, without having to consider concrete instances, is also an advantage of UML method revealed here.

5 APPLICATION INTERPRETED MODEL

The interpretation of the RGEU model under the BCCM must specify, for each information element of the first, which element from the second corresponds. This is somewhat like a contract, by which a part guarantees the requirements of the other. UML defines the concept of realization relationship between classes and initially suggests its use for interface relations, in which a class merely performs a set of operations according with a service needed by another. However, we see no reason for not to use the UML concept of realization to represent the kind of contract that relates classes from distinct models, as in the present situation.

Figure 8 shows a UML class diagram that specifies the correspondence between some RGEU (thin line icons) and BCCM classes (thick line icons) thus

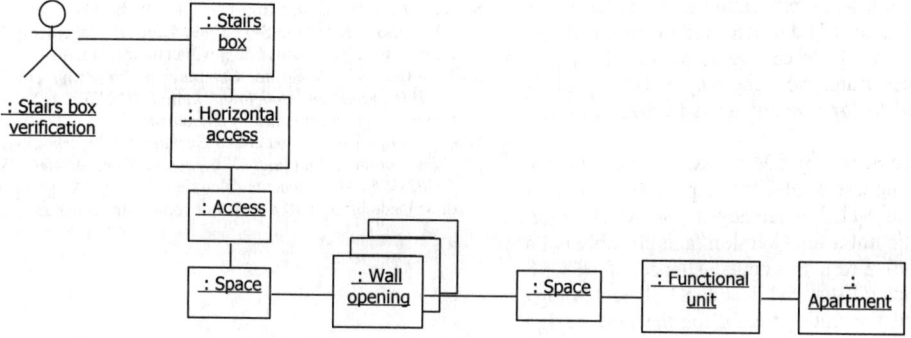

Figure 7. Object diagram related to "access to the stairs box" use case

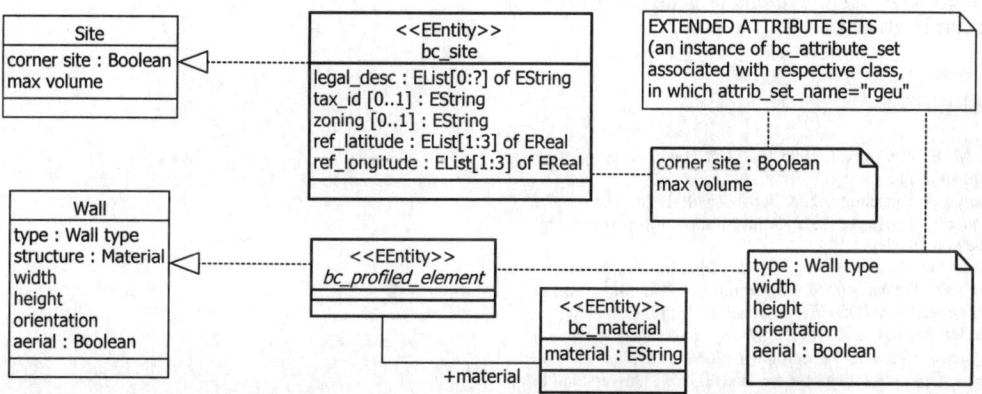

Figure 8. Example of RGEU interpretation specification

related by realization relationships. The use of annotation and defined restrictions also facilitates the specification and communication based on such a diagram.

We look for the presence of each RGEU class attributes in respective BCCM class, or in nearby related BCCM classes. When there are no such attributes, an extended attribute set is defined using the BCCM mechanism offered by the classes "bc_attribute_set" and "bc_attribute_object", distributed by inheritance from the top supertype "bc_product_object".

However, since the information requirements determined for RGEU often involves a well-defined collaboration of a set of classes, as shown by the object diagrams, the above correspondence is not enough. In addition, we need to test and reveal the existence of an equivalent collaboration within the BCCM model. To satisfy this last need, all the UML object diagrams used for the RGEU model validation (example of Fig. 7) are converted according with the possibilities of the interpretation model.

This leads us to an Application Interpreted metamodel in which two kinds of UML diagrams are used: class diagrams for peer to peer correspondence between classes and object diagrams to show equivalent collaboration paths.

6 CONCLUSION

The use of UML to support the development of a STEP Application Protocol has been demonstrated and some advantages put to evidence: (1) being a single language, it's easy to relate distinct views of the same problem; (2) being graphically oriented, it's better suited for the communication of a large set of semantic constructs; (3) since a model may be specified by several diagrams at different levels of abstraction and completeness, it helps analysis and facilitates representation of individual aspects.

We believe that the proposed profile for a UML-EXPRESS mapping is an essential contribute for the adoption of a UML based AP development methodology, because it's not expectable, nor it is necessary that STEP changes it's defined specification method. Besides, such a profile can be automatically implemented in the future, the same way as UML models are converted to (or re-engineered from) Java code, for example.

The choice of the BCCM resources to accommodate the Portuguese RGEU code proved an acceptable degree of STEP efficiency in the AEC sector, even when it's not a final version (and possibly not a complete one). From the actual silence of BCCM, we may conclude that, while the STEP technology has identified the major trends for the new design tools, it's not enough to guarantee its success for all related sectors. Anyway, we think that the future adoption of a final BCCM or the IFC standard is a minor question for our ultimate goal when everything else is assured.

REFERENCES

Bloor, M. S., & Owen, J. 1995. *Product Data Exchange*. London: UCL Press.

Booch, G. & Rumbaugh, J. & Jacobson, I. 1998. *The Unified Modeling Language User Guide*. Rational Software Corp., Addison-Wesley.

D. Appleton Company Inc. 1985. *Integrated Information Support System (IISS): Information Modeling Manual IDEF1-Extended (IDEF1X)*. General Electric Company.

Hernández-Rodríguez, F. 1995. *Modelización de Información Espacial Mediante Tecnología Orientada a Objetos*. PhD thesis, Dpto. de Ingeniería del Diseño, ESI, Universidad de Sevilla.

IAI 1997. *End User Guide to IFC – version 1.0*. International Alliance for Interoperability, downloaded from respective website.

ISO 10303-106:1997. *Building Construction Core Model - Working Draft version T200*. EXPRESS-G diagrams, International Standards Organization TC184/SC4, down-loaded from URL: www.bre.co.uk/~itra/ceic.htm.

ISO 10303-11:1992. *EXPRESS Language Reference Manual (IS)*. International Standards Organization TC184/SC4, Genève.

Nijssen, G.M. & Halpin, T.A. 1989. *Conceptual Schema and Relational Database Design: a Fact Oriented Approach*. Prentice-Hall.

OMG 1999. *Unified Modeling Language Specification – version 1.3*. Object Management Group Inc.

RP 1997. *Regulamento Geral das Edificações Urbanas (RGEU)*. Decreto-Lei n.° 38382, de 7/8/1951, e outros. Governo da República Portuguesa. Lisboa: Rei dos Livros (6.ª edição).

Rumbaugh, J. & Blaha, M. & Premerlani, W. & Eddy, F. & Lorensen, W. 1991. *Object-Oriented Modeling and Design*. New Jersey, Englewood Cliffs: Prentice Hall Int. Ed.

Santos, I.A. & Hernández-Rodríguez, F. 1999b. *Especificación UML de BCCM*. TR-DID-IAS-02-1999, Dpto. de Ingeniería del Diseño, ESI, Universidad de Sevilla (fhr@esi.us.es).

Santos, I.A. & Hernández-Rodríguez, F. 1999c. *Modelo UML del Regulamento Geral das Edificações Urbanas (RGEU)*. TR-DID-IAS-03-1999, Dpto. de Ingeniería del Diseño, ESI, Universidad de Sevilla (fhr@esi.us.es).

Santos, I.A. & Hernández-Rodríguez, F. 1999a. *Perfil UML de EXPRESS*. TR-DID-IAS-01-1999, Dpto. de Ingeniería del Diseño, ESI, Universidad de Sevilla (fhr@esi.us.es).

SofTech Inc. 1981. *Integrated Computer-Aided Manufacturing (ICAM): Function Modeling Manual (IDEF0)*. U.S. Air Force, Wright Aeronautical Laboratories.

Wix, J. 1996-1997. *Computerized Exchange of Information in Construction - Building Construction Core Model (ISO 10303-106) – Documents*. Newsletters 2, 3, 4, 5 and 6, downloaded from URL: www.bre.co.uk/~itra/ceic.htm.

Product and Process Modelling in Building and Construction, Gonçalves, Steiger-Garção & Scherer (eds)
© *2000 Balkema, Rotterdam, ISBN 90 5809 179 1*

Virtual product model – A concept for better information management process

Danijel Rebolj & Andrej Tibaut
Faculty of Civil Engineering, Construction IT Centre, University of Maribor, Slovenia

ABSTRACT: *Today, many researchers working on the information management process in construction recognize the problem of modelling complex product-related data structures. Furthermore, many experts doubt whether an all-inclusive-product-model is a solution for an integrated information environment that should efficiently support the life cycle of a product. It seems that the rich experiences in product modelling gathered during the last decade does not lead necessarily to better models but rather to the awareness that the more complex the product models are, the more rigid and the less usable they become in practice. These recognitions already led to some suggestions for the future integration methods and product modelling, like building smaller domain oriented models and link them using mapping schemes. In this way, however, the complexity of the whole system hasn't decreased, even worse, it grew.*

The paper summarizes some deficiencies of complex product models and then introduces a solution, called Virtual product model, which is based on decomposition of a conventional product model. The concepts, basic components and an example of the Virtual product model are described.

1 INTRODUCTION

From the large international projects that promote use of IT in construction we can learn that technology push approach still dominates over process driven approach. In the first case commercial IT tools and standards are applied to cover the particular sub processes, information is modelled in a standardized way (IFC, STEP) and exchanged between the sub processes. Unfortunately, data exchange between sub processes is based on a "single standard agreement" that promotes unity in data exchange. As such communication is limited to sub processes that "speak" an agreed data modelling language. It seems that such restriction suits most of the project teams in construction. However, in distributed environments where construction project team consists of heterogeneous data sources (loosely coupled contractors) a single standard agreement is a-priori an obstacle in collaboration.

Today many researchers, working in the field of engineering information technology, recognize the problem of modelling complex structures, and many are asking themselves whether an all-inclusive-product-model is a solution for an integrated information environment that should efficiently support the life-cycle of a product. It seems that rich experiences in product modelling in the last decade

lead not to better and better models but rather to the awareness that the more complex the product models are, the more rigid and the less usable they become in practice. These recognitions already led to some suggestions for the future integration methods and product modelling.

Before we continue to analyse the deficiencies of complex product models, let us briefly browse through the short history of product modelling. Probably everything started when the first data interface has been implemented, which has linked the output of one computer program to the input of another. After that, researchers started to develop more sophisticated integration methods. According to the principle we can divide them in the following groups:

- Integration of different stand-alone programs with the help of information interpreters, as for example in the "software fixing" method (Syal et al. 1991). These methods have two main deficiencies: they don't enable fluent information flow, and it is necessary to implement a new interpreter for every new program we want to include.
- The use of a common medium for information exchange between programs. "Blackboard" is one such method (Yau et al. 1991), which enables a fluent information exchange through a common "blackboard". The "Object shell"

method (Rebolj 1993) supports a fluent information exchange as well, however all these methods still require implementation of new interfaces to include new programs.

- The integrated database concept, where all included programs use a common . data repository. There have been many projects, which have developed and used this concept: RATAS (Björk 1989), ATLAS (ATLAS 1992), COMBINE (Augenbroe 1993), COMBI (Ammerman et al. 1994), and in the last years SPACE and OSCON, which set the fundamentals of the technologically advanced integrated environment in civil engineering, described in (Faraj et al. 1999). Among earlier, but less known systems, the CIS, Construction Information System (Rebolj 1990) introduced an integrated geometry-construction database. Many authors published more detailed reviews of relevant projects and systems, including the listed ones (e.g. Amor 1998, Eastman & Augenbroe 1998).

Nowadays the integrated database concept is recognized as the most effective method for integration of computer programs in the life cycle of a building object. The integrated database contains the complete description of a product, therefore such data models are known as product models.

2 APPLICABILITY OF COMPLEX PRODUCT MODELS

Present examples of product models show a tendency to build a unique all-inclusive complex model for a specific engineering field (like shipbuilding, car industry, building industry, road building, etc.). However, none of these attempts has been generally accepted in the civil engineering practice. Rather than that, the past development of building product models led to a question, whether a definition and use of a standard product model has sense at all. To overcome the need to have a single product model some authors have proposed inter-model linking schemes (like in Spooner & Hardwick 1997, and Pfennigschmidt et al. 1997), in this way, however, the complexity of the whole system hasn't decreased, even worse, it grew.

Another problem is arising from the necessity for standard building elements. The history of mankind shows that in communication the only "standard" is the diversity of standards. In other words, it seems most unlikely that the whole mankind would use a single standard language. Even if such a language would exist, it is very likely that soon many dialects would appear, since every individual or group is seeing the same thing in its own perspective.

This problem is even extended in civil engineering and construction, where many different views have to be considered through a product life cycle. Different views are leading to more or less different descriptions (data structures) representing the same entity. Notable progress has been made by the International Alliance for Interoperability with the development of the Industrial Foundation Classes, which can be seen as applicable building blocks (IAI 1996), but which are still not resolving the problem of views (as evident from Yu et al. 2000).

A conflict between the concept of a single integrated model and the need for individuality also showed up. Companies (and individuals) have a strong affection to fully control their own data, which also form the company's "memory" (Larson 1998), a vital part of every company.

Such and similar problems have already been recognized by some authors, who have expressed their hesitation about applicability of complex product models, either between lines (e.g. Graves 1998, and Amor 1998), or directly (as in Eastman & Augenbroe 1998, and Turk 1999). Appending authors' own experiences, the main deficiencies of product models could be summed up into the following essential points:

- Product models are based on clearly defined semantics and demand unique standard basic elements, however, such elements don't exist,
- computers are not (yet) capable to fill up semantic inconsistencies and gaps, which show up in the integration of computer programs - a human is adapting daily to communication patterns with other humans with different mental models, and is capable to reconceptualize parts of information, which don't fit into the whole,
- product models are subjective interpretations, not objective representations of the real, therefore an effective uniform product definition is not possible,
- product models only include parts from the building process and disregard some important views (social, environmental, etc.), which form the process in the real,
- models are restricting creativity due to their complexity and rigidity,
- when implementing prototype models into the real environment they fail due to the inability to capture the rich knowledge and experience of the people,
- although product models are basically open, they get stiff and hardly upgradeable in the real,
- in an integrated database each client's control over his own data is limited.

Eastman & Augenbroe (1998) and Turk (1999) also proposed some solutions to the problems they described:

- product models should be rather small and limited to specific areas; coexistence of more models in the same field is not necessarily bad,

- implementation of middleware tools between applications and models, which will help humans to navigate between the islands of automation,
- gradual implementation of small models into industry,
- development of a richer set of language constructs for model description,
- product and process models should be linked more closely,
- new integration concepts should be tried, which would not reside on integrated semantics,
- it is necessary to allow coexistence of structured information and unstructured data and leave their interpretation to the human,
- programs should not limit but extend the engineers capabilities (virtual reality, telepresence, multimedia, etc.)
- pure information exchange should be upgraded with communication software for collaboration support.

3 VIRTUAL PRODUCT MODEL

3.1 *Concept*

Based on the positive and negative experiences in modelling of building products we propose a concept of the Virtual Product Model (VPM), which could preserve positive, and avoid some negative characteristics of product models.

The virtual product model is represented by a network of loosely coupled particle models,

interconnected by relatively simple but strong rules (like gravity in the macro-cosmos). The neighbourhood of a particle model is in a logical sense defined through a process model, which also determines relations between particles. So the main point is in communication network that interconnects the particles.

Let us recall here the concept, which solved the complexity problems of computer networks. The problems of how to get together many different communication technologies seemed unsolvable, until a cut has been made in decomposing the network into clearly defined functional layers – a solution nowadays known as the 7-layer ISO model (CTRC 1989).

Decomposition has also become a magic word in software development. Huge monolithic systems tend to evolve into open and flexible structures of software components (IEEE 1997).

The virtual product model can be explained as a decomposed product model, consisting of three main layers (Fig. 1):
- particle models or particles (data structures used by applications),
- a process model, which determines the particle interconnection scheme (the "higher sense" of particles), and
- communication network, which is responsible for harmonization of particles and implements the "rules" on them.

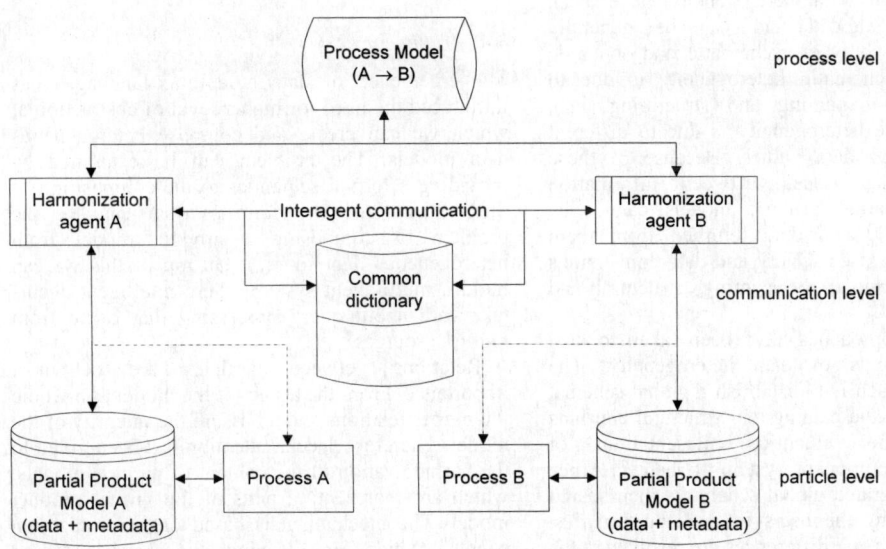

Figure 1. Virtual Product Model basic scheme.

It is believed that such decomposition will decrease complexity of the product model to a manageable level and increase it's flexibility through the autonomy of applications and partial product models - particles.

3.2 *System and data heterogeneity*

In order to implement communication mechanisms for the Virtual product model we have investigated the integration issues that arise in a federation of heterogeneous data sources, possibly storing related information. In an integrated heterogeneous information system, system and data heterogeneities among particle systems are the major obstacles toward information sharing. While system heterogeneity problems are usually resolved by using a common communication infrastructure (e.g. TCP/IP, RPC, CORBA ORB, etc.) the differences in information and data models are more difficult to deal with. It is not possible to force all users to use the same data or information model for different applications since one data model, which is suitable for one application domain, may not be suitable for another. Besides people working in different application areas may have their own preferences on which models to use.

The solution to this problem is either to do pair-wise translations of modelling constructs or to use a common or neutral model, to and from which all data models used by the users are translated. The data model heterogeneity problem needs to be resolved before any successful system integration can be achieved.

Problems with data heterogeneity have been thoroughly investigated and can be generally categorized in two types: schematic and semantic heterogeneity. Schematic heterogeneity is due to different ways of naming and structuring data, whereas semantic heterogeneity is due to different representations of data values. Because of these problems, queries issued between information systems with partial product models cannot be processed directly and data returned from them cannot be readily used. Query and data conversions are needed to bridge the naming, structural and semantic gaps.

Two basic approaches have been taken to deal with the problems of data heterogeneity. The traditional approach is to establish a global schema, which reconciles the naming and structural conflicts and unifies the semantic representations of heterogeneous component systems (Chen & Arbee 1988). This integrated global schema is then shared and referenced by the users to exchange queries. Thus, all conflicts and differences are resolved at the time of schema design and integration. The major drawback of this approach is that the shared integrated schema forces users to view data in the same way instead of the ways that are familiar to them.

3.3 *Mediation versus integration*

In contrast to the traditional schema integration approach, more recent thinking is to "mediate" dissimilar data representations instead of "integrating" them. This mediation approach is typically done by using some mediation rules or specifications, which are used to resolve various kinds of conflicts among particle systems at runtime. A mediated information system allows the users to see data in their own views. They can issue queries based on their own views and receive data in representations that are familiar to them. Furthermore, the mediation approach provides better support for system extensibility and scalability because, unlike schema integration approach, adding new particle systems to the heterogeneous system can be done by changing or adding mediation rules or specifications instead of redesigning or modifying the integrated schema.

Some of the important features of our approach, motivated by the shortcomings of existing technology, include (a) the ability to share data across multiple heterogeneous data sources, (b) the ability to manipulate the meta-data (schema) component of a data source in the same manner as data can be manipulated, and (c) the ability to query besides well-structured data sources (such as relational databases), semi-structured data sources (such as the HTML documents on the World Wide Web).

3.4 *Architecture*

The emergence of data modelling languages has introduced the need for higher level of abstraction at which we can create and compose heterogeneous data models. The problem can be examined by providing a formal semantics to the composition of different data models within a virtual team, i.e. the problem of composing a product model from heterogeneous data sources. On top of this we can build a multi-agent system where intelligent agents take care about query processing that come from multiple sources.

Relations between particles are of most importance. From the aspect of the model as a whole the proper relations should assure the integrity of the model. Therefore special attention has been given to the harmonization of the content of particle models, which are representing parts of the virtual product model. The mechanism is based on harmonization agents, which are leaving the particles their individuality but also bind them to the whole.

Harmonization agents do not require uniform semantics of particle models, but only common basic primitives. It is therefore possible to allow different structuring and representation techniques and standards for particle models. While communicating harmonization agents use their own knowledge about structures, which they gathered and saved in common dictionaries, whereby in insolvable situations agents establish contact with humans. Actually, agents will in the first stage act as assistants, then as advisors and at the final stage as autonomous agents.

The common dictionary is a repository of basic element (term) descriptions in a semantic domain. There is a domain dependent starting set of terms with relations between them, which are necessary for communication start-up between agents. It is supposed that agents will soon come into situations where basic terms and relations won't be enough to exchange views in a new situation. In such cases, agents will have to ask an expert - either a human or another, more "experienced" software agent. For the second case, agent "chat rooms" will be the place, where "inexperienced" agents will have the opportunity to learn, and then improve their "native" dictionary. Building a dictionary automatically from

very simple starting terms should avoid the known trap of defining complex view mapping schemes.

To become a part of the VPM the particles (applications + data structures) have to fulfil certain conditions regarding data representation:

- all exchange data has to be available in an external representation in a text form (which also implies database systems able to communicate in text form),
- a data description (metadata) has to be available,
- data has to be structured in an object-oriented way, using Express, XML, but also non-standard description languages.

Having in mind that it is not necessary to adapt the semantic and the data structure of a particle to integrate it with the product model, even old (but good working) computer programs can be upgraded to suit the conditions. It is, however, a good idea to redesign the data structures to give particles a better ability to interact with others.

On the other hand, the VPM concept, which supports the flexibility and autonomy of applications, is a good accelerator of application's (particle's, component's) self-intelligence and adaptability.

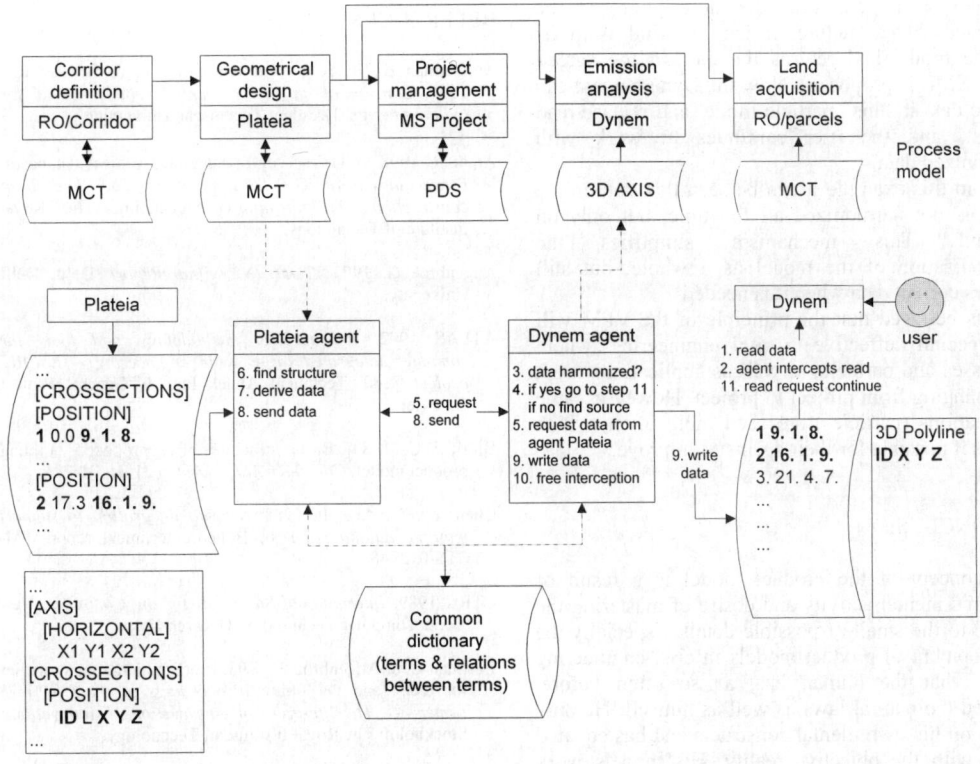

Figure 2. Example of VPM mechanism operations

259

3.5 *Example*

Figure 2 shows the use of the VPM concept on the example of a part of the road life cycle (the road has been in the focus of our research group in the last few years; see Rebolj 1999). A simplified scheme of the process model shows a chain of tasks, with the information about the program(s) and the external data representation used in a specific task. The process model is built, or adopted, for each specific project, because every project can include slightly different tasks, carried out by different programs.

The scenario (shown on Fig. 2) starts with the activation of the task "Emission analysis", which is supported by the program named Dynem. When the user specifies the project name he wants to work on, Dynem tries to read relevant data. The read request is intercepted by the Dynem's harmonization agent, which checks the status of the data in the project model (implemented as a project database). If the requested data is harmonized with the predecessor particles, the agent releases the reading request. Otherwise, it locates the data source in the process model and establishes communication with the responsible harmonization agent. In our case it is the Plateia-agent, responsible for the geometrical design. When Plateia-agent gets the description of the requested data structure, it tries to find it in its particle model and returns the data in the agreed form (XML is proposed). Now the Dynem-agent can update data in "his" particle model and release read request, and the user continues to work with harmonized data.

From this example it can be seen that data in the VPM is not harmonized all the time, but only on demand. This mechanism simplifies the harmonization of the model as a whole, but still assures correct data when it is needed.

It is believed that the principle of the VPM will be especially effective in civil engineering, where processes and partners, as well as applications used, are changing from project to project. However, same applications (particles) are used more often, which makes it possible for their agents to improve.

4 CONCLUSION

The concept of the product model is a result of human's mental activity and desire of mastering the whole to the smallest possible detail. Especially the development of product models in civil engineering shows that the human has, as so often before, ignored the natural laws as well as himself. He only relied on his own mental constructs and has equated them with the objective reality. His models work only under special circumstances, but they are not generally applicable. This does, however, not mean that the concept of product models is useless.

Through the concept of the virtual product model it is believed that it is possible to preserve the independence and flexibility of particles - existing island models and applications, and the simplicity of mastering them, but also to preserve the positive integration effects of complex product models. The reason for this belief lies in the simplicity of used principles and in their closer relation to natural mechanisms (basic laws), which also includes the ability of implicit evolution. The evolution and improvement is supported by harmonization agents, which not only communicate, but through the communication also gain new knowledge and develop adaptability. We believe that our approach would enable more transparent information exchange in construction process regardless of semantic and structural heterogeneity of data. In other words, we have tried to find a mechanism to preserve the advantages of product models while avoiding the shortcomings caused by their complexity, having in mind the words of the German philosopher Oswald Spengler "Everything complex is of short lifetime".

REFERENCES

Ammerman E., Junge R., Katranuschkov P. & Scherer R.J. 1994. *Concept of an object-oriented product model for building design.* Dresden: Technische Universität.

Amor R. 1998. A UK survey of integrated project databases. *Proceedings of the CIB W78 conference The life-cycle of construction IT innovations.* Stockholm: The Royal Institute of Technology.

Augenbroe G. 1993. *COMBINE, Final Report.* Delft: Delft University.

ATLAS 1992. *Architecture, methodology and tools for computer integrated large scale engineering – ESPRIT project 7280.* Technical Annex Part 1, General Project Overview.

Björk B.C. 1989. Basic structure of a proposed building product model. *Computer Aided Design* 21(2): 71-78.

Chen, L. & Arbee, P. 1988. *Schema integration to support multiple database access.* Bellcore technical report TM-STS-012948

CTRC 1989. *International Standards for the Computer.* New York: Computer Technology Research Corp.

Eastman C. & Augenbroe F. 1998. Product modeling strategies for today and the future. *Proceedings of the CIB W78 conference The life-cycle of construction IT innovations.* Stockholm: The Royal Institute of Technology.

Faraj I., Alshawi M., Aouad G., Child T. & Underwood J. 1999. Distributed Object Environment: Using International Standards for Data Exchange in the Construction Industry,

Computer-Aided Civil and Infrastructure Engineering 14(6): 395-405.

Graves G. 1998. Industry requirements for data standards harmonization. *Proceedings of the Global Business Solutions for the new millenium.* CD ROM.

IAI 1996. *End User Guide to Industry Foundation Classes, Enabling Interoperability in the AEC/FM Industry.* International Alliance for Interoperability (IAI).

IEEE 1997. Engineering meets the internet: how will the new technology affect engineering practice? *IEEE Internet Computing* 1(1): 30-38.

Larson M. 1998. AF integrated digital environment. *Proceedings of the Global Business Solutions for the new millenium.* CD ROM.

Pfennigschmidt S., Kolbe P. & Pahl P.J. 1997. Integration von Datenmodellen. *Proceedings of the IKM conference.* CD-ROM. Weimar.

Rebolj D. 1990. Graphic Modelling of Superstructures. *Automatika* 31(1-2): 147-156.

Rebolj D. 1993. *Computerunterstützter integrierter Straßenentwurf in einer objekt-orientierten Umgebung.* Graz: Verlag für die Technische Universität.

Rebolj D. 1999. Integration of computer supported processes in road life cycle. *Journal of transportation engineering* 125(1): 39-45.

Spooner D.L. & Hardwick M. 1997. Using views for product data exchange. *IEEE Computer Graphics and Applications* 17(5): 58-65.

Syal M.G., Parfitt M.K. & Willenbrock J.H. 1991. *Computer integrated design/drafting, cost estimating, and construction scheduling. Housing Research Center Series Report No. 11.* The Pennsylvania State University, Dept. of Civil Eng.

Turk Ž. 1999. Constraints of product modelling approach in building. *Proceedings of the 8th International conference on Durability of Building Materials and Components.* Vancouver: NRC Research Press.

Yau N.J., Melin J.W., Garrett J.H. & Kim S. 1991. An environment for integrating building design, construction scheduling, and cost estimating. *ASCE Seventh Conference on Computing in Civil Engineering and Symposium on Databases.* Washington, D.C.: ASCE.

Yu K., Froese T. & Grobler F. 2000. A development framework for data models for computer-integrated facilities management. *Automation in Construction* 9(2): 145-167.

Product and Process Modelling in Building and Construction, Gonçalves, Steiger-Garção & Scherer (eds)
© 2000 Balkema, Rotterdam, ISBN 90 5809 179 1

The LexiCon: An update

Kees Woestenenk
STABU, Ede, Netherlands

ABSTRACT: The LexiCon provides a structure, populated with objects of interest that can be used as a 'common language' for the construction industry. The structure of the LexiCon can also be used for storing and exchanging data between applications, between participants in construction processes and between owners and users of products resulting from construction activities. The LexiCon has been presented earlier at the ECPPM '98 conference in Watford, UK, and since then at several other conferences and meetings. This paper will give an update on the current status of the LexiCon itself and derived applications, such as SpecExplorer.

1 INTRODUCTION

The LexiCon is one of the source documents for the development of part 3 of the international ISO standard, under development as work item 12006-3 "Building construction – Organization of information about construction works – part 3: Framework for object oriented information exchange", by Working Group 6 of and ISO TC59/SC13. This standard will be restricted to the description of the structure of the framework; the LexiCon itself – or a similar tool – will provide access on the Internet to the population of this framework. It is anticipated that an international body will authorize the contents of the LexiCon.

SpecExplorer is an example of an application using the LexiCon, both its structure and its contents. A first prototype has been demonstrated at the IN-CITE 2000 conference in Hong Kong, January 2000. SpecExplorer is a generic specification tool, capable to store all types of non-geometrical data of any type of facility, covering all stages of its life cycle. It uses the structure of the LexiCon to store the data, and the contents of the LexiCon as templates for the specification contents. The prototype is developed as a blueprint for a series of dedicated, specialized applications, all using the same 'engine' that acts as an object layer between the application and the actual data store. Dedicated applications could for example concentrate on the client's brief, tendering documents, scheduling and facility management.

Other applications might use the LexiCon to provide links with general information, such as building regulations, product information, cost data and quality assessments.

2 THE LEXICON

At the time of this writing Working Group 6 of ISO TC59/SC13 is about to meet to discuss the model that will be at the base of part 3: Framework for object oriented information exchange. The LexiCon served as one of the input documents for WG6. WG6 also wanted the new model to be a subset of the – very large and complex – EPISTLE 3 model. For the co-ordination with other groups (ISO TC184, CIB, ICIS, IAI) that are active in the same area the so called Standing Conference has been initiated by ISO TC59/SC13.

2.1 *The model*

As simplified version of the proposed WG6 model is shown in Figure 1. The LexiCon structure will be adjusted to the final model that will result from WG6.

The model is a meta-model describing classes, where classes represent concepts. All class names start with the prefix "Lex", which might change into another prefix to the preference of WG6. The model itself has been kept as simple as possible. At the root there is an abstract supertype, called LexObject. Associated with this abstract supertype are identifiers like ID and Version, a Version Date, and a number of Descriptors. A Descriptor is derived from the another abstract supertype: LexDescriptor, and can either be a LongName, a ShortName, a Description or a BinaryDescriptor (not shown in Figure 1).

A ShortName is often a kind of abbreviation of a LongName. For example, the LongName 'metre' can be abbreviated to 'm', which

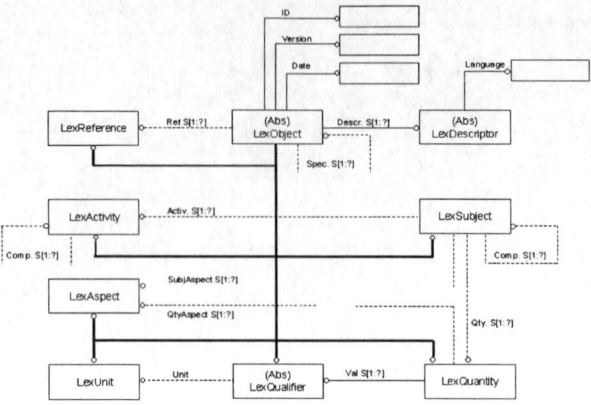

Figure 1: Simplified model

will be the ShortName. A Description provides a textual explanation of the concept to which it is associated. A BinaryDescriptor can be a graphical illustration, or another non-textual description, such as audio or video. All Descriptors have an assigned language. A LexObject may have any number of descriptors in any language, but must have at least one LongName. Hence, a concept has a unique identifier, and can be described in any language with all kinds of descriptors, including synonyms, acronyms and non-textual illustrations.

Also associated with LexObjects are References. These are external 'documents', such as standards, regulations, but also classification systems like SfB or Master Format. Again, any number of References may be associated. This provides a means to compare one document with another, through a common concept. For example, if both SfB and Master Format identify something like 'Floors', these terms could both be associated with the concept 'Floor', if this would be applicable.

A final characteristic of LexObject is that it provides a specialization mechanism, which means that LexObjects can be arranged in a specialization hierarchy, allowing inheritance. For example, a 'Concrete slab floor' is a type of a 'Floor', and inherits the characteristics of that Floor.

Derived from LexObject is LexSubject, which represents the concepts of the objects of interest, which are mostly tangible things like products, buildings and building parts and other structures, but also spaces and non-tangible things, such as software. In addition to the characteristics of LexObjects LexSubjects can be arranged in a composition hierarchy, through the composition mechanism. This means that a Subject may be a part of another Subject and may also contain other Subjects. For example, a 'Doorset' could be composed of a 'Doorframe', a 'Door leaf', 'Hinges' and 'Door furniture', where each of the components are Subjects themselves.

LexObjects are described by LexQuantities, another derived concept from LexObject. A Quantity is a concept that is comprised of a Qualifier and possibly a Unit. Qualifiers and Units are – as LexQualifier and LexUnit – also derived from LexObject. A Qualifier contains the actual 'data' about a Quantity. These data are called Values (not shown in Figure 1), which can be of several types, such as Nominal Value, Range Value, Enumeration Value and Text Value. Another Value type, Tolerance, can be associated with the Nominal or the Range Value.

Quantities may be grouped into Aspects with the LexAspect class. An Aspect can be used for any grouping, such as views and sets of interdependent Quantities. LexAspect has an optional association with LexSubject.

LexActivity is the final concept derived from LexObject. It is meant to provide a home for activities related to Subjects, for example activities to formulate requirements, resulting in a client's brief, or design activities, or activities for building and assembling the Subject, or maintenance activities. An Activity has also a decomposition mechanism.

2.2 The contents

The LexiCon is currently populated with examples of concepts. The intention is to have it populated on an international level. For this web-enabled access tools are under development. STABU has developed an editor and browser that can be used to populate the LexiCon and to browse through its contents (Figure2). This editor/browser will be made compliant with the final model.

In the Netherlands the BAS organization will take responsibility for the Dutch population of the LexiCon, using the Dutch language. BAS is an alliance between most organizations that are involved in construction. As a start, the existing Article classification, developed by electro technical and mechanical contractors organizations (UNETO and VNI),

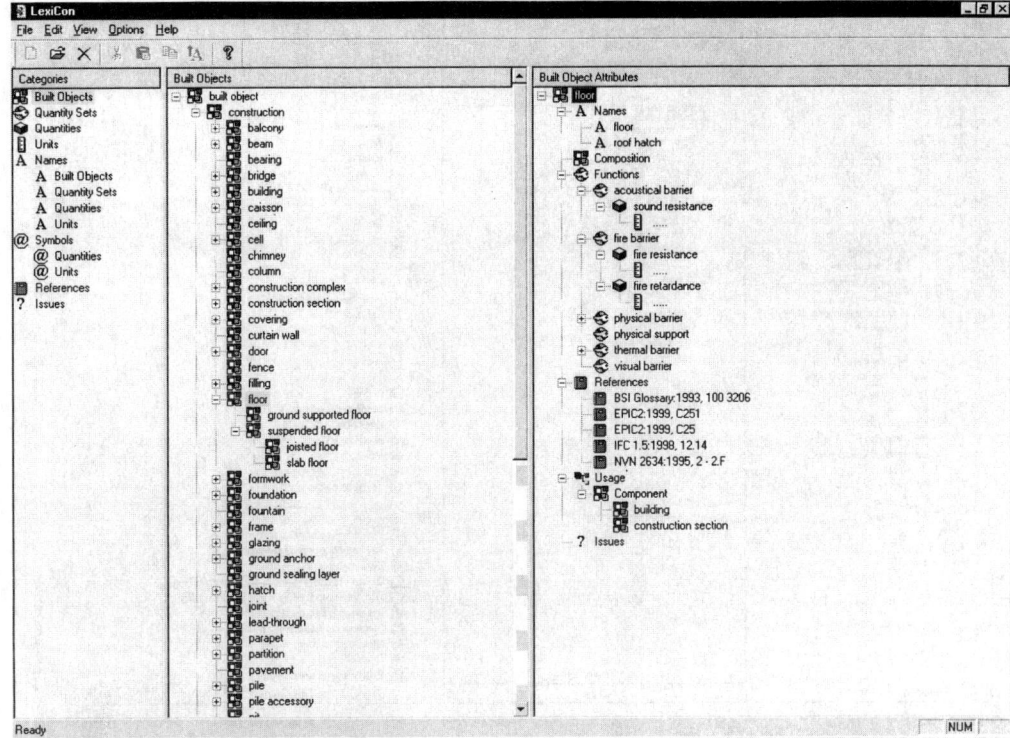

Figure 2: The LexiCon editor/browser

will be moved to the LexiCon. STABU will translate its current specification system into the LexiCon. It is anticipated that other countries will do their part of the population, in their language. At the time of this writing a number of members of ICIS (International Construction Information Society) have expressed their willingness to participate. These are: Construction Information Systems Australia (CIS), Construction Information Limited New Zealand, Construction Specifications Canada (CSC), Norwegian Council for Building Standardization (NBR) and Gemeinsamer Ausschuss Elektronik im Bauwesen (GAEB) from Germany. An organization has to be set up to co-ordinate the international activities.

3 SPECEXPLORER

The SpecExplorer prototype application, in development by STABU, is an example of one of the possible uses of the LexiCon. SpecExplorer provides a tool to maintain project information for the whole life of a construction facility, covering inception, design, construction, maintenance, facility management and decommissioning. SpecExplorer itself stores only alphanumeric data, for geometrical data it needs the support of a CAD tool.

SpecExplorer uses the same structure as the LexiCon, with some additions, and it uses the concepts of the LexiCon as templates for describing the Subjects of a project. Figure 3 gives an impression of the user interface.

SpecExplorer shows in one of its panes the composition of the subject. At the root of this composition tree the project itself is found. A project can be anything, it could be a project for building an airport, or a project for refurbishing a kitchen.

Each node in the composition tree represents a Subject, and each Subject may have an associated Specification, a set of associated Component Quantities, a set of associated Tasks and a set of Notes.

The Specification of a Subject can be copied from the LexiCon Subject. In that case only the project dependent Values have to be added, to make the description project specific. A Value in the Specification can have one of three states: 'Required', 'Proposed' or 'Actual', which can all be part of a Specification, making comparison between these Values possible. A Specification does not need to be an exact copy of a LexiCon Subject; it should be regarded as a derived type, allowing for additional attributes. Additional attributes can be copied too from the LexiCon, as Quantities. If, however, a Specification is associated with several Subjects, this Specifi-

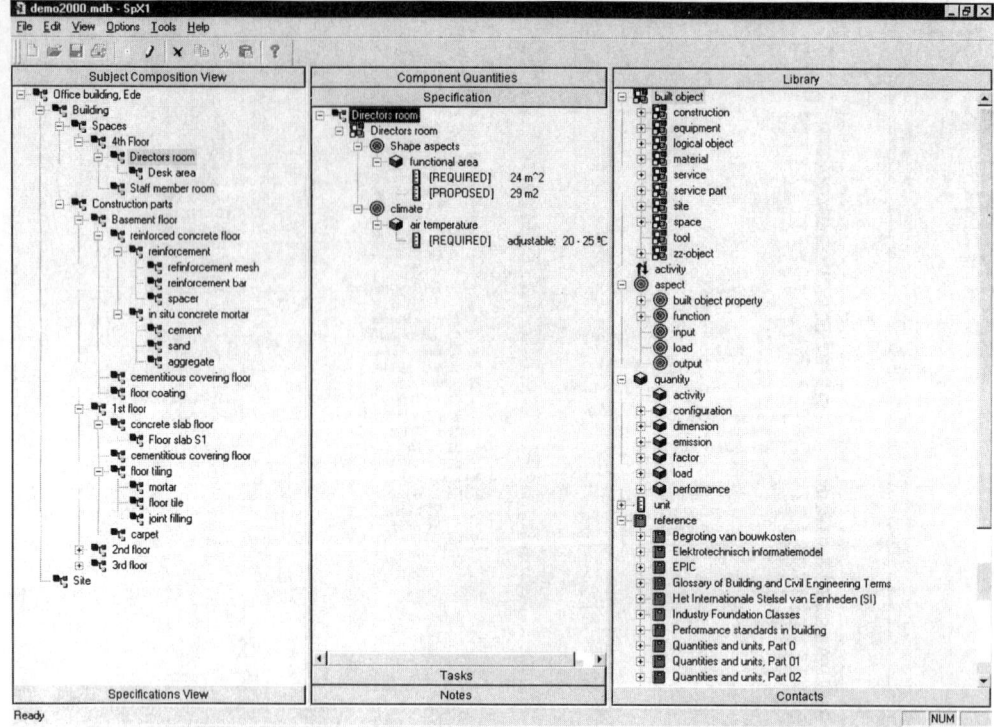

Figure 3: SpecExplorer

cation is the same for all these Subjects, with the exception of Values. Values are specific for a single occurrence of a Subject. The set of Specifications can be viewed in an additional view, showing all Specifications with their associated Subjects.

Component Quantities are composition dependent Quantities. For example, the number of Slabs making up a Slab Floor, where both the Slab Floor and the Slabs are Subjects. The number of Slabs in a Slab Floor is a Component Quantity.

Tasks describe activities associated with a Subject. A task can be anything that has to be done, such as writing the brief, making the design, assembling a set of products to a part, cleaning a part, etc. A task description describes the equipment that is needed, the time that is needed, the participants involved the job and the (contractual) conditions that apply for the job. Whereas a Task can be assigned with every single Subject, it therefore can contain all types of contents. For example, on the project level it describes project related tasks and conditions, such as general contract conditions. By the same token it can also describe the task of brick laying, or cleaning a floor.

Notes provide a means to attach documents, reports, issues and the like to a Subject. Again, these attachments can be on any composition level.

The right panes of SpecExplorer show the Library and the Contacts.

The Library can be the LexiCon, but it could also be a company library derived from the LexiCon, or even another project. The contents of these Libraries contain the knowledge, which can be made available for a project.

The Contacts provide a link with a Contacts database providing information about Participants involved in the project.

4 CONCLUSION

SpecExplorer is one example of using the LexiCon. Many other examples could be given, such as the use of the LexiCon in E-Commerce, or for product information.

The idea of the LexiCon: providing an extensive, language independent list of concepts that are of interest for the construction industry, described in a computer interpretable way, gets more and more interest and support. We hope we can make it work in the near future.

Product libraries – Technology review

James Nyambayo & Ihsan Faraj
Building Research Establishment, Watford, UK

Robert Amor
University of Auckland, New Zealand

ABSTRACT: This paper reviews the development of the standards for access to external product libraries in relation to procurement and product data exchange. Procurement accounts for 70-85 % of the value of a building project and hence it is vitally important for the data exchange required to support these processes to be streamlined. Product data exchange between applications is equally important for streamlining the specification, design and construction and maintenance of a building facility. A considerable amount of effort within the Architecture/Engineering/Construction and Facilities Management (A/E/C/FM) has gone into developing standards for product data exchange. Manufacturers and suppliers have supported the development of procurement related standards. With the developments in XML-based standards the technology is now available for developing cost-effective solutions that meet the requirements of both procurement and product data exchange. The paper evaluates the existing and emerging standards and discusses the issues arising from the development of standards for access to product libraries and possible ways forward.

1 INTRODUCTION

The case for providing electronic access to manufactured products data in a structured way is now well established, both for procurement and product data exchange between applications. A recent survey suggests that 75-85 % of the project cost is attributed to procurement [Clark, et. al 1999]. Product information is available through a number of sources, which includes manufacturers, suppliers, and product information providers. The information is available in a variety of formats ranging from paper-based catalogues and unstructured web pages to structured databases. The problem with the lack of standards is that manufactured product data is not being accessed effectively from within applications.

Within Architecture/Engineering/Construction and Facilities Management (A/E/C/FM), considerable effort has gone into defining standard models for exchanging product data using EXPRESS [ISO 1994], a data definition language. A considerable effort has also gone into developing procurement standards, but these standards have been more widely used in industries such as the manufacturing and process industries. The procurement and product data exchange standards have been developed independently of each other largely because of the different business processes involved.

This paper reviews the data requirements in relation to procurement and product data exchange. It reviews the development of the standards for procurement and model-based product data exchange. The role of XML in the development of these standards is discussed.

2 REQUIREMENTS

2.1 *Procurement related requirements*

The main role of procurement systems is to facilitate the purchase, delivery and installation of products. The requirements of such systems include:
- Product identification i.e. description, classification or functional specification
- Order information
- Negotiation
- Secure financial exchanges
- Contract exchanges
- Tracking of transactions and delivery of products.

Procurement does not require detailed attributes of the products. However attributes in conjunction with classification systems may be required for identification. The rest of the data that is exchanged is related to the procurement financial exchanges and the management of the procurement process.

2.2 Product Data Exchange Related Requirements

Product models within A/E/C/FM define the physical and conceptual objects that make up a building or construction project. The role of the product models is to provide standards for the exchange of product data between applications within the construction industry. The model-based exchange is not complete without incorporating products that reside in external product libraries, e.g. manufactured products. The requirements of an external library identified in [Nyambayo et al 2000] include:

- common searches across libraries
- common structures to the library catalogues through classification systems or product data dictionaries for easy access.
- common external library management systems e.g. brokers or data warehouses.
- common interfaces to the project environment e.g. IPDB systems, CAD and analysis tools.

Generally, product data definitions in these models are detailed. The amount of data exchanged is significantly more than in the procurement process.

2.3 Analysis of Requirements

The need for searching for libraries is common to both systems. Therefore standard catalogue interfaces are essential for both processes. Detailed attributes are not essential to the procurement process while they are essential for data exchange within the design, construction and facilities management processes. Being able to use common data structures will have a positive impact to the management of whole life cycle cost of a facility.

3 DEVELOPMENTS OF PROCUREMENT RELATED STANDARDS

Procurement has traditionally relied on paper-based exchanges, paper-based catalogues, telephone and Fax. A recent survey suggest that most of the transactions within the construction industry still rely on Telephone and Fax [Clarke et. al. 1999]. However, there is evidence to suggest that this is changing as electronic based technology matures and more people become more confident in the new technology. This section reviews some of the existing and emerging standards for procurement.

3.1 Electronic Data Interchange (EDI) Standard

EDI (Electronic Data Interchange Format) emerged as a standard in the early 90s. It is a protocol and data format for exchanging (mainly financial) data. EDI provides a data structure and mechanism for exchanging the information over value added net-

works. EDI transactions are business to business transactions that involve no or little human intervention. Subsets of the UN/EDIFACT standard have been developed specifically for exchanges within the construction industry.

The uptake of EDI has been poor within the construction industry. The barriers to its uptake has been the need for value added networks, which proved to be too expensive for small to medium enterprises involved in the procurement process.

Recent developments have involved taking the EDI structure to the web using XML. It has been noted that a simple conversion of EDI tags to XML would not work without significant changes to the data structure. However some of the basic data structure has been adopted in the XML-based standards under development.

3.2 XML based catalogue standards

One of the most significant developments in the electronic technology has been XML. XML (Extensible Markup Language is a subset of SGML. XML is not a data exchange standard, but a transport mechanism. The most significant development has been the development of XML-based catalogue standards and integration frameworks. The catalogue standard provides a framework for defining product catalogues. Examples of these standards are briefly described in this section.

3.2.1 cXML

cXML (Commerce Extensible Markup Language) standard defines an XML-based protocol and data format for business-to-business transactions, developed by Ariba [Ariba 1999]. It defines electronic catalogues and uses the Internet to exchange documents such as purchase-orders, contracts. It supports parametric searches across catalogues. This standard is supported by Autodesk, XML.Org and Microsoft through its Biztalk [URL1] framework.

3.2.2 xCBL

xCBL (Common Business Library) is a rival protocol and catalogue standard developed by CommerceOne [CommerceOne 2000]. It defines purchase order documents, invoice documents and catalogue content documents. The xCBL is quite similar in structure to the cXML. xCBL is supported by a number of standards development consortia that include:

- RosettaNet [URL2], a consortium of organizations that include Microsoft, IBM, Hewlett Packard, Netscape, whose objective is to develop standards for the exchange of electronic catalogues.

- XML.Org [URL3], a repository for XML schemas that can be used by participating organizations within the same industry to exchange XML based schemas. It is supported by OASIS, a non-profit organization whose objective is to develop and promote the use of XML based standards within and across industries.
- CommerceNet [URL4], a consortium whose objective is to develop XML standards for e-commerce within the financial sector.

Even though cXML and xCBL standards are similar, and have support of common influential industry players such as Microsoft, there is no sign that they will merge.

3.3 ProCat-Gen Catalogues (Genial)

PROCAT-GEN [Cook et. al 1999] is an XML based catalogue system for use within the Global Engineering Network. It is the result of an EU funded project. The catalogue system defines a hierarchical classification structure for construction products and product families. Its structure is more specific to the engineering product data than the xCBL and cXML. It provides for more intelligent searches and organization of information. Its features include views and profiles. A view is a facility to define different classification structures for different users. The access to the views is controlled by 'access control lists' (ACL). Profiles define different sets of attributes for different users. For example the attributes used by an architect may be different from a services engineer. However PROCAT-GEN catalogue standard is not publicly available.

3.4 XML Catalogue Standards summary

The XML-based catalogue standards, in conjunction with other product data access frameworks, have the potential to provide cost-effective solutions for access to external library product data. However, the problem with having with so many standards is that there is no interoperability between them. Both xCBL and cXML standards are supported by the large industry players. PROCAT-GEN is the result of a pilot project and its use is currently confined to the GEN network.

4 CLASSIFICATION STANDARDS

The catalogue standards described in the *section 3* only provide a framework for defining catalogues, but not the product families themselves. Classification systems are required to define these structures.

Classification systems in the construction industry define classes of product families. The classification systems have evolved over a long period of time. Several classification systems exist e.g. CI/Sfb, Uniclass, CAWS, EPIC. According to a survey conducted by RIBA [URL5] CI/Sfb is still the most widely used classification system within the United Kingdom. It is a paper based hierarchical classification system of building products. It has been in existence for over 40 years. In this classification system each product family is represented by a code that is made up of letters and numbers. The structure of the code does not lend itself well to electronic use. The Uniclass system evolved from the CI/Sfb, CAWS CESMME3 and EPIC. The Uniclass system [Uniclass 1997] defines a hierarchical product classification, in addition to management entities, facilities, construction, spaces, civil engineering works, work sections etc. Each product or product family is represented by a letter and number, in a way that lends itself to electronic use. EPIC Version 2 [URL6] released in 1999 is the most recent classification system. The product specification in EPIC is similar to the construction product classification (Table L) of the Uniclass system. They both lend themselves to electronic use. There are indications that EPIC and Uniclass are moving closer together in the way they classify products. They are both built according to the ISO generic classification structure specification, which is why they are compatible.

There are also general classification systems such as the Universal Standard Products and Services Classification (UNSPSC) Code [URL7]. It is a hierarchical classification system for products and services for procurement purposes. It is a result of a merger between the United Nations' Common Coding System (UNCCS) and Dun &Bradstreet's Standard Products and Services Classification (SPSC). The significance of this classification system is that it is supported by VISA and other payment systems. However the classification system is not detailed enough for construction-related products.

The main problem with classification systems is that they are mostly nationally based. The decomposition of products to their constituent components sometimes differs. Some entities defined as products may be elements of larger products in other classification systems. Mapping between these classification systems is not an easy task. The only way forward is possibly the de facto dominance of a few classification systems over the rest or the development of a common vocabulary for describing objects in the construction industry. This concept, known as the Lexicon, is described in detail in [Woestenenk, 1998] and [Woestenenk, 2000].

5 DEVELOPMENT IN PRODUCT DATA STANDARDS

Product data models within A/E/C/FM define elements used in a building and construction project. A number of models have been developed examples of which include: BCCM [Wix and Liebich 1997], ATLAS [Greening and Edwards. 1995], COMBINE [Augenbroe 1995], RATAS [Bjork 1994], COMBI [Scherer 1995], standards developed by STEP and Industry Foundation Classes (IFCs) developed by the International Alliance for Interoperability (IAI). The role of these models is to provide standards for exchange of product data between applications. These models access external product libraries in different ways. This section reviews the model-based access to external product libraries. The models reviewed in this section are: IFC Release 2.X for the construction industry, Part-lib (ISO 135840) [ISO 1997] for the manufacturing industry and the Epistle Class Libraries for the Oil and Process industries.

5.1 Referencing External Libraries using Industry Foundation Classes (Release 2.X)

The Industry Foundation Classes (IFC) model is an integrated model based on ISO 10303 [ISO 1994], representing the physical and conceptual objects used in the A/E/C/FM industry. The model has a number of domains that include Architecture, HVAC, Construction Management, Geometry. A detailed description of the structure of the IFC model can be found in [IAI 1999] and the online documentation on [URL8].

While the model defines most of the products explicitly, it does not attempt to explicitly define all the possible attributes. Instead it uses the property definition to extend the explicit classes. The property definition is a meta-model that defines classes that can be associated to the explicit classes at run-time.

In this model, the manufactured products (defined as property sets) are related to the explicit classes within the model through a relationship class (*IfcRelAssignsProperties*). A property set may contain a list of properties. It may also have a reference to an external library, defined in the *IfcExternalReference* schema. The role of the external library reference schema is to capture the information about external libraries. The property set model is described in detail in [IAI 2000]. An example of its use in practice is described in detail in [Nyambayo et al 2000].

This IFC Release 2.X model supports the use of classification systems. This means that external library information can be organized in hierarchical structures conforming to classification systems such as Uniclass and EPIC. Classification systems can be defined as instances of the *IfcClassificationReference* schema. Furthermore, different classifcation systems can be represented within the model. Products can have a classification system view as well as a model-based view.

The IFC approach does not explicitly define all the product information, but depends on the industry to agree to the structures of their products. The advantage of this approach is that it is flexible. In the construction industry, clients do not have much in-

PropertySet Definitions:

PropertySet Name	Pset_Chair
Typed	True
Typed Class	IfcFurniture
TypeName	Chair
Definition	Definition from IAI: A set of specific properties for furniture type chair HISTORY: ([existing Pset from IFC Release R1.5]) ISSUE:

Property Definitions:

Name	Property Type	Data Type	Definition
SeatingHeight	IfcSimpleProperty	IfcPositiveLengthMeasure	The value of seating height if the chair height is not adjustable
HeighestSeatingHeight	IfcSimpleProperty	IfcPositiveLengthMeasure	The value of seating height of high level is the chair height is adjustable
LowestSeatingHeight	IfcSimpleProperty	IfcPositiveLengthMeasure-	The value of seating height of high level is the chair height is adjustable

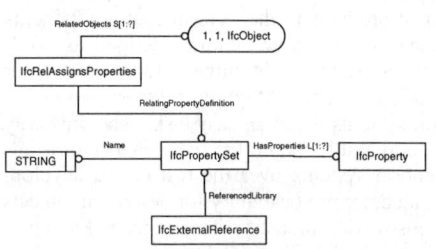

Figure 1: Object extension Model within the IFC Release 2.X (beta) model

Figure 2. Sample Property Set describing a chair

fluence on the procurement process as is the case in the manufacturing and process industries. Supply chains in the construction industry are not established on a long-term basis. They very much depend on the location of the constructed facility. In such an environment, it is very difficult for the industry to dictate the structures of the product information.

The disadvantage of this approach is that it does not solve the interoperability problem. It passes on the responsibility for standardization to the end users. From an implementation viewpoint, it is not easy for end user applications to manipulate product data effectively if they have no prior knowledge of the structure. For example parametric searches are not easy when the parameters have not been standardized.

5.2 *Other STEP- based Industry Initiatives –The Epistle Class Libraries*

The Epistle Class Libraries provide a standard for defining components used in the Oil and Process Industries. The components defined by this standard are used with the Epistle Core model. The concept of the library is based on STEPLib [ISO TC184/SC4/WG3 1997], a class originally developed for use with the AP221 Step Application protocol within the Process Industry. The components are defied as instances of 'class of object'.

In the Epistle framework, standard components have been defined to form the core of the library. The standard components may be organized in hierarchical classification structures. These classifications could be used to develop manufacturer catalogues.

The concept of extending the core model with class libraries is very similar to that of the IFCs where property sets are used to represent the external library products. The main difference is that properties are more closely tied to the objects than is the case in the IFC model. This makes the use of the library, especially the manipulation of the library products by applications, more efficient.

Classification and parametric-based searches can be standardized as all products are predefined. Symbols can be assigned to the classes to enable the use of the product data in CAD and other related applications. A sample of the standard parts defined within the class library are shown in Figure 2.

The Epistle approach is very similar to the IFC approach. However in Epistle Class libraries, the products have been defined and structured into agreed classification structures. These classes are related to the rest of the model by the semantic rela-

Unique class id	Sub-ject area	Object type	Class of item name (narrower term)	Association type
100,355	heat transfer	physical object	air conditioner	can be part of a
100,355	heat transfer	physical object	air conditioner	is a class of
100,006	heat transfer	physical object	air cooled heat exchanger	can be part of a
100,006	heat transfer	physical object	air cooled heat exchanger	is a class of
100,380	heat transfer	physical object	air cooled heat exchanger system	is a synonym of
100,006	heat transfer	physical object	air cooler	is a synonym of
100,380	heat transfer	physical object	air cooler system	is a class of

Figure 2. Partial definition of products within the Epistle Class Library

tionships. Access to these classes is as easy as for any other explicit class. This has been made possible by the influence of the large client base within the process and oil industries. The Epistle class library provides for catalogue structures though its classification model.

5.3 *PartLib (ISO 13584).*

Part-Lib, ISO 13584 [ISO 1995], is a STEP standard used for defining external library products, catalogues and the relevant supplier information. The standard represents an integrated approach to the access to external libraries by:
- providing detailed structures of the product data explicitly using its Library_Content schema (13584-24) and Instance schema.
- defining property definitions using the dictionary schema and instance schema. The model explicitly defines parametric parts data.
- defining individual external libraries using the Library Schema.
- defining standards for supplier information.

STEP data exchange protocols such as STEP physical files (Part 21) and Standard Data Access Interface (SDAI) are used to exchange the product data between external libraries and the project environment.

Part-Lib represents an integrated approach to the access to external libraries. It defines all the structures required to access external libraries. This approach works well in the manufacturing industry where clients have more control over the supply chain. In the construction industry the clients do not have the same influence. This is why the models for the construction industry do not try to specify in detail all the attributes.

5.4 Lessons for the construction Industry

There are a few things that the construction industry could learn from other industries in terms of developing standards for access to external libraries and these include:

- explicitly defining external library products facilitates the development of applications to handle information in these libraries.
- Defining standards for catalogues facilitates searches across many manufacturer libraries

However there are some inherent problems that are barriers to the development and adoption of these standards, and these include:

- too many regional differences in the way product data is classified i.e. there are too many different classification systems.
- differences in the attribute requirements i.e. between users in the same domain and across the different domains. Therefore predefining these attributes as in Epistle and Part-Lib is a challenge.

6 PRODUCT DATA MODEL RELATED FRAMEWORKS

There are a number of frameworks that have been developed to access externally defined data within the construction industry. Unlike the manufacturing and process industries, where systems based on Part-lib and STEPlib are available commercially, the systems available are still largely research prototypes. The lack of progress within the industry is related to the poor adoption of integration within the construction industry. Reasons for the slow uptake of data integration within the construction industry have been discussed in detail in [Amor and Faraj 2000]

6.1 The ARROW System

ARROW [Newnham et al 1997] is a manufactured product warehouse that provides access to product information. The ARROW model is an extension of the IFC model release1.5.1 that represents all product information as explicit classes. The model defines attributes for each of the products explicitly. The advantage of this approach is that explicit classes are more efficient to search against. The problem with this approach is that it is not entirely standards-based, as the explicit extensions to the IFC model do not conform to the IFC standard.

There are no standards that define all the attributes for products within the construction industry. ARROW demonstrates the need for such standards. The practical application of such a system in the absence of detailed attribute standards is not yet realizable.

6.2 The CONNET System

The CONNET system [Amor et al 2000], the product of an EU project, is a service portal that provides access a number of services for the construction industry. The services currently supported are: a construction industry-specific bookshop, news service, product information service, a waste exchange service, a calculation software service and a best practice service.

CONNET has introduced a new dimension to the to the access of information within the construction industry, i.e. the concept of a network of services that provide standard access to information across regional borders. The major advantage of the CONNET approach is that there is standardization in the searches across different countries and a query to one system can be passed on to others.

The barrier to this approach is the different classification systems that are used in different countries. Mapping between different classification systems is still a major issue. Furthermore CONNET does not support the exchange of data between the system and the project environment.

6.3 IAI UK/BRE External Library Access System

The Library demonstrator [Nyambayo et al 2000] is a framework to demonstrate the use of the IFC model Release 2.0 and 2.X in accessing externally defined product data. The project is funded by the Department for Environment, Transport and Regions (DETR) in the United Kingdom. The library access system demonstrated the following:

- use of the property sets to structure external product data.
- how parametric searches on products structured as property sets could be implemented.
- how the library model could be used to represent external libraries, i.e. manufacturer or supplier libraries.
- an overall framework for the access of externally defined products, and how these could be integrated into the design environment

The emphasis of this framework, unlike the ARROW system or the CONNET system emphasized the exchange of product data between libraries and the project environment. In the view of the authors, a combination of CONNET-like services and the External Library Access-like systems is the way forward.

7 LINKING MODEL-BASED PRODUCT DATA STRUCTURES TO XML-BASED CATALOGUES –(EXPRESS TO XML MAPPING)

XML is a character-based syntax for representing structured data objects. It is just a serialisation syntax which means that there is still need to develop standards in order to meaningfully exchange any information. Product data structures already exist in EXPRESS format while catalogue standards exist in XML format. The challenge has been mapping EXPRESS data structures to XML.

The problem with XML is that there are many ways of representing an entity, resulting in many possible ways of mapping EXPRESS to XML. This section reviews the standards for mapping EXPRESS to XML that have and are currently being developed.

7.1 *Product Data Markup Language (PDML)*

The PDML [Shocklee et. al. 1999] has been developed by Product Data Technologies and funded by the US Air Force. It is an early binding to EXPRESS. An early bound XML schema or DTD is specific to the corresponding EXPRESS schema. In early binding, the applications developed to manipulate the XML schema have to have prior knowledge of the schema. And these applications can only be used with that schema.

7.2 *XML Meta-data Interchange (XMI)*

XML Meta-data Interchange (XMI) is a specification for defining XML meta-models. It is a specification developed by the Object Management Group (OMG). A meta-model is a generic model for defining specific models. The mapping provided for by this specification is general and can be used with any model including EXPRESS. However, since it was specifically designed for interchange of UML models, it depends on an UML-to-EXPRESS mapping.

7.3 *Part 28 (ISO 10303-28)*

ISO 10303-Part 28 is an ongoing development by ISO/TC 184/SC4. It defines a late binding to EX-

PRESS i.e. it specifies an XML DTD that can be used to encode one or more EXPRESS schemas and associated data sets. The benefits of late binding over early binding is that applications can be developed to handle any number of schemas and corresponding data sets.

8 WAY FORWARD

The development of the standards for mapping XML to EXPRESS has opened up the real opportunities in the standardization of access to externally defined products. The links between XML-based catalogues and EXPRESS product data repositories is now a possibility.

XML represents a cost-effective way of exchanging information and enables elemental data to be exchanged, unlike the case with Part 21 files. There are several possible building blocks to this solution and these include:
- Repositories for the product data schemas, to facilitate the standard specifications of these standards. Existing repositories such XML.org, Biztalk could be used or more specific ones could be developed for use within the construction industry.
- Brokers or Access frameworks, these will provide standard access and querying mechanism for accessing the external catalogues (see Section 6)
- Standard specifications and classifications of the product data. This is probably the area where the product data standards that have been developed in EXPRESS could be used. By applying the mapping between EXPRESS and XML, XML schemas of the product data could be generated. However, within the construction industry the barriers discussed in Section 5.4 have to be overcome.

9 CONCLUSIONS

This paper presented a review of the data requirements and the development of standards for access to external libraries, for procurement and product data exchange in the construction industry. There are similarities in these requirements and there is scope for a common data structure that meets the requirements of both processes. XML-based catalogue standards and XML to EXPRESS mapping are providing a basis for the development of cost-effective frameworks for the access of external libraries.

10 REFERENCES

Almeida, L., Grilo, A., Rabe, L., Duin, H., 1998. Implementing EDI and STEP in the Construction industry

Amor, R. and Faraj. I., 2000a, Misconceptions of an IPDB, In Amor, R. and Faraj, I: *Proceeding Objects and Integration for Architecture, Engineering and Construction*, BRE, UK March 13-14.

Amor R.,Turk, Z., Hyvarinen, J. and Finne, C. 2000b. CONNET: A Gateway to Europe's Construction Information, In Gudnason, G. (ed) *CIT2000-taking the construction industry into the 21st century*, Reykjavik, Iceland, 28-30 June, 2000

Ariba, Inc. 1999. cXML/1.1 Specification Available on http://ww.ariba.com

Augenbroe, G., 1995. An Overview of the COMBINE Project. In Scherer (ed.), *Proceedings ECPPM'94: Product and Process Modelling in the Building Industry*, Balkema, pp. 547 - 554.

Björk, B. C., 1994. The RATAS project - an example of cooperation between industry and research toward computer integrated construction. *Journal of Computing in Civil Engineering*, ASCE, 8(4), 401-19.

Clark, A.M., Atkin, B.L., Betts, M.P., Smith, D.A. 1999. Benchmarking the use of IT to support supplier management in construction. *ITcon* Vol. 4, July

Cook, G., Czubayko, R., Klemme, R., 1999. PROCAT-GEN – Conformance Specification

CommerceOne, 2000. xCBL 2.0 Release 3.0 Documentation available on http://www.commerceone.com/xbl/cbl/docs/index.html

Ekholm, A., 1996. A Conceptual Framework for Classification of Construction Works, *ITCon*, Vol 1, March

Hemio, T. and Salonen, M., 1999. *Virtual Reality: Human Interface to Product Data*. In

IAI 1999a. Introduction to the International Alliance for Interoperability and the Industry Foundation Classes

IAI, 2000. IFC Release 2.x beta version, available on http://www.iai.org.uk/

ISO, 1994. ISO 10303 part 11, Product Data Representation and Exchange

ISO, 1995. ISO 13584 PLIB TC184/sc4 Parts Library

ISO TC184/SC4/WG11 N101, 1999. ISO/WD 10303-28 Product data representation and exchange: Implementation Methods: XML representation of EXPRESS-driven data

ISO TC184/SC4/WG3 N397, 1997. Guide to STEPlib: Guide for the creation and maintenance of Standard Data International alliance for Interoperability manuals

Newnham,L.; Amor R. and Parand, F., 1997: Gaining Quality Manufactured Product information through Arrow. *IT-CPR-97*, Cairns, Australia, 9-11 July 1997.

NIST GCR 99-781 1999. XML Representation Methods for EXPRESS-Driven Data

Nyambayo, J., Amor, R., Faraj, I. and Wix, J. 2000. External Product libraries: an implementation of IFC release 2.0. *Product Data Technology Europe 2000 Conference*, Noordwijk, The Netherlands, 4- 5 May.

Ofluoglu, S., Coyne, R. and Lee, J., 1999. Managing Building Information on the Web. In Heng Li., Quiping Shen., David Scott., Peter E. D. Love. (eds) . *INCITE 2000: Implementing IT to obtain a competitive advantage in the 21 st Century*.

Shocklee, M., Burkett, B. and Yang, Y., 1999. Product Data Markup Language (PDML) Specification,

URL1, Biztalk Framework on http://www.biztalk.com/

URL2, http://www.rosettanet.org/

URL3, http://xml.org/

URL4, http://commerce.net/

URL5 http://www.ris.gb.com/manuf/sbfagenc.htm

URL6 http://www.epicproducts.org.

URL7 http://www.unspsc.org/

URL8 http://www.iai.org.uk/

Woestenenk, K., 1998. A Common Construction Vocabulary, *In ECPPM'98*, BRE, UK, 19-21 October, pp 561-568

Woestenenk, K., 2000. Implementing the LexiCon for practical use. In *CIT2000: taking the construction industry into the 21st century*. Reykjavik, Iceland, June 28-30. Pp 1049-1057

Geotechnical site investigation: A model based study

Hilary J. Kahn, Alan R. Williams & Nick P. Filer
Department of Computer Science, University of Manchester, UK

ABSTRACT: The work reported in this paper has been done in the context of a collaborative project called KLICON concerned in general with knowledge-management and organisational learning within the construction industry. This specific study is investigating the use of various product modelling techniques and standards to support the capture and use of construction-related knowledge. The goal is to identify how a variety of modelling techniques may be harnessed to provide improved understanding and use of the information available to construction engineers and designers. To provide a focus for the modelling, and so allow us to provide adequate levels of detail in the models, a specific topic was selected as the basis for this work. The topic selected was Geotechnical Site Investigation. The approach used in this work combines IDEF0 activity modelling, EXPRESS information modelling, XML, HTML, the AGS proposed standard, and an ontology being developed at the Building Research Establishment (BRE).

1 INTRODUCTION

The work reported in this paper has been done in the context of a collaborative project, supported by the UK Engineering and Physical Sciences Research Council (EPSRC). This aspect of the work has specifically involved building consultants Ove Arup and Partners, various companies that are part of the Kvaerner Group and the Building Research Establishment (BRE). The broad context for the project is knowledge-management within the construction industry. The specific aspect of this study is investigating the use of a variety of modelling techniques to support the capture and use of construction-related knowledge. The goal is to provide the prototype of a mechanism that can be used to relate project specific information to a generic framework. The eventual goal is to provide support for such diverse functions as engineer training and project audit.

1.1 Domain selection

Geotechnical site investigation was selected as a suitable domain for this development for a number of reasons. It is a domain that is fundamental to much of the work of the construction industry and in particular, it is one that is relevant both to consulting engineers such as Ove Arup and Partners and to construction companies such as the various parts of the Kvaerner Group. It was also important in selecting the domain to ensure that it had sufficient complexity to be worth addressing; geotechnical site investigation is very suitable here because it involves a large number of different stages and interacts at various points with different aspects of construction. A key point for the study was the availability of geotechnical domain expertise; in this case it was mainly provide to us by Ove Arup and Partners. Finally it was also important to have access to data examples from both large and small projects so that the correctness and relevance of the models being developed could be checked.

2 OVERALL SYSTEM

The overall system produced (see Figure 1) links together a generic model of the domain of geotechnical site investigation with

- a simple ontology which provides definitions of terminology in a structured manner
- a model of a format commonly used to transfer geotechnical data between site investigator and the consulting geotechnical engineer
- example data for specific projects supplied by site investigators
- a mechanism for presenting project specific data.

The purpose of this prototype system is to investigate the use of a model driven environment to sup-

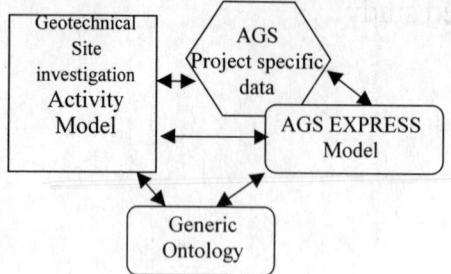

Figure 1: Links between system components

port knowledge re-use. A generic model is used to capture the wide range of activities that are applicable in a given domain of interest. That model is enhanced by links into a general terminology database (ontology) where further explanatory material is provided. Project specific data can then be linked into these generic resources. The project specific information could vary from scanned project documentation to data that can be interpreted or analysed. The prototype illustrates how project-specific data is linked. In this case, a formal model is provided for a format called AGS defined by the Association of Geotechnical and Geoenvironmental Specialists. In addition, project-specific data in AGS format is provided. The formal model is used as the basis for interpreting the AGS project data and generating a meaningful presentation of that data.

3 SYSTEM TECHNOLOGIES

The underlying technologies used in the project are related to specifications or models of some kind and to standard representations.

In this development the following types of modelling or specification are included:

- Activity modelling using IDEF0 (Softech Inc. 1981)
- Information modelling using EXPRESS (ISO10303-11 1994)

In addition, links are made to an ontology under development by BRE. We have restructured this so that it is represented using the Virtual Hypermarkup language and is therefore compliant with the Virtual Hyperglossary (http://www.vhs.org.uk). Other representations used include HTML and XML (Neil Bradley 1999, World Wide Web Consortium 2000). The EXPRESS model, and the HTML and XML examples have all been checked for conformance to the relevant standard definitions.

As an example of a standard format, AGS is used to carry project specific data. An EXPRESS information model of AGS has been defined and an XML

representation of the format created in order to support the presentation of the information.

3.1 Activity Modelling: IDEF0

IDEF0 is a graphical method used to capture the activities carried out within a domain. The IDEF0 approach allows a complex domain to be decomposed into a number of activities, each of which can be further decomposed if required. Activities are characterised by the definition of the external factors that affect them and by the relationships they have to other activities. The contents or decomposition of a given activity is presented as a single page diagram. Hence the activity model as a whole is a hierarchical tree of diagrams. The overall domain is represented by a top-level activity, shown as a box identified as "A0" on a diagram numbered "A-0".

The external factors that affect a given activity are its inputs (I), controls (C), outputs (O) and mechanisms (M), known collectively as the ICOMs.

- Inputs enter the left of the box and represent materials or information that are altered or consumed so as to generate the outputs or perform the activity
- Controls enter the top of the box and alter or control the functionality of the activity and its ability to produce the desired outputs
- Outputs leave the right of the box and are the material or information which result from the activity
- Mechanisms enter the bottom of the box and are the means that support or carry out the activity.

This is illustrated in Figure 2.

Relationships between different activities are of three general kinds. The passing of outputs from one activity to another indicates a flow of information and, in general, a dependency relationship between the two activities. There is also a relationship established between activities if they share inputs, mechanisms or controls. The third relationship is the one established by the decomposition of an activity into its sub-activities.

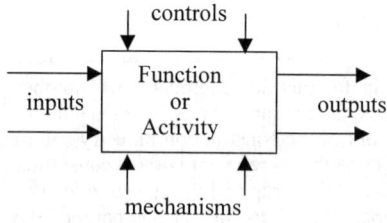

Figure 2: Constituents of an IDEF0 activity definition.

The purpose of an activity model is to define the activities or functions performed in a given domain. It is not a process model and there is no concept of time should be inferred when using or defining an activity model. It would of course be possible to associate the information in an IDEF0 activity model with a separate, project-specific process model.

3.2 Information Modelling

Information modelling is a technique for defining the information that characterises a given domain. In the prototype system described here, information models were written in the ISO Standard language EXPRESS. An EXPRESS model supports the definition of the objects of interest in a domain. These are called entities. Entities are characterised by their attributes, which may be of various kinds. Relationships between different entity types can be represented and the language, which has many object-oriented features, supports the definition of supertype-subtype inheritance hierarchies of entities. Both single and multiple inheritance are supported. Significant facilities for defining limits and constraints such as domain and global rules are also provided.

As an example, consider the excerpt from an information model of the AGS data exchange format illustrated in Figure 3.

```
(*
** A "sample_reference_information" corresponds
** to a SAMP group.
*)
ENTITY sample_reference_information;
  depth_to_top_of_sample : metre_measurement;
  samp_ref : sample_reference_number;
  samp_type : sample_type;
  sample_diameter :
      OPTIONAL millimetre_measurement;
  depth_to_base_of_sample :
      OPTIONAL metre_measurement;
  ...
  chalk_tests : SET [0:?] OF chalk_test;
  suction_tests : SET [0:?] OF suction_test;
INVERSE
  containing_hole : hole
      FOR sample_reference_informations;
UNIQUE
  key_attributes :
  depth_to_top_of_sample, samp_ref, samp_type,
      containing_hole;
END_ENTITY;
```

Figure 3: Excerpt from the AGS EXPRESS model

The definition for *sample_reference_information* includes a large number of characterising attributes, a few of which are shown in Figure 3. Some of the attributes, such as *depth_to_top_of_sample* and *sample_ref* are mandatory attributes; others such as *sample_diameter* are optional. Two of the attributes shown, *chalk_tests* and *suction_tests* are aggregates (e.g. SETs in this case). The UNIQUE clause identifies some *key_attributes* the values of which must be unique in the domain. The INVERSE clause establishes a relationship to another entity, *hole*.

3.3 Virtual Hyperglossary (VHG)

The Virtual Hyperglossary provides an XML-based mechanism (called Virtual Hypermarkup) for representing terminology in an interpretable manner. The top level information object in a hyperglossary is a VHG element. A VHG element consists of three types of component:

- *termEntrys*, which encapsulate the information for a term.
- *links*, which allow subsidiary information such as multilingual glossaries to be associated with terms
- *admin information* which is used to authenticate and describe how the information was collected and organised.

Figure 4 illustrates part of the VHG definition for information relating to soil sampling based on Clayton et al. (1995).

In VHG, different parts of the terminology description are delineated using tags. For example, in Figure 4 the start of an entry is indicated using the markup tag <termEntry> and the end of the entry is indicated by a matching tag, </termEntry>. Similarly, the administration data is delineated by <admin> and </admin>. Tags can also be used to indicate guidance to the presentation of the glossary information, For example, <EM>in situ</EM> indicates that words in situ should be emphasised in some way.

3.4 AGS Format

The format defined by the Association of Geotechnical and Geoenvironmental Specialists is intended to provide a way of communicating geotechnical data in a computer sensible way. The comma separated text data file has been designed to hold only fundamental data such as exploratory hole and test data required to be reported by the relevant British (or other National) Standard or similar recognised documents.

<termEntry>
    <term>Sampling or soil sampling</term>
    <definition>
        <P>Sampling is carried out in order that soil
        and rock description, and laboratory testing
        can be carried out. Laboratory tests typi
        cally consist of:</P>
    <OL>
    <LI>index tests (for example, unconfigured
        compressive strength tests on rock);</LI>
    <LI>classification tests (for example, Atterberg
        limit tests on clay); and</LI>
    <LI>tests to determine engineering design
        parameters (for example strength,
        compressibility, and permeability).</LI>
    </OL>
    <P>Samples obtained either for description or
        testing should be representative of the
        ground from which they are taken. They
        should be large enough to contain
        representative particle sizes, fabric, and
        fissuring and fracturing. They should be
        taken in such a way that they have not lost
        fractions of the <EM>in situ</EM>soil (for
        example, coarse or fine particles)
        and, where strength and compressibility
        tests are planned, they should be subject to
        as little disturbance as possible</P>
    </P>
    </definition>
    <admin>Geotechnics,BRE</admin>
</termEntry>

Figure 4: Example of VHG definition

The format contains tables of information. Each
table consists of a group name, column headings and
data variables. Figure 5 illustrates a small part of an
AGS file. This section is returning results related to
sample reference information. Note that in the real
AGS file each entry must be on a single line.

In Figure 5 the group name is SAMP. The next
entry (taking four lines in the figure) defines the
names of the 10 column headings required for the
group. The lines starting "A10" provide results data.
All the data shown here relates to a single hole, A10.
Each entry includes values in a fixed order. The
meaning of the order can be interpreted in terms of
the column headings defined for the group.

It should be noted that the semantics of this file
format have been captured in the EXPRESS information model for AGS. Hence, the heading
SAMP_TOP in the AGS file is represented in Figure
3 by *depth_to_top_of_sample : metre_measurement*;
the heading SAMP_REF in the AGS file matches

```
"**SAMP"

"*HOLE_ID","*SAMP_TOP","*SAMP_REF",
"*SAMP_TYPE","*SAMP_DIA","*SAMP_BASE",
"*SAMP_DESC","*SAMP_UBLO",
"*SAMP_REM","*GEOL_GEOL"

"A10","1.2000","01","B","0","1.7000",
"MADE GROUND","0","","FILL:"

"A10","1.2000","29","D","0","1.6500",
"MADE GROUND","0","","FILL:"

"A10","2.2000","02","B","0","2.7000",
"MADE GROUND","0","","FILL:"

"A10","2.2000","29","D","0","2.6500",
"MADE GROUND","0","","FILL:"
...
```

Figure 5: Excerpt from an example AGS file

samp_ref : sample_reference_number in the model;
the heading SAMP_DIA in the AGS file format
matches   *sample_diameter : OPTIONAL milli-
metre_measurement;*

3.5 *XML for AGS*

The EXPRESS model of AGS has been used to generate an XML Document Type Definition (DTD)
using a software transform that reads an EXPRESS
model and produces a valid XML DTD mirroring
the structure of the entities and types in the EX-
PRESS model. The XML DTD for AGS has in turn
been used to ensure that XML data generated from a
file of AGS data is structurally correct.

An XML DTD specifies for a set of elements
what other elements may appear within each element. The elements form a tree with, in this case, a
"project information" element at the root and with
elements for simple typed data (such as *integer* or
string) plus enumeration items at the leaves.

The DTD mirrors the structure of the EXPRESS
model. Non structural features of this EXPRESS
model such as the unique and domain rules are ignored as they affect the semantics not the structure
of the DTD and, in any case, an XML DTD has no
way to represent these constraints. Each type and
entity in the model is represented by an element.
Enumeration items are represented by EMPTY elements. Elements representing the attributes of each
entity in the model are required to appear in the
same order as in the model. Optional attributes are
generated as elements which can appear zero or
once, "(element-name)?" in XML. Aggregates generate different XML content structure depending on
the bounds of the aggregate.

XML and EXPRESS have quite different scope

and identifier visibility rules. The name of each element in the XML DTD is required to be unique within the DTD. The EXPRESS model was therefore pre-processed using the STEPWISE "Name Clash Factory" (Kahn et al. 2000, Kahn et al., in press) to find any identifiers which would result in duplicate naming if left unchanged in the XML DTD. In order to resolve name clashes, the names in the model were changed as necessary. In addition, the names of elements representing enumeration items in the model were prefixed with their EXPRESS type name to avoid clashes between enumeration item names and entity names.

```
<sample_reference_information>
    <depth_to_top_of_sample>
     <metre_measurement>
      <measured_value>
       <REAL>1.2</REAL>
      </measured_value>
     </metre_measurement>
    </depth_to_top_of_sample>
    <samp_ref>
     <sample_reference_number>
      <INTEGER>01</INTEGER>
     </sample_reference_number>
    </samp_ref>
    <samp_type>
     <sample_type>
      <sample_type.bulk_disturbed_sample/>
     </sample_type>
    </samp_type>
    <sample_diameter>
     <millimetre_measurement>
      <measured_value>
       <REAL>0</REAL>
      </measured_value>
     </millimetre_measurement>
    </sample_diameter>
    <depth_to_base_of_sample>
     <metre_measurement>
      <measured_value>
       <REAL>1.7</REAL>
      </measured_value>
     </metre_measurement>
    </depth_to_base_of_sample>
    <samp_desc>
     <sample_description>
      <STRING>MADE GROUND</STRING>
     </sample_description>
    ...
</sample_reference_information>
```

Figure 6: Partial AGS XML file of project specific data

Once the XML DTD was created, it was possible to process a project specific AGS file and translate from the AGS format to the more easily processed XML format. Figure 6 shows a small part of the AGS data represented using the XML DTD generated from the AGS EXPRESS model. The section illustrated once again relates to the *sample_reference_information* (see Figure 3) and the SAMP group (see Figure 5).

4 GEOTECHNICAL SITE INVESTIGATION MODEL

Although the current prototype builds relationships between many of the different information representations used, the activity model of geotechnical site investigation is central to the system.

Expert knowledge provided by Ove Arup and Partners was key to supporting the definition of the IDEF0 activity model. In addition, numerous other resources were used, including in particular (Clayton et al.1995).

4.1 *Scoping the Activity Model*

It is essential to realise when creating a model of any kind that the model is a representation of reality, not reality itself. It is never possible to describe all aspects of any form of reality, because in the end, individual perceptions affect the interpretation. Instead, when creating a model the purpose of the model should be understood and the amount of information represented in the model should be as faithful as possible to an agreed interpretation of the reality, within the bounds defined to meet the model's purpose.

In the case of the IDEF0 model created as part of the KLICON project, the goal was to identify to a reasonable level of detail what functions are carried out during geotechnical site investigation, what information is required to perform those functions and what information is generated by the functions. The purpose was to provide the basis for a framework of tools that could link into the model.

The activity model examines the manner in which data and information are used both in the carrying out of geotechnical work and in the application of information gained from geotechnical investigation to the development, construction and maintenance of a project. The activity model considers:

- the evaluation of a site by geotechnical means
- the manner in which geotechnical investigation is used to help:
  - the development of a project
  - the choice of a site

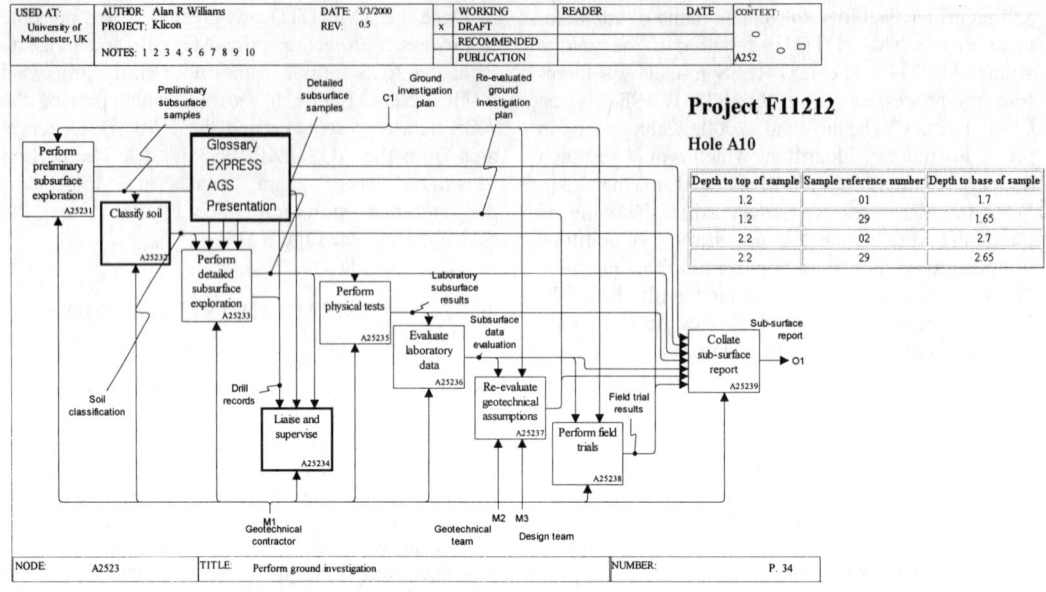

Figure 7: Overall Activity Diagram for Perform Ground Investigation including selection pop-up

- the conceptual design of a project
- the detailed design of a project
- the construction of a project
- the post-construction maintenance and performance of a project

The activity model does not consider, for example:

- the means or criteria by which contractors are appointed
- non-geotechnical criteria for project development, construction or maintenance

The activity model includes topics such as the examination of records to produce a desk study, site walk-over reports and sub-surface investigation, including non-intrusive (e.g. geophysical) and intrusive (e.g. bore sampling) investigation.

An example sheet from the activity model is given in Figure 7. In the full model, further information about each activity may be found by looking on the relevant sub-activity diagram.

On the right hand side of Figure 7, an additional window is shown. In this case, the information provided is for a selection of the samples described in the AGS XML format (see Figure 6 which illustrates just one sample data report line and the source AGS format in Figure 5). The appropriate XML data is transformed by XSL-T into a presentation in HTML.

4.2 Interpreting the model

In this section, additional explanation is given to support the IDEF0 diagram in Figure 7. In the context of a full model, which in this case is a total of 48

diagrams, the overall linkage is easier to follow. The figure also shows the pop-up box that links the IDEF0 diagram to other representations.

The activity "perform ground investigation" is controlled by the "ground investigation plan" generated by A2521, "plan ground investigation". The "ground investigation plan" specifies amongst other information where holes and trenches are to be drilled or dug, what tests are to carried out, the results which may be expected and the standards which must be satisfied.

Most of the activity is carried out by the "geotechnical contractor" who is, normally, subcontracted by the "geotechnical team". The geotechnical and design teams are only involved when the geotechnical assumptions made within the ground investigation plan are not satisfied.

In many cases, holes are drilled to carry out a preliminary subsurface exploration. The preliminary subsurface samples are used to classify the soil and to ensure that the geotechnical plan is approximately accurate.

The "soil classification" and the, possibly revised, "ground investigation plan" control the detailed subsurface investigation i.e. the drilling of holes and the digging of trenches. This investigation results in a set of drill records (this term includes trench digging) and samples taken from the holes and trenches.

The samples are sent away for physical testing at a laboratory. The selection of a laboratory and the shipping of samples and receipt of results have not been modelled.

280

The raw data results received from the laboratory are examined and interpreted to produce a "subsurface data evaluation".

The "subsurface data evaluation" is compared with the "ground investigation plan". If the actual data does not agree with the expectations of the "ground investigation plan", then the "geotechnical team" and the "design team" may decide that the "ground investigation plan" should be modified. As a result, the "geotechnical contractor" may be required to carry out additional or revised investigation.

Field trials of parts of the "construction plan" may be specified as part of the "ground investigation plan". The carrying out of these trials may be triggered by evaluation of the data found during subsurface investigation; i.e. a subsurface result may indicate that a construction technique is probably not suitable. The results of the field trials may control the detailed subsurface investigation.

The raw data from the sub-activities and their evaluation is collated into the "sub-surface report" which forms part of the overall "geotechnical report".

The "geotechnical contractor" responsible for the drilling, etc. is required to liaise with the people carrying out the work and to ensure that they record the data correctly.

4.3 *Activity Modelling Issues*

When creating an IDEF0 model, as with any other modelling techniques, the modeller needs to make some conscious decisions about techniques to use. This leads to some points of issue, which need to be understood.

4.3.1 *Depth decision difficulties*
During the development of the activity model, although the breadth of the model, i.e. the domain it covered, was reasonably well determined, it was difficult to decide how much detail should be specified. These difficulties can be split into two categories: decomposing out of the domain and insignificant detail.

4.3.1.1 Decomposition out of the domain
In the full model, "perform physical tests", which is already at a low level in the diagram hierarchy, could, in theory, be decomposed into details such as "consider test tube criteria" and "determine availability of appropriate test tube". It was decided that such decomposition would be encroaching into the domain of laboratory work. If such details were needed then it would probably be better to handle them by referencing a separate model of laboratory testing.

4.3.1.2 Insignificant detail
In activity "research site geology", the geological records are split into five sub-categories corresponding to the most common types of geological records. These five kinds of record all act as controls upon the same activity, "examine geological records". In previous versions of the model, there were five separate activities corresponding to "examine geological maps", "examine geological publications", "examine air photographs", etc. The presence of these record-type specific activities had the advantage that it would be simple to decompose them. Such decomposition could be used to model any actions specific to examining, for example, geological publications. However, no details of such activities are currently available. Because these separate activities cluttered up the "research site geology" diagram, they were removed.

4.3.1.3 Inputs, Controls and Mechanisms
IDEF0 distinguishes between inputs, controls and mechanisms. At times, it has been difficult to distinguish between them. For example, in activity "research site topography" it is difficult to determine to which category records should be assigned:

Records are definitely not consumed or altered by the activity so should not be inputs

Records do not appear to control the functionality of the activity; certainly they do not do so in the same manner as the choice of site.

Records might be considered as mechanisms. However, they do not appear to be in the same category of mechanism as the teams and contractors.

The decision to model the majority of arrows as controls is arbitrary. The authors welcome suggestions as to a better modelling approach.

4.3.2 *Amount of overlap*
The geotechnical activities that are carried out to support the different stages in a construction project have a large amount of overlap. For example, investigation of records could be carried out at any time from when the project is still being designed to when problems arise with the maintenance of the buildings. Because of the overlap, a completely hierarchical approach to the model could not be taken. Instead, the geotechnical evaluation of the site is modelled within activity "A2". The stages in the construction of the project, 'call' the "A2" activity by issuing evaluation requests for a geotechnical investigation of a chosen site for a particular purpose. The construction stage then receives a geotechnical report about the site.

4.4 *Availability of the model for review*

The IDEF0 model is being prepared as a publicly available technical report and will, it is hoped, be made available on the web as well. Anyone interested in reviewing the model and giving comments to us can, in the meantime, contact us directly.

5 CONCLUSIONS

The prototype system has shown that it is feasible to link various representations of domain data in a way that enhances the value of each form. In particular, the example shows that it is in principle possible to relate generic information, as represented by the IDEF0 model, to project specific data, such as the AGS file.

The prototype has deliberately been built using standard techniques and making use of widely available standard IT tools, such as Internet Explorer or Netscape. It is intended that this approach will make the use of the model-based technology proposed here less daunting.

Future work planned will include making further use of the IDEF0 model so that it can be customised to reflect the activities actually done in a specific project. Further work is also needed to link in additional project-specific data sources.

The precise nature of future developments will depend on the application domain. At present, we are considering developing a means of supporting project audit and project checklists as useful applications that will show the benefit of capturing and retaining knowledge in this way.

ACKNOWLEDGEMENTS

The authors wish to acknowledge the support of the Engineering and Physical Sciences Research Council (Grant Number: GR/M43388/01). They particularly wish to thank Roger Milburn and Colin Curtis of Ove Arup and Partners for their enthusiastic support of this study and for their detailed review of the activity model. Thanks are also due to Francis Cook of the Department of Computer Science for the work he did in supporting the development of the browsing interface included in this prototype system.

REFERENCES

Bradley, Neil 1999. *XML Companion* Longman Higher Education; ISBN: 0201674866

Clayton, C.R.I. , M.C. Matthews, & N.E. Simons 1995. *Site Investigation*, Blackwell Scientific Press Ltd, London, ISBN 0-632-02908-0

ISO 10303 1994. Industrial automation systems and integration: *Product data representation and exchange: Part 11: Description methods: The Express language reference manual*, Reference Number ISO 10303-11:1994, ISO, Switzerland

Kahn Hilary J. & Nick P. Filer 2000. *From Information Model to Efficient Implementation*, to be published *in Information modeling for the New Millennium*, Idea Group Publishing,

Kahn Hilary J., Nick P. Filer, Alan R. Williams, Nigel A. Whitaker & Denis J. Reilly 2000. *Transforming Information Models to Support the Generation of Efficient Implementations*, Proceedings of the 33rd Hawaii International Conference on System Sciences, ISBN 0-7695-0493-0

Softech, Inc 1981. *Integrated Computer-Aided Manufacturing (ICAM) Architecture Part II, Vol. IV Function Modeling Manual (IDEF0)*, Technical Report AFWAL-TR-81 4023, Materials Laboratory (AFWAL/MLTC), AF Wright Aeronautical Laboratories (AFSC), Wright-Patterson AFB, Dayton, Ohio. Federal Information Processing Standards Publication 183

World Wide Web Consortium 2000. http://www.w3.org

Project modelling and management

Product and Process Modelling in Building and Construction, Gonçalves, Steiger-Garção & Scherer (eds)
© 2000 Balkema, Rotterdam, ISBN 90 5809 179 1

Approaches to risk assessment: A pilot study

T. Andersen
Technical University of Denmark, Lyngby, Denmark

F. Madsen
Tryg-Baltica Risk Management, Denmark

ABSTRACT: This paper presents a guideline for contractors risk management on site and a related pilot study performed on seven different sites. The guideline provides a set of three forms which are in compliance with the generic risk management process and other management system such as quality- and environmental management. The pilot study reveal that, management attention and focus on interactions with other management and reporting system, are crucial. Adapting the general risk assessment method where risks are described by likelihood times consequence, is difficult in practice. An easy to access company knowledge archive could provide support to the site manager.

1 INTRODUCTION

In November 1998 a practical guidebook to support the managing of risks at construction sites were issued by the Society of Danish Contractors (Madsen & Andersen 1998)

This initiative was taken after a feasibility study (Falk 1997) showed that risk management could help the Danish building industry to obtain a better revenue – a conservative estimate indicates savings on at least 6 Billion Danish Kroner (800 mill EUR) pr. year.

This paper will introduce the key elements in the guidebook, and the outcome of testing this guide at building sites. Finally, suggestions to improved methods will be outlined.

2 RISK MANAGEMENT

Literature on Risk Management is rather abundant. For example (Wideman 1992) gives a good introduction to general principles on how to manage risks in projects. Risk Management is more or less applicable to all kind of projects (and processes), and the major principles for approaching risks are very similar from one domain to another. Since risk management is generally applicable a "generic" risk management process - as described in e.g. the Australian Risk Management Standard - has been adobted in many domains. This process is illustrated in figure 1.

3 THE GUIDEBOOK

In the guidebook we are focusing exclusively on risk management in the domain of construction.

We have adopted the generic process shown in figure 1, but in a transformed version, which is practical and easy to work with for construction professionals. Moreover we narrowed the scope to construction sites i.e. we attempted to design a risk management process for sites managers. The guidebook has the following main features:

- Its is directed towards, and written directly to, those professionals working on sites.
- The method presented complies with existing quality-, environmental- and risk management standards.
- It draw attention to the need for exchange of risk related information between different parts, e.g. consulting engineers, contractors and end-users.
- And finally it provides a coherent, easy to work with, tool to support stage 2-8 in figure 1. This tool comprises 3 forms, presented in a little more detail below.

3.1 *Suggested approach*

The generic risk management model - shown in figure 1 – has been broken down into 3 distinct forms which together represents a risk register aiming at a smooth risk identification, assessment and monitoring.

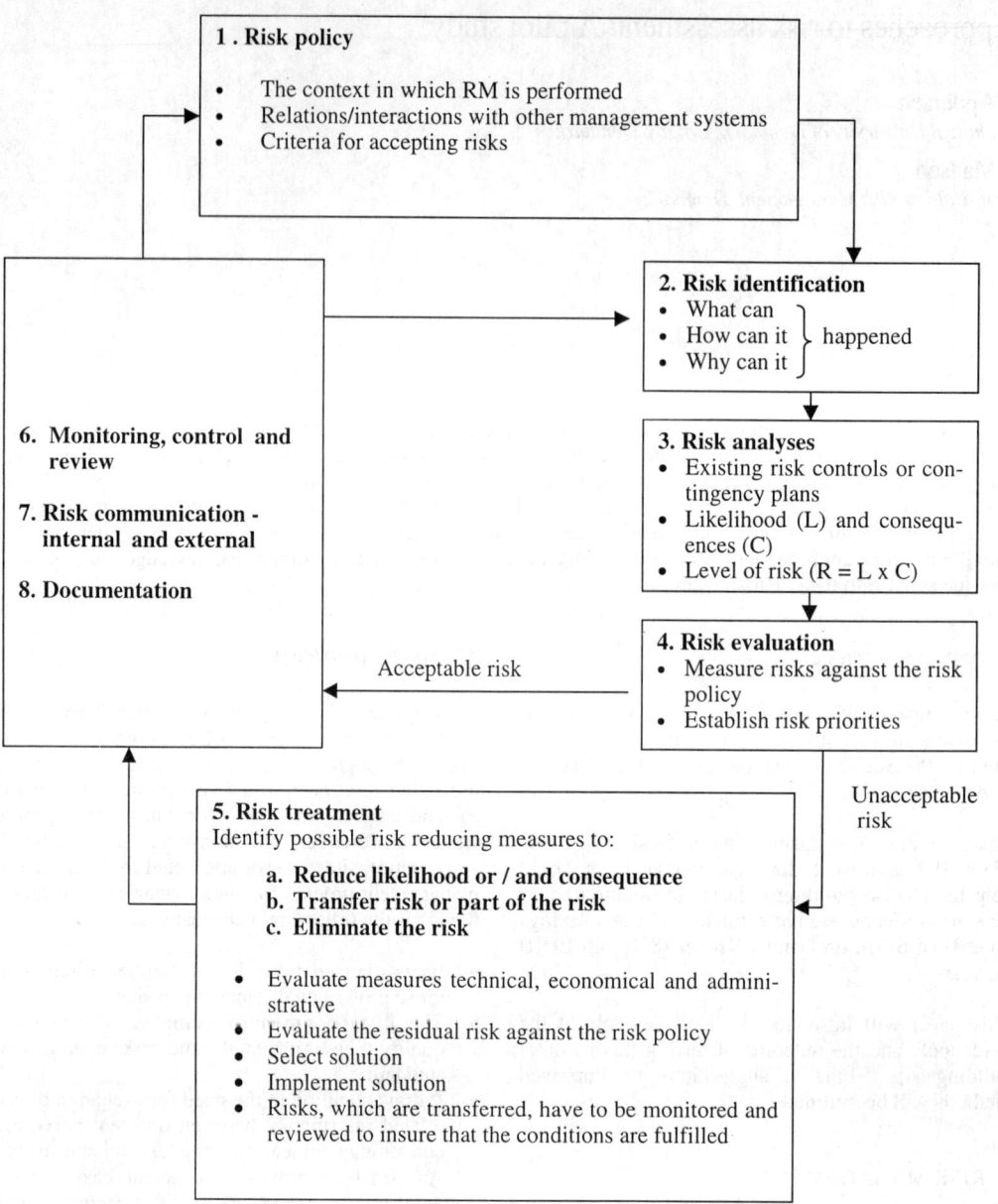

1. Risk policy

- The context in which RM is performed
- Relations/interactions with other management systems
- Criteria for accepting risks

2. Risk identification
- What can ⎫
- How can it ⎬ happened
- Why can it ⎭

6. Monitoring, control and review

7. Risk communication - internal and external

8. Documentation

3. Risk analyses
- Existing risk controls or contingency plans
- Likelihood (L) and consequences (C)
- Level of risk (R = L x C)

4. Risk evaluation
- Measure risks against the risk policy
- Establish risk priorities

Acceptable risk

Unacceptable risk

5. Risk treatment
Identify possible risk reducing measures to:

a. Reduce likelihood or / and consequence
b. Transfer risk or part of the risk
c. Eliminate the risk

- Evaluate measures technical, economical and administrative
- Evaluate the residual risk against the risk policy
- Select solution
- Implement solution
- Risks, which are transferred, have to be monitored and reviewed to insure that the conditions are fulfilled

Figure 1. The generic risk management process.

3.2 *Form #1 - The risk register*

It represents a combination of risk identification, risk analyses and the initial steps in the risk evaluation process. Risks can be related to three different "entry-keys":

1) Construction activity

2) Risk area
3) Events

It is in the hands of the user(s) which approach to follow – and it's possible to use a sort of mixed approach. Any combination which suites the purpose, to identify the most important risks, is fine.

Activity	Area of risk	Events	Likelihood	Consequence	Level of risk	Accept
			(5x5 matrice).	(5x5 matrice)	Likelihood x consequence	Y : No further treatment N : Transfer to Form #2

Figure 2. Form #1, the risk register in principle.

Consequense / Likelihood	Insignificant - 1 < 5000 kr	Minor - 2 5000 – 25,000 kr	Moderate – 3 25,000 – 100,000 kr	Major - 4 100,000 - 1.0 mio kr	Catastrophic - 5 > 1.0 mio. kr
Certain - 5 > 10	5	10	15	20	25
Likely - 4 1 - 10	4	8	12	16	20
Moderate - 3 0,1 - 1	3	6	9	12	15
Unlikely - 2 0,01 - 0,1	2	4	6	8	10
Rare - 1 < 0,01	1	2	3	4	5

Figure 3. Matrix used as part of the risk assessment in the guidebook.

A successful identification of risk basically depends on the involved professionals capability to a structured brainstorm, and not least of what is "in store" to support such a process i.e. personal experience and/ or access to a "corporate knowledge archive".

When filling out Form #1 it is important to include those typical dominating risks already identified during the design and planning stages of the project. Very large risks are usual identified and analysed in detail, by third party or in house experts. It is of paramount importance to establish a systematic passing of risk related data to the construction site. The major part of risk management efforts at the construction site should be related to "small" and "middle size" events, since the large events have already been uncovered. The feasibility study (Falk 1997) indicates that the major part of the savings potential is related to managing these non-catastrophic events.

When risks are identified, next step is the analyses, i.e. to establish a value of each risk. In the guidebook we have selected a "semi-qualitative" method for risk analysis, as a natural choice when there is no solid statistical material to support a pure quantitative analysis.

The term "semi-qualitative" is used because the matrix contains a scoring system, which provides a numerical score for risk level and for consequence.

The first step in evaluation of the risk is to decide whether the risk is acceptable or further treatment plans should be considered.

Decisions are made according with a predefined acceptance sheet where different combinations of the risk level and the maximum consequence place the responsibility for the risk and risk treatment plan at different levels in the organisation.

3.3 Form #2 - the risk treatment register.

Form #2 combines risk evaluation, preliminary cost/benefit analyses with monitoring and review of treatments during the building period and a final evaluation when the job is completed.

Level of risk	Consequence	Responsible	Risk Treatment (Yes / No)
> 15	5	Company management	Y
10 - 15	4	Site management	Y
6 - 10	3	Site risk manager	Y
1 - 6	< 3	Acceptable as is	N

Figure 4. Evaluation of risk. The risk matrix corresponds with predefined acceptance criteria, as outlined in the company risk policy.

Activity:	Risk area:		Events:	Risk treatment:
Likelyhood and consequence: With treatmenet			Likelyhood and consequence: Without treatmenet	
Monitoring:			Review	
Evaluation of treatment:				

Figure 5. Form #2, the risk treatment register in principle

Activity	Risk area	Events	Cause	Related to Objects on site	Expences.

Figure 6. Form #3, the claims / events registration in principle.

The level of risk and the maximum consequence are estimated with, and without, treatment measures.

If treatment does not reduce the likelihood and consequences sufficient, the risk management process have to be repeated until it is possible to meet the acceptance criteria.

3.4 Form #3 - events and claims registration.

Form #3 is a traditional registration of events and claims. The IT application accompanying the guidebook including a predefined terminology for risks, activities and events etc as an initial attempt to ensure consistency in the reported events and claim. Consistent data is a must for future development towards a corporate knowledge archive.

4 PILOT PROJECT:

The guide has been tested at several building sites in form as a simple IT-system primarily offering a more convenient access to the 3 forms-approach. The system employed the 3 forms and the 5*5 matrix presented above.

The site managers (referred to as "pilots") used the system for their local risk management over a period of 6 month at 7 different sites. The main objective was to explore how well suited the three forms were in practice. It soon became clear that filling out Form #1 was a tough challenge for the pilots, and we decided to concentrate on this phase in the test period. Hence, findings from our project are primarily related to the usage of Form #1.

All the involved "pilots" received education in operating the tool and the computer. We established a hot line service to ensure that the pilots could get immediate help – if necessary on site. Moreover, they were thoroughly instructed in the purpose and daily use of the different forms. The procedure was as follows:

First, the pilot should provide a list of risks – either as "risky" activities or as risk areas. We coached this initial process. Next step was the analysis: to estimate likelihood and consequence based on the 5*5 matrix, and finally they should make an action plan for the identified risks.

During the test period, a research assistant were monitoring the progress, and the process were "formally" evaluated two times, halfway through, and after end test. All relevant personal participated in both events.

4.1 *Experiences gained - findings:*

It was difficult to get the project smoothly on the tracks. Site managers are very busy people and time is a very scarce resource. Thus, it is crucial to success that the site people are convinced that they will gain direct profit by attending in the tests. We were moderately successful in getting the right attention, and after a somewhat slow beginning the "risk management" performance on the sites were acceptable and appropriate for our studies.

It should be mentioned that another reason to the delay in the "start up" stages, was an unclear organisation around the pilot project. We have learned that testing at construction sites must have 100% support from, and direct link to, the home office. If not, it is very likely that tests fail. Management involvement is crucial to success!

The major direct findings of our studies are:

A) If an IT-solution is used, it must have more intelligence, that is: be able to give "on line" help particularly in the ID brainstorm stages. This was evident, and clearly outspoken of the pilots.

B) A more stringent breakdown structure supporting ID must be established. It is not, as a starting point, a good idea to promote different entries to identification.

C) It is very difficult for site managers to estimate on "likelihood", whereas "consequence" is more natural. Put in other words: They know what things cost, but not how often things happen. This lies probably in the current nature of construction work, where project management success eventually is measured in hard cash.

We concluded that a revision of Form #1 was needed to comply with the issues raised as a result of our empirical studies.

In general, the pilot project indicated that the principle of the guidebook, i.e. forms dedicated to the construction practice based on the generic risk management process are sound, but revisions and improved tools for analysis are needed.

4.2 *Improvements of method – future development*

As a result of the pilot project Form #1 has been revised, and is about to be implemented in the next version of the mentioned IT application. The new form will mirror the answer on issue B following below i.e. it contains a revised entry to risk identification. Issue A and C is still under investigation:

Issue A is a real challenge, since it points at the key problem for risk management in construction, that is: the outspoken absence of historical data on risks. The only answer to this issue is the provision of a corporate pool of knowledge (Royer 2000).

Issue B points at another well-known problem in construction – the absence of a common, widely acknowledged terminology. In the actual case, we carried out further studies at the contractor and we have established a company "unique" breakdown structure for activities, heavily based on the SfB breakdown standard (Giertz et al. 1975). The structure has three levels:

1) Main-activity (e.g. concrete)

2) Sub-activity (e.g. form work)

3) Specific topic

For each identification (risk entry) the risk must be described. For instance:

- Sub-activity: Form work
- Description: Low temperatures

Issue C is very important to address, simply because it touches upon the core "equation" in risk management theory:

Likelihood times consequence = level of risk

The risk level for a risk area or type of activity provides valuable information for project and company management, and as such it cannot be neglected. We have to incorporate the likelihood measurement in our tool. On a long-term basis, proper historical data will help overcome this obstacle, because we will be able to have access to statistical information providing the likelihood for us. We do not have to guess. In the mean while, the only proper way to solve this problem is education, education and more education, that is: we have to change the attitude and way of thinking for people in construction.

Current and future research will address the issues pointed out here, including more work on IT based solutions and in the area of knowledge management as a general tool to assist corporate risk management.

REFERENCES:

Falk N. et al. 1997: Project Risk Management i byggeriet, Feasibility study for the Danish Ministry of Trade. Technical Report, DTU.

Giertz L M. et al. 1975, SfB-review, technical report, Kokkedal.

Madsen F. & Andersen T. 1998, Risikostyring i bygge- og anlægssektoren, a guide for the construction industry. Society of Danish Contractors.

Royer P. S. 2000, Risk Management: The undiscovered dimension of project management. Project Management Journal, vol.1.

Wideman R. 1992, Project and Program Risk Management, The PMBOK Handbook Series, Vol 6. PMI.

Product and Process Modelling in Building and Construction, Gonçalves, Steiger-Garção & Scherer (eds)
© 2000 Balkema, Rotterdam, ISBN 90 5809 179 1

Neutral object tree support for inter-discipline communication in large-scale construction

G.A. van Nederveen
TNO Building and Construction Research, Delft, Netherlands

F.P. Tolman
Delft University of Technology, Netherlands

ABSTRACT: Communication between disciplines in building and construction can be improved significantly by the proper use of Information and Communication Technology. For that reason, many research groups have been trying to achieve such improvement, especially by using Product Data Technology and STEP, and more recently the IAI Industry Foundation Classes. Unfortunately, the practical results of these efforts, in terms of tools that are used in practice, are still poor. The main reason for this seems to be that most research efforts have been too much top-down oriented, resulting in complex models and long-lasting developments. This paper presents an approach which tries to combine the sound ideas of Product Data Technology and STEP with a strong bottom up development strategy. The approach is based on Neutral Object Trees: hierarchical structures in which the objects that must be designed and built, are structured in a simple decomposition tree. The Neutral Object Tree approach has been used for the development of an Object Tree at the Dutch High Speed Railroad project, which is implemented in a PDM-system that is currently in use.

1 COMMUNICATION IN BUILDING AND CONSTRUCTION

The building and construction industry is facing great challenges. A large amount of work is waiting, including a number of very large projects. In the Netherlands for example: the extension of Schiphol Airport, the second Maasvlakte, and railroad projects such as the Betuweroute and the HSL. In such large projects, but also smaller ones, high demands are put on the control of quality, time and cost.

Traditionally, the building and construction industry is characterized by dynamic partnerships between different disciplines from different organizations. In this situation, communication between different disciplines is a critical success factor. Therefore the aim for better control of control of quality, time and cost, often leads to an aim for better communication between disciplines.

In recent years a number of developments can be recognized that aim at better communication between disciplines, such as:
- New contract types
- Classification and coding,
- Performance approach,
- Systems engineering approach.

1.1 New Contract Types

At the moment new contract types such as Design & Construct and Build - Operate - Transfer are very popular. The idea behind such developments is to make use of each party's capacities in an optimal way (De Ridder 1994). For example by taking care that risks are managed by the party which is best equipped for the job.

1.2 Classification and Coding

Classification is the distinction of (object) classes. Coding is the addition of codes to these classes. The best known example of classification and coding is the SfB-system. Traditionally, classification in building and construction used to aim at building elements. Later on, such classifications are extended with activity classes next to element classes (see for example the Dutch STABU system), with multiple decomposition levels, library structures, etc. In fact, developments such as these go beyond classification methods, and must be regarded as building product modelling developments.

1.3 Performance Approach

The essence of the performance approach is that the objectives of a (building) project are formulated in terms of quantifiable performance requirements, not in terms of prescribed solutions.

1.4 *Systems Engineering Approach*

The systems engineering approach means that a product is seen as a collection of systems that should take care of a certain performance (INCOSE 1998). For example: a space system, a structural system, a heating system.

Developments aiming at better communication such as described above, are closely related. For example: in a Design & Construct project, a key role is played by performance-based specifications; and it is often worthwhile to do this by using systems requirements.

2 INFORMATION AND COMMUNICATION TECHNOLOGY IN BUILDING AND CONSTRUCTION

In communication in building and construction, information and communication technology (ICT) of course plays a key role. But when the state of the art of ICT in building and construction is considered, then it must be concluded that ICT in building and construction practice is still on a rather low level. For example, ICT in design is mainly based on CAD-systems (drawing systems) and exchange of CAD data. Only on a modest level, some integration exists between CAD systems and for example CAE-systems (calculation programs).

This low level state of the art was already recognized in the mid eighties, leading to the initiative to the well-known ISO-STEP standard (officially ISO 10303) (ISO 1993). The STEP standardization initiative aims at electronic communication of product information on a semantic level, based on standardized product models, i.e. standardized information models of product data to be exchanged between participants in engineering environments.

Since the start of STEP in the mid eighties, many follow-up initiatives and projects have started. But the results of all these efforts are rather disappointing for the building and construction industry. For other industries, specific STEP-based standards have been developed, for example for the process industry (STEP AP 221 and the Epistle work), for shipbuilding and for the automotive industry.

But for building and construction a widely accepted standard based on STEP is still missing. This is probably due to the following problems:
- The building and construction sector is fragmented, often small-scale, nationally oriented, and without dominant parties; therefore it is difficult to reach agreement on sector-specific standards.

- The ICT-sector on the other hand, is internationally oriented; therefore it is even more difficult to achieve ICT-support for sector-specific standards.
- Many different ICT-approaches for support of standards exist, and new approaches emerge almost continuously.

A recent development in the area of electronic communication for building and construction is the development of Industry Foundation Classes (IFCs) by the International Alliance for Interoperability (IAI 1997). These IFCs are essentially standardized CAD objects that contain both geometric and semantic product data. The IAI that is developing the IFCs, is a consortium of CAD vendors, such as AutoDesk, Bentley, Nemetschek, etc. The IFC development has some important advantages above STEP, for example the leading role of the software vendors. But also the IFC development is taking place very slowly.

3 TOWARDS BETTER COMMUNICATION AND INFORMATION EXCHANGE

Besides STEP and IFCs there are a few important developments that may help to achieve practical solutions for electronic communication in building and construction.

3.1 *STEP and OO*

One of the problems of the STEP work is its slowness, especially compared to developments in ICT. As a result, several initiatives have started aiming at the application of the latest technologies for STEP work.

One example of this is the support of behaviour using the object-oriented CORBA technology (OMG/CORBA 1998). In the European project VEGA work is done on the application of STEP combined with CORBA in order to support workflow management in industries such as building and construction (VEGA 1998).

3.2 *Minimal Models*

Another trend can be called "the minimal approach". According to this approach data models are kept very small in order to achieve a simple format for data exchange in building and construction. Elaborations of this approach aimed at building geometry (Tarandi 1998), and on integration with Electronic Data Interchange (EDI) (De Vries 1996).

3.3 *View models*

With respect to new concepts for communication the so-called view-approach is important. The view approach, or more precisely the discipline view approach, starts from the observation that different participants in building and construction represent different disciplines, each of which has its own view on design information (Van Nederveen 1993). As a result, each participant has its own specific information requirements, which must be supported by specific information models (so-called view models). For communication in building this means that support of view conversion is a first prerequisite.

The view approach is elaborated in different ways. For building and construction the approach is worked out most extensively in the European project ATLAS (ATLAS 1993). As shown in this project, the big practical issue in the view approach is that the relationships between the various view-specific information models (the view conversion) become too complex, leading to too costly implementation and maintenance of conversion software.

3.4 Internet and XML

A very promising trend in the context of basic technologies, is the ongoing development of Internet technology, especially with XML. However, work in this area is not yet matured. An interesting project aimed at E-Commerce in building and construction using XML is the EU-project eConstruct (Tolman & Böhms, 2000).

3.5 Evaluation

Despite all the work done within and outside STEP, we must conclude that in building and construction very little results have been achieved in terms of tools used in daily practice.

Because of the specific characteristics of building and construction, it seems that the introduction of a new concept for electronic communication on a semantic level can only be successful when a very pragmatic and bottom-up approach is used. At least much more bottom-up than in the various projects in the past.

4 NEUTRAL OBJECT TREES

In order to achieve an approach that is pragmatic, bottom-up, allowing fast implementations and quick returns of investment, it was necessary to reconsider a number of "STEP-habits", and ask ourselves questions such as: "Do we really need complex data models? What kind of relationships and structures do we really need? Can we avoid the difficult subject of geometric models and shape description?". But most importantly, we started really bottom-up, with the development of a model of a single project, without having a reference model or type-model in advance.

The experiences of this work led to the so-called *Neutral Object Tree* approach, of which the main characteristics are listed below:
- An Object Tree is an instance model, it describes the objects of one particular project.
- A Neutral Object Tree is independent of software vendors, building participants and standardization developments, etc.
- Objects are function performers.
- An Object Tree is a decomposition tree (using: contains, or consists of).
- An Object Tree supports a minimum set of relationships.
- For shape description a reference to CAD drawings, or a VRML shape description is specified.
- An Object Tree can be made largely by hand, using cut and paste (think of it as a simple hierarchy as used in the Internet Explorer).

4.1 An Object Tree is an Instance Model

Or in common English: an Object Tree describes a single thing, not a class of things. The White House is an example.

One of the "STEP habits" stated above is to start with a product type model (or reference model, or STEP ARM), a model that describes a certain class of objects, such as buildings, roads or viaducts. Type models have to undergo a standardization process. Once a standard is available ICT vendors have to implement it in their tools. Next Building-Construction companies have to buy these systems. And finally a consortium has to agree on using the standard and tools in a project. A long long way.

Instead, the Object Tree approach starts right away with the creation of the instance model that supports electronic communication. True, the Neutral Object Tree will not be as elaborate as the future product models, but at least we don't have to wait for another decade.

4.2 A Neutral Object Tree is independent of software vendors, participants and standards

This characteristic follows from risks such as being dependent of vendors that might change their strategy, or just might vanish. Or being dependent of organizations that might change (narrow) their strategy and stop their commitment. Or being dependent of slow acting standardization committees with long-lasting procedures. An Object Tree can be created in a matter of days. The real problem is getting everybody on the same line, but that is a problem anyway. The bottom-up approach at least starts at the right place: the bottom.

4.3 Objects are Function Performers

The "building blocks" of the Object Tree are the objects. Objects are regarded as physical things that perform a function. This function can be the realization of a required performance, as specified in a Requirements Specification.

This means that a Neutral Object Tree can be regarded as a solution tree, in which also the functions are specified for which the objects are solutions.

4.4 A Neutral Object Tree is a Decomposition Tree

As the term "tree" already suggests, Object Trees have a hierarchical structure. Now there are many ways in which a hierarchy can be made in a product structure. Nowadays object hierarchies, or trees are usually specialisation trees (following the Object Oriented approach). But the Neutral Object Tree approach aims at simplicity, and therefore it proposes to use one hierarchical principle: decomposition ("consists of").

However, decomposition can still be applied in different ways. For the Object Tree two decomposition principles are used:

- subsystem decomposition, in which an assembly is decomposed into groups of objects that share a location,
- aspect system decomposition, in which an assembly is decomposed into groups of objects that share a specific aspect or role.

Both decomposition principles are needed and are therefore part of the Object Tree approach. But the relationship between elements of these decomposition trees can be very complex. Therefore this relationship is not modelled explicitly in the Object Tree.

4.5 An Object Tree Contains a Minimum Set of Relationships

Relationships (such as "is connected to") between objects can easily lead to a very complex model structure. For that reason a Neutral Object Tree contains only a minimum set of relationship types. First of all, there are the decomposition relationships as described above. Furthermore the only other relationship type is the physical interface. No other functional, logical or any other kind of relationships is used.

4.6 For Shape Description a Reference to CAD Drawings or VRML models is Specified

For shape description a number of methods exist, but again these methods can easily lead to very complicated information models. Once more, a simple solution is chosen: the use of references to CAD drawings. In other words, if a user wants to know about the shape and dimensions of an object, then he should find a function that brings him to the drawing in which he can find what he is looking for. An alternative is to represent each object as a local VRML model and let the browser do its work. In the future Neutral Object Trees will be made viewable over the Internet using VRML, Java3D or X3D.

4.7 A Meta-model for Object Trees

Below the characteristics discussed above are modelled in an EXPRESS-G diagram, see Figure 1.

5 IMPLEMENTATION OF OBJECT TREES

As the Object Tree is basically rather simple, the implementation of Neutral Object Trees does not have to be very difficult either. In fact it is even possible with systems such as Excel or Access, but such systems fall short in support of either data management and maintenance (Excel) or user interface (Access). A better solution is to pick one of the commercially available Product Data Management (PDM) systems, and to tailor the system according to the specific needs of the organization.

In the near future new operating systems and Internet software will further enhance the possibilities for implementation of Object Trees.

6 THE HSL CASE

In the Dutch High Speed Line (HSL) project the approach described above has been applied in the so-called HSL Object Tree. This has resulted in a decomposition structure with of course the entire HSL track as top of the tree. The HSL track decomposes in a few steps into some thousands of HSL objects such as bridges, viaducts, tunnels, sound barriers, cables and ducts etc.

In fact the HSL approach has been even simpler than the approach advocated in this thesis. For example the functional side of objects, the aspect systems decomposition and the definition of physical interfaces is only elaborated in part. The implementation of the HSL Object Tree has been done using Excel, Access, and the PDM system SmarTeam subsequently. Figure 2 shows a screendump of the SmarTeam implementation.

From the HSL project it can be concluded that even a simple approach can easily become too difficult. One might think that the decomposition structure has been defined in a couple of days. In reality this has taken several months, including many meeting hours from many people. Also on the implementation side a simple step sometimes took months to take.

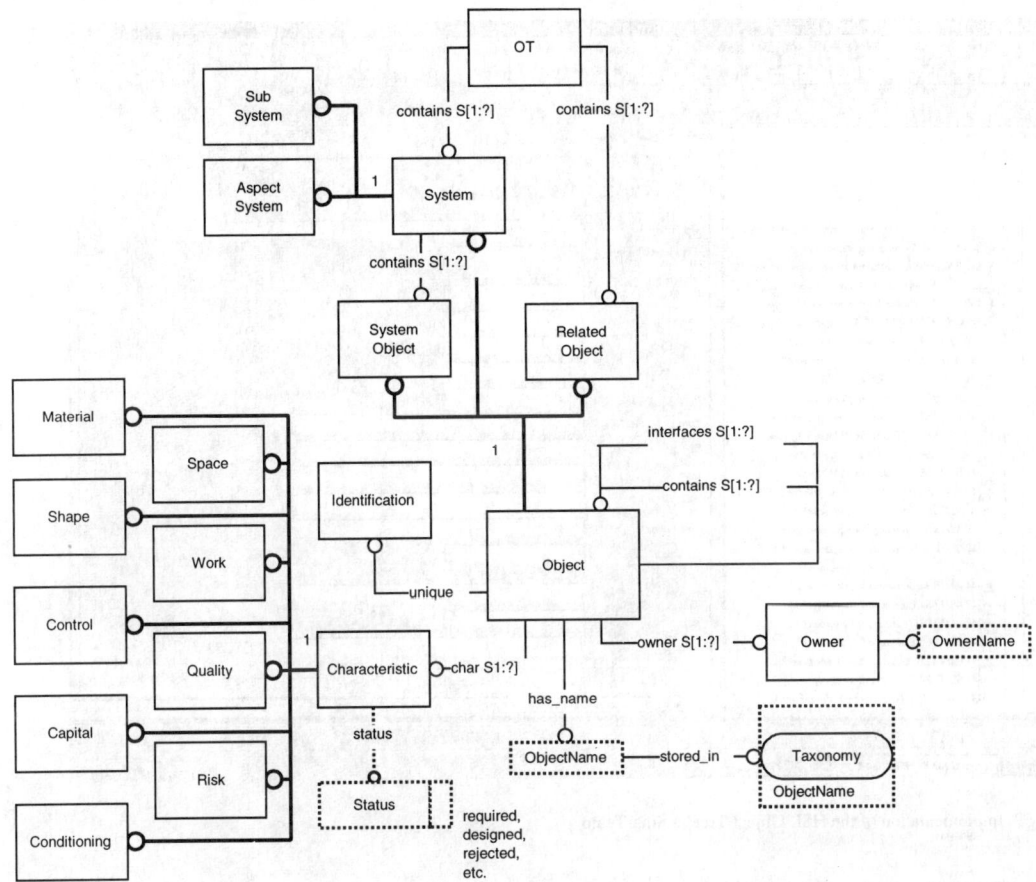

Figure 1. The Object Tree Meta Model

Nevertheless the HSL Object Tree can be regarded as a useful contribution to the enhancement of communication in a large-scale building and construction project based on semantic product data.

7 CONCLUSIONS

The main conclusions are:

1. For improvement of communication in building and construction the semantic representation and exchange of product data is a prerequisite.

2. The development of methods and tools for better communication in building and construction easily fails due to: the complexity of the information models, unfortunate ICT choices, and disappointing developments in standardization.

3. By using Neutral Object Trees a simple yet complete information model for design and engineering can be developed in a short period of time, which can serve as a basis for better communication.

4. Such a Neutral Object Tree must meet the following requirements:
– The Object Tree must be developed bottom-up, i.e. objects (instances) first, classes later.
– The Neutral Object Tree must be *neutral*, that is independent of software vendors, participants and standardization developments.
– Objects must be seen as physical things that perform a function.
– The Object Tree must be a decomposition structure, with both subsystems decomposition (shape driven) and aspect systems decomposition (aspect driven); the Object Tree should not have any other hierarchical structure.
– The Object Tree must support decomposition relationships as described above, furthermore physical interface relationships, but no other types of relationships.
– Shape description must be taken care of by a reference to CAD drawings, or by simple local VRML shapes.

295

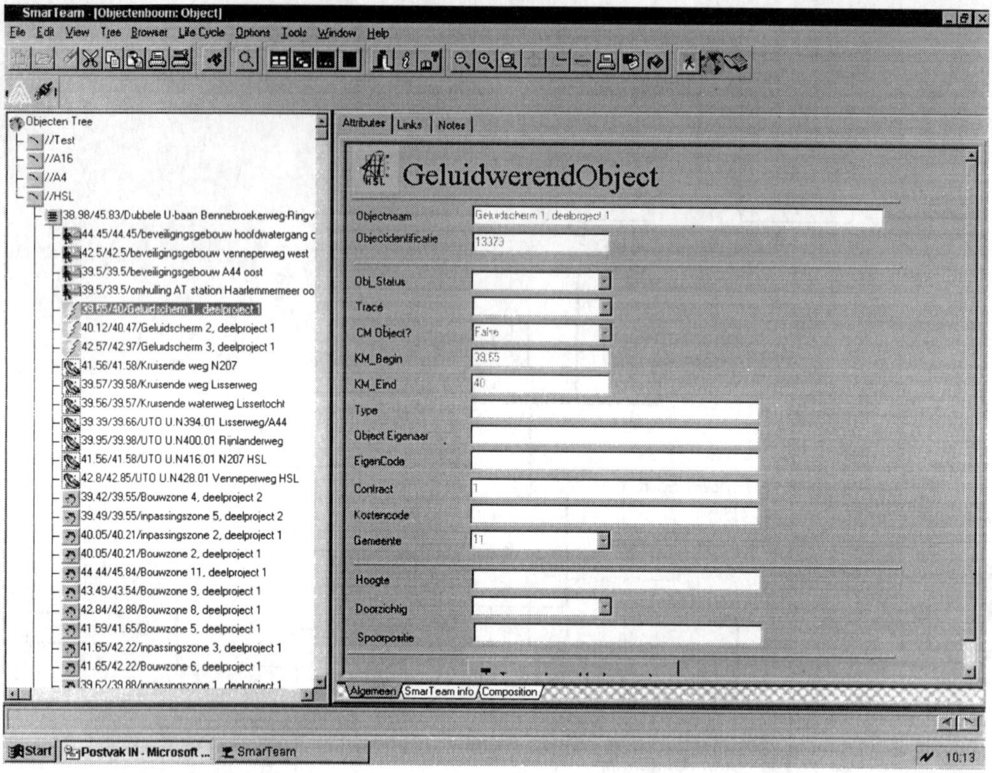

Fig 2. Implementation of the HSL Object Tree in SmarTeam

Such an Object Tree can be developed in a relatively short time. Moreover it can also be implemented in a short time, for example using commercially available PDM systems.

5. The Object Tree can be elaborated further as follows:
− (Further) Development of classification and standardization of object names and object types.
− (Further) Development of libraries of standard objects, but also of standardized resources and processes.
− (Further) Development of management methods using the Object Tree, for example interface management and risk management.
− Generalization of the Object Tree towards a type model to support PDT developments.

ACKNOWLEDGEMENTS

The research presented in this paper is mainly carried out in a PhD-research project that has led to the first author's thesis (Van Nederveen, 2000). The research has been sponsored by the Dutch Technology Foundation (STW) and TNO Building and Construction Research.

REFERENCES

ATLAS Consortium 1993. *ATLAS Public Project Overview*.
De Ridder, H.A.J. 1994. *Design and Construct of Complex Civil Engineering Systems, A new approach to organization and contracts*. Delft University Press.
De Vries, B. 1996. *Communication in the Building Industry*. PhD-thesis. Eindhoven University.
IAI 1997. "Industry Foundation Classes". Release 1.5.
INCOSE 1998. *Systems Engineering Handbook*.
ISO/TC184 1993. "Part 1: Overview and fundamental principles". In *Industrial automation systems and integration − Product data representation and exchange*. International Standard, First edition 1994-12-15. Geneva: ISO 10303-1:1994(E).
OMG/CORBA 1998. "The Common Object Request Broker Architecture (CORBA) Specification", Revision 2.2. www.omg.org/techprocess/meetings/schedule/Technology_Adoptions.html.
Tarandi, V. 1998. *Neutral Intelligent CAD Communication (information exchange in construction based upon a minimal schema)*. Stockholm: KTH.
Tolman, F. & M. Böhms 2000. "Electronic Business in the Building and Construction Industry: Preparing for the New Internet". In *Proceedings CIT 2000/CIB W78*. Reykjavik. (in prep.)
Van Nederveen, G.A. 1993. "View integration in Building Design", in Mathur et.al. (eds) *Management of Information Technology for Construction*: 391-406. World Scientific Publishing Co.

Van Nederveen, G.A. 2000. *Object Trees – Improving Electronic Communication between Participants of Different Disciplines in large-scale Construction Projects.* Delft University.

VEGA 1998. J. Stephens, M. Böhms, M. Köthe, J. Ranges, R. Steinmann, R. Junge, A. Zarli. "Virtual Enterprise using Groupware tools and Distributed Architectures". In R. Amor (ed), *Product and Process Modelling in the Building Industry,* Watford: British Research Establishment.

An approach to a computer aided project management (CAPM) in the building industry

Siegbert K. Heinecke
IBD GmbH, Karlsruhe, Germany

ABSTRACT: The software environment in the construction industry consists of a wide range of individual software solutions. These solutions are often unable to communicate. On the other hand we have a strictly competitive market situation. This requires an efficient project management aided by consistent software and communication concepts. There is a wide gap between wish and reality in the building industry concerning these points. This paper states the requirements to support the building process and project management as a whole by software. An approach to a solution from the view of a building company is presented.

1 THE SITUATION IN THE BUILDING INDUSTRY

The number of enterprises in the German building industry is diminishing. Causes are the stagnating market, significantly reduced public investments and a stiff competition which has increased since the national market has changed into an European one.

The building contractors react by trusts, looking for new spheres of activity such as project development or facility management and by passing work on to cheaper sub-contractors. However the building contractors fail to support this with an appropriate software and infrastructure.

In the building industry the opinion still prevails that only technical know how ensures success and thus it ensures surviving. The factor management is widely underestimated. Behaviours like actively approaching potential customers, which are normal in many other industries, are very rare in the building industry.

Shorter construction times and rising costs determine the building contractors' way of thinking. Instead of eliminating the roots of the evil and providing better and faster information paths, improvisation is rising.

Each building contractor believes that his flows are unique. Therefore no software can satisfy his requirements. Typical arguments are the uniqueness of building projects, project specific partners, long execution times of the projects and the high complexity.

2 SUPPORT BY SOFTWARE: CURRENT SITUATION

Naturally building contractors use software. There are innumerable systems supporting accounting, employee management, purchase, estimation, site-controlling, management of drawings, quantification etc. Normally all of them are isolated systems, incompatible, not supporting data interface standards. Manual transfer of data is often the only chance to be able to do the next process step within the next system.

Where software is concerned building contractors live on islands. The integration of contract givers, planners, suppliers and sub-contractors into one's own data network is considered revolutionary and can hardly ever be found, although it is obvious, that there is a high potential for cost reductions.

A typical example:
- The advertiser sends a printed bill of quantities to a potential contractor.
- The building company enters the positions of this bill manually into her estimation system.
- The bill of quantities is divided into separate bills for the sub-contractors.
- The different sub-bills are printed out and passed on to possible sub-contractors by fax.
- The sub-contractors return their offers by fax, too.
- The building contractor enters the received offers manually into the estimation tool (or maybe a different tool for comparing prices).

Such a method is slow, expensive and error-prone. There are some internet-based attempts to solve this problem, so-called market places for online advertisements. However, the handling of these market places is often incomprehensible to the user, the contents are not up to date and in addition the access to these market places is expensive.

Most software companies also contribute to the fact that the situation changes only slowly. New products are developed which offer more and more functions and can be better adapted to the non-standard needs of individual building contractors, but these new software products are often more proprietary. The aspect of communication is still missing in almost all software solutions. The management of project files and directories, i.e. the direct access to unstructured information of any type, is a further shortcoming.

There have been several attempts to develop a new software under the patronage of a group of building contractors with the aim to create a general product satisfying all their requirements. However such a methodology leads to giant products which are very expensive in development and cannot be completed in an acceptable time. What is more they make excessive demands on the user by too many features.

Another strategy of many software companies is to grow by additional purchase, often by additional purchase of former competitors. Thus a software company can offer a wider range of different solutions for different tasks. The customer however still cannot get an integrated solution.

3 PROCESS ORIENTATION AS A BEGINNING TOWARDS A CAPM-SYSTEM

Process orientation is a matter of course in many industries today. In the building industry the opinion prevails that each project is unique. Therefore it is assumed that processes that are supportable by software can be found in administration only.

Actually one cannot find two completely identical construction projects. However it only depends on the degree of abstraction whether generally valid processes can be identified. In other words, one needs to obtain a general view of the complete construction process first.

The individual process steps must correspond to application modules. These application modules have to leave sufficient flexibility to the user to react to the requirements of a particular project. Naturally these application modules have to communicate with each other. It should be possible for individual modules to be substituted by other solutions and to integrate other available software components where the platform must not be an issue. At present this is not possible in many existing applications.

As a consequence a process oriented software system must consist of modules which

- support the current process step as completely as possible, but without limiting flexibility,
- receive their input from previous modules and pass their output to following modules by defined interfaces
- and are able to exchange information with other specialized systems.

This seems trivial but in practice it is not.

4 REQUIREMENTS FOR A CAPM-SYSTEM

As explained above one requirement is that the application modules are process oriented. A further requirement results from the way many construction companies are organized. Construction companies compete locally. That means company A is not a competitor of company B but the local office of company A in region X is a competitor of the local office of company B in region X and also of smaller, only locally operating contractors in region X.

This decentralization frequently leads to information losses and to unnecessary parallel activities. One demand derived from that is that the different local systems must be able to communicate.

Under the impression of rising costs many building contractors in Germany find themselves in a permanent process of reorganization. The software landscape changes in a similar way. De-central systems become more centralized again. A typical example is the central procurement which leaves fewer responsibilities to the local profit centers or the building sites than they used to have.

If the existing software lacks flexibility, complete new systems have to be bought. Sometimes this implies new hardware, in any case it implies costs for education and loss of effectiveness for a while. So modern software must be able to adapt to structural changes.

The creation of so-called competence centers is a way of reorganization that can frequently be found in the building industry. For example all specialists for bridge construction are concentrated in one or two locations, and all bridge construction projects are controlled and supported from these competence centers. This means that a project manager controls several projects at the same time. If the project manager is not supported by software that allows him to access all up to date information of the different projects from his location, he will spend a substantial amount of his work time in an unproductive way in his car.

5 USE OF EXPERIENCE AS A BASIS FOR A CAPM

Documentation and direct use of experience is widely neglected in the building industry. Knowledge and competence exist only in the heads of some persons or in personal recordings that are not accessible to others.

Undocumented knowledge can lead to loss of jobs or to avoidable costs by ineffective methods. Reference numbers of projects are one form of knowledge. For example they can be used to check the plausibility of an offer.

Knowledge databases should not be assigned to a specific process step: they collect information from one process step and provide it to other steps. Databases that are independent of a process step will be called „satellites" in this paper. Thus our model must be able to be completed by satellites.

6 THE NUCLEUS OF THE MODEL

Our aim is to support the full cycle of building projects from the point of view of a building company, that is from acquisition to warranty. In this paper we concentrate on the documentation and control of all activities in such a process.

We do not assume deterministic workflows because they are only suitable for limited process steps. Instead we allow black boxes, that means parts of the process may be subject to adhoc-decisions (partly deterministic). We are also able to handle the allegedly typical construction process (adhoc workflow only).

The main purpose of the model is to support the control of projects, that means it represents the process chain from the view of a manager. Special functions such as estimation, costs analysis or scheduling are supported excellently by numerous systems available on the market. Here only data exchange between our multi-project process model and project-related software modules has to be guaranteed. Permanent data exchange with commercial application systems is mandatory.

The five fundamental process steps are
- acquisition
- estimation and control of offers
- control of contracts
- documentation and control of the construction site
- warranty after termination of the construction site

The acquisition starts with the first incoming piece of project information and ends with the decision whether an offer is to be created. If an offer is to be created the acquisition information is passed to the next phase. At the same time a project is created in the estimation software, taking all header-information from the acquisition. Acquisition information is archived.

The process of offering starts with the decision to create an offer and ends with the acceptance or rejection of the offer by the customer. In the positive case data are transferred to the contract control module. In any case the offer information is archived.

If the offer is accepted contract control (including contract negotiation) starts. Contract control

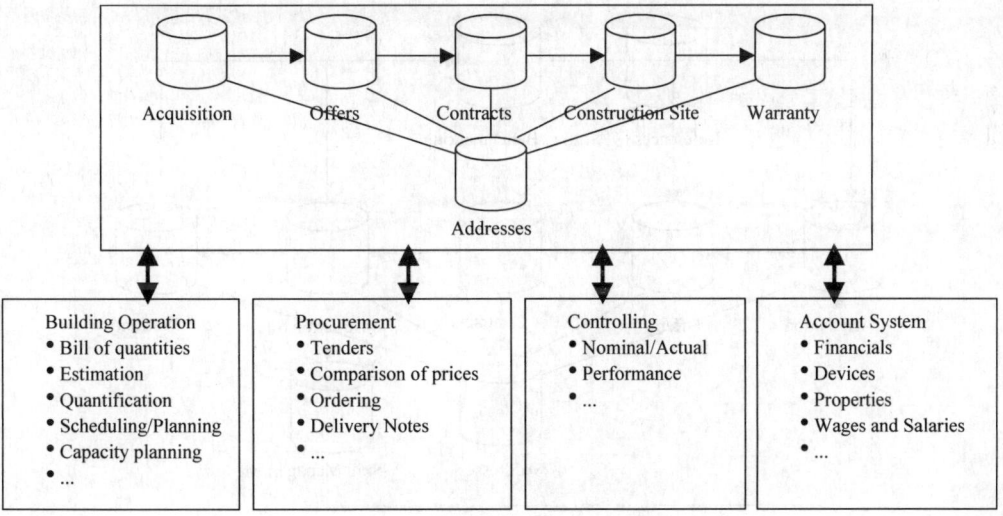

Figure 1. The nucleus of the model

301

ends with the completion of the building. With the beginning of construction on the site an additional module for the support of the site becomes active. This module also ends with the termination of the site.

The contract control module concentrates on the contractual part as well as on the administration and evaluation of sub-contractors. The construction site module serves the documentation of all activities on the site, for example a building diary.

When the construction of the building is finished, the contract and site documentation are archived and transferred to the warranty phase. Warranty concerns two points: the warranty of the main contractor towards the owner and the control of warranties and defect removals by the sub-contractors. This process ends with the expiration of the legal or contractually agreed guarantee period. The project documents are archived (or destroyed).

7 THE SATELLITES OF THE MODEL

Within the general process steps described above, numerous sub-processes run which can be improved substantially by supplying suitable information. As an example let us consider the acquisition process again. One step of this process is pre-qualification.

The main problem with pre-qualification is the compilation of the correct documents, for example finding exactly those reference projects suitable for the request of the owner. Here the satellite „references" serves well. This database is automatically supplied with basic contract information by the contract control module. By manually adding supple-

mental text, photos, etc to the contract information in this database one gets a reference pool that can be easily used for pre-qualification.

A further step in acquisition is the selection of projects. For capacity reasons not every project information can result in an offer. One must be able to identify the interesting projects easily. Certain boundary conditions such as project scope, special technical features, environmental conditions, etc are to be evaluated. And naturally there has to be a person in the company who has already gained experience concerning the technical requirements of the project.

It is easy to verify the last point if the project experience of all employees are collected in an appropriate data base.

A further satellite that is based on project experience is benchmarking. This module passes information to the acquisition phase and offering phase. Project reference numbers are collected and can be used for a first estimation or for plausibility checks.

There are many more helpful satellites such as claim management, risk management, management of drawings, etc. However, they will not be discussed in this paper.

8 THE ASPECT OF COMMUNICATION IN THE MODEL

One of the requirements of the model is the integration of external partners, that is the step from intranet to extranet. The integration of external partners often fails because one partner is not willing to invest or he has already decided for another platform

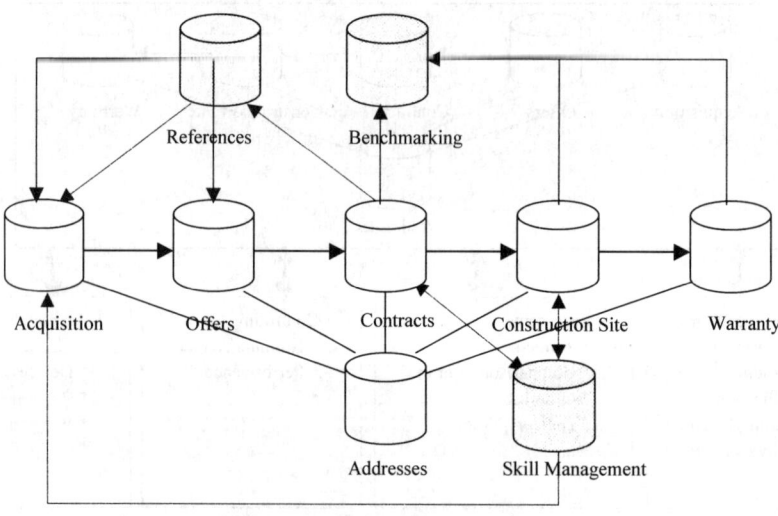

Figure 2. The satellites of the model

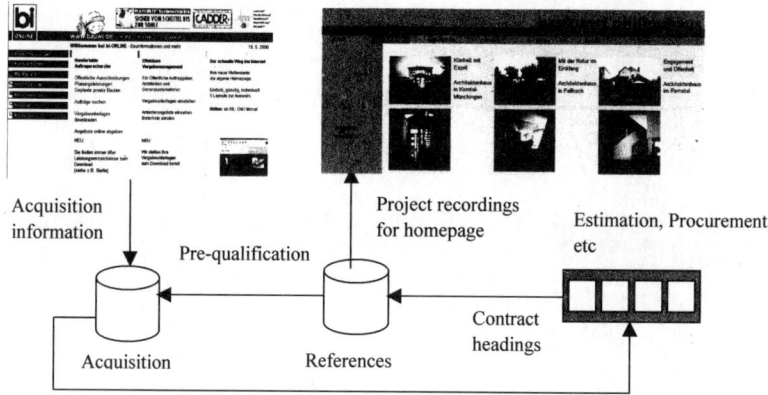

Figure 3. Interaction with the internet

that is not compatible to the preferred platform of the building company. Thus a system that cannot be accessed by internet via a browser cannot be successful.

Once the technical problems are solved the benefit that is achieved by the integration of the external partners becomes obvious very fast. Databases for management of drawings are a typical example of extranet applications.

Apart from this external aspect of communication more transparency is achieved internally. The evaluation of sub-contractors can serve as an example. These evaluations are mostly done by the site-manager, i.e. often on the construction site. However these evaluations are rarely collected in a central database. It is more common to have an individual database at each location and not to make the information accessible to other locations. Thus a sub-contractor who has worked poorly in region A and is closed for further jobs can easily get a job from another location of the same building contractor. The use of a communicative application helps to collect such information and to avoid such mistakes.

A further benefit of our system results from the interaction with other internet applications. For example the acquisition database can fetch project information automatically from internet providers of bills of quantities. Reference projects can not only be used for pre-qualification but also for the company's presentation on their homepage.

The technology is basically available. The implementation has been obstructed by two things so far:
– Software companies are not interested in committing themselves to agreements about data exchange interfaces.
– Many decision makers in the building industry are not in favor of transparency in their processes.

The market pressure on the building contractors and on the software companies is now forcing both sides to embed functional bound sub-systems into modular process-oriented systems.

The relevance of information technology in the building industry is rising because the companies change their focus from execution to management. We would like to support this by our approach.

9 CONCLUSION

A system that satisfies all requirements mentioned above is not realizable on only one platform in reasonable time and at reasonable costs. By combining suitable platforms a model can be implemented which comes very close to the requirements without losing the necessary flexibility.

Product and Process Modelling in Building and Construction, Gonçalves, Steiger-Garção & Scherer (eds)
© *2000 Balkema, Rotterdam, ISBN 90 5809 179 1*

Integrated process simulation modelling system – A proactive decision tool for life cycle project management

H. K. Doloi & A. Jaafari
Project Management Research Group, Department of Civil Engineering, The University of Sydney, N.S.W.,
Australia

ABSTRACT: This paper puts forward a new application of process simulation as an integrated tool specifically designed for holistic evaluation of project functionality within a life cycle project management framework. The authors describe a methodology for development of the aforementioned simulation tool, dubbed as DSMS (Dynamic Simulation Modeling System), devoted to project management decision making. The simulation model can act as a dynamic vehicle for optimizing technical and operational functionality at both development and operational phases of projects. The project life cycle objective functions are the basis for decision making throughout the project's life. Thus the simulation approach is considered as an added facility in the quest for optimizing decisions to resolve progressively market and external uncertainties associated typically with the project's environment.

1 INTRODUCTION

This paper discusses the application of simulation model to evaluate the feasibility and viability of capital projects throughout the project's life cycle. An integrated computer-based simulation tool will provide a basis for scenario analysis of the project in real time. Project processes can be simulated and optimized in both planning and operation phases dynamically, and impacts on the life cycle objective functions (LCOFs) estimated.

The purpose of this paper is to introduce a holistic approach to managing the project deliverables by focusing on the business objectives in the early phase of the project. This approach will provide a platform for real time project definition based on technical, functional and operational aspects of the project. Simulation modeling is introduced as a unique management tool for effective front-end planning of projects. It helps determine the optimality of decisions on operability, functionality, quality or performance issues and reduces the overall uncertainty associated with the project (Luk 1990). This approach supplements traditional static approaches by using simulation modeling for managing the functionality of the project scope.

2 SIMULATION MODELING AND STRATEGIC DECISION ANALYSIS

Continuous definition and improvement of a project, as a viable business entity is becoming an important challenge for project management in a competitive global environment. Continuous improvement in one way can help increase the efficiency of current operations through better utilization of facilities and develop strategies to do the right things at the right time in project's life cycle. To deal with the growing project complexity, organizations increasingly rely upon the benefits of new technologies (Heindel & Lasten 1996). Technologies need to be incorporated in the end result of a project (i.e. a product, a system or a facility) as well as being applied in the management of the project (i.e. technology to support, planning, control, communication and decision-making).

Key strategies in the global marketplace include the development of high-quality products at lower cost and faster commercialization. The main objective of the project definition process is to maximize the chances of successful project realization, which defines the project concepts, selection of alternatives, definition of technical contents and scope of project, and determination of financial and commercial requirements (Doloi & Jaafari 2000).

To cope with the influences of market uncertainties and global competition, the authors have put forward a simulation model as an additional tool for optimal management of projects. Modeling of the technical and operational functionality of the project (i.e. end results) in each production environment and simulating the operation phase in a dynamic fashion are considered useful as aids to decision-making and project definition. Simulation enables not only making the target facility optimization in terms of functionality, but also the total investment cost required can be forecast more accurately (Cano et al. 1998). Furthermore, the simulation tool will help fine tune the existing facilities in order to accommodate competitive market pressures in internal and external environments. Continuous project definition based on LCOFs supported by a simulation modeling approach will provide a significant improvement in planning and management of projects.

3 THE INTEGRATED FACILITY ENGINEERING PROJECT

The Integrated Facility Engineering (IFE) project, currently under development at the Department of Civil Engineering, of the University of Sydney, is a generic system that will aid management of capital projects in an integrated project management environment. Target values set for LCOFs are used as decision criteria to guide decision-making. It has been conceptualized and designed to facilitate the uptake of life cycle project management (LCPM) methodology for the delivery of projects (Jaafari & Manivong 1998). LCOFs comprise the following:
- the project's financial status and its profitability;
- the operability, quality or performance of the facility; and
- the project short and long term liabilities, including occupational health and safety (OH&S)
  risks throughout project life, environmental impacts and third party liabilities.

The IFE system comprises the following integrated modules:

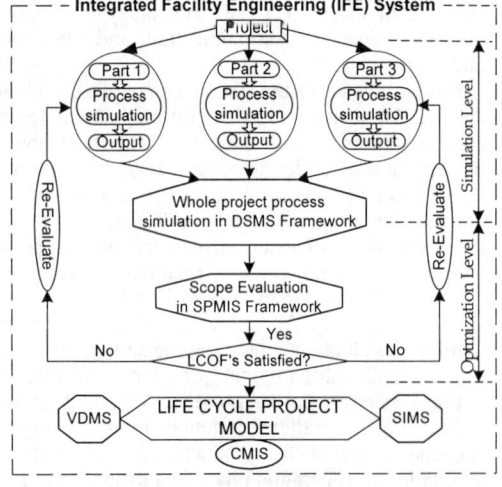

Figure 1. Simulation model integration in the IFE System

- A Smart Project Management Information System (SPMIS) to facilitate the analysis of project management functions (Jaafari & Manivong 1998);
- A Visual Design Management (VDM) system to assist in visualization/schedule simulation and management of the design process (Chaaya & Jaafari 1999);
- A Construction Management Information System (CMIS) (Jaafari et al. 2000);
- A Dynamic Simulation Modeling System (DSMS) to enhance the strategic decision analysis for project's viability (Doloi & Jaafari 2000); and
- A Soft Issues Management Systems (SIMS) to evaluate soft functions such as community and stakeholders' issues (Jaafari & Vlasic 1999).

The IFE system has a unified project databank that establishes a multi-access Intranet configuration, allowing information entry at the point of information generation, distributed access to the system reports and general client-server functions to aid information integration, expedite communication and decision processes. The work breakdown structure determines parts and products of the project. Target values set for LCOFs are used for the evaluation of decisions or alternatives to locate optimal solutions. Figure 1 shows the integration of all the modules in the IFE framework.

4 DYNAMIC SIMULATION MODELING SYSTEM (DSMS)

The DSMS module adds a simulation capability to the IFE system for decision evaluation. The main purpose of DSMS is to facilitate the optimization of the end facility, particularly reliability, throughput times, stocks, facility utilization and optimization versus LCOFs. A comparative study of the literature also provides evidence that the throughput times, buffer sizes and interdependencies of resources cannot be determined with conventional methods. The analytical approach to the management of these characteristics is difficult due to the following factors:
- the increasing number of components and cross relationships with each other make the project substantially complex. It becomes more and more difficult to define the system mathematically; and
- the presence of uncertainties on such key issues as market dynamics, plant breakdown, waste and rejects, etc. demands a simulation model of the facility for testing against different scenarios, each simulating probable conditions.

Figure 2 illustrates a strategic life cycle view of a project. It also shows the part, which is the focus

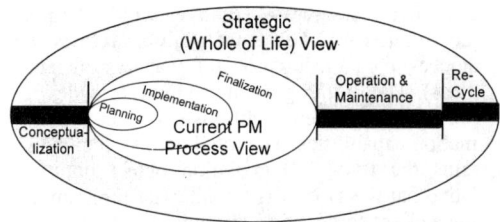

Figure 2. A generic life cycle view of the project

of the current project management approach. Modeling of the technical and operational functionality of the end deliverable supports decision-making and project definition from a strategic (whole of life) perspective as opposed to current PM approaches. The current PM approaches concentrate on the delivery process and associated functions of contractual scope, time and cost as the decision criteria. Morris (1998) has discussed similar principles i.e. concerning an instinctive business sense associated with the project delivery. Economic analysis reflecting the final customer's or investor's life cycle costs associated with the project deliverables are important for optimal decision making, particularly in the early phase of projects (Jaafari 1997), (Jordanger 1998). This is because solutions devised and commitments made at the early phase fix a major part of the project cost.

The DSMS module is geared to feed the output information into the IFE system in order to reflect, analyze and forecast the impact of the same on the LCOFs. The re-evaluation of the project definition throughout the project life cycle will also be emphasized in the LCPM approach.

5 BENCHMARKING OF DSMS FOR FRONT END LIFE CYCLE PROJECT MANAGEMENT

Dynamic simulation modeling assists the performance and sensitivity studies on changing operating conditions and input-output characteristics. It allows evaluating a broad range of system conditions to monitor the consequences of altering plant processes. The benefits of an effective simulation model should include:

1. *Optimization of processes*:- the simulation model will test multiple solutions for process layout. The model allows recognition and elimination of bottlenecks, optimization of queue sizes, observation of system behavior, and generation of valuable statistics on each alternative before the final investment decision;
2. *Impact evaluation for facility/plant design changes*:- the model will act as a test-bed for examination of facility/design changes, including how the revised process can satisfy the demand

or how the changes affect the LCOFs. The model facilitates the evaluation of the facility performance and capacity utilization based on future expansion and growth of current product lines. It will compute the sensitivity of operations to specific uncertainty variables;

3. *The ability to assess contingency and trial handling procedures*:- the model should allow the possibility to better serve customers with faster delivery rates, increased quality through quicker feedback and corrective action, what if analysis to evaluate control strategies for material flows and storage capacity;
4. *Evaluation of the effects of market variations to project scope and facility operating conditions*:- the model defines the scope of the project and evaluates the sustainability and profitability in real time based on market fluctuation. It facilitates the capability forecasting alternative operating scenarios to cope with market shifts and provide optimal solution;
5. *Proactive decision analysis with life cycle consideration*:- proactivity, as the name implies, has a specific meaning in the context of life cycle project management. It refers to the real time management of the project so that the specific target values set for LCOFs at the outset can be met or exceeded (Jaafari 1997). The model should be able to simulate the processes and facilitate a proactive approach for continuous assessment and reassessment based on expected or probable events or problems. Furthermore, model capability should ensure that value addition is being achieved and the changes made to the project justify the end deliverable as a viable business entity.
6. *Systematic modeling, recording, storing, validating, retrieval and general data management in conjunction with the overall project management information system (PMIS)*:- the SPMIS provides a hierarchical modular framework for the project which will be initialized by the DSMS to set up the relevant processes and sub-processes of the project. The model will have the necessary degree of interoperability, compatibility and interface facilities to integrate with the main information system, or other standard software. Object-oriented database systems provide the common platform for input-output data management; and
7. *Improved staff training capabilities*:- the model should offer the facility as a learning vehicle to train staff in a realistic environment simulating extreme and other unusual conditions. It must be compatible with the latest available computer technology and facilitate distant learning via the Internet.

In short, an idealized simulation tool of the type stated will facilitate construction of a digital model

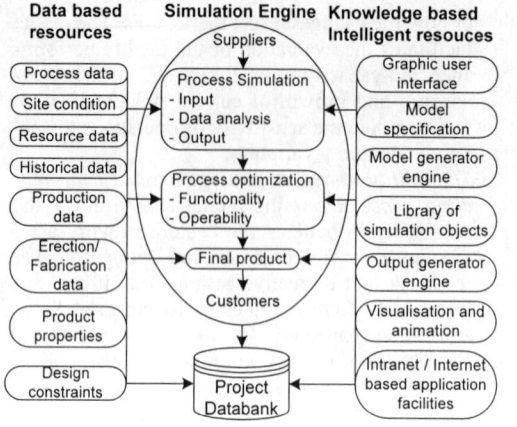

Data based resources	Simulation Engine	Knowledge based Intelligent resouces

Figure 3. Broad conceptual architecture of DSMS

of a project that will act as an experimental test-bed for decision making throughout the life of the project. It will furnish the capability to test diverse scenarios in order to derive the most viable solution in the project's life cycle.

6 DSMS BROAD ARCHITECTURE

Figure 3 illustrates a conceptual architecture of the simulation model. The DSMS comprises four broad sub-systems:- 1) data based resource management system, 2) simulation engine 3) knowledge-based intelligent resources and 4) database management system.

1. *Resource management system*:- the main purpose of this system is to operate and manage the overall resource data. The information on process data, site condition, resource data, historical data, production data, product properties and design constraints are needed prior to the simulation modeling. Different scenarios are combined with various constraints during the simulation model construction.

2. *Process Simulation Engine*:- the simulation engine facilitates systematic sequencing of different processes and sub-processes containing events occurring in the system. The simulator manages and runs the relevant processes with input and output requirements. It provides a set of operations that can be called or invoked by the simulation model. Functionality and operability offered by different scenarios are observed and optimized based on the model behavior and simulation run statistics.

3. *Knowledge-based intelligent system*:- the knowledge-based intelligent system facilitates evaluation of real life scenarios via the digital computer model of the project. The multi-objective evalua-

tion engine assesses alternatives within the project framework. The input analyzer refines the stochastic data using probability analysis (Jaafari 1988). The output generator engine provides reports and documents with visualization and animation capabilities. The intelligent system maintains the track of simulation object libraries. Note that the DSMS is a multi-user program using a client/server configuration.

4. *Database management system or repository system*:- the database system acts as a motherboard in the IFE system. All program modules reside in and interact with this system. The project model is created via SPMIS. The process model is created via DSMS framework for simulation. The model data is stored in the database. This system manages the infrastructure/facility condition, process information, operation conditions and resource mobilization data imported from the Data Base Management System (DBMS). The use of object-oriented DBMS extends the capability to accommodate multimedia data objects such as text files, image files, spreadsheet files, CAD drawings and project information.

7 DSMS SPECIFICATIONS

7.1 General Description

Figure 4 shows a hierarchical and modular structure utilized in DSMS. This concept will enhance the program's capability in simulating truly varied design alternatives (Zeigler 1987).

The following steps depict a brief description of the DSMS development process:

- Develop a project information model with hierarchical breakdown structure;
- Identify the major constituent parts of the project. It parts comprise complex systems, break these further down into major constituent subsystems;
- Develop process models defining processes and operations involved in each part vis-à-vis system and input all the information via graphic interfaces;
- Develop resource libraries defining the various resources available for the aforementioned systems;
- Define the sequences and interconnectedness of all operations. These will provide links with dependent and independent activities in the process model;
- Run the process simulation engine at part and project levels and produce design solutions. Analyze the outputs for possible optimization;
- Feed outputs to the project model for further life cycle evaluation. Functions can be linked across

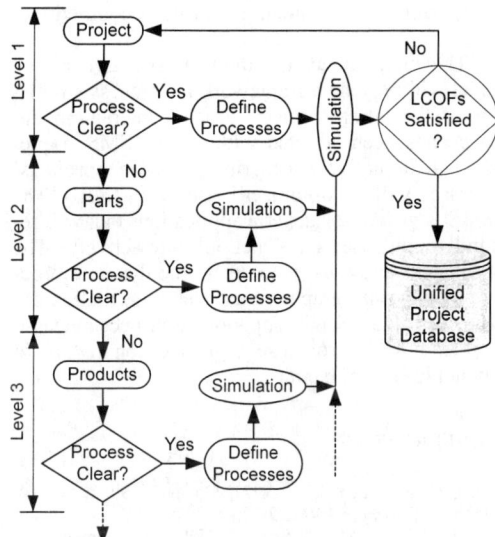

Figure 4. The 3-level hierarchy structure in DSMS

interactions with one another. Each process maintains its own list of activities. This approach utilizes object-oriented modeling. High level processes are represented as a class and instances of the process are represented as object (Garrido 1999), (Adkins & Pooch 1977).

7.3 IT Outlook

As with the rest of the IFE system, the development of the DSMS will be within the object-oriented environment using C++/Visual C++ programming languages. The DSMS will be designed to set up automatically project based process models against instructions from the users and using information extracted from the IFE system. Interactive Windows will facilitate users to determine the association between each object class in the DSMS. Data exchanging and dynamic linking of libraries will allow integration of the DSMS into the IFE system. Furthermore, DSMS's interfacing capabilities will provide an added facility for entering and modifying system input, reviewing facility need, redefining project scope, producing reports and maps.

the life cycle phases within the SPMIS environment to get a comprehensive overview of the project's status in real time. This objective-based approach will allow verifying the impact of any change(s) on project scope (see Figure 1);

- If LCOFs are not satisfied at the project level, request re-submission of the corresponding processes for alternative solutions; and
- Incorporate the accepted solutions into the project model. The outcomes are produced into reports, barcharts, piecharts, tables, histograms, time plot variables as well as visualization and animation.

7.2 Process Interaction Approach

As stated earlier, processes are the collection of events, activities and delays with respect to time. Processes are the set of abstract data structures and entities carrying out the operation sequences within the system. Thus, the behavior of the system is represented by the set of interacting processes. The event list used with this approach is composed of a sequence of event nodes. Each node contains the respective event time and process.

The process can be in one of the several states:

- *active*, when its activities are being executed;
- *ready*, when the process is waiting to start;
- *idle*, when the process is not active; and
- *terminated*, when the process has exhausted its action and not going to be active again.

The process interaction approach in the simulation model represents the dynamic behavior of a group of processes carrying out the operations and

8 PROCESS SIMULATION IMPLEMENTATION

Figure 5 shows the process-event interaction mechanism proposed for DSMS. The event list contains all event objects of the processes occurring in the system. Each event contains event ID, event time and the process to which the event belongs. The simulation engine manages the processes in the event list by carrying out three main tasks: process placement, process removal and process rescheduling. The engine keeps track of simulation clock and decides which process is due for immediate activation. The event time for this process becomes the new value of the simulation clock.

The code segments for the sequence of activities (phase) of events represent the processes as a co-system. The simulation engine activates or reactivates these co-systems one at a time. The model of the system abstractly describes simultaneous activities of the real system. To accomplish the execution of co-systems, the engine facilitates the following sequences:

1. Get the next event from the event list;
2. Advance the simulation clock and activate the respective co-system;
3. Carry out the corresponding activities in the operation;
4. Update the reactivation pointer to indicate the next set of activities and place the corresponding event in the event list;
5. If process is conditional, get the next events and place into the corresponding co-system;
6. Abandon the current co-system;

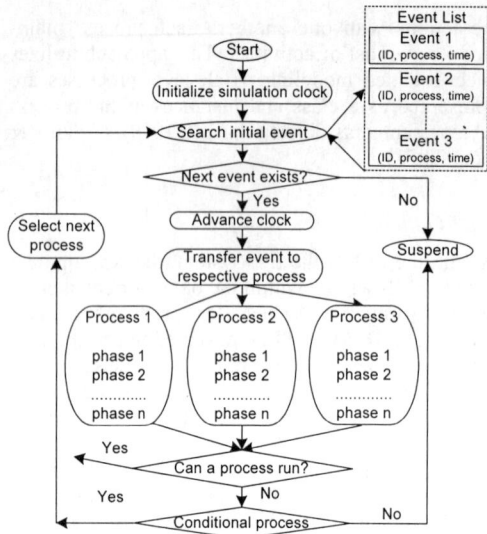

Figure 5. Process interaction mechanism in DSMS

7. Get the next event, activate or reactivate the corresponding co-system and facilitate activities;
8. Repeat the cycle until the event list is empty or the preset simulation time is over.

9 FUTURE DIRECTIONS

In order to verify and validate the proposed model, the authors will conduct a field (case) study. The model will focus on manufacturing, construction and or services type projects. The case study will provide a thorough understanding of how simulation techniques can be used effectively to optimize decisions on projects. It is postulated that the DSMS can be used as an interactive learning tool via the Internet.

10 CONCLUSION

In this paper the authors argued that process simulation in an integrated project information framework can be a valuable tool in terms of optimizing project decisions vis-à-vis life cycle objective functions. Projects are considered as value-driven business undertakings. The simulation technology can be used to improve the project's base line value and determine project investment decisions optimally.

A review of the existing systems against the idealized system supports the proposal for DSMS development (Doloi & Jaafari 2000). The project's functionality and operability can be simulated reflecting anticipated future market shifts at an early stage of the project. This will permit greater understanding of the project's capacity to respond to mar-

ket dynamics and maintain its business competitiveness.

The hierarchical and modular structure of the model will provide a framework for process simulation of the whole project. Process interaction approach has been adopted within the DSMS framework. Alternative scenario vis-à-vis projects' processes will be optimized based on LCOFs. The DSMS is geared for generic applications that will respond to the user's instructions interactively. The DSMS will allow users to create the dynamic process models using graphic interfaces. Information derived from DSMS will support the project management team for improved strategic decision making in a dynamic environment.

11 REFERENCES

Adkins, G. & Pooch, U.W. 1977. Computer Simulation: A Tutorial. *Computer* 10(4):12-17.
Cano, J.L., Saenz, M.J. & Sanz, D. 1998. Development of a Project Simulation Game. *Transaction of the 14th World Congress on Project Management.* International Project Management Association IPMA:2. Slovenia: Finland.
Chaaya, M. & Jaafari, A. 1999. Integrated Design Management within a Life Cycle Project Management Paradigm. *Second International Conference, Construction Process Re-engineering, 12-13 July 1999.* Sydney, Australia.
Doloi, H. & Jaafari, A. 2000. Towards a dynamic simulation model for strategic decision making in life cycle project management, *Project management journal*, PMI, under review.
Garrido, J.M. 1999. Practical Process Simulation using Object-Oriented Techniques and C++. *Artech House.* Boston: London.
Heindel, L.E. & Kasten, V.A. 1996. Next generation PC-based project management systems: the path forward. *International Journal of Project Management* 14(4):249-253.
Jaafari, A. 1988. Probabilistic unit cost estimation for project configuration optimization, *International Journal of Project Management.* 4: 266-234.
Jaafari, A. 1997. Concurrent Construction and Life Cycle Project Management. *ASCE Journal of Construction Engineering and Management* 123(4): 427-436.
Jaafari, A. & Manivong, K. 1998. Towards a smart project management information system. *International journal of project management.* 6(4): 249-265.
Jaafari, A. & Vlasic, A. 1999. Integration of Soft Issues into a Life-Cycle Project Management System. *Second International Conference, Construction Process Re-engineering, 12-13 July 1999.* Sydney: Australia.
Jaafari, A., Manivong, K.K, & Chaaya, M. 2000. The story of VIRCON in simulating and teaching professional construction management, *INCITE 2000.* Hong Kong.
Jordanger, I. 1998. Value-Oriented Management of Project Uncertainties. *14th World Congress on Project Management, June 1998.* Slovenia: Finland.
Luk, M. 1990. Hong Kong Air Cargo Terminals to work in Synch because of Simulation *Applications. Industrial Engineeing* 11:42-45.
Morris, P.W.G. 1998. Why Project Management does not always make Business Sense. *Project management* 1:12-16.
Zeigler, B. P. 1987. Hierarchical, modular discrete-event modelling in an object-oriented environment, *Simulation* 49(2):219-230.

Application of artificial intelligence methods

Multi-agents architecture for product development activities

P.Ghodous
LIGIM (Laboratory of Computer Graphics, Image and CAD/CAM Modeling), University Lyon I, France

S.Hassas & S.Pimont
LISI (Laboratory of Information Systems Engineering), University Lyon I, France

M.Martinez
PRISMa (Advanced Manufacturing and Information Systems for Production Laboratory), AIP, INSA, France

ABSTRACT: The product development is a complex engineering process that requires not only different sources of knowledge but also different types of knowledge e.g. physical, mathematical and experiential. In consequence, facilitating the interference of many experts of different disciplines and their communication is an important research problem. Actually the majority of systems and models focus on reasoning activities of product developers. However they don't reflect really the environment in which the product developers work. The multi-agents systems of distributed artificial intelligence permit to develop the coordination models, the communication among experts, multiple representations and concurrent reasoning. In this paper, we present a multi-agent system architecture which facilitates the work of product developers and provides them the advantages such as modularity, efficiency, reliability, creativity....To verify our architecture we give some examples in the domain of thermal power plant design.

1 INTRODUCTION

Today a modern engineering environment is considered as one that emphasises on communication and involves not only many designers but also many people concerned by manufacturing, quality control, marketing,..., working concurrently on different subparts of a more global design project (Figure 1).

The concurrent engineering approach is considered as one of the key concepts that enables the improving the quality of products, reducing the development time, reducing the cost and create the new concepts. Using this approach, every one contributing to the development of product, from definition of needs, conceptual design to manufacturing, maintenance and recycling.

Over the last two decades, there have been numerous efforts directed towards the development of computerised product models. The early research for this development was concentrated on the problem-solving techniques. The recent trends concentrate on representation issues. The evolving nature of engineering design and the diversity and complexity of engineering knowledge require the knowledge representation schema to be very flexible.

In this paper based on this new concept, we present an architecture which incorporates the current works in distributed artificial intelligence and product representation standards (uniform product representation and effective communication mechanism).

2 PRODUCT DEVELOPMENT ENVIRONMENT

To explain the environment of product development, firstly, we present the definition of the product and the different existing classifications.

A product is a physical object that can be realized by manufacturing or natural process (ISO10303, Part 1). The products can be classified by their discipline, by their function and by their structure. Generally there exists a complex relationship between structure and function of the product. This complexity is one of the origins of the difficulties to formalize the design process and product development.

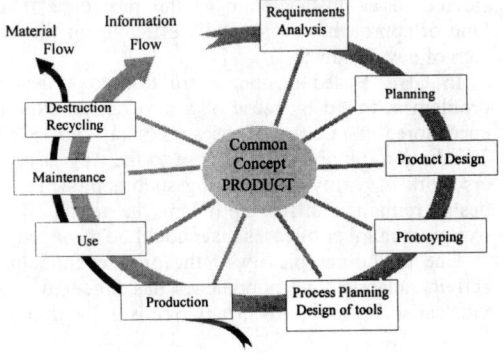

Figure 1.Product Development Environment

Each product is born by a need, it is developed and it is manufactured. The number and the sequence of the phases of the product actually depends on the discipline of the product, the management methods of the industry and also the culture of the society.

The life cycle of industrial product is complicated. Normally, there exists a great number of persons with the different knowledge and expertise engaged in the different activities during several years. They can be located in different places. There exist the different products and different versions of product. Each product is normally composed of components. Actually in many industries the life cycle of a product is a sequential process and It is difficult to have the concurrency between the different phases of the life cycle.

The traditional life cycle can be composed of: Requirements Analysis, Planification, Conceptual design, Detailed design and Analysis, Evaluation and prototyping, Design of tools, Process planning, Production management, Production, Use, Maintenance, Destruction and recycling.

Each of these activities is a complex aggregation of many activities, which uses diverse sources of knowledge, information and data.

By the result of study of product development we came to these conclusions :
- The major problem in product development analysis is how to acquire the knowledge of experts.
- The knowledge used in the development process is not always comprehensible.
- The knowledge is modified and is evolved during the analysis.
- The knowledge is not always formalizable.
- There is always an information that is hidden.
- There exists varieties of tools, file formats and data bases, in other words, there exist heterogeneous systems.
- The other types of data such as inventory, accounting,.... exist which are also important in the design process.
- Generally the design activity is the modification of existing designs.
- The tools for versioning and configuration do not give the reasons of the choice or the alternatives in the design. They do not give neither the intentions of design process.
- The designer is not capable to view or query the design process and its specifications globally.
- The process of design has a passive nature.

We can deduct that all the problems come from two major problems : the fact that the engineering data is not interconnected and the fact that the interaction between the participants are not represented and correlated . To resolve this problem two approaches exist :

1- Proposing a project organization in the way that all the related agents operate concurrently. The specialists in different domains interfered in the project join together (meet) and give their opinion. Unfortunately the meeting of the implied persons of the project is not sufficient to solve completely the problem.

2- Integrating in a computer system all the knowledge related to different expertises and give the possibility to the persons who interfere in the product development access to these information and incorporate.

In this paper, we propose a solution for the second approach.

3 CONTRIBUTION OF DISTRIBUTED ARTIFICIAL INTELLIGENCE

Distributed Artificial Intelligence (DAI) presents an interesting solution for computer support of product development based on concurrent engineering (Tong, C. and Sriram, D. 1992)(Gero J.S and Sudweeks F 1994) (Gero and Sudweeks 98).
Many models using intelligent approaches have been proposed to resolve the complex problem of product modeling activity:
- Problem decomposition models
- Prototype models
- Cased based models
- Distributed models

In the problem decomposition design, the idea is to decompose a complex problem into a sequence of simpler sub-problems for which solutions are given as a sequence of elementary actions allowing the achievement of design of the final product. This approach is based on a hierarchical planning of the sub-objective in a top-down sequential way [Brown and al., 1985].

In the prototype design, modeling supposes the existence of a general model represented by a prototype. The design prototype notion (Gero, 1985) provides a structure which is closer to the past design experience. In the ideal case, the design is considered of an instanciation of the prototype. This kind of approaches is not really efficient for the design of new products.

In CBR Based design, a solution to a design problem is found by reuse of a past design experience stored in a library of cases (Oxman, Vao 1996). This bottom-up approach is close to the way designers work in reality. Automating such approaches in design remains a difficult problem, due to the effective great number of cases that should be managed.

Due to the complexity of the product modeling activity, classical AI approaches has failed to provide satisfying solutions. Many trends in the domain of product development, modeling or manufacturing are now developed by new approaches allowing the combination of different kinds of mechanisms and reasoning.

Agent-based systems seems to provide a new perspective for the achievement of this task (Shen et al., 1999).

Multi-agents offer a paradigm for overcoming the complexity of the problem currently associated with constructing of the computer aided systems for design. The idea is to decompose the overall system into a number of semi-autonomous agents that interact with each other and with the system users, within a decentralized control regime.

The advantages of multi-agents approach can be classified in :

-The natural distribution of the design problem, that it requires the interaction between distributed (even geographically) experts, each having his own viewpoint (so distributed viewpoints).
-Robustness of multi-agents systems, with regard to fault tolerance because of their decentralized control.
-Modularity of the system design, allowing for the application of "divide and conquer" approaches to overcome problems complexity
-Efficiency provided by balanced and competence targeted tasks sharing and use of concurrency.

In another part, in the domain of product development activity we have come to these conclusions :

- The recent researches are concentrated on the representation of the knowledge related to a domain in order to use the problem solving methods.
- The major problem with the models studied is the lack of flexibility which does not permit the evolution of the model in the time.
- Most of these models are concerned with a specific domain and a special subject which make the approach of concurrent engineering and achievement of a better communication hard and inefficient.
- Most of the models are not normalized and they don't use the recent works in the domain of standardization.

We believe that the multi-agents paradigm offers an interesting approach to come up with new solutions to the problem of product modeling activity and in the domain of intelligent design in general (Campbell et al., 1999)(Campbell & al., 2000). We will describe in the following section the multi-agents architecture we propose, and the different kind of knowledge and reasoning mechanisms used.

4 DEFINITION OF CONCEPTS RELATED TO PRODUCT DEVELOPMENT ENVIRONMENT

To present our multi-agent system architecture firstly we define the necessary concepts related to product development (Rosenman & Gero, 1996) (Howard et al. 1992) (Umeda Y. et al 1990) (Shimoumura Y. et al. 1995) (Van Nederveen et al. 1992). Our study concentrates on definition of requirements (needs) and preliminary design phases, however our proposed architecture can be also applied to the other product development phases.

The design proceeds from a conceptual description of a problem (need) to a concrete description of an artefact as a solution to the problem. There is a clear shift from a semantic to a syntactic description. The categorization of the different concepts involved in the design process, namely purpose, need, function, behavior and structure, has the advantage of clearly identifying the role that each category plays in the design process. It allows for decomposition of the design process into clear, identifiable and manageable roles that identify the flow of and the type of information in the design process.

Different design disciplines, during the course of the design process need to collaborate but have different views of a design product according to their functional concerns. These views translate into different models of a product, which need to be accommodated in any comprehensive description of a design product. It is only by accommodating the functional concerns of the different agents and representing the functional properties of products that useful representations for design and design processes, which require collaborative participation, can be effected.

Some models used for product description are shown in Figure 2 (Ghodous et al. 1998).

The structure is the physical object itself, it is "what is".

Behavior is "how does", behavior is thus a description of the object's actions or processes in given circumstances.

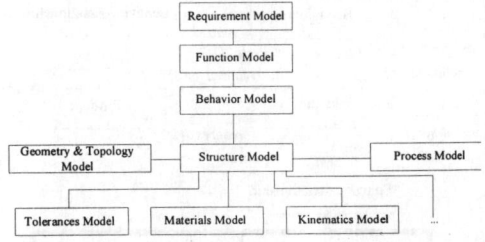

Figure 2. Some models used for product description

315

Function is "what does", is the result of the behavior, i.e. as its product or effect, so that function is closely related to behavior, the latter being the mechanism by which results are achieved.

Requirement is "the needs that it is established before the design of a product begin". One type of requirement can be purpose.

Purpose is "why does" or "what for"; Purpose only exists when related to human values of utility. Humans relate to design objects through their purpose. A design object has an intended purpose and may have unintended or realized purpose. As new behaviors and functions of an object come into focus, purposes other than those intended can be discovered.

Structure is the state of the object, in a given physical environment, and, in that environment, exhibits certain behaviors. These behaviors effect various physical functions. These functions are interpreted according to human values within a particular socio-cultural environment as enabling certain purposes to be fulfilled.

To define these facts (Information Models), due to the reasons of normalization, we use the STEP standard (ISO10303). STEP is an international standard for representation and exchange of product data. We represent the concepts by EXPRESS-G formalism. EXPRESS-G is a graphical language of EXPRESS developed by ISO 10303 STEP(ISO10303-11). The EXPRESS-G basic notations used in figures include entities (rectangles); super-type/subtype relationships (thick solid lines); required attributes (normal lines); relationship for optional attributes (dashed lines). Additionally, the direction of an attribute is symbolized by an open circle, where the circle represents the "many" side of a "one to many" relationship.

The function model is shown in Figure 3 (Ghodous et al. 1999).

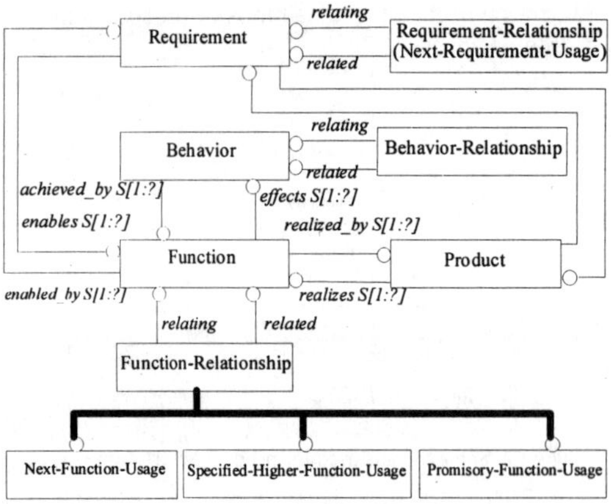

Functions are represented with Function entity, which represents both main functions and subfunctions. The relations between functions are represented by Function-Relationship entity, which represents all kinds of functional relationships. The Next-Function-Usage and Specified-Higher-Function-Usage represent functional decomposition. Promissory-Function-Usage represents the functions that we need to define, but actually we don't know where should we place them in the functional tree. The examples of instanciation of these models are expressed in the section 6.

5 A MULTI-AGENTS SYSTEM ARCHITECTURE

In a concurrent approach for the modeling process, different participants (customer, designer, process planner, producer, user) interact through a product development environment, according to different viewpoints attached to a design disciplines. A system which vocation is to help for the achievement of this task, needs to represent the participants diversity as well as the different disciplines. We suggest to view the activity of product modeling as a problem described in a multidimensional space, where dimensions represent the aspects to take into account in this activity such as different design disciplines, different participants having different viewpoints, different stages in the whole activity, and different requirements.

To visualize the problem space we have represented different axis (Figure 4). Each axis represents a considered dimension. On this figure we can observe the viewpoints of the different participants expressed according to different disciplines.

Requirements expressed by a customer can be projected, at a given level of activity, according to a given discipline.

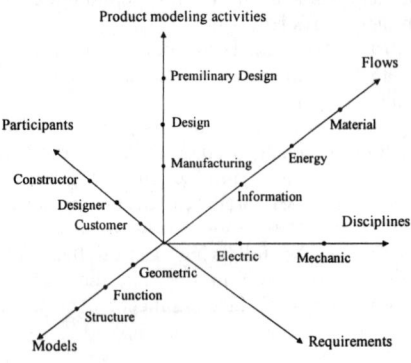

Figure 4. The problem space

Figure 3. Functional, Behavioral and Requirement Model

According to this representation, a customer expresses his requirements in an ad hoc global way. The expressed requirements are then dispatched through the different disciplines defining what we call the *Requirement Model*, so as each expert in a given discipline, can take into account the customer requirements with respect to his domain of expertise, during the modeling activity.

We suggest a multi-agents architecture, where each participant is represented by an agency. An agency is a multi-agents system for which each agent represents the activity held in a design discipline. The use of agencies allows for the restitution of a global view, to a given participant (customer, designer,..), whereas agents in a given agency, represent disciplines taken into account for a given participant. We can notice that for the *Customer Agency*, only requirements are represented and manipulated. For the *Designer Agency*, we represent not only the expression of customer requirements and knowledge necessary for their dispatching through disciplines, but also knowledge necessary to elaborate the functional model, and parts of the functional model by disciplines (Figure 5).

By zooming the *Customer Agency*, we can observe (see figure 6) as many agents as disciplines

considered. Agents interact through a blackboard that contains the general expression of requirements as entered by the customer. Each agent, representing a given design discipline, has a base of knowledge (KR) allowing him to extract an expression of requirements according to his discipline. Requirements that are not related to any discipline are conserved in the general blackboard to be used later during the modeling process.

The other agencies have the same structure. As we can see on the *Designer Agency* (Figure 7), the agency is composed of two parts: a requirement expression part with the same structure as the *Customer Agency*, and a functional model elaboration part which is composed of as many agents as design disciplines. In Figure 7, we only represent three agents corresponding to three disciplines. Each agent has a knowledge base, which allows him to elaborate the functional model. To do so agents make also use of the *Requirement Model*, and generate a tree according to three flows: energy flow, information flow and material flow. The *Requirement Model* is manipulated and can evolve during the whole activity of product modeling. The functional model is elaborated progressively and is accessible to the different participants.

5.1 *Knowledge manipulated by agents*

Each agent makes use of different kind of knowledge: General knowledge related to his design discipline, called the Domain Knowledge, and knowledge representing his experience in the given discipline, called Experience Knowledge. In our system, agents are able to reason in order to elaborate the functional model. To do so two mechanisms of reasoning have been considered: a deduction mechanism which is implemented by a rule based system (RBS), and a reasoning by analogy, which is implemented by a cased based reasoning system (CBR). To be able to choose a reasoning mechanism during the elaboration of the functional model, agents use Reasoning Knowledge.

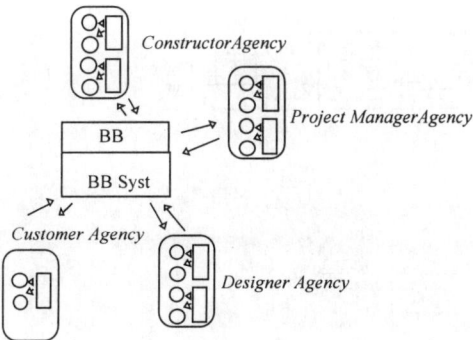

Figure 5. The multi-agents architecture of the system

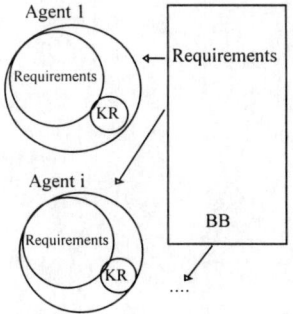

Figure 6. Customer Agency

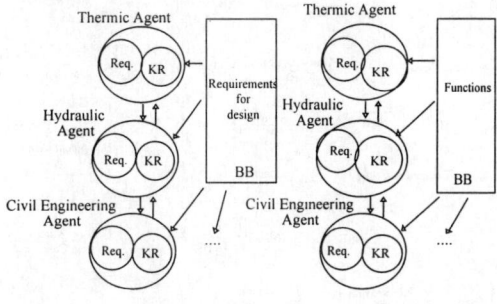

Figure 7. Designer Agency

5.2 Organization of the multi-agents system

Agencies are organized around a blackboard system. Each agency communicates its intermediate results to other agencies through the shared blackboard. For example, when the Designer Agency elaborate the functional model, it communicate it to other agencies through the blackboard.

5.3 Organization of the blackboard

In each agency, as well as at the global architecture level, the blackboard is composed of two parts: the data/result workspace part (DW) and a workspace of collaboration part (CW). The DW is accessible by all agents of the agency. The CW is organized into areas: Questions Areas (QA), Coordination Area (CA), Conflicts and Negociation Area (CNA), and Interaction Area (IA).

5.4 The data/results exchange workspace

This workspace is accessible to all agents of the agency. During the activity of modeling agents put into/get from this workspace the initial data and in-

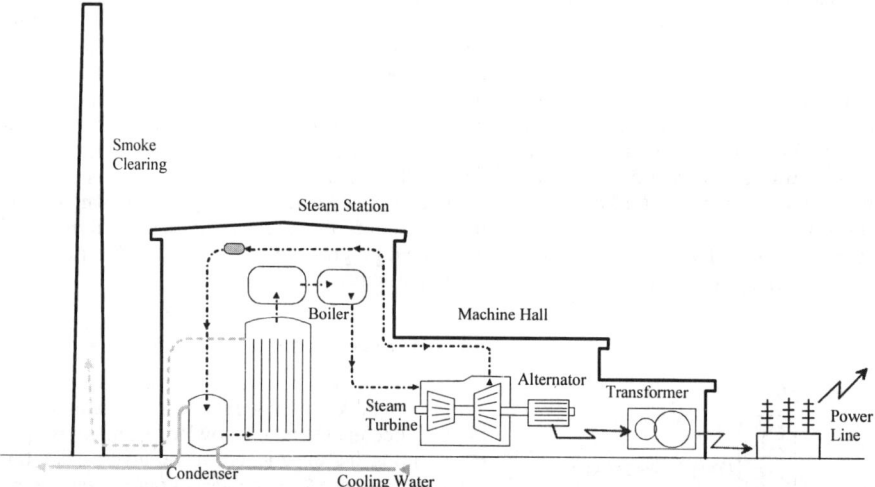

Figure 8. A thermal power plant

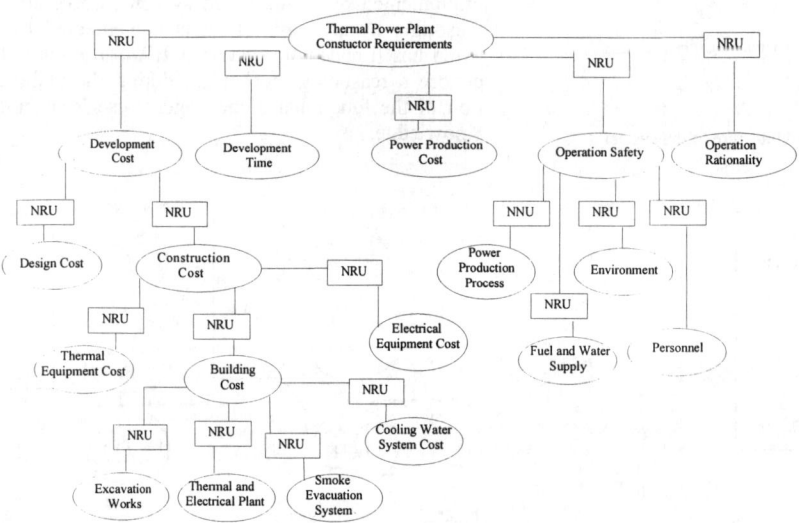

Figure 9. An Instance of Requirement Model from Constructor Agent Point of View

318

termediate results of their activity of reasoning. At the beginning of their activity, initial data corresponds to the results obtained by the reasoning on initial requirements.

5.5 The collaboration workspace

Each area of the workspace is accessible to all agents of the agency. The QA contains questions asked to all the agency, by any agent. The CA contains information and strategies of coordination (when needed). The CNA contains conflicts encountered between agents and strategies of negociations to solve them. The IA contains messages addressed by a given agent to one or several specified agents. Each agent has his own addressed area inside the IA.

5.6 How do agents work?

Each agent of the Design Agency begin to solve the sub-problem (of the global problem) related to his domain of expertise. He begins from the initial data (expression of requirements) and reason to elaborate the functional model. When the agent is unable to make a decision, and falls into a situation where he needs to ask a question to go further, he puts his question into the QA.

The human participant to the modeling activity can access any of the areas described above. He can take part of the collective activity by entering new data or responses to questions asked, for which no agent could answer. To do so they interact with the system through defined interfaces.

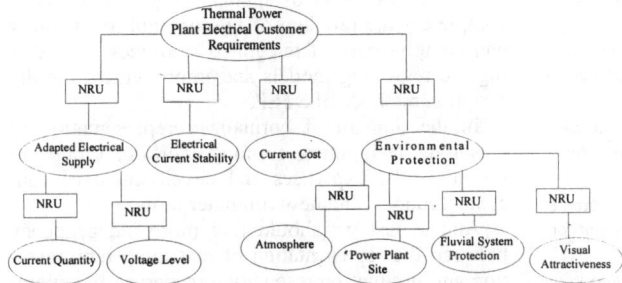

Figure 10. An Instance of Requirement Model from Electrical Customer Agent Point of View

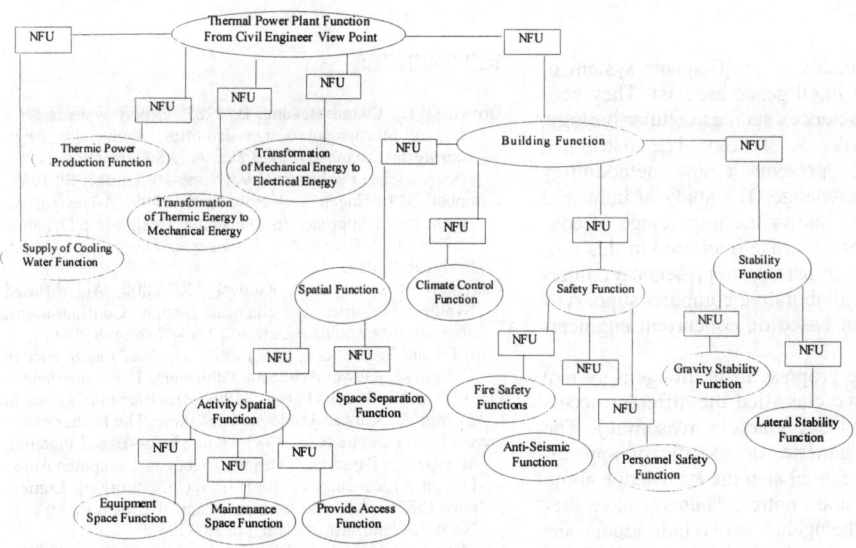

Figure 11. The functional instance viewpoint of civil engineer for thermal power plant

319

6 APPLICATION

To validate this architecture, we show the example of thermal power plant design (Figure 8). In the following, we describe in detail some viewpoints on requirement model and functional model.

We present two instances of requirement model from a Constructor Agent View Point (Figure 9) and Electrical Customer Agent view point (Figure 10).

One instance of the functional model described in figure 3 is represented in figure 11. This instance represents the viewpoint of civil engineer agent on functional model.

Concerning the reasoning techniques, for the requirement to function transformation we use the rule based reasoning, for the function-behavior transformation, we use the case-based reasoning. We show some examples of rule-based reasoning:

if Activity is pollutant *then* Add Subtype Space Separation function *to* function Spatial.

If the temperature gradient is greater than temperature threshold *then* Add Subtype function Space separation to function Spatial

If cooling fluid is river water *then* Add Subtypes functions Harnessing, Headbay, Afterbay, Forced Circulation *to* function Supply of Cooling Water

If frequency of personnel operation is greater than 1 time a day *then* Add Subtype function Provide Access to function Activity spatial

If material *or* energy flow *between* activities *then* Add Subtype function Adjacent Placement to function Activity Spatial

7 CONCLUSIONS

The application domains of multi-agents system of distributed artificial intelligence are vast. They concern the variety of sciences such as cellular biology, sociology or cognitive psychology. The distributed artificial intelligence presents a new methodology for modeling the knowledge. The study of industrial product development shows the importance of considering the different expertise interfered in this process. The multi-agent paradigm presents an interesting solution for collaborative computer support of product development based on concurrent engineering.

In this paper, we propose an multi-agent system architecture. We have classified the different necessary types of knowledge (facts + reasoning). The knowledge can be generic, or specific of application ; We have represented also the knowledge about the communication and control. Then we have presented the agents. The agents can be individual (person, software, .. ;) or collective (group). We have introduced the notion of agency to represent these two types of agent associated to each participant in product development. The knowledge related to each agent are explained and we have classified the different types of cooperation among them.

Due to the reason of normalization, we have used the recent research work on standard product representation and we have represented the product information models by STEP. These models constitutes the facts models of our multi-agent system architecture.

To validate our architecture, we have studied the case of thermal power plant design. We have presented the examples of different agents viewpoints.

Our further researches depends on the new works in different domains :

In the domain of cognitive modeling, the new research for a better understanding of engineering as a process and our ability to model it.

In the domain of standards the new works on STEP methodology, the stability of techniques used for the development of application protocols, solve completely the problem of interoperability, stability and completeness of integrated resources, considering the reasoning models and improvement the description methods of STEP.

In the domain of normalized representation of reasoning techniques, forming working groups to formalize the expertises and development of conceptual models and new computer tools.

And at last we should find more independency between the representation of reasoning representation and product representation to permit the evolution of these models in time. In this way, the modification of reasoning model can not influence the product model and vice versa.

REFERENCES

Brown D. C., Chandrasekaran B. 1985, Expert systems for a class of mechanical design activities, *Knowledge Engineering in Computer-Aided Design*, J.S.Gero (Editor), Elsevier Science Publishers B.V. (North-Holland) IFIP, 1985.

Campbell M.I., Cagan J., Kotovsky K. 1999, A-Design: An Agent based Approach to Conceptual Design in a Dynamic Environment, *Research in Engineering Design* 11:172-192, Springer-Verlag

Campbell M.I., Cagan J., Kotovsky K. 2000, Agent-Based Synthesis of Electromechanical Design Configurations, *Journal of Mechanical Design*, 22:61-69, March 2000.

Gero J.S and Sudweeks F. (eds.) 1994. *Artificial Intelligence in Design'94*, Kluwer Academic Publishers, The Netherlands.

Gero J.S and Sudweeks F. (eds.) 1998. *Artificial Intelligence in Design'98*, Kluwer Academic Publishers, The Netherlands.

Gero J.S. and Coyne R.D., 1985, "Knowledge-Based Planning as a Design Paradigm", Design Theory in Computer Aided Design, Proceedings of the IFIP WG5.2 Working Conference 1985, Tokyo, Yoshikawa, H. and Warman, E. A., eds., North Holland, Amsterdam, pp. 261-295.

Ghodous P. and Vandorpe D. 1998, A systematic Approach for Product and Process Data Modeling Based on the STEP Standard, *Computer-Aided Civil and Infrastructure Engineering*, 13:189-205.

Ghodous P. and Vandorpe D. 1999. Product's Multiple Views Modeling using STEP Standard, *Swiss Conference of CAD/CAM'99, Neuchâtel University, Switzerland, February 1999* 122-127.

Howard H.C., Abdallah J.A. and Phan D.H. 1992. Primitive-composite approach for structural data modeling, *Journal of Computer in Civil Engineering*, 6(1): 19-40.

ISO 10303, STEP Product Data Representation and Exchange, International Organization for Standardization, Subcommittee 4, NIST, http://www.nist.gov/.

ISO 10303-11, STEP Product Data Representation and Exchange, "Part 11, Description Methods: The EXPRESS Language Reference Manual", International Organization for Standardization, Subcommittee 4, NIST, 1999.

Oxman R., Vao A. 1996, CBR in design, *AI Communications*, 9:117-127.

Rosenman, M.A., Gero, J.S. 1996. Modelling multiple views in a collaborative environment. *Computer- Aided Design Special Issue, Artificial Intelligence in Computer-Aided Design* :28(3):193-205.

Shen W., Norrie D.H. 1999, Agent-based systems for intelligent Manufacturing: A state of the art Survey, *Knowledge and Information Systems, an International Journal*, 1(2): 129-156.

Shimoumura Y. et al. 1995, Representation of Design Object Based of the Functional Evolution Process Model, *ASME Design Engineering Technical Conferences*, Vol. 2 , MA, USA.

Tong, C. and Sriram, D. 1992. *Artificial Intelligence in Engineering Design*, Vol. I, II, III, Academic Press, San Diego, CA, USA.

Umeda Y. et al 1990., Function, Behavior and Structure in *AIEGN'90 Application of AI in Engineering* , 177-193 , Computational Mechanics Publications and Springer-Verlag.

Van Nederveen G. A. and Tolman F. P. 1992. Modelling Multiple Views on Buildings, *Automatic Construction Journal*, Vol. 1, 215-224.

Woodwark J. and Piegel L. (eds.) 1996, *Computer-Aided Design, Special Issue: Artificial Intelligence in Computer-aided design*, 28(3).

Woodwark J. and Piegel L. (eds.) 1996, *Computer-Aided Design, Special Issue: Computer-aided concurrent design*, 28(5).

321

An approach to a knowledge-based design assistant system for conceptual structural system design

R.J. Scherer & A. Gehre
Technische Universität Dresden, Germany

ABSTRACT: Design is a highly innovative and creative process as well, which is strongly based on empirical knowledge and less on mathematical knowledge. Nevertheless, a design system has to unify both. Innovation is much more a human habit than a formularizable property, which can be automated and taken over by a computer system, however. Therefore the design tasks, i.e. design knowledge provision, planning and cognizance have to be shared between the user and the design system. To cope with this the planning operator method has been chosen as the backbone of the design assistant system. There the empirical design knowledge from experience is capsulated in operators and the more complex empirical design knowledge about the design process is represented in the relationships between the operators. The relationships are flexible, dynamic and represented by rules and are instantiated by the use of a knowledge engine. The design assistant system provides design alternatives at each design step and based on his expert knowledge, experience and innovation, the user decides upon them and therefore decides upon the design path in the design space. The design assistant system is based on an object-oriented product model with multi-inheritance, which allows dynamic evolution of the structural system from principal decision down to detailing and dimensioning including rollbacks. The design process is structured into three layers. First, on the strategic layer principal conceptual design decisions are represented, which needs a highly flexible product model to map all the back and forth steps in the design process. On the tactic layer the detailing and pre-dimensioning are carried out. The result a conceptual structural system represented as an object-oriented product model will be obtained, ready for a structural analysis with an arbitrary analysis tool. The last layer is called the reactive layer, which is necessary to cope with conflicts, revealed trough the structural analysis, or introduced by late or external modifications or neglected constraints, simplifications and approximations.

1 INTRODUCTION

The early phase of the design of a structural system as well as design in general, is dominated by knowledge of experience, empirical rules, rules of principles and rules of thumb. This provides – as an essential property – the human designer with the necessary flexibility and degrees of freedom in such a natural way that he can incorporate his individual personality and innovative potential. This, and only this degree of freedom led and leads to an enormous variety of building types and is the gateway to evolution. However the big handicap in practice is the missing empirical knowledge and experience at the right time for the right problem, which is even worse for newcomers but also holds for experienced engineers.

Furthermore, design is a decision making based on hypotheses, expectations and speculations as well. As a consequence a design result is not a proved result in the mathematical sense. A conceptual design of a structural system is either over- or under-designed, normally and has to be corrected and optimized through re-design processes known as design iterations.

Especially this highly sophisticated while not mathematically formularizable and very important starting phase of the whole design process is not or insufficiently supported by existing software systems on the market. The proposed design assistant system should help fill this gap.

The knowledge-based design assistant does not have the objective to do creative design by itself but should support and assist the real designer to stay and remain on the innovative and creative level of design, whereas the assistant systems provide all necessary information, prepare decision-making by working out alternatives for the following-up design step as alternatives in order to trigger the innovation potential of the designer, i.e. to inspire him. This requires a high human-computer interaction [1] The designer should be freed from stupid work hindering him from entering innovative levels continuously. The intuitive use of the design assistant system based on artificial intelligence methods increases once more the productivity of the designer and avoids long learning and training time for the proper application of the system.

As the final result of the design process, the dèsign assistant system provides the user with an object-oriented product model of the structural system ready to be used for the structural analysis by any tool on the market, provided that the tool has a product model interface. It is intended to start with the IFC V2.0 basic model [2] and extend it by a short-form structural system model which should be merged into the IFC V3.0 version [3].

2 THE PRODUCT MODEL

The functionality of the proposed architecture of the design assistant system is strongly dependent upon the use of an advanced object-oriented product model, much more than conventional structural analysis software will be, because flexibility and intentions are the guiding forces of design, which can only be representable by a semantically high and sophisticated data structure with flexible and powerful relation mechanisms. It cannot be represented by mathematical relationships only. The object-oriented product model has to allow multiple views, i.e. aspect models (or short forms in the naming convention of the IFC) and mappings between each other. Based on that the start-up architectural model, the structural system model will be continuously refined, specialized and detailed during the design process (Fig. 1) and at the end the data in the form of product data can be forwarded to the tools of other disciplines, where the corresponding aspect model, the short form of the data can properly be selected and adopted by the tool.

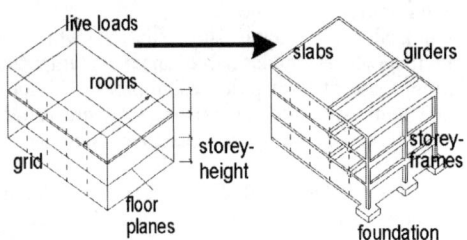

Figure 1. Continuously evolving specialization and detailing of the building product model.

Such a concept opens the possibility of an integrated environment incorporating the design, the detailing and the construction processes in building construction and civil engineering. The product model contains instances and relations. Instances are instantiated objects from generic object classes. They contain attributes (or slots) and methods. The

class structure will support multiple inheritances, which make possible to create new object classes in a very flexible way. For instance, the product model under development should allow creating a new object class during the current design process by merging (binding) two parent object classes into one new object class. A case example for instance is the detailing of a column to a steel column through binding the parent object class column from the aspect model "architecture" with the aspect model "steel structural system".

It has to be mentioned that the application of multiple inheritance leads to very complex taxonomies in the product model, which are not in the scope of current product model development, neither in the IAI [2], nor in STEP [4].

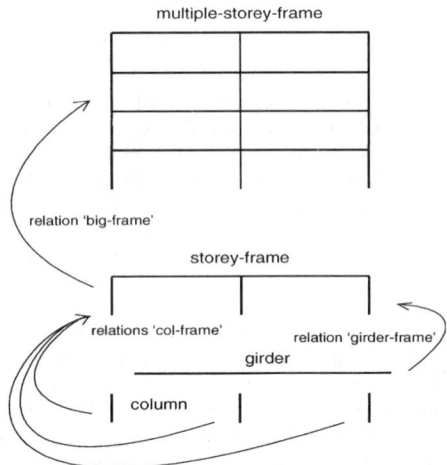

Figure 2. Structuring in formation in the product model based on relations of the type 'part-of'.

Relations of the product model developed for the design assistant system are used for semantically high expressed connection between instances. They are represented as objects as well. These constructs are necessary in order to provide the demanded flexibility and interpretability of the model and ensure consistency of the model after modifications, which will appear internal much more than externally recognizable by the user. For instance, a lot of model evolutions will be realized through model modifications. This will allow configuring the model evolution efficiently and flexibly as well.

A typical example for such a flexible relation is the well-known part-of relation, which describes the decomposition of building element groups. They introduce the structuring into the product model by describing the relation between the instances. Without such relations the instances of a product model would be only an unsorted list of instantiated objects. In Figure 2 this is illustrated by a frame system. Every column 'knows' – expressed by a relation – which storey frame it belongs to and every storey frame 'knows' – due to a relation –which storey it belongs to.

3 THE DESIGN PROCESS

3.1 Architecture of the system

Design means much more than computation with digits, like structural analysis. Design activities incorporate such aspects like perception, recognition, imagination, intuition, presumption, expectation, learning, planning, and problem solving.

Keeping this in mind, it will become immediately plausible that the representation of the design process cannot be done with the classical methods and techniques based on numerical mathematics applied for instance in CAD and structural analysis tools. On the contrary, the decision made by the designer must be represented on a semantically high level in order to be properly transformed into the final structural analysis model. This is what is needed to support the early phases of structural system design.

Furthermore, for the integration of the shared tasks, the innovation and the empirical knowledge between the user and the system, respectively, a communication on a high cognitive level must be realized, which demands transparency and communication of not only end results but also the intermediate results as well and much more than that the intermediate, step-wise decisions made. The design assistant system does show a quite higher communication demand and a higher complexity of the data structure than conventional tools show. To cope with these requirements properly the design process is mapped onto three different levels, where different methods of artificial intelligence are applied [5, 6.]. They determine the internal processes, which are not viewable for the user. He will see only one interface but a highly interactive one.

On the strategic level, design decisions are made which are of general nature. Their level of abstraction is high. On this level, conclusions are to a high extend based on forecasting and expectations. The focus is on general, system-wide design. On the tactic level, the degree of freedom for decision-making is much more narrowed due to the pre-established decisions made on the strategic level. The focus is on detailing and on local design.

On the reactive level, conflicts, which are to be detected after the structural analysis and dimensioning, will be coped with. Conflicts are naturally arisen from the fact that experience and empirical knowledge applied to the preliminary design process is vague, simplified, fuzzy, and must be approximated to fit the new situation to be applied. It is a reaction to already achieved design facts and often resulting in locally limited modifications. If not, the methods on the reactive level will trigger a design step on the tactic level for re-design and there a design step on the strategic level may be triggered as well.

All in all, the design process can be considered the search of visible solution in a more general space of possible design solutions. It is obvious that such an ill-conditioned problem will lead to a combinational explosion if this search was carried out through exact combinational methods. Heuristic methods are needed which limit the combinationally possible solutions drastically. The heuristic methods may be on a general level pre-formularized and provided by the design assistant system but the more specific and therefore powerful ones have to be provided by the user as part of his expert knowledge, experience and innovation. At the end, this means that additional knowledge besides the basic knowledge from text books and standards is provided for the search either trough the system or by the user, which reduces complexity and increases control over the complexity of the search process. From the user's view the design strategy can be considered a projection of most likely solution paths in the design space. Tactics means the selection of specific search knots, whereas on the reactive level search knots are only locally modified. Of course if there, modifications are not locally limited re-design actions are triggered on the tactic level or even on the strategic level.

3.2 Implementation of the architecture

3.2.1 Strategic level

The top level of conclusion in the design assistant system is represented through design operators which follow the concept of planning operators as specified as an artificial intelligence method [7, 8]. They contain the pre-conditions, so-called design aspects, which determine the applicability of the operator as well as the expected transformations of the design state. Viewed from the strategic level, the transformations are newly expected aspects, which are generated according to the design step chosen by the user.

The design operators are distinguished in extendable and elementary operators. The extendable operators are completely located on the strategic level. They contain, in addition to the above-mentioned aspects, conditions, which represent goal expectations. These expectations are step-by-step getting into reality during the following-up application of operators for both extendable and elementary ones, where their kind of instantiation strongly depends upon the particular design problem and the intervention of the designer, of course. The principles are given in Figure 3. The chosen scenario there shows the systematic specialization of the product model, i.e. the top down design process, starting from the highest abstraction level,

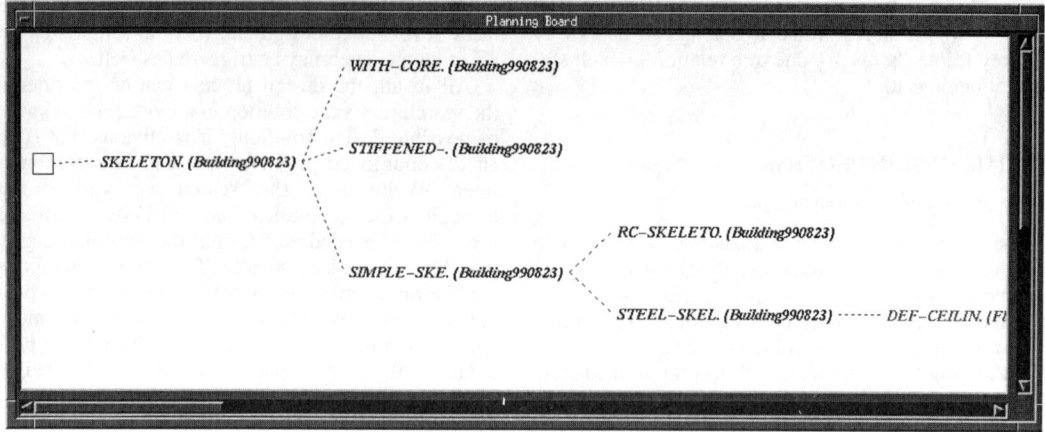

Figure 3 Prototype of the planning window showing a sequence of extendable operators from skeleton to defined ceiling..

there the 'skeleton' (a structural system type) operator down to next level represented by three alternatives, from which the designer already chose the solution path 'simple skeleton', which means non-stiffened frame system. There, again the design assistant system had generated two design alternatives, which had been provided to the user, where he selected 'steel skeleton' and therefore decided that steel will be used as the construction material.

In a very short time scale the designer decides upon the main principle design basics and instantiates the principle product model part of the structural system aspect model. Because this process strongly follows the decision-making path of a real designer, no additional training is necessary for the use of the design assistant system. The operation of the system is mainly intuitive in the natural way.

3.2.2 Tactic level

At a certain abstraction level the potential of further specialization of the product model is saturated and the extendable operators reference one or more elementary design operators. Those will extend the principle design actions of the design operators by tools for either detailing or dimensioning, which are instantiation steps with no or minor modifications and evolutions of the instantiated object structure. This marks the transition from the strategic to the tactic design action. An expandable operator may not only expand to one single elementary operator but his strategic design goal may only be reached if several elementary design operators are applied which, of course, are based upon each other in a certain way. The end result is not predetermined, because by each elementary operator the designer is asked for his particular design decision, which is an impact on the design solution path followed in the general design space.

Figure 4 shows a snap shot of such a design scenario on the tactic level, as it will be visualized in the planning window of the design assistant system. In the particular case shown there, the expandable operator 'RC-skeleton' triggers the goal that all levels of the building are to be dimensioned. The according expectation is that this is possibly remaining on RC and not shifted to steel, for instance and to stay on 'simple-skeleton' system and not to introduce additional stiffening subsystems. As a consequence three subsequent tools 'define ceiling', dimension ceiling' and 'dimension frame' make up the only useful way to do this job. The word 'triggering' as used above was only applied to describe this process in simple words. Triggering of a sequence assumes that this is a deterministic, prescribed process, which is here not at all the case. The design assistant system internally concludes about the choice and sequence of the finally suggested operators on the basis of intelligent conclusions applying rules and an inference mechanism, however. A given sequence would contradict the basic philosophy of the design assistant system of a free interaction between formalized, non-innovative parts, represented by the system and innovative parts provided, say introduced by the design.

In the case of triggering a sequence of elementary operators a further method is internally applied, which is the method of a design focus as for instance introduced by [9].

Based on the design focus, which is deduced from the design aspects extracted from the instantiated design data, i.e. the product model as instantiated on the strategic level, the actions of the designer are somewhat directed in such a way that the design assistant system provides only tools which are visible in the actual – and as an expectation – for the further design state to reach the design goal

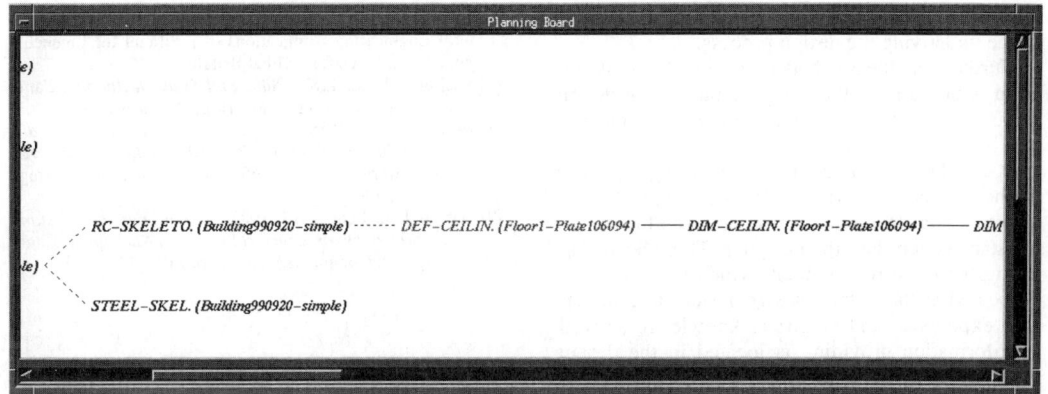

Figure 4. Prototype of a planning window showing a design sequence of elementary operators expanded up to the design state of dimensioning of frame.

as set up from the expandable operators on the strategic level. With the method of the design focus two important objectives are fulfilled. First, consistency of the design process is achieved and second, the possibility of mal-use of the design assistance system is strongly reduced.

3.2.3 Reactive level

The design actions on the above described strategic and tactic level serve for the decision upon important design basics and for the generation of instances and relations between them with the subsequent dimensions of all the instantiated elements.

However, this is not sufficient for a design system, even if – at a first glance – it may look like. Design is a process based upon knowledge of experience and empiric processes and which has to be applied to unknown, unforeseeable situations, properly adopted and combined with different patterns of knowledge of experience, say cases of best practice. There the design system has to cope with uncertainties, inherent in the knowledge of experience, with fuzziness in the particular application, with necessary simplifications and approximations to apply the knowledge to the particular problem. Further on, the designer may wish alterations in order to generate alternative solutions after a finished design process, and so on. A lot of external, not foreseen impacts may happen. This all may be subsumed to conflicts. They arise after a completed design or they are detected after the structural analysis is carried out in order to verify, as a natural process, if the expectations of the design are fulfilled or an overload has occurred here and there. A typical example may be that separately designed building parts may end up with overlapping elements along their boarded plane.

The solution of such a conflict is a reaction to already given facts. Usually such conflicts are

solvable through modifications applied to a locally limited area. Therefore the method of constraint propagation is selected to cope with this kind of problems. The constraints are located to categories of design instances and to particular design actions. Conflicts, which are detected, are solved either by the support of the system by supposing solution alternatives from which the designer may select one or directly by intervention of the designer. Theses solutions are reactions and modify the design product model, which again can trigger new conflicts. A domino affect may be possible. Therefore not only simple reactions but also a chain of reactions has to be considered and represented by the methods applied. Recursive algorithms are applied which of course need to be controlled if endless loops are possible, which may occur for these kinds of design problems.

4 SUMMARY

The early design phases are only insufficiently supported by available software. The reasons may be that a shift of paradigms is necessary changing from structural system analysis problems to structural system design problems, i.e. changing from mathematical methods to artificial-intelligence methods. It is imaginable that new, innovative concepts and methods are needed and introduced in the engineering world. The proposed design assistant system is based on several innovative assumptions.

First, the design process is subdivided in three different levels, the strategic level, the tactic level, and the reactive level, where on each level different methods of artificial intelligence are applied. This homorphic approach is of upmost importance because a single artificial intelligence method is unable to mirror the real complexity and intelligence underlying the design process.

Further on, the method of design focus is applied which guides the designer more goal-driven in his actions and save a lot of time in the interaction between the system and the designer.

As a bracing method the planning operator methods is chosen, which allows a highly flexible system and a strong interaction between the design assistant system and the designer. This allows fulfilling the basic requirement, which is: innovation is located at the human designer side, formularizable experience and empirical knowledge as well as information providing is located in the design assistant system. There is a tight co-existence of human and machine, which has to be achieved here.

Further on, an object-oriented product model on a very high semantic level including multi-inheritance is needed, otherwise it is not possible to achieve the necessary flexibility and the representation of the evolutionary character of a real design process.

ACKNOWLEDGEMENT

The support of the DFG (German research foundation) during 1997 through 1998 under contract Sche223/18, which allowed to develop the principle architecture and select the principle AI methods, is very much acknowledged. The ongoing support (2000-02) of the EU under the programme IST, contract IST-1999-11508, project ISTforCE is very much appreciated and will give us the possibility to prove the system as a whole, develop the knowledge and develop a real prototype configured and implemented as a web-based agent.

REFERENCES

[1] Card, S.K.; Moran, T.P.; Newell, A. 1983. *The Psychology of Human-Computer Interaction*. Lawrence Erlbaum Associates, Hillsdale, N.J.

[2] International Alliance for Interoperability (IAI); Liebich, T., See, R. (eds.). *IFC Object Model Architecture Guide*. Final, Oakton/Virginia: PDF file, 1999.

[3] Weise M., Katranuschkov P., Scherer R.J. 2000. *A Proposed Extension of the IFC Project Model for Structural Systems*, Proceedings of the 3rd ECPPM in Lisbon, Balkema Rotterdam, Netherlands.

[4] ISO 10303 -1, -11, -21, -22. 1994-96. *Product Data Representation and Exchange* - Parts 1, 11, 21, 22, ISO TC184/SC4, Geneva., Switzerland.

[5] Hauser, M. 1998. *A cognitive architecture for knowledge-based assistance of the conceptual design of structural system* (in German). PhD thesis, Institute of Applied Informatics in Civil Engineering, Technische Universität Dresden, Germany..

[6] Hauser M., Scherer R.J. 1997. *Application of intelligent CAD paradigms to preliminary structural design*. Artificial Intelligence in Engineering 11 (Special Issue: Structural Engineering Applications of Artificial Intelligence), pp. 217 - 229, Oxford, Great Britain.

[7] Suchman, L. A. 1987. *Plans and Situated Actions*, Cambridge University Press, Cambridge, Great Britain.

[8] Hetzberg, J. 1989. *Planen – Einführung in die Planerstellungsmethoden der Künstliche Intelligenz*, Reihe Informatik, Bd. 65, BI Wissenschaftsverlag, Mannheim, Germany.

[9] Garrett J.H., Fenves S.J. 1987. *A knowledge-based standards processor for structural comonent design*. J. Engineering with Computers, Vol. 2, pp. 219 – 238.

Product and Process Modelling in Building and Construction, Gonçalves, Steiger-Garção & Scherer (eds)
© *2000 Balkema, Rotterdam, ISBN 90 5809 179 1*

A model for pre-qualifying contractors based on the artificial neural networks technique

F. Khosrowshahi
Faculty of the Built Environment, South Bank University, London, UK

ABSTRACT: The success of the construction phase of a project is closely tied with the quality of the main contractor assigned for the job. Therefore, the selection of the contractor is one of the most important decisions that clients will have to address. The importance of this issue is evident for all clients however the focus of this paper is on local authority clients in the UK.

A new model for contractor pre-qualification is proposed which is based on the use of artificial neural networks. This alternative model is easy to use, practical and objective. The model is based on the analysis of the variables pertaining pre-qualification and the use of data collected from local authorities in the UK. The results suggest that the proposed model is capable of producing highly accurate predictions about contractors' position in relation to pre-qualification, thus, it is a viable and more practical alternative to the current methods for determining the success-or-failure state of contractors to pre-qualify for a particular project.

1 INTRODUCTION

Pre-qualification is an important part of procurement strategy. The quality of the selected contractor is highly dependent on the quality of those pre-qualified for selection. Nonetheless, the practices leading to the short listing and pre-qualification varies from ad-hoc non-structured techniques to fully structured multi-criteria methods.

An orchestrated pre-qualification exercise can be a laborious and costly exercise, thus, many clients are inclined to resort to superficial forms of investigation and analysis, resulting in eventual selection of the contractor for the wrong reasons such as misrepresentation (e.g. unjustified low bids) or inappropriateness (e.g. lack of expertise, low financial backing, etc.). Sometimes, those companies that do exercise a form of structured pre-qualification, their decisions are based on outdated information that is compiled over the years but not updated regularly. These sources tend to compile all sorts of information, which can overwhelm the user of the information.

Contractors obtain about 70% of their work through competition. An understanding of the mechanics of pre-qualification can be just as beneficial to the contractor as it is for the client. Large proportions of tenders (more than 80%) submitted by contractors are doomed to fail (Bresnen et al. 1987). Therefore, any advance indication of the likely chance of success can help those contractors with very little chance of success to avoid excessive costs of preparation of the tender. Furthermore, a prior knowledge about the relevant importance of pre-qualification variable, as perceived by the client, can help the contractor present themselves in a more plausible and appealing fashion.

These complexities have resulted in wide spread use of heuristic approaches rather than optimisng techniques (Moslehi et al. 1991; Skitmore et al. 1992).

The aim of this paper is to propose a new decision support model for contractors pre-qualification for use by local authorities in the UK and for contractors who are likely to bid for local authority works. Accordingly, for a given project, the model can be used to identify whether the particular contractor is suitable for pre-qualification or not.

The focus of the current paper is on the practices of the local authorities in the UK. The local authorities are a major client of construction works (Merna & Smith 1990). As clients, they have certain characteristics that distinguish them form private clients. They are the guardian of public money and accountable for public funds while they are compelled to emphasise issues such as health and safety and fairness to contractors and suppliers.

The large number of input variables and the availability of a large quantity of data suggest that Artificial Neural Networks (ANN) is the most appropriate AI technique applicable to this problem. This method is capable of dealing with most complex multi-variate and non-linear problems.

2 PRE-QUALIFICATION

There is no universal method for the pre-qualification process. This lack of standardisation is also evident across local authority practices in the UK. Numerous ad-hoc methods exist across the industry many of which rely on heuristic approaches, however, these approaches have not proven to yield satisfactory results (Holt et al. 1993). There are also deterministic methods: the Internal Audit section of the Treasury Department has developed the Building Contractors Appraisal System (BCAS). This system has been in use since 1988 by a major UK Local Authority. This method relies on the use of Z-Scores which compares the solvency profile of a given firm to profile-distributions generated from previous cases of success and failure (Deakin 1972 & Taffler 1976). A similar approach has been adopted in the USA, the results of which are comparable to those produced by BCAS (Kaka and Khosrowshahi, 1998). These models use a set of financial ratios and focus on the financial standing of the potential pre-qualifier.

The process involved in awarding contracts consist of project packaging, invitation, pre-qualification, short listing and bid evaluation (Hatush & Skitmore 1997a). Therefore, the activity preceding pre-qualification – invitation – relies on the availability or identification of the list of potential contractors. Normally, the client has an approved list of their own, or has access to an approved list. In the absence of an approved list, the client obtains this information through a detailed questionnaire and issues an invitation for participation. This process will help the client to provide a short list of usually four to eight contractors who are then invited to tender. This screening process requires evaluation of each contractor on the basis of the available information that needs to be gathered (Hunt et al. 1966). Often, this information is collected in form of a series of selection criteria. If the information exists in a different format, then it is decomposed into a list of selection criteria. The examination of the studies in this area indicates the absence of a universal set of criteria that are considered as being influential for making pre-qualification decisions. The focus of these studies varies from technical ability of the contractor (Dennis 1993), managerial and financial aspects (Hunt 1966), financial ratios (Deakin 1972; Mason & Harris 1979; Kangari 1992; Langford et al. 1993; Khosrowshahi & Taha 1996), to experience and size of the company (Moslhi & Martinelli 1990).

In all these studies, each variable contributes to a different degree towards the final assessment. The final selection is based on the standing of each participant. In some cases, a minimum set of requirement is also applied. For instance, some or all criteria must score at least a minimum level. Marsh (1987) reports on one such case in Japan where the contractor must score an overall 80% prior to being considered.

The assignment of the importance level of each variable is evaluated in many ways: Holt et al. (1993) and Holt et al. (1994) used a probability rating system as well as relative index ranking; Hawwash (1991) identified a weighting system for the variables, Russell et al. (1988) did the same through the use of Z-score profiling, and Hatush & Skitmore (1997b) separated the most important and the least important criteria by applying the Delphi technique to elicit information form experts. Holt (1998) proposes a multi-attribute utility theory in order to quantify the subjective elements. This was carried out by examining past projects and identifying the success factor of each selection criteria by measuring the desirability or satisfaction that was achieved.

The current paper indiscriminately considers many variables related to pre-qualification with the intention that the clients (local authorities) would identify the most influential variables. These variables will then form, the basis the input variables of the neural net model.

3 ARTIFICIAL NEURAL NETWORKS

The technology was pioneered as early as 1940s (McCulloch & Pitts 1943, Hebb 1949), before the Preceptron concept was designed and applied for image recognition (Rosenblatt 1958). Further advancements took place during the 1960s (Widrow & Hoff 1960) which was followed by the consolidation of the theoretical foundation of the method during the 1970s (Anderson 1972). It was not until after the introduction of multi-layer Perceptrons in late 1980s (Rumelhart, Hinton & Williams 1986) before ANN assumed a more serious place in the research as well as commercial arenas. The progress in this filed is ongoing (e.g. introduction of the Back Propagation network by Shubnikov 1997). ANN has bee successfully applied to many areas within construction and its further potential has been highlighted by Moselhi et al. (1991) and Flood & Kartam (1993).

Artificial neural net (ANN) is a superior technique for dealing with complex systems that cannot be properly explained by a clear set of rules and where the problem requires a degree of fault tolerance. The choice of the method is also influenced by the availability of large amounts of data relating to a large number of variables the relationship between which cannot be simply described mathematically. This is particularly true for pre-qualification deci-

Table 1. Score of pr-qualification variables

	Qualifying Attributes	Avg.	Total no'	≥4 ★	Level of importance*				
					5 no/%	4 no/%	3 no/%	2 no/%	1 no/%
1	General experience	4.48	42	0.90	24(57)	14(33)	4(10)	0(0)	0(0)
2	Image of the organisation	2.58	38	0.16	1(3)	5(13)	16(42)	9(24)	7(18)
3	Age of the organisation	2.79	38	0.24	0(0)	9(24)	13(34)	15(39)	1(3)
4	Financial standing and record	4.90	42	1.00	38(90)	4(10)	0(0)	0(0)	0(0)
5	Quality assurance registration	2.59	34	0.21	3(9)	4(12)	9(26)	12(35)	6(18)
6	Health and safety record	4.38	42	0.81	24(57)	10(24)	8(19)	0(0)	0(0)
7	Personal and social contact of staff	2.26	35	0.26	0(0)	9(26)	5(14)	7(20)	14(40)
8	Previous business relationship	2.83	35	0.20	1(3)	6(17)	16(46)	10(29)	2(6)
9	Post business relationship.	3.79	42	0.67	6(14)	22(52)	13(31)	1(2)	0(0)
10	Reputation for completion on time	4.12	42	0.86	11(26)	25(60)	6(14)	0(0)	0(0)
11	Reputation for high quality service	4.07	42	0.83	10(24)	25(60)	7(17)	0(0)	0(0)
12	Reputation for low contract price	3.08	38	0.39	4(11)	11(29)	13(34)	4(11)	6(16)
13	Personnel / team's expertise	3.71	42	0.57	8(19)	16(38)	16(38)	2(5)	0(0)
14	Depth of technical resources	3.62	42	0.50	7(17)	14(45)	19(45)	2(5)	0(0)
15	Recent experience of similar project	3.67	42	0.57	11(26)	13(26)	11(26)	7(17)	0(0)
16	Value of the project in hand	3.52	42	0.62	6(14)	20(24)	10(24)	2(5)	4(10)
17	Recommendation by consultants	2.54	26	0.19	2(8)	3(38)	10(38)	3(12)	8(31)
18	Flexible attitude towards contractual form	2.25	32	0.19	2(6)	4(16)	5(16)	10(31)	11(34)
19	Friendly co-operation	3.09	35	0.34	1(3)	11(43)	15(43)	6(17)	2(6)
20	Office organisation and efficiency	3.64	42	0.60	5(12)	20(36)	15(36)	1(2)	1(2)
21	Location of firm	2.29	35	0.17	1(3)	5(23)	8(23)	10(29)	11(31)

* Score: 5 = essential; 4 = very important; 3 = important; 2 = quite important; and 1 = not important.
★ Sum of scores for levels of importance 4 & 5 (i.e. attribute is considered very important to essential).

sions where subjective issues introduce noise in the data. Also, ANN is suitable for the problem in hand because it involves pattern recognition, abstraction and generalisation (Maren et al. Pap 1990).

Typically, a neural network produces an output pattern by operating on an input pattern. The network contains a number of layers of neurons (processing elements), each accessing and processing data simultaneously. Each operation by a layer of neurons is referred to as a cycle [e.g. a three-layered network has three cycles]. All neural networks have a form of learning faculty. They are distinguished on the basis of the way the processing elements (PE) are connected, their transfer function (rules for the way neurons fire to other neurons and itself), and training laws (determination of the relative importance of individual interconnections to a neuron's input). To this end, several types and architectures have been developed.

The network topology used for the purpose of this research is based on the supervised multi-layered feed forward network - the back propagation network. The choice is influenced by its superiority (Shubnikov 1997), its popularity (Maren et al. 1990), and its suitability for generalisation – identification of significant similarities in the input variables and the negation of irrelevant data (Freeman 1992). With this method, training can be undertaken with only a small sample, thus the current sample size of 42 is more than adequate. Also, unlike Kohonen networks, the back propagation is suitable for situations where there is small number of outputs - one, in the case of pre-qualification decision.

4 THE SAMPLE

A questionnaire was forwarded to all 379 London Boroughs, County Councils and District Councils in England. This resulted in 42 useful responses constituting 11% of local authorities. The questionnaire included all possible variables relating to pre-

Table 2. Variables ranked & calculation of rho†

	Attributes Ranked	Avg. score	Rank >3	di²	% Ranks
1	Financial standing and record	4.90	4.90	0.000	100.0
2	General experience	4.48	4.63	0.024	90.5
3	Reputation for completion on time	4.12	4.31	0.035	85.7
4	Reputation for high quality service	4.07	4.29	0.046	83.3
5	Health and safety record	4.38	4.71	0.106	81.0
6	Post business relationship	3.79	4.21	0.184	66.7
7	Project value	3.52	4.23	0.500	61.9
8	Efficient organisation	3.64	4.20	0.310	59.5
9	Personnel / team's expertise	3.71	4.33	0.383	57.1
10	Recent experience in similar project	3.67	4.46	0.627	57.1
11	Depth of technical resources	3.62	4.33	0.510	50.0
12	Reputation for low price	3.08	4.27	1.411	39.5
13	Friendly co-operation	3.09	4.08	0.995	34.3
14	Personal social contact	2.26	4.00	3.038	25.7
15	Age of organisation	2.79	4.00	1.465	23.7
16	Quality assurance registration	2.59	4.43	3.387	20.6
17	Previous business relationship	2.83	4.14	1.727	20.0
18	Recommendations by consultants	2.54	4.40	3.465	19.2
19	Flexible contractual form	2.25	4.33	4.340	18.8
20	Location of firm	2.29	4.17	3.538	17.1
21	Image of organisation	2.58	4.17	2.521	15.8
			Total =	28.612	

{Spearman Ranking Coefficient **rho** = 1 -(6 Σ di²) / (n (n² - 1)) = **0.981** [n=21]}

qualification. Indeed, respondents were invited to add to the list if necessary. The initial purpose of the questionnaire was to exploit the knowledge of the respondents to distinguish between the important and unimportant variables. This paved the path towards the development of a simple model which accommodated the important variables only. These variables formed the basis of the input for the neural net model.

The questionnaire consisted of 21 variables and the respondents were asked to score the importance of each variable when they considered contractors for their previous tenders. The scores ranged from 1 to 5 representing 'not important', 'quite important', 'important', 'very important' and 'essential'. This labeling was designed as such in order to draw a clear line between the important and not–so-important variables. Accordingly, the threshold used for separating the important variables was set at 4+ which included 'very important' and 'essential'. To this end, the respondents had been advised that the scores of 4 and 5 were intended for the qualifiers and the lower scores were for the runners-up. Table 1 shows the list of all variables together with their average score, total number, those above the 4 and 5

threshold, and the ratings that each score attained. The latter is given in % form, because in some cases the respondents did not attend all questions. Also, the total number was used in the calculation of the average score in the following fashion which is an example for the 'Personnel / Team's Expertise':
[(8x5) + (16x4) + (16x3) + (2x2) + (0x1)] / 42= 3.71

In Table 2, The variables are ranked in terms of their level of importance by focusing on scores above 3 only. The Spearman Ranking Coefficient (rho) was used to calculate the percentage ranks and determine the level of correlation between ranking of important variables (those above 3) and the overall ranking of variables (average score). The rho of 0.99 suggests that the two values are highly correlated and the respondents are clear about their selection of important variables. The dividing line which separates those variable that at least 50% of respondents believed them to be important was found to be conveniently placed in the middle of the list at variable number 11. Therefore, the first 11 variables were selected as the input variables of the neural net model. These are listed in Table 3 where the range, and standard deviations are also given.

Table 3. Important attributes; qualified and disqualified statistics

No	Qualifying Attributes ⅄	Average score of qualified contractors	Range	S.D. ★	Average score of disqualified contractors	
					Designated ✳	Actual +
1	Financial standing and record	4.90	4 – 5	0.297	4.3	4.42
2	General experience	4.48	3 – 5	0.671	3.1	3.24
3	Reputation for completion on time	4.12	3 – 5	0.633	2.8	2.99
4	Reputation for high quality service	4.07	3 – 5	0.640	2.7	2.88
5	Health and safety record	4.38	3 – 5	0.795	2.7	2.93
6	Post business relationship	3.79	2 – 5	0.717	2.3	2.50
7	Project value	3.52	1 – 5	1.110	1.2	1.47
8	Efficient organisation	3.64	1 – 5	0.821	2.0	2.18
9	Personnel / team's expertise	3.71	2 – 5	0.835	2.0	2.17
10	Recent experience in similar projects	3.67	2 – 5	1.052	1.5	1.72
11	Depth of technical resources	3.62	2 – 5	0.825	1.9	2.07

⅄ Attribute scored ≥ 4 (rated as very important to essential) by at least 50% of respondent.
★ Standard deviation (S.D.) for the distribution of respondents' scores for each attribute (variable).
✳ Designated score of disqualified contractors based on average score of approved contractors - 2 S.D.
+ For data generated for disqualified instances. See Table 4. Numbers 38 - 74, for each variable.

5 THE NEURAL NET MODEL

Basically, the neural net model consists of the input layer, processing and the output layer. Two sets of independent input-output data are required for training the network and for testing the network.

There is only one PE in the output layer – decision to qualify or disqualify a contractor. Theses two possibilities are represented by 1 and 0 respectively.

The input variables are the 11 PEs listed in Table 3. The input data are the data collected for these variables.

A shortcoming of this work, which had been envisaged at the outset, relates to the fact that only information about qualifiers and potential qualifiers could be obtained, as local authorities do not have policies for non-qualifiers. Therefore, this information had to be driven from the existing data. This was carried out through the examination of the pattern of the existing data from which additional data could be generated. However, in order for the new pattern to represent disqualifiers, it had to have a mean lower than the pattern the qualifier data. Initially, the two patterns were separated sharing a boundary line. This was undertaken by shifting the mean of the disqualified population by 6 standard deviations (SD) from the mean of the qualifiers. For a normally distributed data 95% and 99% of all data lie within 2 and 3 SDs across the mean respectively.

The Range and Standard Deviation columns in

Table 3 show that the frequency distribution of most variables are widely spread suggesting that a clear boarder-separation of two populations is not a viable assumption. Furthermore, for certain variables, a disqualified contractor may have the same score as the qualified contractor but fail to qualify because of poor performance in other variables.

It is evident that the two patterns have an overlap area suggesting that here are no black and white circumstances and the gap between the two groups is somewhat narrow. Subsequently, it was decided to separate the means of the two groups by 2 standard deviations. The resulting scores (designated) are given in Table 3.

The resulting statistics of the disqualified data facilitated the use of the standard Z-table in order to tally the whole population, from which, a pattern of data relating to disqualified contractors was generated. In order to complement the existing 42 data for qualifiers, another 42 data were generated to represent the disqualifiers. Form each sample 37 items were randomly selected for the training of the model. The remaining 10 data were used for testing the model.

6 THE ANALYSIS

Having defined the PEs of the input and output layers and their corresponding data, the Neural Desk program was used to carry out the training process.

333

Table 4. The performance of the model

Sample No [+]	Output		Error	Standard Error	
	Actual [+]	ANN [*]	Actual - ANN		
1	1	0.99999	0.00001		Qualified
2	1	0.99996	0.00004		
3	1	0.99998	0.00002	0.0000277	
4	1	0.99999	0.00001		
5	1	0.99999	0.00001		
6	0	0.00001	-0.00001		Disqualified
7	0	0.00001	-0.00001		
8	0	0.00001	-0.00001	0.0000129	
9	0	0.00001	-0.00001		
10	0	0.00001	-0.00001		

[+] Output: 1 = qualified; 0 = disqualified.
[*] Artificial Neural Network output.

By means of trial and error and calculation of the system errors, it was identified that seven PEs in the hidden layer produced the least system error.

The training was carried out with the 'error tolerance' set to 7000 iterations. Initially, in order to examine its replicability, the model was tested against the training data,. With standard errors of 0.000489 for the qualified and 0.00134 for the disqualified instances, the model proved to demonstrate extreme accuracy for replication purposes. However, the real test is against the independent set of test data. The results of the test, shown in Table 4, reveal that the predictive accuracy of the model is highly satisfactory: the standard errors are 0.0000277 and 0.0000129 for qualified and disqualified respectively. This is further consolidated by the R^2 of 0.99 for both tests.

7 CONCLUSIONS

The importance of pre-qualification as an stage in contract award practices was highlighted and its relevance to the local authorities in England was discussed.

The work initially identified a long list of selection criteria that could be used for pre-qualification. A questionnaire was designed to separate the important

attributes from the less important attributes. The former formed the basis of the input variables, which were used to evaluate the state of the output - success or failure of the contractor to qualify. Here, due to the nature of the problem and the type of available data, the artificial neural network technique was recognised as an appropriate modeling tool. To this end, the back propagation neural net was used with 11 processing elements in the input layer, 7 processing elements in one hidden layer and 1 processing element in the output layer.

In order to improve the performance of the model, additional data were generated through the separation of the mean of the disqualified contractors from the mean of the qualified contractors by 2 standard deviations. Subsequently, a new pattern of disqualified data were produced from which 42 data were generated and added to the existing 42 data relating to the qualified data.

Having trained the model, its replicability was tested by using the training data. Also, the predictive accuracy of the model was tested against an independent set of test data. Both tests produced highly satisfactory results. The independent test produced a R^2 of 0.99 and standard errors below 0.0000277. The results clearly suggest that the model is a viable tool for predicting the pre-qualification state of contractors.

REFERENCES

Anderson, J. A. 1972. A simple neural network generating an interactive memory, Mathematical Biosciences, 14, 197-220

Bresnen, M. J., Haslam, C. O., Beardsworth, A. D., Bryman, A. E. & Keil, E. T, 1987. Performance on site and the building client, CIOB Occasional paper 42, CIOB, Ascot.

Deakin E. B., 1972. A Discriminant Analysis of Predictors of Business Failure, Journal of Accounting Research, 10, pp167-179.

Dennis, L., 1993. Handbook of engineering management, Butterworth - Heinemann Ltd, Oxford.

Flood, I. & Kartam, N., 1993. The use of artificial neural networks in construction, in Proceedings of the Conference on Organisation and Management of Construction - The Way Forward, CIB W65, Trinidad, pp., 81 - 89.

Hatush, Z. & Skitmore, M., 1997/a. Criteria for contractor selection, Construction Management and Economics, 15, pp. 19 - 38.

Hatush, Z. & Skitmore, M., 1997/b. Evaluating contractor prequalification data: selection criteria and project success factors, Construction Management and Economics, 15, pp. 129-147.

Hawwash, K., 1991. Selection of contractors and tender analysis, Management of contracts and projects, Project Management Group, UMIST.

Hebb, D. O. 1949. The organisation of behaviour, John Wiley, NY.

Holt, G. D., 1998. Which contractor selection methodology?, International Journal of Project Management, Vol. 16, no. 3, pp. 153 - 164.

Holt, G. D., Olomolaye, P. O. & Harris, F. C., 1993. A conceptual alternative to current tendering practice, Building Reseach and Information, vol. 21, no. 3, pp. 167 - 172.

Holt, G. D., Olomolaye, P. O. & Harris, F. C., 1994. Factors influencing U.K. construction clients' choice of contractor, Building and Environment, vol. 22, no. 2, pp. 241 - 248.

Hunt, H. W., Logan, D. H., Cobertta, R. H., Crimmins, A. H., Bayard, R. P., Lore, H. E. & Bogen, S. A., 1966. Contract awards practices, Journal of the Construction Division, Proceedings of the ASCE, 92, pp. 1 - 16.

Kaka, A. & Khosrowshahi, F., 1998. Investigation of the accuracy of current practices of contractors' pre-qualification using case studies, Proceedings of International Conference on New Information Technologies for Decision Making in Civil Engineering, Montreal October, vol. 2, pp. 1235 - 1244.

Khosrowshahi, F. & Taha, E., 1996. A Neural Net Model for Bankruptcy Prediction of Contracting Organisations, Proceedings of ARCOM 96 Vol. 1, Sheffield, pp. 200 - 209.

Maren, A., Harston, C., & Pap, R. 1990. Handbook of neural computing applications, London Academic Press Ltd, 220 - 221

Marsh, P., 1987. The art of tendering, Gower Technical Press.

McCulloch, W. S. & Pitts, W. 1943. A logical calculus of the ideas immanent in nervous activity, Bulletin of Mathematical Biophysics, %, 115-133.

Merna, A. & Smith, N. J., 1990. Bid evaluation for UK public sector construction contracts, Proc. Instn Civ. Engrs, Part 1, Feb., pp. 91 - 105.

Moselhi, O., Hegazy, T., & Fazio, P., 1992. Potential applications of neural networks in construction, Canadian Journal of Civil Engineering, 19, pp. 521-529.

Moselhi, O., Hegazy, T., & Fazio, P., 1991. Neural networks as a tool in construction, Journal of construction engineering Management, ASCE, 117, pp. 606-625

Moselhi, O. & Martinelli, A., 1990. Analysis of bids using multiattribute utility theory, in Proceedings of the international Symposium on Buiulding Economics and Construction Management, Sydney.

Rosenblatt, F. 1958. The Perceptron: a probabilistic model for information storage and organisation in the brain, Psychological Review, 65, 386-408

Rumelhart, D. E., Hinton, G. E. & Williams, R. J. 1986. Learning internal representations by error propagation. In Parallel Distributed Processing, editted by Rumelhart D. E., McCelland J. L. & PDP Research Group (Eds.., chapter 8, 1986, 318-364.

Russell J. S. & Skibniewski M. J., 1988. Decision Criteria in Contractor Prequalification, ASCE Journal of Construction Engineering and Management, 4 (2), pp.148-164.

Shubnikov, E. I. 1997. The main models of neural networks, Journal of Optical Technology, 64, 989-1003.

Skitmore, M., Thorpe, T., McCaffer, R., & Couzens, A., 1992. Contract bidding decision support system. Interim and final reports, The University of Salford, UK.

Taffler R. J., 1976. Finding Those Firms in Danger, City University Business School, London, Working Paper No. 3.

Widrow, B. & Hoff, M. E. 1960. Adaptive Switching Circuits, Institute of Radio Engineers, Western Electronic Show and Convention, Convention Record, Part 4, 96-104.

Modelling and visualization in design

Product and Process Modelling in Building and Construction, Gonçalves, Steiger-Garção & Scherer (eds)
© 2000 Balkema, Rotterdam, ISBN 90 5809 179 1

Automation and integration of 2D and 3D geometric models of bridge decks

A.Z. Sampaio
Department of Civil Engineering, Technical University of Lisbon, Portugal

A. Recuero
Eduardo Torroja Institute, CSIS, Madrid, Spain

ABSTRACT: The bridge deck shape has a complex configuration resulting of the overlapping of two longitudinal morphologic effects: the deck deep evolution and the layout of road geometry. Because of the great interference between these components, the elaboration of the 2D and 3D deck representations, usually required in a bridge design, is a hard task.

Is this paper is proposed a methodology to automate deck representations based on a deck geometric modelling scheme. The concepts here presented were implemented in a computer program oriented to box girder deck bridges. The designer describes the deck shape, in a simple and direct way, creating a deck geometric database. Using this database, as a mean of integration, is possible to automate the deck 2D and 3D representations usually required in the bridge design process.

This programme is then a great support in a bridge design namely on the preliminary and graphical documentation stages.

1 DECK SHAPE GEOMETRIC MODELLING

The surface involving a deck can be seen as generated by a deck cross-section that moves along its longitudinal axe. This geometric modelling scheme is based on the concept of sweeping construction of models (Anand 1993, Woodwark 1986): when a polygon moves along a trajectory, analytically defined, it generates a surface.

In the usual applications of this procedure the configuration of the generator element is not affected (Anand 1993). However, on the deck case, the shape of the generator cross-section and its transversal orientation are modified, point to point along the deck, due to the influence of two longitudinal geometric components, acting in simultaneous:

The morphologic evolution of the cross-section shape along the deck, established by the bridge designer;

The layout of the road geometry, defined to the extension of the road where the bridge is inserted.

1.1 Bridge deck database

To generate a deck as a sweeping model the geometric description of all components is needed. For the box girder solution, there were selected adequate descriptive methods to define the exact geometric characteristics of a real case for each component. Using them, it is possible to create a geometric database of the deck. The database is formed with the generator cross-section shape, the longitudinal morphology evolution and the layout of the road geometry (Figure 1).

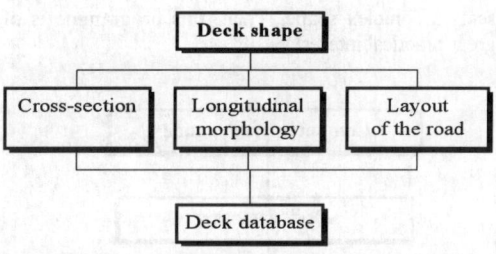

Figure 1. Geometric database of the deck

The set of the descriptive methods deal with the geometric components and the characteristic data usually managed by the designer. So, the use of those descriptive methods results natural.

1.2 Sweeping modelling process

By analysing the way of how the several deck shape components interfere with each other, on the definition of the real deck configuration, it was possible to establish the algorithms that allows automatically the generation of cross-sections along the deck. The algorithms use the database as a mean of integration.

This generating procedure determines cross-sections with shape, position and spatial orientation correctly defined (Figure 2).

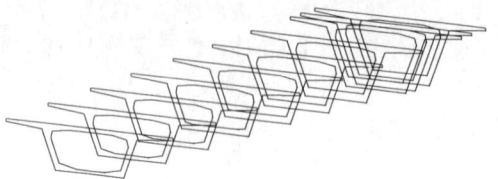

Figure 2. Deck sweeping model

1.3 *2D and 3D deck representations*

Those cross-sections are used on the automate procedure of definition of the deck geometric models usually required on a bridge design:

The cross-sections drawing presented on its won support plane;

The deck longitudinal section drawing;

A deck 3D face model.

1.4 *Computer programme*

The methodology here presented was implemented in a computer programme oriented to the box girder deck (Sampaio 1998a, Sampaio 1998b). It is one of the most frequent bridge solutions and it usually presents a complex shape. Then, this programme is of great practical interest.

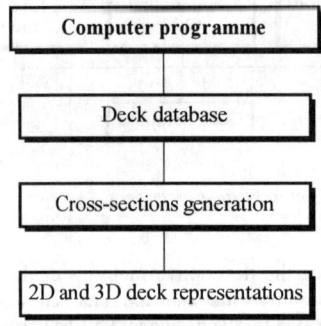

Figure 3. Modular structure of the computer programme

The programme is structured in three sequential modules. Each one corresponds to a geometric modelling step (Figure 3): the creation of the deck geo-

metric database, the process of cross-sections generation along the deck and, finally, the automatic elaboration of different kind of deck representations.

In follow, is presented the concepts inherent to the development of each module.

2 CREATION OF THE DECK DATABASE

The module that allows the creation of deck databases is structured in three sub-modules. Each one allows the description of one deck shape morphologic component:

The generator cross-section configuration;

The deck depth and the webs and slabs thickness evolution along the longitudinal axe;

The geometry of the layout of the road components (horizontal alignment, vertical alignment, superelevation and superwidth).

The characteristic data of those components are structured in independent files and, later on, they are used on automate procedures.

2.1 *Cross-section shape*

To describe the cross-section of box girder decks it was established the parameterised shape presented in Figure 4. This method is adequate to define families of shapes, such as different solutions of a cross-section type.

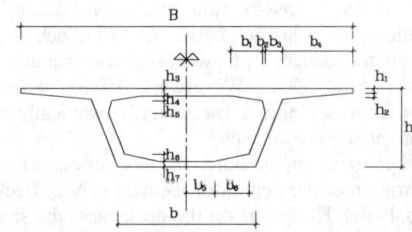

Figure 4. Parameterised shape established for box girder cross-section solution

This parameterised form has sufficient details to permit the description of a great number of real cases of that structural solution. The associated parameters correspond to the dimensions normally used on this kind of drawings:

$$\{B, b, h, b_1, b_2, b_3, b_4, b_5, b_6, h_1, h_2, h_3, h_4, h_5, h_6, h_7\} \quad (1)$$

The procedure that allows the generation of cross-sections along the deck is based on the definition of analytical expressions for each longitudinal deck edges. To establish that, it must be admitted a constant number of vertices included on the cross-section outlines and an identical numeration of those vertices. It was identified 18 vertices related to the parameterised shape, as shown in Figure 5.

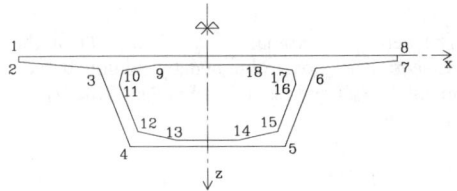

Figure 5. Numeration of the vertices included in the outlines of a cross-section

The cross-section module transposes the dimensions data to a format more suited to the geometric transformation that, later on, the generator element is going to be submitted. This other format is a vertex co-ordinates array refereed to a local orthogonal axes system (x and z, Figure 5).

The vertex co-ordinates are listed in a file. The sequential order must correspond to the vertex numeration shown in Figure 5. It respects the topology of the vertices included in each cross-section outline.

2.2 Deck longitudinal morphology

The deck longitudinal configuration is characterised by periodicity of shape limited by their spans. The central spans of decks are usually symmetric and two types of morphology compose the lateral ones. As a descriptive method it was established that: first, each deck span is divided in two segments and, then, its exact shape is described (Figure 6).

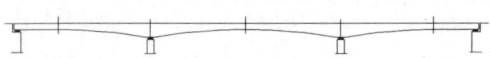

Figure 6. Deck divided in segments of regular morphology

The bridge designer to define the deck shape establishes the analytical functions related to the deck dept variation and to the webs and slabs (top and bottom) thickness variation. The deck dept variation modes considered are:
· Constant, linear and parabolic variation (Figure 6);

And for the webs and slabs thickness variation there were considered only:
· Constant and linear increment.

The developed procedure that allows the geometric characterisation of these four sub-components is based on the definition of parameterised generic shapes (for each type of variation). In Figure 7 is presented the generic parameterised form defined as a span segment established for the parabolic dept deck variation. The parameters related to the generic shape characterise, directly and completely, that type of variation. They are sufficient to represent the trajectory defined by each cross-section vertex when it sweeps a deck segment.

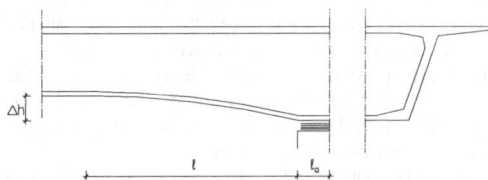

Figure 7. Parameterised sketch for parabolic deck variation

To define the real shape of a deck, all segments of the deck must be described. For that: first, is classified the type of variation presented for each sub-component in all segments and, then, the user must indicate the adequate values to the geometric parameter associated to each generic parameterised sketch. The data are organised by deck segment and sub-component and listed in a file to be used later on.

2.3 Layout of the road geometry

Finally, is described the geometry established for the layout of the road corresponding to the location where the bridge is going to be inserted. To characterise the sub-components of the layout: horizontal alignment, vertical alignment, superelevation and superwidth, the geometric parameters and data are the normally used by the designer. This information is usually included in the preliminary design documentation of the new bridge.

The data are structured in regular geometric segments (a circular arc, a clotóide arc, a straight line, ...) and listed in independent files for each sub-component. Those files complete the deck geometric database.

3 CROSS-SECTION SHAPE DEFINITION

The generator cross-section (defined as a list of 18 vertices, Figure 5) modifies its shape, orientation and spatial location, when it moves along the deck. Two steps compose the cross-section generation procedure:

- First, a series of cross-sections is defined along the deck. Their shapes are determined due only to the deck dept variation and the slabs and webs thickness variation. The deck in this stage is admitted straight and horizontal;
- Later on, each cross-section is adapted to the layout of the road geometry. For that, the vertex array of each cross-section is submitted to successive geometric transformations.

3.1 Cross-section generation along the deck axe

The shape of a cross-section included in a deck segment is obtained using the trajectory of each vertices of the generator cross-section when it moves along the segment. Each trajectory is the result of the influence of four longitudinal sub-components (deck dept and webs and slabs thickness variations), acting in simultaneous over the segment.

However, each final trajectory can be obtained by addition of individual trajectories correspondent each one to a sub-component. It is admitted that each sub-component acts in an independent way over the segment. The analytical function of an individual trajectory is defined using the respective geometric parameters associated to the generic parameterised shape. In a concrete case the parameters take the correspondent dimension values included on the deck geometric database (previously created).

Let us determine, for example, the shape of the cross-section S included in a segment submitted to a deck dept parabolic variation (Figure 8) and to a bottom slab thickness linear variation (Figure 9). For the vertices of the initial cross-section, S_0, involved in each type of variation, increments of co-ordinates, Δx and Δz, are obtained in the respective routine. Next, the increments are added to the vertex co-ordinates list of the initial cross-section in order to obtain the final shape of the S cross-section.

Then, for the bottom slab vertices (vertices 4, 5, 12 to 15, Figure 5) of the cross-section, there are obtained the increments $\Delta x_1()$ and $\Delta z_1()$, due only to the deck dept variation (Figure 8).

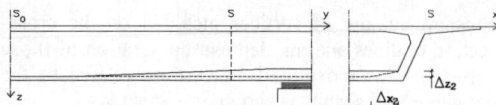

Figure 9. Increments due to bottom slab thickness variation

The vertex co-ordinates list $x()$ and $z()$ of the S cross-section is, now, obtained by adding those increments to $x_0()$ and $z_0()$ co-ordinates of the S_0 initial cross-section:

for $i = 1$ to 18
$$x(i) = x_0(i) + \Delta x_1(i) + \Delta x_2(i)$$
$$z(i) = z_0(i) + \Delta z_1(i) + \Delta z_2(i)$$
(2)

In these expressions, the increments $\Delta x_k()$ and $\Delta z_k()$ related to vertices not affected by any kind of variation, are zero. The final vertex co-ordinates are listed in a file and, later on, used on the adaptation of the cross-section to the required layout of the road geometry.

3.2 Adaptation of cross-sections to the road geometry

Using directly the files of the road geometry included in the database, the programme determines the geometric characteristics of the road for each sub-component, at the kilometric point that localises each cross-section on the road.

To incorporate the superwidth data (SW) over the cross-section, the extension of one (Figure 10) or both cross-section cantilevers have to be incremented:

for $i = 1$ to 2
$$r(i) = x_0(i) + SW$$
$$z(i) = z_0(i)$$
(3)

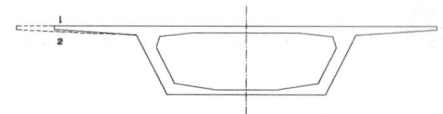

Figure 10. The extension of one cantilever is incremented by adding the superwidth

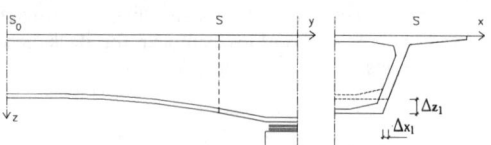

Figure 8. Increments due to deck dept variation

For the intern vertices (12 to 15, Figure 5) there are determined the increments $\Delta x_2()$ and $\Delta z_2()$ due to the linear increment of the bottom slab thickness (Figure 9).

Next, the cross-section is adapted to the superelevation (SE). The cross-section is, then, submitted to a bi-dimensional rotation as a rigid body (Figure 11). The initial cross-section co-ordinates array, $x_0()$ and $z_0()$, is multiplied by a rotational matrix defined on the cross-section plane x,z (Foley & Dam 1991):

$$\begin{bmatrix} x(1) & z(1) \\ x(2) & z(2) \\ \dots & \dots \end{bmatrix} = \begin{bmatrix} x_0(1) & z_0(1) \\ x_0(2) & z_0(2) \\ \dots & \dots \end{bmatrix} \times \begin{bmatrix} \cos\theta & sen\theta \\ -sen\theta & \cos\theta \end{bmatrix} \quad (4)$$

with,

$$\theta = artg(\text{SE}) \quad (5)$$

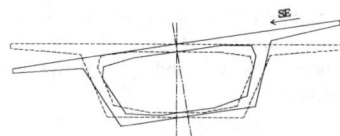

Figure 11. Cross-section rotation by adding the superelevation

Later on, the cross-section is localised and oriented on the cartographic (Figure 12) referential, using the geometric data (M, P and azimuth) of the horizontal alignment and the elevation value of the vertical alignment, obtained at the cross-section kilometric point (KP).

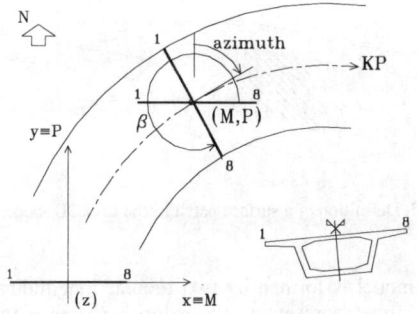

Figure 12. Cross-section adapted to the horizontal alignment geometry

The cross-section vertex co-ordinates $x_0()$, $y_0()=0$ and $z_0()$ are affected of a spatial transformation matrix (Foley & Dam 1991) in order to obtain the final position of the cross-section, $x()$, $y()$ and $z()$:

$$\begin{bmatrix} x(1) & y(1) & z(1) & 1 \\ x(2) & y(2) & z(2) & 1 \\ \dots & \dots & \dots & \dots \end{bmatrix} = \begin{bmatrix} x_0(1) & 0 & z_0(1) & 1 \\ x_0(2) & 0 & z_0(2) & 1 \\ \dots & \dots & \dots & \dots \end{bmatrix} \times T \quad (6)$$

with,

$$T = \begin{bmatrix} \cos\beta & sen\beta & 0 & 0 \\ -sen\beta & \cos\beta & 0 & 0 \\ 0 & 0 & -1 & 0 \\ M & P & elevation & 1 \end{bmatrix} \quad (7)$$

and,

$$\beta \doteq -azimuth \quad (8)$$

In this way, the cross-sections generated along the deck and transformed by adaptation to the super-width and superelevation geometry, stays correctly located and oriented in a spatial referential (Figure 2). These cross-section files are used on the automatic generation of deck representations.

4 DECK GEOMETRIC MODELS

The developed programme includes a drawing module that allows the automatic creation of the deck representations usually required on bridge design processes:
·   Cross-sections drawing;
·   Plane projection of deck longitudinal section;
·   Deck 3D face model.

The geometric models are generated at the drawing file DXF format (Jones & Martin 1991). The procedures to execute automatically the distinct representations are based on the selection of a series of graphical entities. In a concrete case the graphical entities, that form a drawing, are defined using the co-ordinate data of the cross-sections generated along the deck.

The DXF structure of a graphical entity is constant. Only the numeric values that particularise an entity are distinct. So, the drawing module includes a routine for each entity type used in any deck representation. The geometric parameters of each entity are variables. The routine concretises the variables with the values that identify its representation and list the entity in the DXF format (Jones & Martin 1991) into the drawing file in creation.

4.1 *Drawing of a cross-section*

The drawing of a series of sequential cross-sections, defined with their exact configurations (with super-width and superelevation incorporated), is a usual type of deck plane representation included in the bridge design graphic documentation.

The graphical entity, polygonal line (named POLYLINE on the AutoCAD system, AutoCAD 1997), is used to define cross-section outlines. The polygonal line vertices are concretised with the coordinate values included in the respective cross-section file (Figure 13).

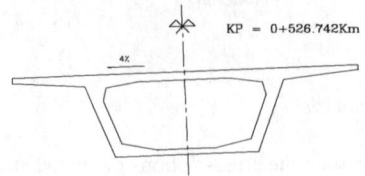

Figure 13. Cross-section drawing

The drawing DXF file is also complemented with the graphical representation data of the axe (LINE) and symbol (POLYLINE) of symmetry and with the inclusion over the cross-section of the respective kilometric and superelevation values (TEXT).

4.2 *Definition of the longitudinal section*

The orthogonal projection of a deck longitudinal section is another usual type of drawing required on the bridge design.

To define this drawing is used one polygonal line (POLYLINE) for each longitudinal edge visualised on the deck longitudinal section. The number of vertices of each polygonal line corresponds to the number of consecutive cross-sections used to compose the longitudinal section of a selected deck segment. Vertical straight lines (LINE) represent cross-sections intersected by the cut surface (Figure 14).

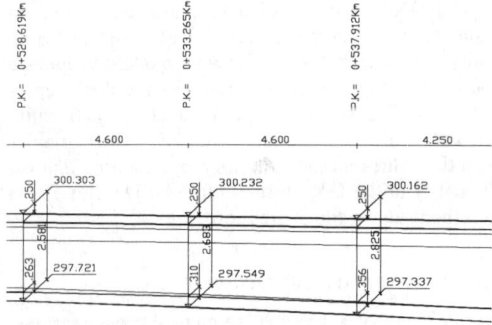

Figure 14. Detail of a deck longitudinal section

The longitudinal section representation includes data related to each cross-section: the kilometric value, the elevation value over the deck longitudinal axe and the dimensions of the deck dept and the thickness of both slabs. The respective values are obtained using the vertex co-ordinates and longitudinal location of each cross-section. Those values are automatically insert into the longitudinal deck representation.

4.3 *Elaboration of a deck 3D face model*

It was developed an algorithm that allows the automatic execution of a deck 3D model, defined in the DXF format.

The model is formed by a series of surface patches (3DFACE) defined by four vertices. Each entity is limited by identical par of vertices of sequential numeration, i and $i+1$, belonging to two consecutive cross-sections, n and $n+1$ (Figure 15). The four vertex co-ordinates, $x_s()$, $y_s()$ and $z_s()$, of the 3DFACE entity are defined by,

$$\begin{aligned}
x_s(1)&=x_0(i,n) & y_s(1)&=y_0(i,n) & z_s(1)&=-z_0(i,n) \\
x_s(2)&=x_0(i+1,n) & y_s(2)&=y_0(i+1,n) & z_s(2)&=-z_0(i+1,n) \\
x_s(3)&=x_0(i+1,n+1) & y_s(3)&=y_0(i+1,n+1) & \\
& & & & z_s(3)&=-z_0(i+1,n+1) \\
x_s(4)&=x_0(i,n+1) & y_s(4)&=y_0(i,n+1) & z_s(4)&=-z_0(i,n+1)
\end{aligned}$$
$$(9)$$

where the co-ordinates, $x_0()$, $y_0()$ and $z_0()$, are data included in the cross-sections file used to compose the model.

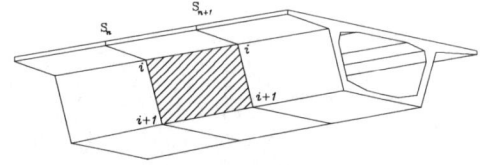

Figure 15. Definition of a surface patch in the deck 3D model

The model is formed by two tubular longitudinal surfaces: one representing the exterior shape of the deck and the other its interior. The effect of opacity required for the deck top cross-sections is obtained defining an adequate triangulation with a set of 3DFACE elements (Figure 16).

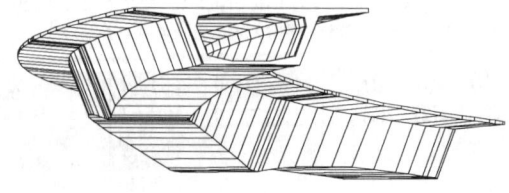

Figure 16. Projection of a deck 3D model

Is possible to apply algorithms over the bridge deck face model to simulate colour, material patterns and the incidence of different types of light. Then, this model, presenting a realist image, is of great interests in the conceptual design phase and in the presentation sessions of the new bridge.

5 EXAMPLE

It was characterised a real case of bridge deck (GRID 1995). The data obtained by the descriptive process, using the developed computer programme, form the geometric database of the deck.

The cross-section, presented in Figure 17, was characterised using the parameterised generic cross-section included in Figure 4. The values given to the parameters correspond to the dimensions represented in Figure 17.

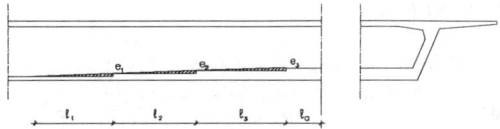

Figure 18. Generic parameterised sketch for bottom slab thickness linear variation

The layout of the road geometry is also described. Directly using the created database it was possible to generate a series of cross-sections localised at constructive joints and other specific points required in the definition of longitudinal deck representations.

The drawing included in Figure 13 represents one of the generated cross-section and, in Figure 14, is presented a detail of the longitudinal section of the selected deck segment. Of the some deck segment it was possible to obtain the DXF file corresponding to the 3D face model projected in Figure 19. The model is formed with the cross-section visualised in that projection.

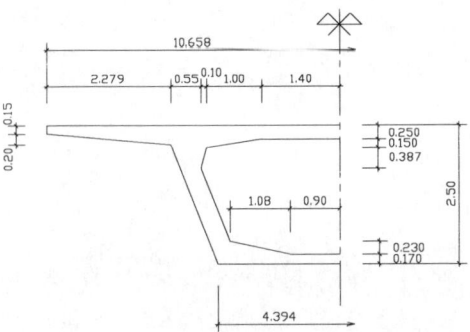

Figure 17. Initial cross-section

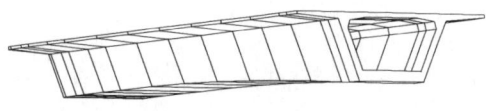

Figure 19. Axonometric projection of a deck segment

6 CONCLUSIONS

The proposal descriptive model allows the automation of the geometric phases, related to the deck element, in a bridge design process. The model uses as a mean of integration a geometric database representative of the real deck shape.

The principal advantages of the use of the developed programme are:

· A considerable reduction of time consuming inherent to the elaboration of the deck graphical documentation usually included in a bridge design;

· The possibility to visualise, in a quick way, projections of a deck 3D model from different points of view. The model is formed with cross-sections correctly defined in shape, orientation and position. This model takes advantage over the traditional approximate modes of defining deck 3D models;

· The deck shape redefinition, needed on the conceptual design phase, is easily actualised on the deck geometric database. Then, any modification of the initial shape is rapidly visualised (in the form of plane drawings and perspectives). It conduces to a optimised definition of deck shapes.

The deck segment, where the initial cross-section is inserted, is characterised by a parabolic evolution of its dept. The parabolic curve is defined with the values shown bellow corresponding to the parameters associated to the longitudinal parameterised shape represented in Figure 7.

$$\Delta h = 1.0m \qquad \ell = 32.25m \qquad \ell_a = 1.5m \qquad (10)$$

Along the same deck segment, the bottom slab thickness is incremented in a linear way. The exact mode of variation of the slab thickness is defined by the values presented bellow corresponding to the parameters related to the generic parameterised sketch established for this type of evolution (presented in Figure 18).

$$e_1 = 0.23m \qquad e_2 = 0.15m \qquad e_3 = 0.0m$$
$$\ell_1 = 22.65m \qquad \ell_2 = 8.5m \qquad \ell_3 = 1.1m \qquad (11)$$
$$\ell_a = 1.5m$$

REFERENCES

Anand, V. 1993. *Computer graphics and geometric modelling for engineers*, John Wiley & Sons, Inc.

Woodwark, J. 1986. *Computer shape*, London: Butterworths.

Sampaio, A. 1998. *Geometric modelling programme oriented to box girder decks* (in Portuguese), report IC, AI12/98, I.S.T., Lisbon: Portugal.

Sampaio, A. 1998. *Bridge decks geometric modelling* (in Portuguese), PhD Thesis, I.S.T., Lisbon: Portugal.

AutoCAD - User manual, Release 14 1997. AutoDesk, Inc.

Jones, F. H. & L. Martin 1991. *The AutoCAD database book - Accessing and managing CAD drawing information*, Ventana Press, 4ª ed., E.U.A.

Foley, J. & A. Dam 1991. *Computer graphics - Principles and practice*, Addison-Wesley Publishing Company, 2ª ed., U.S.A.

Funchal fast way design, 1ª phase - Ponte da Quinta design (in Portuguese) 1995. *GRID – Consultas, Estudos e Projectos de Engenharia, Lda*, Lisbon, Portugal.

Pros and contras of the use of virtual reality in architectural design

U. Kunze
Department of Civil Engineering and Architecture, University of Applied Sciences Dresden, Germany

ABSTRACT: This paper discusses the pros and contras of the application of VR computer systems in architectural design practice and compares several modelling techniques.

1. TODAY'S PRACTICE IN ARCHITECTURE OFFICES

It is well known that computer systems are widely used in all areas of society today; they are accepted as tools to facilitate work in all types of offices.

In architecture offices in Germany, computer technique is mainly applied in the final stages of the design process to produce 2D drawings, tender documents and cost calculations, that means to support the monotonous part of the job. Compared to other creative fields like media design or publicity, architects are still very conservative in using computers in the creative part of design and often work with traditional means and methods like paper drawings and classical building models.

The use of CAAD systems in architecture offices has very much increased for the last ten years: While in 1990 only 10% of the offices used CAD, currently 80 % have a CAD system in use. But most of them don't fully treat the opportunities of CAD and 3D-modelling and produce computer drawings on a low level.

Virtual Reality (VR) simulations can only be carried out by 10% of the architecture offices, mostly not in a good quality. But the need of visualisation performances for architecture is estimated to 80% and it is still increasing. [3]

Due to the special knowledge, abilities and equipment that are required to carry out high quality computer visualisation and animation, most architecture offices can not effort to do that themselves. Here is a gap for smaller specialised companies to perform 3D modelling and rendering as a service.

2. USE OF VIRTUAL REALITY IN ARCHITECTURE

2.1. 3D-Modelling versus 2D-Techniques

The creation of a 3D-model of a building by means of a computer system offers some advantages compared to the presentation of objects by 2-dimensional floor plans, site-views and section drawings.

Once created, the 3D-model includes all needed geometric information about the object and you can very easily derive any perspective view you would like to produce by help of the computer system. It is possible to look at the designed object from outside and inside from any desired angle of view, to walk through or fly around. This leads to a better understanding of what is designed, especially for people who are not familiar with reading 2-dimensional drawings.

Picture 1: Example of a 3D building model [1]

Furthermore, 3D surface or solids models are the basis for the creation of coloured rendered pictures or videos as described in the next part of the paper. Therefore, 3D-models are well suited to prepare discussions with the owner of the building under construction or with authorities or to produce presentations to take part in architectural competitions.

By means of 3D-models, the spatial imagination of the designer is also supported. In this sense, computer-based 3D-modelling can be seen as a design aid, especially in the case of geometrically complicated structures in the space. Here, probably conflicts can more easily be recognised than in 2D, for instance complicated penetrations of building parts and technical equipment or intersection of oblique beams or roof faces.

A 3D building model can contain much more information than geometry. Today's CAAD systems allow to build product models from which you can derive information for other design phases like cost calculation, project or facility management.

But in addition to the pleasant effects, 3D-modelling brings along a lot of problems that you won't have with 2D-techniques.

Architects often don't build 3D-models because the effort is much more higher than the effort to produce a 2D-drawing: It takes longer to create a whole building in 3D; there are much more sources of error and special knowledge, skills and abilities are required in the 3D world.

3D-models can not so easily be changes like 2D-drawings, because simple modifications of one part of the object often affect the whole 3D-model.

Another fact against the use of 3D-systems are higher costs for the architecture office concerning the price of the computer programs and particularly the cost for qualification of the personal caused to expensive courses and the long time that is needed until a person is familiar with the capabilities of a 3D-system.

For simple understanding of the structure and function of traditionally constructed buildings which consist of several floors, floor plans, side views and section drawings are often sufficient.

2.2. Virtual Reality versus simpler 3D-Presentations

Computer based photorealistic visualisation is an effective method to support the design process, particularly to facilitate the assessment of the optical effect of design objects in advance. In architecture, the choice and combination of material and light sources is very important to carry out

Picture 2: VR simulation of a building [1]

Picture 3: Simple mass models [1]

good building designs, especially at the final stages of the design process. By means of a rendering system, the designer is able to play with these design components, judge design results immediately and can easily perform changes. [2]

But sometimes the enormous capabilities of material simulation are a danger to get lost in details in early phases of design, where it is more important to find solutions for general problems than to produce realistic pictures of intermediate stages of design

Concerning the process of developing the shape of buildings, virtual reality presentations are not necessary. Simpler 3D-models as for instance shaded mass models without materials or pure wireframe presentations are often better suited to support the architect in early phases of design.

The question what are the right tools for an architect to produce 3D-models and to render realistic images and animations can not generally be answered. Is it better to work with a CAAD system like ALLPLAN by Nemetschek or speedikon by IEZ or should one use a general system for modelling and rendering like 3D STUDIO MAX or VIZ or Lightscape by AutoDESK or ALIAS by wavefront?

In case of conventional buildings, it is recommended to start 3D-modelling in a CAAD system, to transfer the data to a highly specialised rendering system and do the material and light simulation, the camera definition, the rendering and animation in this system, because the quality of rendering tools in traditional CAAD systems is seldom very high.

2.3. Computer Techniques versus traditional Design Tools in 3D-Modelling

Traditional models of buildings and hand-made drawings are unique works. They can only be reproduced by photography or by photocopy in the case of drawings. Changes can hardly be performed; in worst case a new model has to be built.
Here, computer models offer great advantages, because they can be easily copied and modified as often as you wish.

Film and video production is very complicated with traditional architectural models, you need very special equipment as endoscope and camera and it costs a great effort to finish a whole film which is in fact not very realistic.
As already mentioned, the easy production of animation sequences which simulate architectural walkthroughs or flights is one of the main reasons to build computer models. The designer only has to define cameras and motion paths and a frame rate for the desired animation, the computer system calculates all the needed pictures of the movement due to these parameters.
Of course, a video of high quality needs some more effort and special knowledge in post processing as well as additional technical devices.
However, computer technique is the only one that offers the possibility of real virtual reality; by means of special equipment like a head-mounted display one can imagine to be inside a virtual building.

On the other hand, with computer systems only the visual sense is active. In architecture, this is of course the most important sense, because people perceive rooms, light and shadow, colours and textures by their eyes.
But man has more than the visual sense, and sometimes it can be important to touch and feel a building part, even with materials. It is not possible to do that in VR computer models, today.
Classical models allow the observer to take it into the hand and to touch it, which can be a necessary method in the design process.

Furthermore, in traditional 3D model building, the architect gets a better understanding for the construction, load bearing and the static behaviour

as an essential part of the whole design, because a wooden or steel and glass model has to stand on a surface, to carry loads, not to buckle and to fulfil other constructive requirements as well as the planned building. So the designer can develop a feeling for the constructive situation.

In normal 3D modelling computer systems, physical or constructive aspects are not taken into account. Therefore, it is possible to produce any "nonsense" of geometry which has no load or where one part is on the same place like another and which can not be built in reality.
That might be seen as an advantage for creativity to be able to form everything that is in your mind, as described in [3]: "Computers are fascinating, because they allow us to make our dreams come true. The visions of our minds are flying and hardly to restore. It seems to me that there are no concrete limits in the production of computer renderings. The only restriction is located in our dreams and in the potential of our fantasy."

But it can also be seen as a disadvantage or danger in architectural design, because in this area the designed objects are a unit of shape, function and construction and have to follow physical laws if they should become reality.

3. EXAMPLE: DESIGN OF A SHOPPING CENTRE

This part of the paper describes an example of the design of a shopping centre in the German town Erfurt which was carried out in a diploma work by my student Tobias Ruhland at the department of Civil Engineering and Architecture at the HTW Dresden University of Applied Sciences in March 2000.

First ideas of the design came to paper by hand sketches. Even in the age of computer technology, the idea in a mind of a person finds its natural way out of mind via the hand in form of pictures, drawing sketches and verbal notes.

Second step of finding solutions for the design task were 2D-plans in form of computer drawings to fulfil room programs and to be a basis for 3D-models.

Very early, the student stepped over to simple 3D-models in order to have first room imaginations and to compare possible alternatives of shaping and placing the building parts into the town planning situation (see pictures 3 and 4).

Picture 4: Shopping Centre by T. Ruhland [1}

Picture 5: Interior of the designed shopping centre [1}

Because the final studies work of our students has to be produces in only 3 months, he had strictly to decide what to model in detail. Concerning a complex project like a big shopping centre; you can not model all building parts like complicated beams and girders in the roof area, trees or people: Many of these details were abstracted as boxes and the final form is produced during the rendering process by opacity mapping.

This way, the student could reach high reality in the rendered images despite of low modelling effort.

As a final result of the work, a video was created which shows the design stages and the ideas of the architect concerning the new shopping centre in Erfurt.

4. CONSEQUENCES FOR A/E/C EDUCATION

Computers are the ideal interface between the ideas of an architect and the other partners in the building process, to show and discuss design proposals and alternatives to exchange data and to easily perform changes.

Therefore, CAD and visualisation techniques should be a normal part of A/E/C education in the compulsory courses for architecture students, today. 3D-modelling has to be trained in early computer courses in order to be used in design later on. Special higher level virtual reality courses should be offered to interested architecture students as optional subjects.

Furthermore, new trends in computer applications have to be considered in A/E/C education such as Internet-based technologies in VR to support concurrent engineering. The "virtual design studio" is the future form of train together in design classes including forms of distance learning. It offers the opportunities to design classes to work on individual projects but exchange ideas and methods during the design process by using computer technology.

Due to the advantages described in this paper, computer-based 3-dimensional design and visualisation will more and more dominate traditional manual drawing and modelling techniques in architecture offices without totally replacing them. Both methods have their own advantages and limits and shall exist side by side and be used according to the special needs in each case.

REFERENCES

[1] Ruhland, Tobias: *Sense of Architectural Visualisation. Remarks to the Diploma Work „Shopping Centre at the Hirschgarten in Erfurt".* HTW Dresden – University of Applied Sciences. March, 2000.

[2] Kunze, Undine: *Application of Visualisation Tools in Architectural and Engineering Design. In: Computing in Civil Engineering.* Proceedings of the Forth Congress. Philadelphia, 1997.

[3] Seichmann, Michael: *Super-Realism in Architectural renderings. In: Virtual Architectural Models.* Köln, Taschen, 1999.

Rapid prototyping for architectural models

Thomas Kvan
Department of Architecture, University of Hong Kong, SAR, People's Republic of China

Ian Gibson & Ling Wai Ming
Department of Mechanical Engineering, The University of Hong Kong, SAR, People's Republic of China

ABSTRACT: Rapid prototyping (RP) technology has developed as a result of the requirements of manufacturing industry. There are a number of other application areas where RP has been used to good effect and one of these is architectural modelling. However, such application areas often have different requirements from what is offered by the current technology. This paper describes work carried out by the authors to investigate potential applications for architectural modelling, as well as an attempt to explore the limits of the technology. It will go on to discuss how the technology may be developed to better serve the requirements of architects.

1. ARCHITECTURAL REQUIREMENTS

Much has been written about the role of drawing in architectural design [Robbins 1994, Schön 1992, Goldschmidt] but the role of the model as a design tool has not been extensively reviewed. Likewise, the demands of architectural design on rapid prototyping have been little explored [Ryder 1999]. An examination of the role of models in design is an appropriate starting point for a discussion of the role of rapid prototyping as a model making technique. This review does not pretend to be an exhaustive review of models in design but presents a survey of the current situation and identifies areas in which developments need to take place.

1.1 Models in architectural design.

Models are used in architectural design for several purposes [Ratensky 1983]. Early in a design cycle, sketch or study models will be created to examine particular aspects of a design idea. Such models are often assembled rapidly and crudely for it is the immediacy of the feedback, which is sought. At later stages in a design cycle, more carefully assembled detailed models may be created to present ideas to colleagues, clients or decision-making bodies. Such presentation models are often highly articulated as they strive for verisimilitude.

In both these conditions, it is common that the models are used to depict form in order to communicate design intent [Eissen 1990]. Often the physical models will be used to supplement drawn images [Ratensky 1983]. It is often noted that models reveal aspects of a design more legibly than drawings to both laymen and those trained to read the media [Koepke 1988]. In this role, models are used at all scales, ranging from town planning to explanation of particular building components [Eissen 1990]. In particular, complex mass-void relationships or spatial sequences are more easily communicated in models. From this we can identify that models play a role as a communication tool.

In addition to representing form, models can be used to examine processes as well. Models can be used to assist in testing constructibility {Burry 2001] or integration of services into complex or tight structural spaces. Models can be disassembled to reveal the components of a building or the spaces within, or they can be abstractions that reveal to the viewer particular properties not hidden when the real object or complete model is viewed [Hohauser 1984].

In a related pedagogical application, models are used to develop spatial thinking [Janke 1968]. Certain geometries can be examined in model form more easily than other means. By assembling and disassembling forms and by representing ideas that are too complex to imagine, models help the designer to understand the spatial implications of their drawn two-dimensional decisions.

As the science of architecture has evolved, so too models are used to examine particular behavioural aspects of a design, such as its structure, lighting, acoustics or ventilation [Cowan 1968]. It is feasible to test models to failure and learn from these the likely behaviour of the final completed structure. As such, models are a safer and more cost-effective means of examining possible design solutions. From testing models, experimental results can be obtained that validate or extend theoretical understanding of structures.

Finally, models can be appreciated in their own right [Hubert 1981] and as such plays an important role in the process of understanding a design even after a building is completed. Where buildings are not (or cannot be) built, models reveal uniquely worlds to appreciate [Pommer 1981].

1.2 Limitations

Models are always representations of the completed object, for if they are not, the model maker has built the final construction. Architectural models, however, are typically not simply faithful replications of built works; they are always interpretative. Even when full scale, models represent only part of the final design, either by omitting structural properties, finishes or elements. The level of detail, articulation or accuracy in finish is limited by the skill of the model maker and the time available. Sketch models typically sacrifice accuracy for the sake of speed. As these models are used for decision making in the design process, it is incumbent on the design team to recognise the compromises made in the representation of the design idea and to take these into consideration as subsequent design decisions are made.

If models are intended for experimentation and verification, it is essential that relevant properties of the model be considered to ensure similitude in these properties. The model maker must consider the similarity of behaviour with the completed object and the translatability of the results from testing a model [Cowan 1968]. On the other hand, since models are not exactly the same as the finished built object, either by virtue of being smaller, cheaper, lighter or otherwise different,

models can allow the designer to try designs that will fail in the real situation.

1.3 Representation

Models are an abstraction in representation. As such, the choice of materials is important, for these will influence the interpretation of the viewer. One of the more important constraints of model making is therefore the difficulty in achieving forms, texture, colour or schedule with the materials and model making techniques chosen.

Some model makers or designers will use real materials in their models, insisting that the model is the real object. There are, of course, obvious problems of literally translating materials from a full-scale object to one at a reduced scale. The choice of materials will often depend upon workability, ability to work at the appropriate size, representational properties or experimental behaviour as well as safety in working. Longevity of the model is also an issue - a model intended for international exhibition and archiving will be made of a different palette than one intended to quickly explore a possible form.

More often, substitutions of materials are made to achieve representational likeness. As has been identified above, we can distinguish three types of model: preliminary, experimental and final. Preliminary are sketch models made quickly to approximate the shape or volume. Such preliminary models may be made from any materials at hand. Preliminary models may be refined into preliminary scale models, which exhibit greater control over the dimensions. Experimental models are made to approximate particular properties of the intended design, the particular property selected depending upon the aspect of the design to be tested or experimented.

Numerous texts present model-making techniques [Hohauser 1984, Ratensky 1983, Pattinson 1982, Koepke 1988 among others]. In these, discussions can be found of the wide variety of tools and techniques to manipulate materials to aid in representing design ideas. Here we find instructions on sheet materials that can be cut, bent, etched, formed or milled. Detailed elements can be made using rubber (latex, synthetic, or silicon) or plaster casts for cold moulding or using more traditional metal casting techniques. It is noted that finishes are often applied, with such application taking up to one third of the time required to assemble the model [Hohauser 1984]. From this we note that a well-established body of knowledge exists for achieving representational success. All of these note, however, that model making is a time consuming and demanding process which makes it difficult to use as a design aid. The designer often cannot afford the time or have the skill to make any but the most simple preliminary models [Janke 1968]. The compromises made in sketch design models are such that they are of limited use. It is here that RP promises to be of great benefit in design.

1.4 Digital models

With the development of computer-aided design systems, it has become possible to supplement or replace analogue models with digital models. Physical models can be come a by-product of the design process rather than a laborious activity for its own end. A digital representation of a architect's building design can be turned into a physical model for a final presentation from the same data set as used to produce an animation on screen.

Digital representations offer benefits beyond those of physical models. It has been noted that physical models are most often used as tools to communicate the visual appearance of a design concept. As has been noted, a distinction can be drawn between such visualisation models and a digital building model [Day 1997]. A visualisation model is a representation of what it might look like, a building model is the description of the building, its components and its assembly. As such, data associated with a digital model may be used to direct the model making itself, allowing models to be produced in which, for example, the colour of a component represents its structural state.

RP in model making offers an unusual opportunity as it melds the digital with the analogue. Digital representations can be manipulated in ways that physical models cannot. For example, a designer of an auditorium executing a Boolean operator on two spaces will find it is easier in digital models than in physical. Likewise, digital models have lead to the use of algorithmic design in the generation and exploration of design solutions [Duarte and Simmondetti 1997]. Linked to RP, this offers the opportunity for algorithmic model making.

2. RAPID PROTOTYPING

Rapid Prototyping (RP) is represented by a range of technologies that are capable of taking virtual models created in CAD systems and fabricating them in a physical form. This process is done automatically, generally regardless of the model geometry, and without the need for special tooling or fixtures. Complex, 3-dimensional contours are quantized in the form of stacks of 2-dimensional layers of finite thickness. If these layers are very thin, then the parts made will be close enough to the final desired model to suit a range of applications.

The development of RP follows the increasing use of computers for design. As forms become more complex, so it becomes more likely they will be designed on computers. These complex forms, however, make it more difficult to visualise, test, manufacture, and generally develop products using conventional technology. The aim of RP is therefore to reduce the time and skill required to perform this process and to provide a means for integrating the use of the design data through a variety of stages in the product's development.

Burns [Burns 1993] categorises RP technology according to basic process used. Extending this classification to include the most recently commercialised RP technologies leads to the following: -

Selective photocuring: Commonly known as Stereolithography (SLA, commercialized by 3D Systems [Hull 1988]), this process has numerous variations. They use a special kind of resin, called a photopolymer, which has the property of turning solid under the influence of light of a certain colour. A scanned laser or a masked lamp delivers light to selected regions on a surface of the liquid to turn it solid in the shape of a single cross section of the desired object. This is repeated for successive layers to form the whole object.

Selective sintering: A laser is scanned across a thin layer of thermoplastic powder to selectively melt and fuse the powder particles to form a single cross section of the object. This is repeated to form successive layers of the whole object. This process is most widely known in the form of

Selective Laser Sintering (SLS) from DTM Corporation [DTM 1995].

Deposition: With these processes material is fed through a mechanism to form the object. They can be subdivided into three distinct types: -

- *Continuous:* A thermoplastic material is melted and fed through a nozzle, which is moved on a robotic arm to lay down the molten material in desired locations. This is commercialised as the Fused Deposition Modelling (FDM) process from Stratasys [Stratasys 1991].

- *Drop on powder:* An adhesive liquid is deposited in a controlled pattern over a thin layer of powder, selectively joining the powder particles to form a single cross section of the object. The process is repeated to form successive layers of the object. A number of systems have been licensed from the inventors at MIT known as the 3D Printing (3DP) process [Michaels et al. 1993].

- *Drop on drop:* Melted thermoplastic material is deposited in a controlled pattern through a droplet deposition mechanism. A number of systems have been commercialised using this method. Examples include the Sanders Modelmaker [CAD/CAM Publishing 1994] and the Genisys machine from Stratasys [Stratasys 1999].

Adhesion of cut sheets: This is a hybrid subtractive/additive process. The contour of each cross section of the desired object is cut, and the cut patterns are stacked and bonded to form the object. Alternatively, the stacking and bonding may take place first, with cutting afterwards. The original and most widely known commercial system using this technique is Laminated Object Manufacturing (LOM from Helisys), but there are a number of other systems now available.

It is clear that, with so many different methods available, no single system dominates all others. Choice of system depends on a number of factors related to application and dependent on how the machines are constructed, how they are controlled, and how the final parts appear to the user.

2.1 System construction

A number of systems use laser technology. Although the lasers are not particularly powerful, they are expensive and difficult to maintain which restricts them to high-cost, high-end machines. Lasers can take advantage of galvanometric scanning mirrors to effect high speed building of individual layers. Laser mirrors need only move short distances and can therefore scan very quickly. Deposition heads on the other hand, like those used on the FDM machines, have much higher momentum associated with them and therefore require much more energy to move around. Also, in order to get the best results, FDM must extrude material in a continuous filament and therefore fabricates using a complex fill pattern that also slows down the build. However, even droplet deposition methods like 3DP, which can use a raster scanning approach, will still be slower to scan a single line than laser based systems because of the mass of the print head. However, the ZCorp machine [Zcorp 1999], which uses this approach, overcomes this problem by having a multiple nozzle printhead that can deposit many lines in a single pass. The LOM system uses a laser and a travelling mirror (rather than scanning mirror) approach. This is because LOM only requires the laser to separate the part from the waste material at the perimeter. The traverse time of the laser is therefore significantly reduced and is suitable for larger, bulkier models.

Machines are often described as office, laboratory or shop floor environment. This is mainly due to the materials being used and how they are processed within the machine. Some materials require fume extraction that makes them generally unsuitable for placing in an office. Other processes require a carefully controlled environment to ensure the process is not disturbed or the materials do not degrade. Examples of this include SLA, which requires low vibration and humidity and SLS, which requires a continuous nitrogen gas supply.

The more complex systems are, the more expensive they tend to be for low volume production. This often means they are also more difficult to use, particularly if they require fine-tuning in order to get optimum quality from the models. Training and technical support are therefore important aspects surrounding choice of technology and this is also reflected in the price of the machines.

2.2 System control

System control relates to part accuracy. This not only includes positioning of the scanning mechanism, but also penetration depth and feature width. Some systems can go to very fine layer thickness, which would limit the stair-step effect. Others have very fine deposition heads or beam widths, which results in fine feature in the build plane and wall thickness.

In addition to this, parts can shrink and warp, particularly for processes that invoke a phase change in the material (which is nearly all of them). This can be minimised with precise control of the process parameters associated with the material. With SLA for instance, stresses can be minimised by ensuring even distribution of the laser across the resin surface. SLS and FDM control the material temperatures during build to avoid excessive material shrinkage during cooling.

Whilst control of the mechanism is likely to be just as precise for the powder processes as the liquid processes, the granularity of the material and the random distribution of the particles is always going to limit the final result. Powder particles can only be so fine before they become extremely difficult to handle and so layer thickness and feature definition of liquid based systems like SLA and Sanders are likely to be the better.

2.3 Part appearance

A number of researchers have discussed the development of colour RP technology [Gibson & Ling 1999]. To date, there are only two systems that commercially provide multi-colour capability (although the Sanders Modelmaker has been demonstrated as having 4-colour capability, it has yet to be commercialised). The FDM system has the most versatile capacity to build bi-colour models from a range of coloured ABS plastics. SLA can make bi-colour parts, albeit limited to a red and yellowish/transparent resin developed by Zeneca [Medical Modelling Corp. 1999]. SLA

resin has this particular benefit of being transparent, which is unique amongst RP materials.

All parts made using RP technology exhibit a characteristic 'stair-step' effect, which results from the quantized, layer based approach. Whilst efforts are being made to reduce this (Sanders and SLA machines can now go better than 0.05mm layer thickness), the foreseeable future shows that parts will generally require manual finishing after processing within the RP machine. The powder-based machines also exhibit a characteristic granular texture.

All RP parts are weaker than if they were made in the same material using a conventional manufacturing process. Some processes result in particularly weak parts, either because the selected material or the bonding method used is weak. The thermal processes (SLS and FDM) result in the stronger parts, suitable for functional applications.

3. EXAMPLE MODELS MADE USING RP

A number of models have been constructed using RP methods. Figure 1 shows parts that formed a space frame construction. These models, whilst geometrically simple, would be quite difficult to construct because of the required accuracy. Originally, the model was attempted using the FDM process. The small diameter of the vertical poles made this a very weak structure. When the supports were removed, the model was unable to stand its own weight and collapsed. After building using the SLS process, the model was strong enough to handle for design analysis.

SLS Duraform nylon was used to create the next model (Rotunda, figure 2). This was also an assembly, specifically made to allow the designer

Figure 1. This space frame model was fabricated in 3 parts using SLS Duraform material and finally assembled to form the complete model.

Figure 2. Various views of the rotunda model

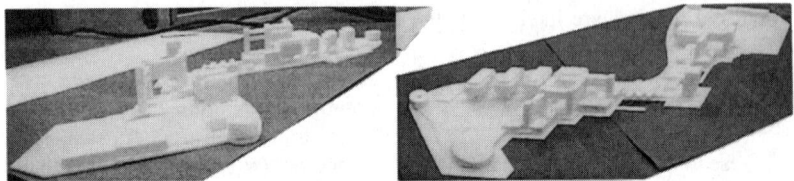

Figure 3. Component model of a landscape, consisting of 6 separate sections, jigsawed together

to show the relationship between different room locations. The building consists of load bearing walls, unlike column and strut in Figure 1. The component method also made it more suitable for the SLS process since each part was of manageable dimensions and could fit into the build envelope easier. Shell structures are best suited to most RP methods since this is their primary application area (injection moulded parts). Had this part been more solid in nature, it would have been appropriate to build it as a hollow component. This would have limited distortion during the build process as well as make it quicker to build.

RP is generally not well suited to building large landscape models. This is because most machines have a comparatively small build envelope of approximately 250mm cubed. Larger machines are available, but the time and cost of building may then become prohibitive. Figure 3 shows a

development model for a possible new university campus made using SLS Duraform. This again was a component-based model, specifically divided to allow the parts to fit into the machine. The dividing lines were chosen to minimise the effect on the final model.

4. DISCUSSION

The examples above show that architectural models can be fabricated using RP. However, there are a number of points arising from our experience that illustrate the difficulties faced by architects wishing to make use of RP: -

- *Appearance*: the models are homogeneous in appearance. It is not possible to highlight specific features in the model using changes in material, colour or texture, something commonly achieved in other model making techniques employed, thus limiting significantly the communicative role of RP models alone.

- *Finish*: RP models are typically rough finished and require further treatment to obtain an appearance of crafted work. Even those architectural models not presented as works of art are expected to have a reasonable level of finish.

- *Time to build*: although called 'Rapid' Prototyping, the process takes a number of hours to complete the models (the landscape model took around 5 days to complete all the sections). Applications requiring quick mock-up designs can not make use of RP, although newer 'concept modellers' may change this.

- *Skill factor*: while architects use computers extensively in design, there is little experience in controlling RP devices to obtain the subtle differences in emphasis employed in other forms of model making to communicate desired design ideas.

An additional factor is choice of technology. Often more than one RP technology is used to construct different components of a model. Sometimes this is through necessity, due to the required strength or the level of detail. However, choice may also be for aesthetic reasons. Parts made using SLS exhibited good detail and strength, but FDM or SLA parts may be preferable where a less granular finish is required. If the landscape model were placed on a more hilly terrain, LOM would have been preferred to produce the terrain because of its speed of build, with the buildings still made from SLS plastic. The ability to choose different technologies for components presupposes each machine is available, an important economic factor.

As noted in section 1, design models are used to communicate design intent or identify design implications. These models are more than simply faithful scale representations of finished objects. The omission of depiction or simplification of representation has to be made carefully. The homogeneity of RP models force simplification but the designer is, as yet, unable to control this as well is in traditional model making materials. There are many instances in which architects wish to use models with few highlights or which have few surface characteristics in order to emphasise form. For example, early sketch models are often made of white card. Similarly, a building being presented in an urban context may be depicted by a carefully detailed and multi-coloured model

while the surrounding masses are monolithically depicted in a homogeneous material. Thus, the homogeneity of RP output may be detrimental or beneficial.

There may be other reasons why RP has yet to make much of a presence in the world of architectural design. Expense is one possible reason; RP systems are still expensive and largely inaccessible to the practitioner. Overall cycle time, after incorporating the computer design time, is another. As has been identified, RP offers greatest potential in producing sketch models. At this time in the design cycle, physical representations are of use to examine geometrical and other formal properties. As has been demonstrated by some practitioners [Novitski 1994], forms can be examined in digital models that are difficult to achieve in physical models made by traditional techniques. RP can overcome this gap and link the digital and physical once again. This is perhaps the greatest promise for such systems in the realm of architectural design.

4.1. Colour in RP

It has already been mentioned that architectural models are often deliberately made from a range of materials in order to emphasise specific features within the model. Also, as the designs are finalised, more realistic colouring is necessary to allow the design to be fully evaluated within the proper context of its surroundings. This makes it sensible to select a RP system that has a multiple material capability or by selecting more than one RP system for large (and costly) models. It may also be appropriate to use conventional modelling techniques to create large, simple geometry landscapes and combine this with RP for creating the more complex and detailed (e.g. building) components.

It is clear from the models shown that one method for enhancing the models is to create them in colour. FDM in particular would be useful in this sense since it has a range of coloured ABS plastic materials and indeed can make bi-colour models. However, a greater colouring capacity would also be extremely useful. Ling and Gibson describe in their paper [Ling & Gibson 1999] a method for incorporating full colour into the SLS RP process. This is done by placing a colour printer mechanism inside the SLS machine. Initial tests indicate that this is capable of multiple coloured component manufacture but it is uncertain on how detailed the colouring would be. Low resolution colouring would make it possible to create regional colour variations, but if the colouring resolution is sufficiently high it may be possible to create detailed featuring within the part. Parts may eventually be made with the windows and decorative feature incorporated directly into the building models. The technology requires substantially more development, however, since architects are particular about the purity of the colours used and the degree to which the colour can be controlled.

5. CONCLUSIONS

This paper has discussed the different types of RP system that are available commercially in terms of their general advantages and disadvantages. However, when considering architectural applications, it is clear that there are further complications since the requirements are different from the conventional application fields of RP. In particular, speed of build, build envelope, and the general homogeneous appearance of RP models are unacceptable for many architectural models and modellers.

However, it can also be noted that architects are unwilling to use new technologies for the building of models unless benefits can be found over old technologies. Some examples have been shown and it has been discussed that for certain effects, RP modelling can be particularly useful. The most exciting developments appear to lie in exploring construction manufacturing techniques. It is also evident that RP technology can be used more creatively than usual by combining models made with different materials and processes, or in conjunction with conventional techniques to overcome some of the shortcomings of the technology.

It is also clear that the requirements of the more creative designers have been somewhat overlooked, with efforts in applying RP limited to replacing traditional model roles. The development of new CAD interfaces, where designers can interact in a more intuitive manner, and functionally graded RP machines, where models can be created with non-homogeneous characteristics (like coloured parts) go some way

to making this technology more accessible to the masses. As we experiment with RP devices in teaching contexts and learn to control the devices, we ·expect to find opportunities in RP representational techniques that we can exploit to new design or communicative effect.

REFERENCES

Burns M. (1993), Automated Fabrication-improving productivity in manufacturing, P T R Prentice Hall, pp. 14-84

Burry, M. (2001), Rapid prototyping, cad/cam and human factors, Automation in Construction, forthcoming

CAD/CAM Publishing (1994). IBM demonstrate new rapid prototyping system, Rapid Prototyping Report 4(3)

Cowan H. J., Gero J. S., Ding G. D. (1968), Models in Architecture, Elsevier, Amsterdam

Day A. (1997), Digital Building, Butterworth-Heinemann, Oxford

DTM Corporation (1995), Product Brochure, Sinterstation 2000

Duarte J. P., Simondetti A. (1997), Basic Grammars and Rapid Prototyping, EG-SEA-AI Workshop '97, Applications of Artificial Intelligence in Structural Engineering

Eissen K. (1990), Presenting architectural designs, Architecture Design and Technology Press, London.

Gibson I., Ling W.M. (1999), Incorporating Colour into Rapid Prototype Models, Proceedings of the 8th European Conference on Rapid Prototyping and Manufacturing.

Goldschmidt, G. (1994) On visual design thinking: The vis kids of architecture. Design Studies, 15(2), pp158-74

Graves G., and Graves D. W. (1997), Box city : an interdisciplinary experience in community planning, Center for Understanding the Built Environment, Prairie Village, Kansas.

Hohauser S. (1984), Architectural and interior models, Van Nostrand Reinhold, New York.

Hubert C. (1981), The ruins of representation, in K. Frampton & S. Kolbowski (eds.), Ideas as Models, Rizzoli, New York.

Hull C. (1988), (3D System Inc), Stereolithography: Plastic Prototypes from CAD Data Without Tooling, in Modern Casting v78 n8, pp.38

Janke R. (1968), Architectural models, translated by James Palmes Thames & Hudson, London.

Koepke M. L. (1988), Model graphics: building and using study models, Van Nostrand Reinhold, New York.

Ling W.M., & Gibson I. (1999), Possibility of Colouring SLS Prototype using the Ink-Jet Method, The Future of Rapid Prototyping, published in the Journal of Rapid Prototyping vol.5(4) pp152-3.

Novitski B. J. (1994), Freedom of form, Architecture.

Medical Modelling Corp. (1999), http://www.medicalmodeling.com/clearview.html

Michaels S., Sachs E.M. & Cima M.J. (1993), Metal parts generation by three dimensional printing, Proceedings of the Fourth International Conference on Rapid Prototyping.

Pattinson G. D. (1982), A guide to professional architectural and industrial scale model building, Prentice-Hall, Englewood Cliffs, N.J.

Pommer R. (1981), The idea of "Ideas as Model", in K. Frampton & S. Kolbowski (eds.), Ideas as Model, Rizzoli, New York.

Ratensky A. (1983), Drawing and modelmaking, Whitney Library of Design, New York.

Robbins E. (1994), Why architects draw, MIT Press, Cambridge, Mass.

Ryder G., et. al. (1999), Rapid Design and Manufacture Tools in Architecture, Automation in Construction, in print.

Schön D. A. & Wiggins G. (1992), Kinds of seeing and their functions in designing, Design Studies, 15:2, pp158-174

Stratasys Inc. (1991), Fast, Precise, Safe Prototypes with FDM, in Solid Freeform Fabrication Symposium Proceedings, the University of Texas at Austin, Austin, Texas, pp115-122.

Stratasys (1999), http://www.stratasys.com/

ZCorp (1999) http://www.ZCorp.com/

Modelling and visualization in construction

Product and Process Modelling in Building and Construction, Gonçalves, Steiger-Garção & Scherer (eds)
© *2000 Balkema, Rotterdam, ISBN 90 5809 179 1*

Integrated construction scheduling, cost forecasting and site layout modelling in virtual reality

J. Mahachi, L. Chege & R. Gajjar
Division of Building and Construction Technology, CSIR, Pretoria, South Africa

ABSTRACT: Construction sites are markedly inefficient in terms of where and how materials are stored and handled, the planning of a site layout, communicating the construction process to the interested parties, and adopting efficient construction sequencing or assembly of building components. Currently, construction managers, clients, planners and engineers must conceptually visualize a site-layout plan, how materials and machinery have to be handled, and how the construction process will unfold. If unplanned, the ratio of material handling to production can become high, with obstacles hindering the flow of materials and substantial double handling of materials occurring. Such inefficiencies have a major impact on the overall productivity and cost-effectiveness of a construction project. This paper reports on the development of integrated construction management tools in a virtual reality environment. A project management scheduler is integrated with cost-forecasting techniques, site-layout techniques using genetic algorithms in an interactive environment, where the user (construction planner, clients etc) can visualize and analyse the construction processes.

1 INTRODUCTION

Despite the rapid development of Information Technology (IT) and its applications in various industries, several processes in the construction industry are still based on experience and ad hoc intuitive decisions. It is commonly acknowledged that the construction industry is adopting advanced information and communication technologies, with a view to improving competitiveness, performance and delivery, at a far slower rate than other comparable industries are. This is due to fragmentation of the industry and the non-repetitive nature of its operations compared with other industries, such as manufacturing.

Studies and interactions with the construction industry of South Africa have shown that a substantial financial loss (in terms of materials, productivity, etc.) occurs due to the lack of integrated and consistent systems of planning and optimising site construction activities (Mahachi *et al*, 2000). In recent years, the cost of construction in South Africa has escalated to a level where commercial buildings give only a 9% return on investment, which does not make the construction attractive from the investors' point of view. In order to survive, the construction industry needs to improve its performance and competitiveness by cutting life-cycle costs by as much as 30 to 35%. Clearly, such radical improvement cannot be achieved overnight or without making fundamental changes to the way the industry works and

thinks. Mahachi *et al*, (2000) identified major problem areas in which IT and systems integration can make a significant contribution to improving construction site processes and minimising costly errors and repetitions, as being:

- scheduling and sequencing of processes,
- time-space scheduling,
- cost estimation and forecasting, and
- the organisation and layout of construction facilities.

The construction industry now recognises the need for adequate construction planning as a means to improve its products and services. However, at present it has no efficient methodological mechanism, such as the manufacturing industry has, largely due to the complexity of the environment where, for instance, materials do not flow through a series of well-defined workstations. Future innovation and technological progress will therefore be facilitated through integrated information technology solutions. In particular an integrated model using Virtual Reality (VR) promises to solve many of the problems faced in the construction industry. With the use of VR, construction site layout, materials movement and other processes can be modelled before construction starts on site, or as the construction process unfolds. This paper describes integrated technologies and software that can be used in the construction industry. It is anticipated that the tools developed will help to shape and influence the transition of the industry into the age of the information society, and will play a major role in its success.

2 CONSTRUCTION SCHEDULING

Construction planning has in the past been performed using a number of scheduling techniques. These techniques were primarily based on scheduling of events and activities on arrow networks, precedence diagrams, or "line of balance" methods. Arrow networks made use of several arithmetical computation methods, namely the Critical Path Method (CPM), Ladder Construction and the Programme Evaluation and Review Technique (PERT). Precedence diagramming involved using activity-on-node diagrams illustrating the sequence and precedence of activities. Line of balance methods have their prime application in situations where there are predominantly repetitive elements as typically found in multi-storey construction.

Each planning technique and activity scheduling method has certain advantages and disadvantages. Definitions and illustrations of these planning techniques have been comprehensively documented in other publications, hence detailed descriptions are omitted from this text.

The above concepts have been briefly touched on in order to introduce a new paradigm where entities that are considered during the planning and scheduling are modelled in an object-oriented environment and simulated in virtual reality. A new approach toward modelling the construction process is using intelligent objects modelled in an object-oriented environment. The advantage of such a technique is that objects such as cranes and other plants have generic attributes that are common to all within their class. These objects can be built with varying degrees of intelligence, which allows them to interact with each other and the temporary facilities on site. Once all objects present on a construction site are modelled, these are stored in a knowledge database (object warehouse), from which they can be retrieved when required in the construction schedule.

In this work, a new approach to construction scheduling was developed by integrating a project scheduler with virtual reality. Primavera P3 was used as the project scheduler as previous studies (Mahachi et al, 2000) had indicated that the software is widely used in South Africa. The objective of such integration was to be able to simulate the construction schedule in real-time (i.e. using VR), allowing the project planner to see the progress of the construction as it would occur on site.

Information regarding time taken to complete activities, the resources required, and building components (columns, beams, slabs, etc.) are extracted from the scheduler and these are built up in virtual reality using the intelligent objects stored in the knowledge database.

Advantages of this technique are that changes imposed on the scheduler can be seen in virtual reality. Alternatively, if the project planner makes changes in virtual reality, these will be related back to the construction schedule and delays or timesaving can be picked up in the schedule. This interaction makes it possible to note the effect that changes to the schedule have on related activities and on the "actual" construction process, typically creating the scenario – "this is what it will look like if...".

The importance of the previously mentioned scheduling techniques should not be understated. Rather, by using project schedulers interfaced with virtual reality and placing intelligent objects in this environment, the project planner now has a powerful tool where his/her schedule can be "visualized" interactively in real-time by all interested parties. Progress monitoring is an added benefit of this integrated environment where members of a project team on and off site can "see" where the construction should be and what it should look like.

3 CONSTRUCTION SITE LAYOUT MODELLING

Construction practitioners typically sketch the layout of temporary facilities at different locations in time, on the site-arrangement blueprint. This sketch is rarely updated as construction progresses and, since many changes are likely to occur, the drawing becomes less valuable. Construction site layout tools that handle 2-D or 3-D arrangements are available. The use of computer graphics has received considerable attention over the last decade. 3D graphics, for example, have been used for construction planning, allowing the planner or owner to simulate the project at various stages of the execution. However, tools that handle spatial arrangements that change over time, in "real time" are not available, due to the complexity of the problem. Existing computerised models do not allow the user to alter a model when the results are different from what they expect. Part of this work involved the development of algorithms for the generation of optimal layouts of site construction. The algorithms developed were a hybrid of Genetics Algorithms (GAs) and the heuristics construction improvement technique, CRAFT (Computerised Relative Allocation of Facilities Technique).

Genetic Algorithms are search algorithms guided by the principles of natural evolution. They work with a population of strings which is evolved over time. A genetic string represents a candidate solution to the optimisation problem. In the algorithm that was developed, a float value chromosome representation was used, with each chromosome representing the coordinates of the centroids of the temporary facilities. The objective function was to minimise the transportation costs between temporary and permanent facilities. The fitness function evaluates the quality of a solution. Highly fit individuals have a better chance to appear in the next generation during the selection process. A penalty-based transformation method was used to convert the constrained problem into an unconstrained problem by penalising unfeasible solutions. New solutions were created by using genetic operators such as crossover and mutation. The crossover operator combines two chromosomes at a time, and generates offspring by combining both chromosomes' features.

When 80% of the chromosomes have converged, the iteration process is stopped. At this point the Genetic Algorithm outputs the "optimized" centroid positions of the temporary facilities.

CRAFT is a well-established improvement procedure which can be used to improve a layout obtained from another method, or improve an existing layout. From the initial layout generated by the genetic algorithm, CRAFT then improves on this layout by interchanging the locations of the facilities, using 2-way and 3-way interchanges. The process continues until there is no benefit in interchanging the facilities.

CRAFT and the genetic algorithm are then integrated in a virtual reality environment, as shown in Figure 1. The user, for example the construction planner, can modify the output from the GA and his/her layout is then improved upon by CRAFT, and the loop continues in an interactive manner until the user's constraints have been satisfied, and a cost reduction accomplished.

4 CONSTRUCTION COST FORECASTING

In order for the construction industry to become competitive, there is a need for it to have the ability to forecast construction costs accurately, and hence provide better decision making for the clients. Dawood and Molson (1997) define a cost forecast as the process of predicting future expenditure of project resources at early stages of project design development and before construction commences. This is an important aspect because the clients' decision on whether to embark on a project depends largely on budgetary requirements. Clients are becoming far more aware of where money is being spent and like to be kept informed of changes in the cost of the project at all stages of the procurement (Bates and Dawood, 1997). Part of this work involves developing techniques for cost forecasting, as well as the best way to report the outcomes in a way that can be easily understood by any interested party, in particular the client.

Several forecasting techniques have been used in other industries. However, because of the unpredictability of the construction industry, little work has been reported on cost forecasting (Dawood and Molson, 1997). In our work, it was decided to use a technique that integrates both the qualitative, subjective parameters and the quantitative techniques in a more holistic manner. The model considers all the historical data in order to perform the quantitative analysis. Using a technique similar to the one suggested by Dawood and Molson (1997), the cost forecasts obtained for each activity are displayed in an interactive manner, in a virtual reality environment. From the output the construction manager will therefore be in a position to control the costs by comparing the cost forecast against actual costs.

5 INTEGRATED DYNAMIC MODEL

The proposed integrated model, which is still under development is shown in Figure 1. In this model, the interactive environment is provided in virtual reality. 3D CAD models, which consist of 3D surface and solid objects, are transformed into a library of intelligent objects. A C++ "communication module" was developed to manage the interface between the Primavera project management scheduler and VR, using dynamic data exchange/object linking and embedding (DDE/OLE), existing within the engine of Primavera (RA).

For a specific date, the OLE scans the Primavera schedule for any tasks that have been completed or are in progress for that date. Via another link to VR, the program instructs the graphical package as to which objects are to be displayed, including information on their status. Each object is characterized by its location, orientation and dimensions. A library of "intelligent" objects comprising slabs, columns, walls, etc., and their parametric representations is therefore part of the integrated system, and these objects are used for the creation of any activity image with the desired composition, dimensions and orientation.

Current developments are investigating the interface of the project scheduler with the site layout. Information about the resources and current status of the project is extracted from the scheduler and used to develop the initial site layout plan, using a "construction" optimisation technique (a GA). The initial layout thus produced by the GA is displayed in "real time", in the VR environment. A construction planner then modifies the layout, in an interactive manner. The output of the planner is then improved upon using CRAFT. Both the site plan and the physical progress of the construction can be viewed by inter-

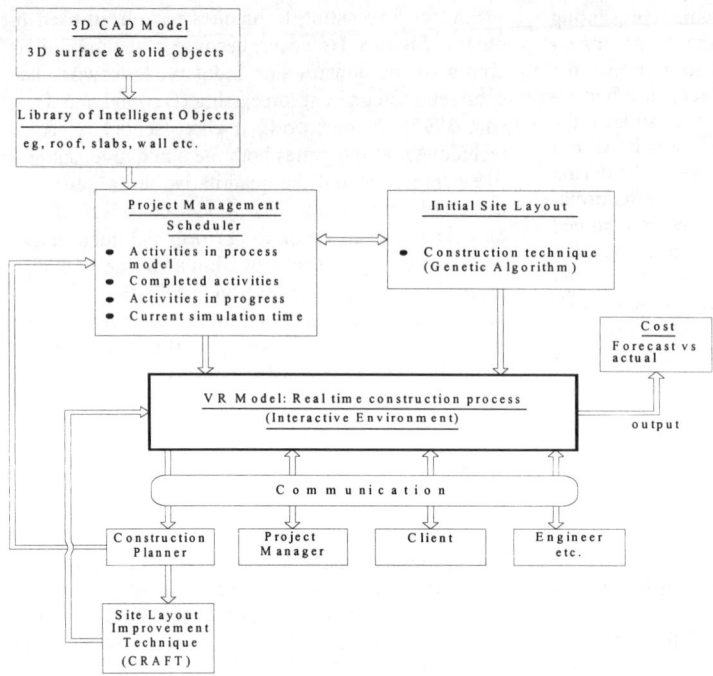

Figure 1. Integrated construction process model

ested parties such as clients, engineers, project managers, etc. A major output from the VR model is the cost. Information about the forecasted costs vs the actual costs can be displayed in real time.

6 DISCUSSIONS AND CONCLUSIONS

The South African construction industry, like most construction industries around the world, will benefit from actively participating in developing new technologies and embarking on the process of adopting them in order to improve its operations and competitiveness to meet the challenges of the 21st century. The conventional and currently used scheduling tools (e.g. GANTT charts and network techniques) are not particularly effective for communication between clients, schedule planners and their users. There is therefore a need to develop integrated, interactive, visualisation tools that enhance viewing of the planned project and site layout, as this will look at various stages during construction. It is anticipated that this project will develop a new approach to modelling construction processes and products and, through the work of the International Alliance for Interoperability (IAI), the tools will have the capability of being linked to any project management software. In terms of future research, it is anticipated that a better representation of the planned against ac-

tual or the deviation between planned and actual performance will be visualized more distinctively in a virtual reality environment.

REFERENCES

Bates, W. & N. Dawood 1997. The effects of IT on construction cost forecasting. *Computing in civil engineering 4th Congress, Philadephia*: 543-550.

Dawood, N. & A. Molson 1997. An integrated approach to cost forecasting and construction planning for the construction industry. *Computing in civil engineering 4th congress, Philadephia*: 535-542.

Mahachi, J. 2000. Construction site layout improvement through a hybrid genetic algorithm. *Paper to be presented at a conference on construction applications of virtual reality: current initiatives and future challenges, University of Teesside, UK.*.

Mahachi, J., Goliger, A.M. & Naidoo, S.R. 2000. Survey of the materials handling and site layout pre-planning practices in the SA construction industry. *Boutek internal report, BK140, Report number BOU/I163, CSIR, Pretoria.*

Product and Process Modelling in Building and Construction, Gonçalves, Steiger-Garção & Scherer (eds)
© 2000 Balkema, Rotterdam, ISBN 90 5809 179 1

Process to Procession: Visualising construction processes with Procession's 3D information visualisation

S. North
University College London, UK

ABSTRACT: The construction industry traditionally uses two dimensional visualisation techniques to analyse project progress. This usually takes the form of an 's-curve', commonly generated by one of three performance measurement systems: earned value analysis, completion analysis or performance trend indices. Analysing project progress in this manner only provides 'single view' analysis of data trends. Deeper interpretation of a project might require studying many individual s-curves. Indicative patterns can go unrecognised in a mass of unfiltered data. This research delivers a conceptual, three-dimensional framework for the interpretation of non-physical construction industry processes. Procession is an information visualisation software tool based on this conceptual framework. It delivers a more comprehensive representation of project progress, as a three-dimensional data surface. Data is either exported from Microsoft Project or from a single project database environment, utilising STEP (Standard for The Exchange of Product model data). Procession is targeted at social housing project clients, providing an 'at-a-glance' indication of project 'health'.

1 INTRODUCTION

The rationale for Procession's three-dimensional framework for construction planning is suggested by previous work on multi-dimensional morphological frameworks, in the field of systems engineering (Hall 1969 and Pohl 1994). Figure 1 shows the conceptual framework, with 'deviation parameters', 'deviation level' and 'tasks' as its three dimensions (North 2000). Deviation level represents units of time or money, which deviate from the flat terrain of the project baseline.

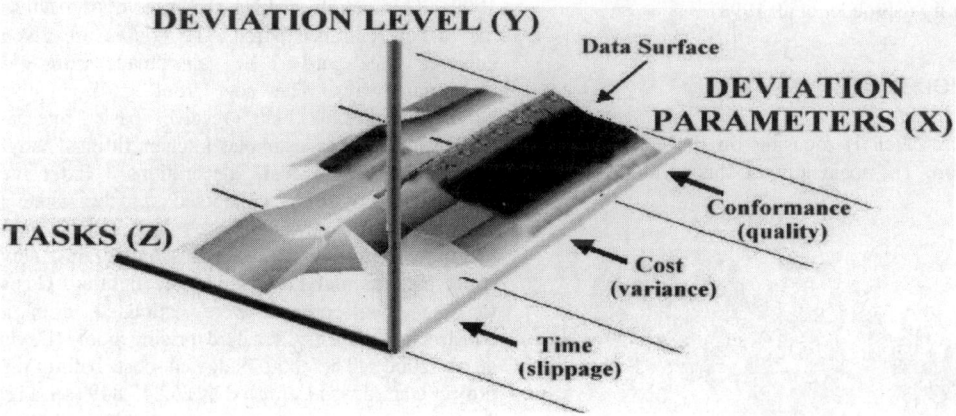

Figure 1. A 3d framework for construction planning, source: North 2000.

The three framework dimensions map to Procession's data surface, as follows:
• The X axis maps to the deviation parameters dimension
• The Y (height) axis maps to the deviation level dimension
• The Z axis maps to the tasks dimension.

The values for X and Z define a two-dimensional plane, with Y providing the height scalar values.

Therefore, a project with no deviations from the project baseline will produce a flat data surface. Procession's data surface is actually a 'carpet plot', generated by a scalar algorithm. The three dimensions are achieved from a two-dimensional set of points, which are warped in the direction of the surface normal. The amount of warping is controlled by the scalar value. The point set is determined by the three-dimensional framework's 'tasks' and 'deviation parameters' dimensions. The scalar (or height values) are provided by the framework's 'deviation level' dimension (see Fig. 1 A 3d framework for construction planning).

Procession is a stand-alone information visualisation application (Card et al. 1999 and Chen 1999) developed for the Microsoft Windows 95/98/2000/NT platform. It is capable of opening data files both from a local computer and from an Internet web server. Procession is an MDI (Multiple Document Interface) application, which is to say that it can have several visualisation windows open simultaneously. Procession was developed in the C++ language, utilising MFCs (Microsoft Foundation Classes).

Procession's three-dimensional graphics functionality is provided by VTK (the Visualization ToolKit), an 'open source' system providing a C++ class library (Shroeder et al 1998).

2 METHODOLOGY

Current research is focusing on the evaluation of Procession. The main aim of this is to determine

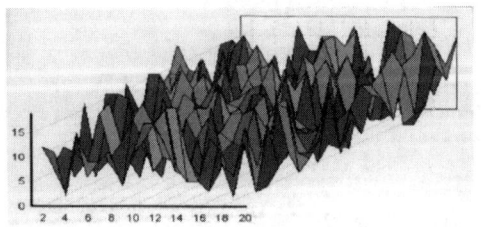

Figure 2. An example of a navigable 3d data surface, source: author.

whether clients perceive Procession's progress reporting information to be more useful than currently utilised approaches. As a secondary aim, it is hoped that the evaluation will prove informative about the suitability of the conceptual three-dimensional framework. In order to evaluate Procession, comparative research will be conducted, making use of interview protocols and rapid prototyping. This evaluation is intended to determine a selected construction client's level of satisfaction with the quality and format of the project progress information provided.

In order that Procession's informational provision improves on the current model, a requirements capture stage is being undertaken. A planning data set has been obtained of a 'live' construction project. The researcher used this data as a skeleton to create simulated projects, with different sequences and outcomes. The purpose of this is to provide a testing ground for Procession, which must be unfamiliar to the construction client (there would be no motivation for information requests). The fictional project scenarios are based on a real nineteen-month project to refurbish flats and houses on an ex-local authority estate, in an inner-city area. The 'live' project objectives were: to provide new and clearly defined street patterns, to enlarge the individual gardens and to remove crime-ridden pedestrian routes. The total estate consists of five hundred and seventeen housing units (individual houses and flats). The 'live' project focuses on two hundred and fifty-two of these units (one hundred and twenty three houses and one hundred and twenty-nine flats).

Work on the houses will be conducted with most of the tenants in situ. A unique aspect of the project is the 'porch turnarounds' on the houses (Smit 1999). This entails making the back of the house into a new entrance porch. The houses are given enlarged back gardens by gains made from old pedestrian ways. The new front gardens have parking spaces. Flats receive basic internal refurbishment. For example: kitchen fittings, vinyl kitchen flooring and decorations. Extensive landscaping is being conducted on the estate's communal areas. This includes turf laying, tree-planting, re-routing of gas mains and drains, new street patterns and improved estate lighting. Costs for the 'live' project were estimated from a construction industry standard pricing book (Davis et al. 2000). The total budgeted cost before the project started was estimated as £6,247,849 (see Fig. 3, 'Live' project summary task budgets before commencement).

In order to generate project scenarios with a high level of realism, it was decided to identify real

myProject baseline budget at summary task level	
SITE_SET_UP	
SCAFFOLD_AND_TEMP_ROOFS_AND_REFURB_EXISTING_ROOFS_ON_SOME_HOUSES_AND_FLAT_BLOCKS	£20,907
'PORCH_TURNAROUNDS'_ON_HOUSES_AND_INTERNAL_REFURBISHMENTS_ON_FLATS_GROUP1	£2,585,299
'PORCH_TURNAROUNDS'_ON_HOUSES_AND_INTERNAL_REFURBISHMENTS_ON_FLATS_GROUP2	£412,588
COMMISION_AND_CLEAR_AFTER_HOUSE_AND_FLAT_WORKS	£532,999
LANDSCAPING_TO_INDIVIDUAL_HOUSE_AND_BLOCK_GARDENS	£2,091
LANDSCAPING_ENTIRE_SITE_AND_COMMUNAL_AREAS_AND_INITIAL_HORTICULTURE	£1,260,290
CLEAN_RESIDENTIAL_AREAS_FOR_SECTIONAL_COMPLETION	£94,535
SERVICES_TO_COMMON_AREAS	£2,509
SITE_HARD_LANDSCAPING_AND_INITIAL_WORK_ON_ROADS_AND_PAVEMENT	£916,083
MAIN_HORTICULTURAL_WORKS	£133,190
FINAL_WORKS_ON_ROADS_AND_PAVEMENTS	£98,316
CLEAR_CLEAN_AND_FINAL_SITE_HAND-OVER	£184,860
	£4,181
	£6,247,849

Figure 3. 'Live' project summary task budgets before commencement, source: author.

'unacceptable' risks within the original budget and schedule. The two most serious risks were then used as a basis for 'problematic' issues, which were then fictionalised into the two scenarios. The identification of unacceptable risks was achieved using the risk analysis software tool Riskman Professional (see Sect. 3, Transforming raw project data into visual structures). When running through the research scenarios, construction clients taking the role of development workers will receive project progress reports at seven quarterly milestone intervals.

Each of the simulated projects has had quarterly milestone files produced, to represent each of the progress reports throughout the project. These files will be stored on the researcher's laptop computer, in the CSV (Comma Separated Value) format. The milestone files compress a project of nineteen months into as many minutes. For this research, it has been decided that 'real time' simulation of a project is not practical. For example, a construction client's need for information concerning the progress of 'on-site' projects is likely to make up a vital, but proportionally small part of their workload. As the requirement for this information may happen at unpredictable times, it does not seem a sensible use of resources to attempt direct observations of the construction client's behaviour on a 'live' project.

The prototypes will be evaluated by several different construction client volunteers.

The interview questions will consider the quality of information provided by the construction clients' current methods of progress reporting, allowing for later comparison with the prototypes. Construction clients will also be asked about their general familiarity with computers. After working with each prototype, construction clients will describe their perception of information quality. One protocol section will relate to functional aspects of the software prototype. Its purpose is to assess not the quality of the information that Procession presents, but the usability of the software. As such, reference will be made to standard guidelines for user-interface design.

In order to verify construction client's *actual* level of informational understanding, the data for each of the simulated projects will include one or more 'problematic issues' (i.e. an apparently trivial deviation factor that might result in catastrophic project failure). These issues will be specifically fictionalised to be more obviously detectable in Procession, than with traditional reporting methods. The protocol will contain questions relating to the construction clients' awareness of these hidden issues.

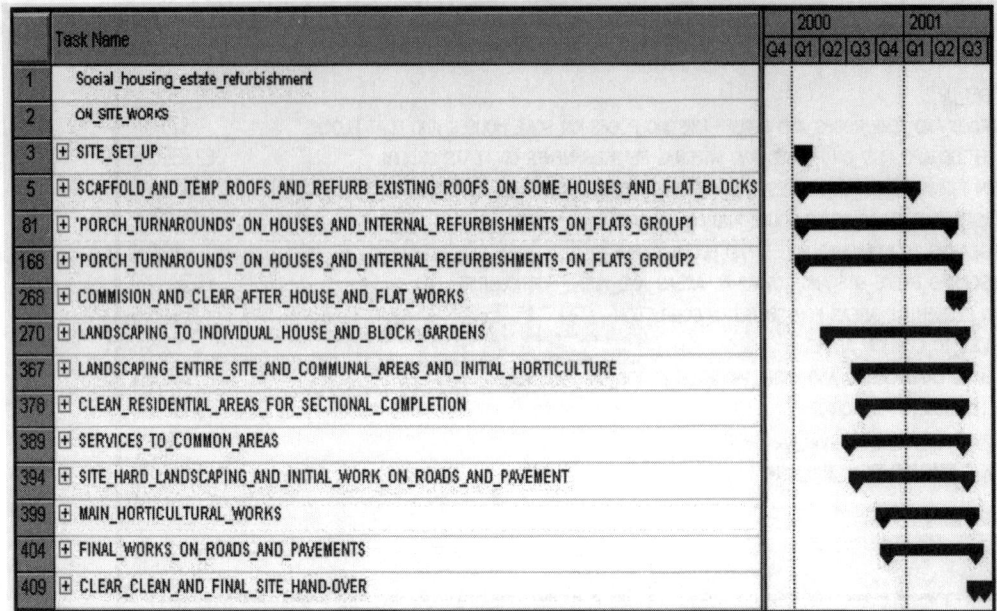

	Task Name	2000				2001			
		Q4	Q1	Q2	Q3	Q4	Q1	Q2	Q3
1	Social_housing_estate_refurbishment								
2	ON_SITE_WORKS								
3	⊞ SITE_SET_UP								
5	⊞ SCAFFOLD_AND_TEMP_ROOFS_AND_REFURB_EXISTING_ROOFS_ON_SOME_HOUSES_AND_FLAT_BLOCKS								
81	⊞ 'PORCH_TURNAROUNDS'_ON_HOUSES_AND_INTERNAL_REFURBISHMENTS_ON_FLATS_GROUP1								
168	⊞ 'PORCH_TURNAROUNDS'_ON_HOUSES_AND_INTERNAL_REFURBISHMENTS_ON_FLATS_GROUP2								
268	⊞ COMMISION_AND_CLEAR_AFTER_HOUSE_AND_FLAT_WORKS								
270	⊞ LANDSCAPING_TO_INDIVIDUAL_HOUSE_AND_BLOCK_GARDENS								
367	⊞ LANDSCAPING_ENTIRE_SITE_AND_COMMUNAL_AREAS_AND_INITIAL_HORTICULTURE								
378	⊞ CLEAN_RESIDENTIAL_AREAS_FOR_SECTIONAL_COMPLETION								
389	⊞ SERVICES_TO_COMMON_AREAS								
394	⊞ SITE_HARD_LANDSCAPING_AND_INITIAL_WORK_ON_ROADS_AND_PAVEMENT								
399	⊞ MAIN_HORTICULTURAL_WORKS								
404	⊞ FINAL_WORKS_ON_ROADS_AND_PAVEMENTS								
409	⊞ CLEAR_CLEAN_AND_FINAL_SITE_HAND-OVER								

Figure 4. 'Live' project summary tasks and their scheduled durations before commencement, source: author.

3 TRANSFORMING RAW PROJECT DATA INTO VISUAL STRUCTURES

In Section 2, the overall evaluation methodology was introduced. This section will now discuss in more detail the transformation of the original 'live' project data into a visualisation format suitable for Procession. Early on in the research sequence, a decision had been made to make Procession compatible with the market-leader project management package, Microsoft Project. Therefore, the 'live' project data was obtained in this format. The original project durations, dependencies and schedule were left unchanged.

In project management terminology, 'resources' are the materials and labour costs assigned to a specific task. Resources had not been assigned to the tasks in the original file and so assumptions were made about example resources that could be applied (i.e. the sub-tasks, materials and labour required to achieve the stated tasks). As the durations were pre-determined, resources based on hourly paid rates were not applicable. Instead, the labour and material costs were obtained from an industry standard pricing book (Davis et al. 2000) and calculated on a per house, block or site basis. In the initial version of the project budget, sums were applied as 'fixed cost' resource elements. Testing revealed that Earned Value (see Sect. 4, Traditional performance

measurement) fields required for Procession seemed not to be functioning as expected when this approach was adopted (i.e. BCWS and BCWP values not present). Instead, a new baseline file was created with the 'cost' values cut and paste as 'fixed costs' directly attached to tasks (i.e. there were now no resources). Allocation of resources to tasks was done as units, rather than as a percentage. Number of available resource units was set to an arbitrary high figure (ten thousand), to make sure that sufficient were available. Each 'fixed cost' was set to 'accrue at prorated'. 'Actual Costs' were calculated automatically by Microsoft Project. A 'Baseline' was saved. The baseline provides a record of all budget values before project commencement. These are then used for later analysis during the project lifecycle.

In order to generate possible scenarios from the data set, risk analysis (in the form of Monte-Carlo simulations) was applied to the Microsoft Project file. The Microsoft Project compatible software tool Riskman Professional (see www.riskdriver.com/riskmantool/) was used to identify potential project outcomes.

Monte-Carlo simulations can be thought of as "statistical simulation methods, where statistical simulation is defined in quite general terms to be any method that utilizes sequences of random numbers to perform the simulation" (CSEP 1995). First

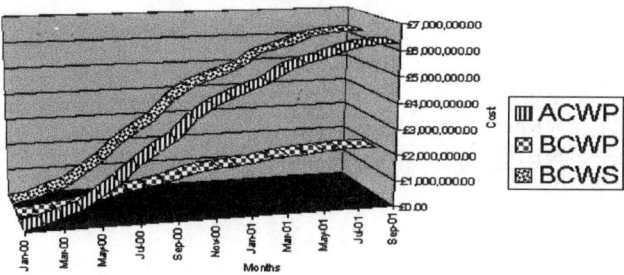

Figure 5. Scenario 1–entire project earned value s-curve, source: author.

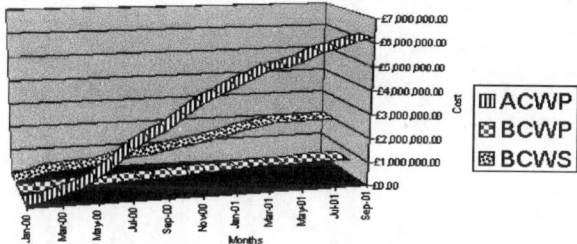

Figure 6. Scenario 2–entire project earned value s-curve, source: author.

described during World War II's Manhattan Project, Monte-Carlo simulations are now widely used, both in hard science and for predicting games of chance (CSEP 1995). Riskman uses Monte-Carlo simulations to analyse the likelihood and impact of specified risks. A project consists of tasks, related by precedence constraints (in project management terminology, a PERT network). In a simplified model, each task has a duration and a set of allocated resources. After completing a large number of simulation cycles, Riskman calculates each task's criticality to successful project outcome. Riskman outputs its results as Microsoft Excel charts and in its customised Microsoft Project 'Simulation View'. In the next section, this paper will go on to describe the results achieved when Riskman (and traditional project management visualisation techniques) were applied to the 'live' project data used for this research.

4 TRADITIONAL PERFORMANCE MEASUREMENT

The traditional method for reporting on project progress is the s-curve. This is generally utilised to analyse performance against budget in one of three ways. Most common of the three is 'earned value analysis' (see Figs. 5-6). This provides the budgeted cost of completing the tasks actually finished at an observation point. This sum is the 'earned value'. Comparing this with the amount actually spent, is a good indicator of project performance. There is an acronym for each of the three curves plotted against each other for earned value analysis (Forsberg et al. 1996):

- ACWP (Actual Cost of Work Performed. Actual cost of performing the completed tasks)
- BCWS (Budgeted Cost of Work Scheduled. The planned budget for the scheduled tasks)
- BCWP (Budgeted Cost of Work Performed. The planned budget for the tasks that were actually completed- the earned value).

371

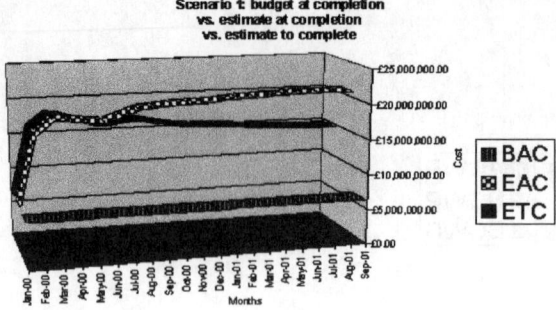

Figure 7. Scenario 1–entire project completion cost s-curve, source: author.

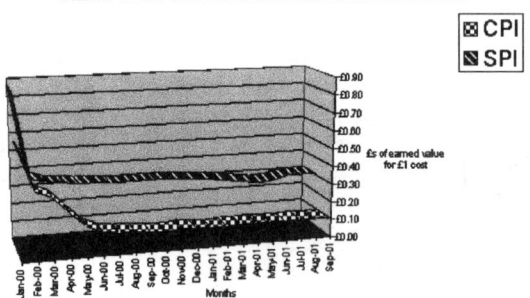

Figure 8. Scenario 1–entire project performance s-curve, source: author.

In Completion Cost analysis (see Figs. 7 and 9), an s-curve plots three lines against each other to compare the original budgeted cost of the project and the revised cost, from the point of observation. There are three acronyms for these curves (Forsberg et al. 1996):

- BAC (Budget At Completion. The planned budget)
- EAC (Estimate At Completion. The cost at completion revised as project progresses)
  =ACWP+((BAC-BCWP) x (ACWP/BCWP))
- ETC (Estimate To Complete. The cost required to complete the project as at time 'T')
  = (BAC-BCWP) x (ACWP/BCWP).

Performance Indices provide a third approach to progress analysis. In this method, two curves are plotted (see Figs. 8 and 11). One represents cost and the other schedule. For both plots, a flat horizontal line at £1 indicates that £1 expended is earning £1 of completed project. Values below £1, suggest that the project is either over-budget or behind schedule. In addition to providing 'to date' progress, the performance indices also indicate the underlying

future trends in the project. The two curves represent:

- CPI (Cost Performance Index)
  = BCWP/ACWP
- SPI (Schedule Performance Index)
  = BCWP/BCWS.

5 CONCLUSION

The Monte-Carlo simulations revealed two summary tasks which were categorised as having attached risks of an 'unacceptable' level. The first of these was the summary task 'scaffold and fit temporary roofs, refurbish existing roofs on some houses and flat blocks' (see Figs. 3-4) which had the following unacceptable risks associated with it:

- Constraints on availability of external scaffolding components, impact the whole product
- Requirements provided to scaffolding subcontractors are ambiguous
- Sub-contracted scaffolding task disturbs project progress

372

- Sub-contracted scaffolding task is badly managed and followed up
- Workloads were under-estimated to gain contract.
- Required security level for the project is higher than usual because some tenants are remaining in occupation. Organisation of key access to occupied properties becomes expensive and time-consuming
- Project is innovating in its field (porch turnarounds- see Sect. 2).

The second identified summary task was 'site hard landscaping and initial work on roads and pavements' (see Figs. 3 and 4), which had the following unacceptable risks associated with it:

- Feasibility study on impact of road and service re-routing is missing
- There is a low awareness of possible technical problems
- When estimating costs, some activities are not quantifiable or have no reference in archived projects
- Project risks were not assessed or are difficult to assess
- Economical and political context has an influence on project and can be changed by unilateral decision at a high level. This particular project has to take particular regard of tenant opinion
- Project environment is evolving
- Technical difficulties are a real challenge
- Detailed assessments of cost and time scale are not compatible with project budget and schedule. Planning does not take into account this fact and relies on false hypotheses.

The two unacceptable risks were fictionalised into scenarios (with embedded 'problematic' issues). In scenario 1, the scaffolding sub-contractor failed to remove house scaffolding according to schedule, extending the duration of the roofing tasks and delaying the commencement of the porch turnaround work by almost three months. Scenario 2 sees the estate vehicle access restricted and most of the tenants remaining in residence. When the initial work on roads and pavements began in the third quarter, the situation became unmanageable. All works relating to the communal estate areas (roads, pavements, hard landscaping and horticulture) started to slip and their durations extended. As discussed in the last section, traditional visualisation techniques were used to analyse the entire sequence. A series of two-dimensional s-curves were produced. BAC (Budget At Completion), EAC (Estimate At Completion) and ETC (Estimate To Complete) were plotted against each other up to milestone 7, as a cost/time s-curve. The expected total cost of the project (before it started) was £6,247,849. By the end of scenario 1's planned project time-scale, £6,169,170.78 had been spent and in order to complete (behind schedule), an additional £14,533,790.31 was required. The total project cost would then have spiralled to £20,781,638.82 (see Fig. 7). In scenario 2, 100% of the £6,247,849 predicted costs had been spent by the end of the planned project time-scale and in order to complete (behind schedule), an additional £56,751,522.90 was required. The total project cost would then have spiralled to £62,920,693.68 (see Fig. 9).

CPI (Cost Performance Index) was plotted against SPI (Schedule Performance Index) as an s-curve to identify trends within the project progress. For scenario 1, the CPI and SPI showed that both the schedule and the budget trends were deteriorating rapidly before the first milestone. In the second quarter, both budget and schedule stabilised. By the third milestone, a slight improvement was seen in both curves, however they quickly dipped back and continued at approximately £0.30 for the rest of the schedule. At completion, the project was equally behind schedule and over budget (see Fig. 8). Scenario 2 saw both the schedule and the budget plans deteriorating rapidly before the first milestone. In the second quarter, the budget stabilised, while the schedule continued to slip. By the third milestone, the schedule was stabilising and it continued to mirror the budget curve, with a differential of approximately £0.20. This continued until the scheduled completion date, with only a slight budget recovery in the last quarter. At completion, the project was still very behind schedule and over budget. The final CPI value reveals that £0.10 of value was being earned for every £1 spent (see Fig. 10).

BCWS (Budgeted Cost of Work Scheduled) was plotted against BCWP (Budgeted Cost of Work Performed) and ACWP (Actual Cost of Work Performed) to produce an s-curve of Earned Value (see Section 4). For scenario 1, expenditure stayed almost as expected (ACWP and BCWS tracked each other throughout) but earned value (BCWP) slowly deteriorated throughout the sequence. Therefore, the work completed by the planned completion date had cost far more than budgeted and the schedule had proven to be over optimistic (see Fig. 5). Throughout scenario 2, ACWP almost followed the budget in the baseline schedule, but increased in task durations. This resulted in the scheduled expenditure

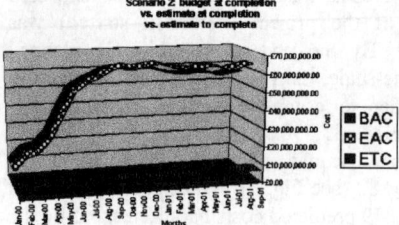

Figure 9. Scenario 2–entire project completion cost s-curve, source: author.

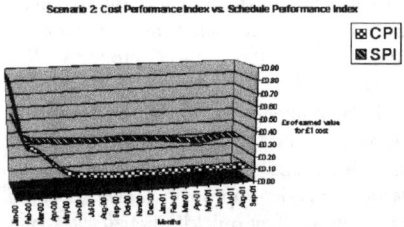

Figure 10. Scenario 2–entire project performance s-curve, source: author.

(BCWS) levelling out to match a longer time-scale. Despite slow improvement through the sequence, the earned value (BCWP) was so low compared with the ACWP, that the situation became irretrievable (see Fig. 6).

The sequence outcome for scenario 1 was:

• Project had run out of time and budget to complete. No cost overruns had been incurred to date but more funds would be required to continue.

The sequence outcome for scenario 2 was:

• Project had run out of time and budget to complete. No cost overruns had been incurred to date but funds totalling many times the original budget would be required to continue.

Finally, two sets of seven milestone Microsoft Project files were generated, one for each scenario. To simulate progress at each milestone, changes were made to the tasks, in terms of their durations and percentages complete. For each scenario, it was necessary to work through the milestone files chronologically. This is because Microsoft Project always passes revised estimates onto the next milestone in the sequence. CSV data files for Procession were exported from each of the Microsoft Project milestone files (i.e. fourteen files total). In the next stage of the research, construction clients will apply the Procession initial prototype to the scenario milestone files.

6 ACKNOWLEDGEMENTS

This work was supported by VIRCON (The VIrtual CONstruction Site) a collaborative project between University College London, Teeside University, The University of Wolverhampton and eleven construction companies, funded by EPSRC award number GR/N000876.

The author wishes to thank Dr Graham M. Winch for his advice and guidance in the course of this research.

7 REFERENCES

Card S.K., Macklinlay, J.D. & Shneiderman, B 1999. *Readings in Information Visualization-using vision to think*. USA: Morgan Kaufmann.

Chen, C. 1999. *Information Visualization and Virtual Environments*. USA: Springer-Verleg.

CSEP, "Introduction to monte-carlo methods", The Computational Science Education Project, http://csep1.phy.ornl.gov/CSEP/MC/MC.html, USA, 1995.

Davis, Langdon and Everest, *Spon's Architects' and Builders' Price Book*, one hundred and twenty fifth edition, Davis Langdon & Everest (Editors), E. & F.N. Spon, UK, 2000.

Forsberg, K., Mooz, H. & Cotterman, H. 1996, *Visualizing Project Management*. USA: John Wiley & Sons.

Hall, A.D. 1969. Three-dimensional morphology of systems engineering. *IEEE Transactions on System Science and Cybernetics*. SSC-5(2): 156-160.

North, S. 2000. Procession: using intelligent 3d information visualisation to support client understanding during construction projects in R.F. Erbacher, P.C. Chen, J.C. Roberts & C.M. Wittenbrink (eds), *Visual Data Exploration and Analysis VII*: 356-364.

Pohl, K. 1994. The Three Dimensions of Requirements Engineering: A Framework and Its Applications. *Information Systems*. 19(3): 243-258.

Shroeder, W., Martin, K. & Lorensen, B. 1998. *The Visualization Toolkit*. Prentice Hall: USA.

Smit, J. 1999. Reclaiming the backlands. *Building Homes*. September issue.

Joined up construction: CONCURing the information maze

H. van de Belt
TNO Building and Construction Research, Delft, Netherlands

Maria Nikolaenko
VTT Technical Research Centre of Finland, Building and Construction, Finland

David Leonard & Jeff Stephens
Taylor Woodrow Construction, London, UK

ABSTRACT: The previous round of research in STEP related projects aimed at developing semantically precise exchange formats. Now the current industrial need is to build on that with much more focus on the control and management of information within processes. There are also several emerging technologies supported by current research activity, which require industrial path finding implementation. Within the Brite-EuRam project CONCUR, a consortium of European construction interests is working on industry deployment to demonstrate the application of previous foundation research and present new developments. The paper presents an overview of some of the work being undertaken in CONCUR, its relationship to other projects, and its contribution to standardisation initiatives such as ISO/STEP and the International Alliance for Interoperability (IAI).

1 INTRODUCTION

The Building and Civil Engineering industry is one of the largest in Europe but it is an industry which is organisationally complex and fragmented, with more than 95% of companies being small to medium sized enterprises and operating in a project-centred "virtual enterprise" culture. This is also the case with the larger, multi-disciplined businesses. In the main, the organisation and execution of projects follow very traditional patterns with corresponding traditional information sharing and exchange methods, resulting in faltering and patchy adoption of any new processes, including product related knowledge approaches.

Until recently, there were no significant drivers of industry change. In the last decade however, this situation has begun to change markedly. Clients and facility operators now demand better quality, faster and cheaper built facilities incorporating more complex technology. At the same time, governments have considerably increased the regulatory constraints on safety, waste and energy consumption.

Other industries facing the challenge of working in dispersed and concurrent business environments are developing integration strategies based on electronic information sharing and exchange using open international standards. Product and Production modelling technology, as featured in ISO-10303 (STEP), UN/EDIFACT and now the IAI IFC, is generally perceived as a key enabler of business integration.

The CONCUR Consortium – comprising:
- contractors:
  Taylor Woodrow, Fortum Engineering and Skanska;
- research and development institutions:
  VTT and TNO,
- the specification system developer STABU,
- universities:
  Technical University of Delft and KTH (Royal Institute of Technology, Sweden)

proposes to develop, implement and deploy integrated environments for creating and managing shared and distributed project information. Systems are being deployed in live construction projects and are based on formal neutral data models of the project information shared and exchanged by software applications (tools) in the Building and Civil Engineering industry. The models help to:
- integrate design, engineering and construction support tools currently used by the industrial partners, but also new, innovative commercial applications,
- improve internal integration of the industrial partners,
- implement and demonstrate concurrent design and engineering in distributed, multi-partner projects and
- implement, evaluate and deploy electronic information exchange and sharing, focusing on (1) the downstream delivery of comprehensive tender information for the construction stage and (2) the

upstream availability of alternative technical solutions to the inception (client brief) stage.

This paper gives an overview of a number of the results of the CONCUR project:

- Information Management System – The use of an IMS to control both documents and model information,
- Browser/Viewers – The demonstration of web based browsing and viewing of information models,
- Model Merging - The merging of IAI IFC Part 21 STEP files and models,
- Property Sets – The definition of property set libraries for defining specifications and products, and their use with IFC building models,
- Derived Attributes – The use of EXPRESS to calculate derived attributes from IFC geometry information,
- Use Cases – Examples of using this technology in industrial use scenarios with industrial strength information.

2 INFORMATION MANAGEMENT SYSTEM

The Information Management System (IMS) is a supervisory system and repository server controlling the Project Database. It is managing the exchange and sharing of distributed information (both documents and model information). One of the main dimensions of the IMS is the level of privacy. Company-IMS systems are systems that are only used internally within that company. The Project-IMS is set up for project needs and it is common for all project partners. The main function of Project-IMS is to make project data available to project participants and partners, keep track of transactions and control access rights. IMS systems can be grouped by their function capabilities:

1 EDM (Electronic Document management) systems are generic – managing electronic unstructured documents (files)
2 PDM (Product Data Management) systems are product specific – managing electronic documents and structure
3 PDT (Product Data Technology) systems are model based – managing pre-defined building objects (Express driven data)

Figure 1 illustrates the main feature of the CONCUR IMS implementation. The core of the IMS is a commercial PDM system and a database. The end-users may access a project IMS either directly (left) or via an in-house IMS system (right).

The basic assumption is that the IMS is mainly used for information sharing between dispersed actors in construction projects. The main problems when setting up a "Virtual Enterprise" arise from the fact that projects consist of temporary co-working

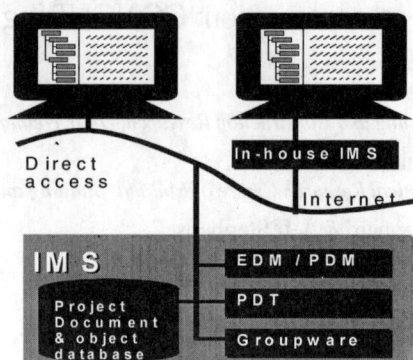

Figure 1 Implementation of IMS

partners working together for limited time. The characteristics of the Project-IMS must consider many aspects: legal, property rights, technical, security, human and management. The IMS should be pre-configured for generic conditions in construction projects rather than for specific end-user companies. Once the project IMS has been set up the most usual operations are as follows:

1 Original documents/objects are stored in the internal company IMSs of participating companies.
2 Only the information, which is to be shared with other project participants, is to be stored in the IMS.
3 Project participants prepare information using their internal systems, which are not accessible to other participants. According to the agreement between actors, information is made available to others by uploading it (check in) to the project IMS.
4 Other actors may view the information from the project IMS and make comments on it using red-lining. The author and other participants can view comments.
5 All transactions are recorded: who stored what information and when, who has seen it and when etc.

3 CONCUR IMS

The CONCUR IMS is a combination of a full-scale document management server and a product model server. The CONCUR IMS provides the following functionality:

1 Uploading and downloading of files
2 Full control of product the model server
3 Access to product data using an XML interface

3.1 Document management

All information that is uploaded to the system is stored in the document management system with the metadata the user has provided. Uploaded files can be seen as replacements to the paper documents so it is extremely important to be able to keep track when and by whom files have been inserted in the system and who has downloaded them and when. It has to be possible to get access to predefined numbers of revised documents to be able to check what information has been available at any specific time during the project. The CONCUR IMS prototype implementation has been built on the top of the Project-Wise document management system from Bentley Systems (http://www.bentley.com). Java servlets have been used to provide the required functionality. The AddDocument servlet is executed when new documents are inserted to the system. Metadata provided by the user is used to locate the document to the right place in the system. The GetDocuments servlet provides access to documents in the system. A core idea in the development has been that the end user does not have to or even want to know what kind of document management system he is using. He is only interested in being able to upload and download documents. An example of the project IMS user interface is shown in Figure 2.

3.2 Product Data Management

If the uploaded file is a STEP physical file (ISO 10303-21) it can be uploaded to the product model server. The model can be loaded in the system as a new model or it can be appended to an existing model. The user can also remove whole models from the system or just separate instances from specific models. This is only possible if the user has the required permissions to do so.

The Express Data Manager, a product from EPM Technology AS, (http://www.jotne.com) is used in the prototype implementation of the CONCUR IMS to handle all EXPRESS driven functionality. Users get access to the available methods using the Java servlet provided for this purpose. The Java programming interface of Express Data Manager is used to execute the user defined tasks.

The product model server using an XML based interface provides access to the product data at object level. The implemented solution is an extension to the ISO 10303-28 specification (Hemio 2000). Elements like 'Database' and 'Repository' have been added to the original specification and 'Attribute' is changed to be an element in the XML file instead of an attribute. Also data type information has been added so that it is possible to instantiate the described object using only the XML file. This implementation is a late bounding implementation so it is applicable to all schemata.

An interface to the CONCUR IMS can be a virtual reality model. Because all documents and also product data objects can be accessed using simple URL requests it is simple to generate a VR-model, which can be used as a graphical user interface. By clicking an object in the virtual reality browser a user can send a URL query to the provided server and get back a list of documents attached to that object or an XML describing the values of attributes of that specific object.

One of the tools available to generate the VRML file is the VTT-ProMoTe software, which was initially developed in the ESPRIT project ToCEE and extended in the CONCUR and ProCure projects.

There are several research projects going on at the moment, which will extend and upgrade the functionality of the CONCUR IMS, for example the ProCure project is adding a virtual reality interface to the IMS.

When documents are uploaded to the project IMS (Figure 3) an html form is used to request the meta data from the user. The CONCUR PDM schema defines allowed attributes for the metadata. After fill-

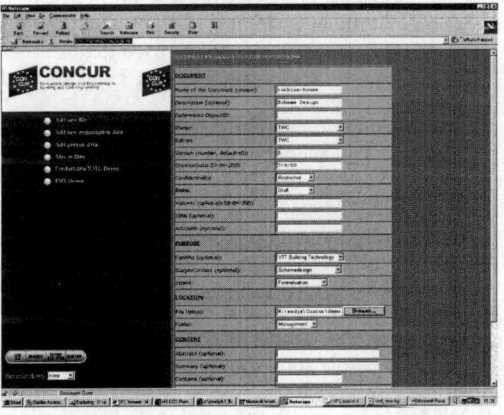

Figure 2 Project IMS User Interface

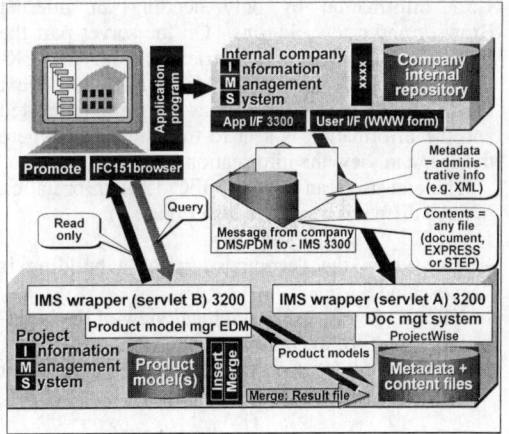

Figure 3 Project IMS architecture

377

ing the form the user sends it to a Java servlet, which stores files and metadata into the ProjectWise database.

4 PROMOTE BROWSER

EXPRESS schema and STEP data browser VTT-ProMoTe (Product Model Technology):

Introduces the concept of a generic schema and data browser. The schema browser provides the generation of ready to compile Java classes, the ability to create reproduced schema as plain text and as an HTML file, as well as browsing the content of EXPRESS schema in many views.

Data browser enables the viewing of STEP product data as an entity hierarchy and as a data content hierarchy based tree, if it is supported in the schema. The software supports access to distributed models in client/server mode over the Internet. A further feature is the creation of virtual reality models which can be used as a 3D user interface to product models and related documents over the Internet. VTT-ProMoTe is to be further developed so that it can access product data using the XML interface provided by the EDM product model server.

5 IFC 1.5.1 BROWSER

The Ifc151Browser is a web application that provides an 'on the fly' translation facility from IAI's IFC 1.5.1 models to VRML and HTML. TNO and CSTB (VRML-generation) developed the base product in the Esprit VEGA Project. The product has been matured, adapted and extended in the CONCUR project.

The Ifc151Browser consists of two parts: a client part and a server part. The client part of the Ifc151Browser makes it possible to view the IFC 1.5.1 information by only needing an Internet Browser and a few plug-ins. On the server part the IFC 1.5.1 information is retrieved from the EX-PRESS Data Manager of EPM Technology. Next the information is processed as the user requested and the information is sent to the client part where the user can view the information.

The end-user can view the IFC 1.5.1 information in three different ways (see also Figure 4):

- Tree view; the decomposition of a building is presented.
- VRML view for a 3D graphical view of the building or parts of the building.
- HTML view for more detailed information about the building or parts of the building.

Furthermore, the user can attach or detach specifications defined using IfcPropertySets, ask for the

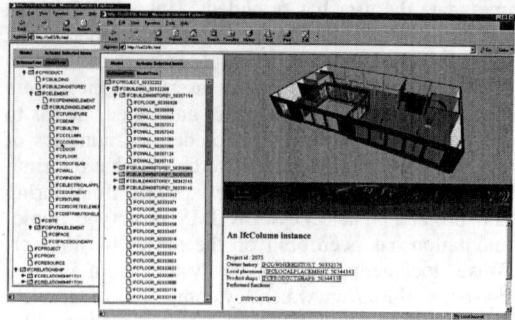

Figure 4 – Ifc151Browser user interface (client part)

derived information (see section 7) create comma-separated files with various information and so on.

6 IFC 1.5.1 MODEL MERGING

Information is changed continuously during the design of building objects. Sometimes the shape changes, other times costs are added or HVAC systems are defined. Within this process, files often transfer the information from one application to another. Every software application handles its own specific discipline and is sometimes not capable of maintaining the information of other disciplines. While an application adds information, other information is lost. This results in several files describing the same building object, but every file containing different information.

A model merging mechanism is therefore needed to overcome the problem of having multiple files describing one building object. Within CONCUR, IFC 1.5.1 is used so a model merging mechanism has been defined and implemented for the IFC 151 EX-PRESS schema. The Model Merging Mechanism is defined in such a way that the model merging process is a separate process from the normal way of working. Without changing any software application, the user can work like he/she used to do.

In this model merge process three kinds of models can be distinguished:
1 The original or source model (the existing information). This model will be called Model A.
2 The updated model. This model will be called Model B. Model B is in fact Model A with changes made by one software application.
3 A new model with parts of model A and parts of model B. This is the merged model and will be called model Z.

Figure 5 below shows how current information is distributed from one application to the other. In the CONCUR project, all software tools that are supporting IFC 1.5.1, are able to import and export

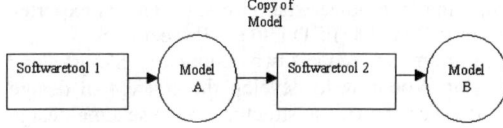

Figure 5 Distribution of information between software tools

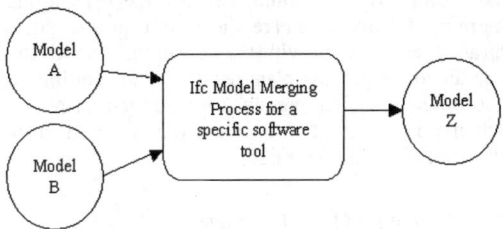

Figure 6 Model merging process

STEP Physical Files. Models A and B will therefore be STEP Physical Files.

The model merging process will be specified using EXPRESS-X. Originally EXPRESS-X was the language to describe the conditions for conversion between two EXPRESS schemas. By slightly changing EXPRESS-X, the conversion is made more generic and by that more useful for model merging as well.

By using permissions in the model merging process on the IFC 1.5.1 entities, certain entities are extracted from Model A and other entities of Model B and created in Model Z (see Figure 6). The specification of the permissions of the entities will differ for every software application.

As long as applications lose information and software applications do not store and retrieve information using a direct interface (such as SDAI) to a database, the need for model merging will remain.

After implementing and testing the model merging mechanism, more research will be done to check if this model merging mechanism is also suitable for concurrent working between users who are working with files.

7 DERIVED ATTRIBUTES

An end-user likes to think of his buildings in terms of length, depths, volumes and the like for his rooms, walls, doors etc. He wants to abstract from the representation details like bounding boxes, polyloops and trimmed curves that ideally should be transparent to him. However, most CAD systems can only handle exactly these shape representations. Even if the CAD systems can handle more abstract notions like height, depth, width etc. they are often

not communicated beyond the application itself i.e. imported and/or exported.

By extracting the property values from the shape representations of IFC 1.5.1 it is possible to use both the representations and the property values of shapes. The IFC 1.5.1 EXPRESS schema is extended with extra procedures and functions to extract length, width, height, depth, thickness, area, volume, perimeter or surface area from the shape representation. Several IFC 1.5.1 entities like IfcWall, IfcFloor, IfcWindow and IfcColumn are extended with derived attributes. A derived attribute represents a property whose value is computed by evaluating an expression. (ISO 10303-11:1994 (E)). The derived attributes refer to the procedures/functions that extract the needed property values.

Extending the standard EXPRESS IFC 1.5.1 schema means that however the EXPRESS schema itself has been changed, the exchange of the information by STEP Physical Files will remain exactly the same. Therefore no software applications have to be changed. The derivation of property values takes place in the EXPRESS Data Manager of EPM Technology, which will 'on the fly' derive the requested property value by executing the necessary procedures and functions.

Below, a small piece of the adapted EXPRESS schema is presented for IfcWindow:

ENTITY IfcWindow
SUBTYPE OF (IfcBuildingElement);
  GenericType: IfcWindowTypeEnum;
-- All derived attributes added for the need of property values
DERIVE
  derivedWidth :   IfcPositiveLengthMeasure
                           := GetWindowWidth (Is-
                           ContainedBy, ProductShape
                           );
  derivedThickness: IfcPositiveLengthMeasure
                           := GetWindowThickness
                           (IsContainedBy, Product-
                           Shape);
  derivedHeight :   IfcPositiveLengthMeasure
                           := GetWindowHeight (Is-
                           ContainedBy, Product-
                           Shape);
  derivedTotalArea :   IfcAreaMeasure := GetTo-
                           talWindowArea (IsCon-
                           tainedBy, ProductShape);
  derivedTotalVolume: IfcVolumeMeasure := Get-
                           TotalWindowVolume (Is-
                           ContainedBy, ProductShape
                           );
  derivedTotalPerimeter: IfcPositiveLengthMeasure
                           := GetTotalWindowPerime-
                           ter (IsContainedBy, Pro-
                           ductShape);
  derivedTotalSurfaceArea: IfcAreaMeasure := GetTo-
                           talWindowSurfaceArea (
                           IsContainedBy, Product-
                           Shape);

WHERE
    WR61: SIZEOF(QUERY(Temp <*
SELF\IfcObject.TypeDefinitions |
        NOT(Temp.TypedClass = 'IfcWindow'))) = 0;
END_ENTITY;

The property values of the derived attributes will always remain up to date as they are extracted on the fly. For example, if the geometry of a building element is changed, the associated derived attributes will also change since the procedures/functions will be re-executed in the EDM.

As some applications are only able to handle property values instead of shape representation, extra functionality is added to the Ifc151browser so the property values are exported to a comma separated value file. A comma separated value file is still a commonly supported file by applications.

8 USE CASES

During the design phase of a project the overall costs need to be evaluated at various stages of the design process. For example at the Inception stage, Concept Design stage, Scheme Design stage, Detailed Design stage etc. At each stage the costs will get more accurate and the design will probably change to keep within the overall budget for the project.

The Concur Project Information Management System (Project IMS) is used to hold project information for both documents and product data. One important use case is for merging different IFC models that are being created during the design process.

8.1 Project IMS Use Case for model merging

The Project IMS Use Case shown in Figure 7 has been defined to support the process of model merging. In the Concur deployment trials Fortum Engineering develop a conceptual design of a Turbine

Building in a power station plant. This is exported as an IFC model (FE1) into the Project IMS.

Fortum then request two contractors: Skanska and Taylor Woodrow to develop the conceptual design of the Turbine House structure to the scheme design stage. The scheme designs are developed using the ArchiCad and Allplan AEC CAD systems and two further IFC models are stored in the Project IMS (SK1 and TW1). Fortum then use the IMS model merging facility to merge the turbine house structural scheme design with the conceptual model that contained the process plant design for the equipment and pipework. One preferred merged model for the scheme design solution is selected and stored in the Project IMS as model FE2.

8.2 Typical End User Use Case

Use cases have been defined by the end user partners in the Concur project for their deployment trials. An example for Fortum Engineering is shown in Figure 8 and covers the following stages in the design of a powerplant:
- Clients brief
- Choice of the type of plant.
- Configuration
- Conceptual Design
- Schema Design

The various inputs and outputs from the processes are identified along with the applications that are used in the deployment trials.

8.3 Taylor Woodrow Deployment Trials

Taylor Woodrow's deployment trial architecture is shown in Figure 9. At the inception of a project various layouts/configurations are compared using an Inception Support System application. This can be used to check the feasibility of various concepts for the overall project and to be able to compare costs as the layout/configuration of the different concepts are varied. The project layouts can then be exported as IFC part 21 files and imported into a

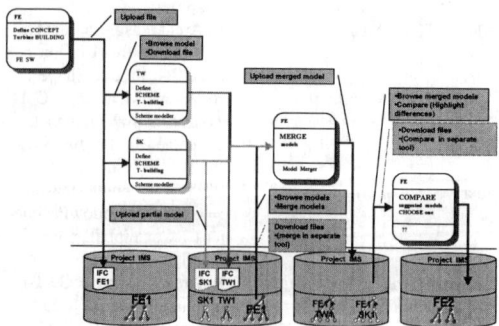

Figure 7 Model Merging Use Case using the Project IMS

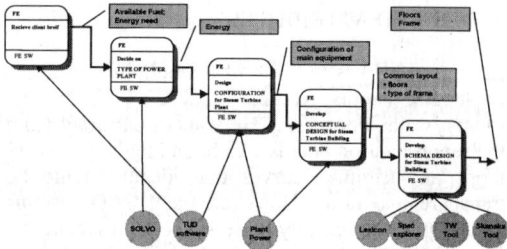

Figure 8 Fortum Engineering Use Case

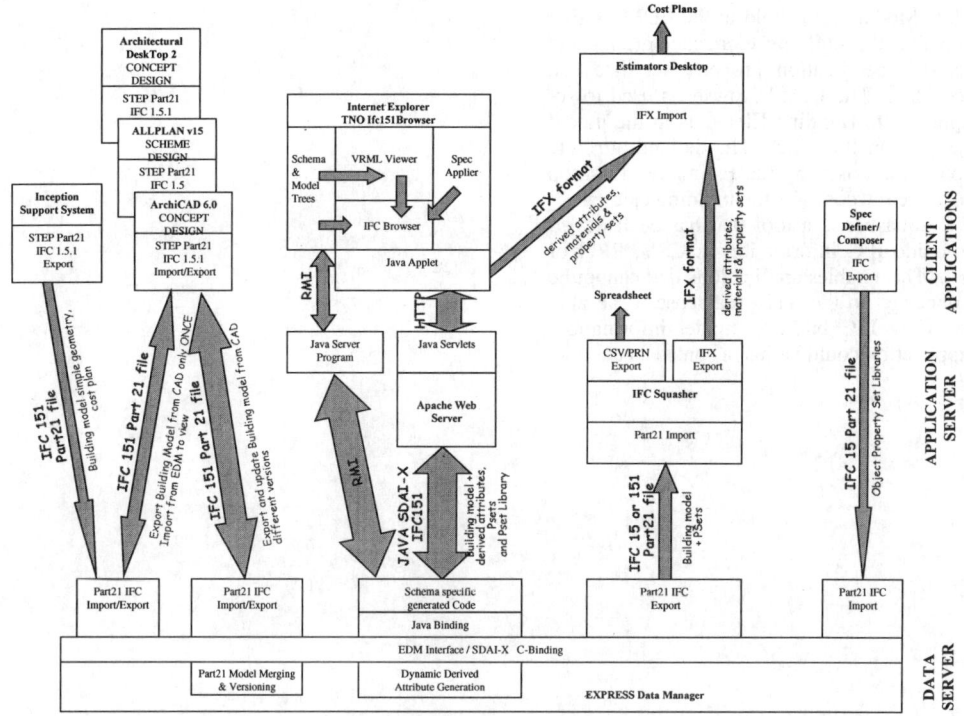

Figure 9 Taylor Woodrow Deployment Trial Architecture

STEP database repository, for example the EPM EXPRESS Data Manager (EDM).

The IFC models can then be imported into AEC CAD systems: ArchiCAD, ALLPLAN or Architectural Desktop in order to develop the Conceptual Designs into Scheme Designs. Once the Scheme Design has been developed the use of IFC Model Merging ensures that information from different disciplines is not lost when the Scheme Design is merged with the original conceptual design.

The TNO Ifc151Browser provides end-users, who have not got a CAD application, with access to the various IFC models in the Project IMS using any standard web browser. The Ifc151Browser application displays schema or model trees, which can be used to select and view or browse any of the building elements in the IFC model.

An IFC model that has been exported by a CAD system will mainly contain building element geometry and material type information. In order to calculate the cost of building elements the specification for these elements needs to be defined. The SpecDefiner/SpecComposer application is used to define templates for the specification properties and then create actual specification property set libraries, which contain instances of the actual specifications

that will then be associated with individual building elements.

The SpecApplier is integrated into the TNO Ifc151Browser and is used to apply specification information to building elements in the IFC model. The application provides a user with the means to associate IfcPropertySets, selected from a specification library, with individual or groups of building elements. The specification information that has been added to the building model objects is then available for other applications to re-use.

Ideally, non CAD based applications would be able to query the IFC model object information and be able to calculate various properties like: volume, area and surface area (derived attributes). However, to generate such information the application must have a good knowledge of how the building element geometry is held in the IFC model. To avoid applications from having to calculate such derived attributes the EXPRESS Data Manager generates the attributes on the fly. This information can be viewed using the TNO Ifc151Browser or exported in a Flat File format (IFX).

The Estimators Desktop is used to develop Cost Plans and is able to abstract building elements and their associated dimensions, areas and volumes etc.

381

from the IFC Model that is held in the EDM. Other information like the building element material and any associated specification property set data can also be accessed. The Ifc151Browser is used to select the appropriate Building Elements in the model and then export an IFX File. The derived attribute information is then used by the Estimators Desktop to help calculate the costs for the building elements.

The IFC Squasher is a tool that has been developed to provide IFC information in CSV, PRN or IFX format. This enables applications that cannot be directly connected to the STEP database to be able to read or view IFC building model information. Such an application could be just a spreadsheet.

IT in construction

Requirements for navigation through drawings on wearable computers by using speech commands

Jan Reinhardt & Raimar J.Scherer
Technische Universität Dresden, Germany

ABSTRACT: Wearable computers will strongly impact the work on a construction site in the near future. These computers are firstly small in size and secondly the persons using them should not be hampered in doing their basic work by the use of mouse, joystick or even keyboard, which are used as input facilities. Oftentimes the only way to put data in the computer under site conditions, is a speech recognition interface. However, the application of voice as the data input media requires new methods and technologies for the navigation through drawings displayed on a wearable computer on site. This paper describes technologies, methods and algorithms that were introduced to support navigation through a drawing by using speech commands. The context for this problem was given by a project of Carnegie Mellon University and TU - Dresden that focuses on the usage of wearable computers to efficiently measure the progress of a construction project. Being confined to have only a small range of speech commands, it is necessary to optimize the applied command algorithms in a way that the users can navigate to the desired element as quickly as possible.

1 INTRODUCTION

Wearable computers will strongly impact processes on the construction site in the future. Having information on construction elements, specifications, check lists or even immediate contact to an expert on site traveling time and resources can be saved. Moreover detailed, structured information can be provided on the construction site. But not only to provide information, also to collect data on site, wearable computers will have a great impact. Progress Monitoring and inspection tasks for instance can be made much more efficient by using wearable computers (Garrett 2000).

Having brought the computer on site helps little if the process of data input on site does not address the special requirements the construction site imposes. A keyboard will not be an option for a computer that is used on site, as the user would have to set down the computer to put in data. This paper discusses how wearable computers for construction sites can be equipped with functionalities that allow the user to fulfill the desired tasks on site.

2 WEARABLE COMPUTERS

Wearable computers are an emerging technology that will strongly influence data procession and communication processes on site.

Many people already take advantage of mobile computers such as notebook computes. It is a fact, that these computers have contributed to a higher productivity, because, the notebook computer is available to the user more often than a desktop computer. The use of a notebook computer is not an option for the rough environment of a construction site. Notebook computers have to be set down to input and retrieve data. A computer that can be used on site has to be robust, small in size and unobtrusive to the user.

There are two major categories of computers that meet this criteria. Firstly there are the emerging palm top computers. There is a growing amount of software available for palm top computers. Auto-Desk has come up with a program that displays drawings on the palmtop. Nevertheless palmtop computers have a fairly small screen size and are therefore only to a limited degree usable to depict drawings or complex user dialogs. For more complex dialogs and
to display drawings, a new generation of WIntel based wearable computers can be used.

These computers, on which this paper sets the focus, can be worn on the belt or a vest and therefore meet the requirements, the site context imposes. The interface to the user can be a hand held or a head worn display. Even though there are hand held keyboards available for these computers, efficient data input will only be possible by using the hand held touch

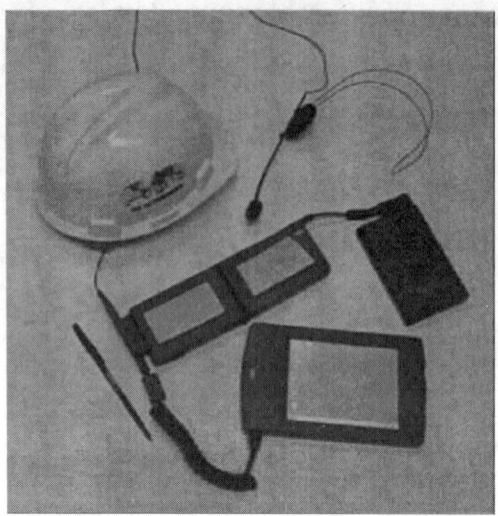

Figure 1, Wearable Computer VIA II

screen and a speech interface. Therefore the user interface of the program to the user is of crucial importance.

There are applications available, that add a speech functionality to standard software applications. Tests have shown, that this added speech functionality does not work very well, in many cases. Applications that use a speech functionality that is embedded in the context o the program will work much better.

The next section will discuss, how this embedded speech functionality can be build in a software.

3 DELIVERING A SPEECH FUNCTIONALITY TO A PROGRAM

For the project, this paper is based on, a speech engine had to be linked to a java program.

A speech engine recognizes spoken text and synthesizes speech. This functionality had to be implemented in a program, that enables the user to access information and put in data on site.

Java provides a Speech API, java.speech, that connects a Java program to external speech processing software. In our case we used IBM's ViaVoice speech recognition, which turned out to be fairly robust and accurate for this particular application. The java.speech API provides a speech engine that has two major sub interfaces: Recognizer and Synthesizer. Whereas the Recognizer transforms spoken text to written text, the Synthesizer generates speech text from written text. Both features are incorporated in ICMMS.

To obtain the text of spoken words from the Recognizer, a grammar had to be defined.

The Java Speech API supports two types of grammar that can be applied by a speech engine. A Rule Grammar is a user defined grammar that is used to recognize key words and phrases that, for instance control a program. The second type of grammar is the dictation grammar, which is based on a predefined grammar provided by the speech engine's vendor. A Rule Grammar is much more accurate in recognizing spoken text than a dictation grammar as it selects the recognized words and phrases from a limited source of predefined grammar rules. Rules are sets of tokens that define the source from which the speech engine can choose phrases. Only if the spoken text coincides with one of the rules, the engine will recognize it. It is possible to use more than one grammar at a time at a speech engine.

There are different ways to create new Rule Grammars. The easiest way probably is to create a text source that complies with the Java Speech Grammar Format (JSGF) and use the loadJSGF method of the Recognizer to load the text into the Recognizer. Storing the JSGF text in a file that is accessed by the loadJSGF method, the file could have the following content:

```
grammar construction;
<job> = architect{architect}|engineer
{engineer}|owner {owner};
public <role> = I am the {job} <job>
public <bye> = good bye {bye};
public <pile> = The pile has to be
moved {dir} <direction>
public <direction> =
up{up}|down{down}|left{left}|right{rig
ht};
```

It is easy to recognize the pattern on which the JSGF is based. The single phrases can be joined together. A possible phrase that the recognizer would understand would be for instance "The pile has to be moved left". The words inside the brackets are tags that can be recognized by the speech engine when recognizing the specific phrase.

The second, more difficult way to create a Rule Grammar is to create it from within the code. The advantage of this method is that the grammar can be altered or appended while the program is running .

A Rule Grammar contains a collection of Rule Tokens. Rule Tokens can consist of other Rule Tokens that are linked to each other by Rule Alternatives or Rule Sequences, which are children of the java.Speech API RuleToken class (Java Speech).

Our application uses a Rule Grammar to recognize speech commands that control the program, do data input and navigate through the drawing. When the user selects a certain element on the drawing, the Rule Grammar is updated, to ensure that the com-

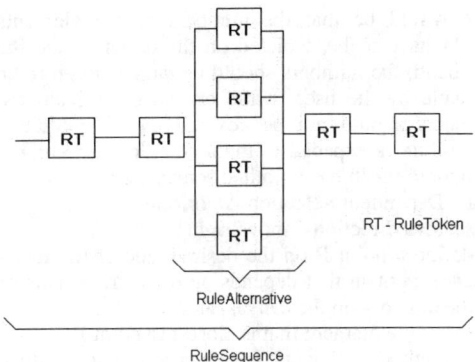

Figure 2, Rule model of a Rule Grammar

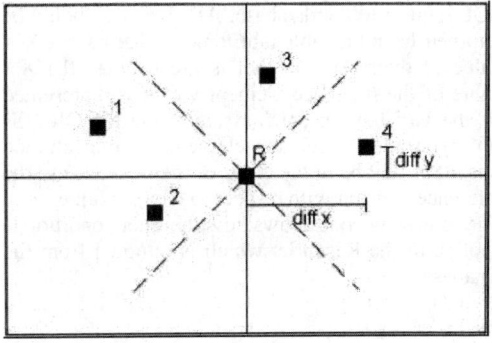

Figure 3, Selection process of an element, using speech

mands that are specific to the selected element are available in the Rule Grammar.

The Recognizer, which uses a Rule Grammar very much contributes to the high level of performance of the speech interface of our application.

Although the Rule Grammar delivers much more accurate results for command recognition, a Dictation Grammar is also needed to recognize comments the user may wish to make on the selected element. Words and phrases can also be added to a dictation grammar, but as the source from which text can be chosen is much broader in scope, the accuracy and speed of speech recognition is lower than when using a Rule Grammar.

4 NAVIGATING THROUGH A DRAWING BY USING SPEECH

Another important use of a speech interface for the application is to provide a functionality to navigate through a drawing using only speech commands, such as:
* Move left
* Move right
* Move up
* Move down
To define the relative positions of the elements in a drawing, each element has specific points that represent the location of the element. A good point to use is the centroid of the element. The centroid can easily be calculated and is probably very close to the point the user naturally associates with the location of the element.

Having reduced the elements to centroid points, all centroid points located in a certain direction relative to the reference point (e.g. right, up), can be determined . If the elements are sorted, the one centroid point that is the closest to the reference centroid point can be determined.

Often the centroid point of an element that is the closest to the reference point, with respect to the desired direction, is not the one the user naturally associates with requested movement. For instance element 3 of the drawing shown in Figure 3 is the closest element to the reference element R if the requested direction is "right". Nevertheless, the user naturally associates element 4 with the direction right, even though it is not the closest element to the reference element to the right.

For this reason an additional filter condition has been introduced, which only considers elements, that are within the triangle that is defined by the broken lines of the desired direction.

For example the condition for a "Move Right" request would be:

diff x > diff y

Applying this condition, element 4 will be selected. For a Move Up request, the filter condition would be:

diff x < diff y

Element 3 will then be selected. To determine the elements that are located in a certain direction relative to the reference element, the program takes advantage of SQL requests, which return only the elements that meet the specific conditions of the SQL request. For example to get a list of elements that are right from the reference element the following java and SQL code can be used:

```
String queryStr;
.....
if(direction=="right"){
    queryStr = "SELECT * FROM tabEle-
    ments WHERE centroidX>
    "+(currentX+0.000001)+" ORDER BY
    centroidX";
}// END if(direction=="right
```

In this example, currentX and currentY represent the coordinates of the reference element. The variable queryStr contains the String that represents the

387

SQL request to the database. All elements should be returned from the table tabElements that have a X - value of their centroid that is greater than the X - value of the reference element which is represented by the variable currentX. By applying the ORDER BY centroidX literal, the elements in the returned ResultSet will be in the order of their distance to the reference element with respect to the X - Value .

The following code shows how the filter condition is applied to the ResultSet which is returned from the database.

```
ResultSet rs;
Vector reply;
...
if((direction=="right")||(direction=="
  left")){
  if(diffY <= diffX)
    {reply.add(new Integer(
    rs.getInt("elementID") )); }
}//END if((direction=="right...
```

The Vector reply will contain the IDs of the elements that are to the right of the reference element and meet the filter condition which says that the distance of the reference element to the desired element has to be smaller with respect to the y - Value than the distance with respect to the x - Value. The first element that meets the filter condition is the new selected element. The quoted code is embedded in a loop that runs until the first element meets the filter condition or the ResultSet has run out of elements.

This, of course, is a very simple algorithm to select the desired element. It will take several steps to move the focused element from the upper left part to the lower right corner of a drawing.

There are also alternative algorithms that could be applied:

Multiple Step Algorithm:

1 Select a direction ("move right")
2 Tell the computer the number of elements to move to the desired direction

This algorithm is easy to implement as it just skips the first n element IDs that are returned from the SQL request and meet the filter condition.

The algorithm also would be very efficient to select elements that are located in a regular order (e.g. grid of columns).

Number Selection Algorithm:

1 Assign numbers to the elements that surround the reference element
2 Tell the computer the number of the desired element

Using this algorithm would make it easier to select elements that are not clearly in one of the four major directions from the reference element. Also navigation on layouts that have a rather random distribution of elements would be improved as it takes only one step to select the desired element. The disadvan-

tage would be that the numbers of the elements would have to be depicted on the drawing. On the one hand, the numbers should be large enough to be readable by the user, while on the other hand the elements should not be covered by the numbers. This issue is especially important for the wearable computer which has a limited screen size.

Scale Dependent Selection Algorithm:

1 select a direction ("move right")
2 define a point P on the desired side of the reference element that depends on the current scale of the drawing on the canvas, and
3 select the element that is closest to point P

This method will work best in conjunction with a powerful zoom function. The selection depends on the current scroll and zoom settings. As scroll and zoom settings have to be adjusted in many cases anyway, there is only one more step ("select the direction") to select the desired element. The disadvantage of this algorithm is that scroll and zoom settings have to be very accurate to select the right element. Nevertheless, it is a good algorithm to move the selected element over larger distances across a drawing.

The most efficient way to navigate through a drawing is very much dependent on the specific characteristics of the drawing. All algorithm have advantages and disadvantages. Powerful speech navigation strategies therefore should allow users to select the most suitable navigation algorithm.

5 APPLICATIONS OF SPEECH CONTROLLED WEARABLE COMPUTERS

Progress Monitoring for construction projects is becoming is increasingly important, as owners demand shorter periods for the delivery of their projects. Construction companies have great interest in knowing where problems can occur and if the actual construction progress is eventually behind the planned construction progress.

Traditionally, Progress Monitoring is done by marking the actual progress on the construction site in a layout. Back in the office the person who does progress monitoring transfers the visual information from the layout in a progress figure and compiles charts that depict the construction progress over time.

This process can be very time consuming and has to be repeated several times during the duration of a construction project. For this reason, Progress Monitoring is often enough neglected, which may result in unrecognized delays and coordination problems of the construction project.

In a joined project of Carnegie Mellon University and Technische Universität Dresden, a prototype of

Figure 4, Flat Panel Display of Xybernaut Wearable computer

a wearable computer based Progress Monitoring System ICMMS (Integrated Contract Management and Monitoring System) was developed. ICMMS supports the processes of collecting progress data on site and the procession of this data in the office. The screen of the wearable computer that is used on site shows a drawing of the construction project. To input progress, the user selects a building element, that is depicted on the screen, selects an activity that is carried out at this particular element and puts in a progress figure. This update process can be done by using a stylus and a touch screen. In many cases, where hands free operation of the wearable computer is required, the user is reduced to input construction progress by using the speech interface, which incorporates the methods and components that have been discussed in this paper.

Doing Progress Monitoring with unobtrusive, lightweight wearable computers, that have a user interface, that is tailored for the specific task and environment, data collection and procession time can be saved. ICMMS, that provides such an interface, can make information about the status and problems of a construction site available much earlier than doing Progress Monitoring traditionally by hand (Reinhardt 2000).

6 SUMMARY AND FUTURE DIRECTIONS

Mobile computing has shown in many fields, how productivity can be increased, if data is put in and retrieved from the computer in the place where it is needed. Even though this is also true for construction industry in general, computers are rarely used on construction sites.

The emerging product group of wearable computers and also palm top computers can help to make the advantages of mobile computing also available on the construction site. Nevertheless, the software that is run on these wearable computers has to respond to the specific context, the wearable computers are used in. For applications that depict a drawing on the screen of a hand held display, special ways of navigation through a drawing have been examined. Besides identifying drawing elements by clicking on them with a stylus, speech can be used as an interface to interact with the drawing. Commercially available speech recognition tools can be linked to new applications. It is important, that the designed application is structured in a way that follows the logic of a speech dialog between user and computer. Navigating through a drawing is of special interest for research, as efficient selection and control processes for the user interaction with a drawing can become very complex.

As the introduction of wearable computers in new fields, such as construction sites, advances, the topic of "keyboard - less" interaction with the computer will gain importance.

Of course the navigation methods stated in this paper are a first step in the field of speech controlled navigation through drawings. Further research will be necessary to improve these methods. It would be also interesting to explore, in how far speech can be used to navigate through a virtual 3D model of a building. It is also conceivable to support the navigation process by external tools, such as a GPS receiver. The system could automatically select the building element on the screen at which the user on site is standing.

Further fields of research are how peripherals, such as digital cameras can be linked to wearable computers and how the communication process between office and construction site can take advantage of this.

Nevertheless the main objective of further steps should be, to improve the interaction between user and computer in different field contexts.

REFERENCES

J. Reinhardt, J. H. Garrett, Jr., R. J. Scherer. 2000. The preliminary design of a wearable computer for supporting Construction Progress Monitoring. *Proceedings of the Internationales Kolloquium über die Anwendung der Informatik und Mathematik in Architektur und Bauwesen (IKM).* Weimar: IKM.

J. H. Garrett, Jr. & J. Sunkpho. Issues in Delivering Mobile IT Systems to Field Users *Proceedings of the Internationales Kolloquium über die Anwendung der Informatik und Mathematik in Architektur und Bauwesen (IKM).* Weimar: IKM.

Java speech Application Programming Interface description. available as *http://java.sun.com* /products /java-media/ speech/.

JavaTM Speech API Programmer's Guide. available as *http:// java.sun.com/products/java-media/speech/ forDevelopers/jsapi-guide/index.html.*

Progress Monitoring with wearable Computers. available as: *http://www.ce.cmu.edu/~janr/icmms/icmmt.*

Product and Process Modelling in Building and Construction, Gonçalves, Steiger-Garção & Scherer (eds)
© *2000 Balkema, Rotterdam, ISBN 90 5809 179 1*

Activity object-oriented simulation strategy for construction simulation system

H. Zhang & C. M. Tam
Department of Building and Construction, City University of Hong Kong, SAR, People's Republic of China

Jonathan J. Shi
Department of Civil and Architectural Engineering, Armour College of Engineering and Science, Illinois Institute of Technology, Chicago, Ill., USA

ABSTRACT: This paper describes the activity object-oriented simulation strategy for the activity object-oriented visual construction simulation system (AOOVCSS) developed with object-oriented approach. Using the activity as the object, the activity object-oriented simulation strategy incorporates the object-oriented approach with conceptualizing of the real construction operation, both from the points of view of developers and modelers. So not only the developers may be assisted during developing period, but also the system users will be assisted in modeling with the graphical activity-based network. Besides, the activity object-oriented simulation algorithm corresponding to the activity object-oriented simulation strategy is used to carry out the simulation experimentation. Instead of checking all activities for each simulation time unit, as AS (activity scanning) simulation algorithm does, it dynamically activates and checks only the related activity only when involved entities are being released by another activity object, so it will speed up simulation experimentation.

1 INTRODUCTION

The simulation techniques for construction areas have been studied for decades, during which some simulation tools for construction areas have been developed, such as CYCLONE (Halpin 1977), STROBOSCOPE (Martinez and Ioannou 1994). Even though the simulation methods have been proven to be an effective tool for improving construction process planning (Halpin 1992), the troubles or difficulties involved in learning relative simulation tools and building simulation models have been widely experienced by all levels of users. Most simulation tools require additional and especial programming method to model complex operations, so it is time consuming for users to learn and be acquainted with the simulation languages. Besides, most simulation tools in construction require multiple kinds of elements to model operations (such as the normal element, constrained element and idle element in CYCLONE), which certainly increases the tediousness, complexities, error-proneness and time in modeling.

The activity-based construction (ABC) modeling and simulation method (Shi 1999) was proposed targeted to solve above problems. Unlike most simulation tools requiring multiple modeling elements, the ABC method requires only one kind of element (activity) in modeling, so that the ABC model resembles the critical path method (CPM) diagram in appearance and can be easily accepted by users in construction areas. However, the feasibility of the ABC concept, e.g., how to model the complex operations with the graphical activity-based element and without the requirement of programming, has not been substantiated. In addition, the ABC method adopts the three-stage simulation algorithm (check start-up conditions of activities first, then advance simulation and finally release simulation entities) to advance the simulation experimentation, which requires checking all activities for each simulation time unit as the AS (activity scanning) simulation algorithm does. This scanning method certainly slows down the simulation experimentation when modeling the complex construction operations (require more activities).

The use of object-oriented approach has the potential for developing software that is relatively easy to maintain, can be re-used and contains close abstractions of real world concepts (Graham 1994). Because of these advantages, there has been phenomenal growth in the research and development of object-oriented simulation (OOS) tools since the mid-1080s (Chell A. Roberts, Yasser M. Dessouky. 1998), such as the process-based object-oriented simulation (J. A. Joines, 1992) and three-phase based object-oriented simulation (Pidd, M. 1995). Besides the advantages in developing simulation systems, the object-oriented simulation assists simulation users in conceptualizing the real world in abstract representation and bridging the gap between the system being studied and the simulation modeling (Chell A. Roberts, Yasser M. Dessouky. 1998).

This paper introduces the activity object-oriented simulation strategy, which is proposed during developing the activity object-oriented visual construction simulation system (AOOVCSS) based on the ABC concept and object-oriented approach. The activity is regarded as the object and the object-oriented analysis is incorporated in the activity object-oriented simulation, which views the real problem under study as being composed of some activity objects during both the development and application period of this system. Accordingly, the understanding of abstract representation for the construction operations being modeled can be enhanced when using graphical activity-based modeling method. In addition, the dynamical binding mechanism of object-oriented approach has been applied into the activity object-oriented simulation algorithm, which dynamically initiates only the related activity objects and results in checking and advancing behaviors. Therefore the speed of simulation experimentation will be increased. The AOOVCSS developed under activity object-oriented simulation strategy is targeted to simplify the modeling process, to get the simulation system quick and easy to use yet to offer sufficient modeling power.

2 ACTIVITY-BASED CONSTRUCTION MODELING CONCEPT

Unlike other simulation systems for construction operations, such as CYCLONE and STROBOSCOPE, which use multiple modeling elements to model real systems, the ABC simulation model requires only one kind of element, i.e., the activity, to model construction operations. The ABC graphical activity network model is similar to the critical path method (CPM) diagram in appearance, with which most users in construction are familiar.

The three-stage simulation algorithm was suggested to carry out the simulation experimentation of the ABC model, which describes the dynamic behavior of a construction operation through controlling the status of activities during simulation. This algorithm includes a) select activity, b) advance simulation, and c) release

simulation entities. All activities are equally and completely checked against their logical or other start-up conditions and resource requirements and the system will select activities satisfying the execution condition. Then the starting time of the selected activity is updated, or the simulating time is advanced, and finally resources or other entities are relocated to the required positions and the available time for these entities is also updated.

In the ABC method, simulation entities are divided into two categories: (1) resource entities and (2) processing entities. Resource entities refer to the tangible resources required by activities, such as trucks or crew. Processing entities represent information to activate the following activity whose starting-up is dependent on the completion of the preceding activity.

3 OBJECT-ORIENTED SIMULATION

Object-oriented approach, many concepts of which were originally derived from the object-oriented capabilities of Simula, provides a currently popular way of designing and implementing computer programs. The main features of object-oriented approach include data abstraction, encapsulation, polymorphism and hierarchical ability. The use of object-oriented approach has the potential for developing software that is relatively easy to maintain, can be re-used and contains close abstractions of real world concepts. Since the mid-1980s, there has been phenomenal growth in the research and development of object-oriented simulation tools. However, it should be noticed that a simulation system written in an objected language is not certainly the object oriented simulation. Only the simulation system designed and implemented based on object-oriented analysis method (Hill, D.R.C. 1996) can be called as object-oriented simulation (OOS). Besides the advantages mentioned above which benefit the system developers, the object-oriented simulation has great intuitive appeal in applications.

The method of object-oriented analysis was used to incorporate the object-oriented approach with the analysis way of real problem under study. Object-oriented analysis remains very close to the problem to be dealt with, for the model which results from it is composed of the objects and processes peculiar to the real system being studied. Therefore, the object-oriented simulation makes the simulation modeling process fundamentally different from conventional simulation modeling. One major modeling distinction is in the way the modeler views and constructs a system model. The object-oriented simulation views the real world as being composed of objects and bridges the gap between the problems being studied and the simulation modeling, which is beneficial to the molders or users.

The procedure for object-oriented analysis includes five tasks:

1. Decompose the problem being studied.
2. Identify objects of the decomposed problem.
3. Classify inheritance and hierarchy relationships among objects.
4. Define attributes (data and methods) of objects.
5. Establish the communications among objects.

The object-oriented analysis deals with the real problems abstractly by denoting the essential characteristics of an object and ignoring the non-significant aspects. The object-oriented analysis is followed by object-oriented design and implementation.

Among the object-oriented discrete event simulation (process-based or three-phase based concept) programs developed under object-oriented approach, the entities that are made to change state in the simulation time according to some defined rules are normally chosen as the major objects composing the problem under study.

4 ACTIVITY OBJECT-ORIENTED SIMULATION STRATEGY

The object-oriented analysis is used to analyze the activity-based modeling during the development of the corresponding simulation system when using object-oriented approach. Instead of using the entities as the major objects, the activities that compose a construction operation are chosen as the major objects, so that provides a link between abstract conceptualizing of the problems with object-oriented analysis method and the graphical activity-based modeling. In addition, the corresponding activity object-oriented simulation algorithm is thus proposed to replace the AS based three-stage simulation algorithm with the aim of speeding up the simulation experimentation.

4.1 *Object-Oriented Analysis of Graphical Activity-Based Modeling*

The graphical representation formalism of a simulation model serves as a very useful framework with which the modeler can analyze and conceptualize the problem and as a communication medium among the people who are involved in the project. It is therefore desirable that the graphical representation formalism be a simple, high-level system abstraction so that it enhances conceptualization of simulation problems and the understanding of the implications of different modeling strategies (Ozden, M. H. 1991). A construction operation is actually composed of a series of activities, some of which may be divided into sub-activities again. Therefore, a construction operation can be modeled through decomposing the operation into sub-operations or activities that can be easily conceptualized and represented. Obviously, the graphical activity-based is a simple and high-level abstract representation formalism of a real construction operation, because it can be easily understood and provides modelers with clear conceptualization of the problem in question.

When the object-oriented approach is applied to develop the graphical activity-based modeling system, it is obvious that the object-oriented analysis is very applicable to analyzing the modeling process with graphical activity-based. Based on the procedure of the object-oriented analysis, the procedure to analyze the graphical activity-based modeling can be transformed into:

1. Decompose the construction operation under study into a series of activities that can be easily represented and defined.
2. Identify the activity object and corresponding activity class at a level of abstraction from the decomposed activities.
3. Classify the hierarchy and inheritance that may exist in the existing activity objects.
4. Define the attributes of the identified activity class, which give detail representation of the activity class and determine its operation.
5. Establish the relationships among the various activity objects that provide the communication method, including logical relationship and releasing-receiving relationship of involved entities.

If using above method to analyze the modeling of an earthmoving operation with activity-based network, the activity objects that compose the operation can be identified as in Figure 1.

This procedure of activity object-oriented analysis not only contributes to the design and implementation of the simulation system, but also benefits users in building models because the system views the problem being studied the same way as the object-oriented analysis does.

393

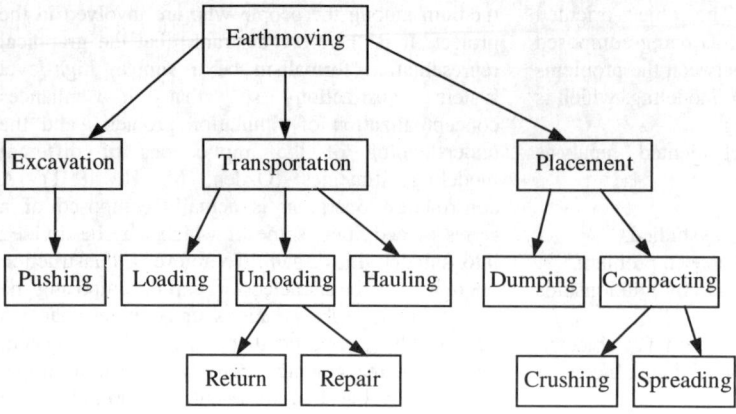

Figure 1. Activity objects composing the earthmoving operation.

4.2 *Definition of Attributes of Activity Object*

The activity-based graphical model should be experimented through dynamical presenting the status of activities by tracing the discrete-event of involved entities (resources or processing entities). So the activity-based construction simulation is still a kind of discrete-event simulation. However, unlike other object-oriented discrete-event simulations that take the entities as the objects, the activities that compose the construction operations are considered as the objects for the activity-based graphical simulation when using object-oriented approach.

Each activity object is created through casting (mapping) from the existing activity class. The activity class is defined by the associated attributes. These attributes are classified into two types-factual and behavioral. The factual attributes describe the activity class by the entities, current state, parameters characterizing its capabilities. The behavioral attributes are essentially procedural functions that characterize how the involved operations take place in certain circumstances. Table 1 shows the details of these attributes.

The list of Related-Activities of the factual attributes describes the relationship among activity objects. The relationship includes the logical relationship (e.g., preceding or following activities) and the entity releasing-receiving relationship (e.g., activities that will receive the released resources or processing entities from current activity).

The factual attributes are to be used by the operations (behaviors) of each activity object. All the behaviors of activity objects take place according to the principle, which is actually a simulation algorithm that advances the simulation through dynamical describing status of activities by tracing the simulation entities.

4.3 *Activity Object-Oriented Simulation Algorithm*

The simulation experimentation is carried out according to a certain simulation algorithm. There are three kinds of algorithm for various simulation systems: a) activity scanning, b) event scheduling and c) process interaction. The ABC three-stage simulation algorithm also belongs to the activity-scanning algorithm, which requires equal and complete checking all activities for satisfied operation conditions. As discussed before, the simulation experimentation will be slowed down because a large amount of activities need to be scanned and more complicated start-up conditions need to be checked when the complexity of operation being studied increases. Based on the characteristics of the activity-based modeling and the dynamic binding mechanism of the object-oriented approach, the activity object-oriented simulation algorithm is proposed to conduct the simulation experimentation of the activity object-oriented model, and speed up the simulation experimentation.

The execution of an object-oriented system is completed through initiating a series of objects by passing messages, which may be sent by other objects. The initiation of an object results in either the provision of selected factual attributes or the initiation of a specific behavior. As for the activity object-oriented system, the message coming from an activity object releasing entities will initiate the activity object that receives the released entities and

Table 1. Attributes of activity object.

Factual Attributes	Behavioral Attributes
Activity Name	Checking start-up conditions
Geometric position	Random sampling
Duration	Computing duration for selected statistical distribution
Representations of start-up conditions and end-off effects	Updating starting time of selected activities
	Updating available time of resource entities
Simulation time	Advancing simulation time
Related Activities: * m_List_ReceiveEntities	Relocating resources to required positions
	Deciding releasing direction

initiate its behaviors, e.g., checking, updating starting time, and advancing simulation time

In an activity-based modeling operation, when an activity finishes its action, only the related activities that receive the released resources or processing entities (determine the logical relationships) may be more likely satisfied with the start-up conditions than other activity objects. So it is suggested here that the related activities be checked first after the completion of an activity. The satisfied activities will begin their operations, the unsatisfied activities will not be checked until some of them receive released simulation entities (resources or processing entities).

On the other hand, the dynamic binding mechanism of the object-oriented approach permits the initiated object to be instanced (created through mapping) from predefined class at run time. For instance, an activity object listed by the pointer (e.g., *m_List_ReceiveEntities) can be instanced at run time. Therefore, the activity object that receives the released entities may be instanced at run time at which the involved entities are being released by another activity object, so that the behaviors concerning checking start-up conditions, advancing time and releasing entities, etc. can be initiated at that time. After the releasing behavior, the instance of this activity object will be deleted. This iterative process will continue until receiving the message for end. This method is called as activity object-oriented simulation algorithm here (see Figure 2). The initiating of other behaviors means initiating those behaviors (methods) that are used to assist in modeling complex construction operations, e.g., sampling, computing of stochastic duration, and deciding of releasing direction, etc.

Compared with the AS or ABC three-stage simulation algorithm, this simulation algorithm is able to speed up the simulation experimentation from two points:

1. It requires checking only activities that may receive the released entities, other than checking all activities.
2. It begins checking operation only at the time at which the involved entities are being releasing, other than at each simulation time unit.

5 AOOVCSS BASED ON AOO SIMULATION STRATEGY

5.1 Visual Modeling Environment for Activity-Based Model

Based on the activity object-oriented simulation strategy discussed above, the activity object-oriented visual construction simulation system (AOOVCSS) is developed in Visual C++ language. The AOOVCSS is furnished with the visual-modeling environment, which provides the users with the communication between the graphical activity-based modeling and the real construction operations under study. This visual-modeling environment conducts system users to decompose the operation being modeled into a series of activity objects, as the object-oriented analysis does. So the users may get enhanced understanding of the graphical activity-based modeling method and process.

The modeling interface (see Figure 3) is used to display the graphical models, and on which the graphical modeling is carried out. The activity object-associated dialog box (see Figure 4) is used to assist modelers in deciding and inputting required parameters or definitions for each activity object. These parameters or definitions will be automatically changed into factual attributes and behavioral attributes as described in Table 1, and then be restored in a database so as to be used during executing the AOOVCSS.

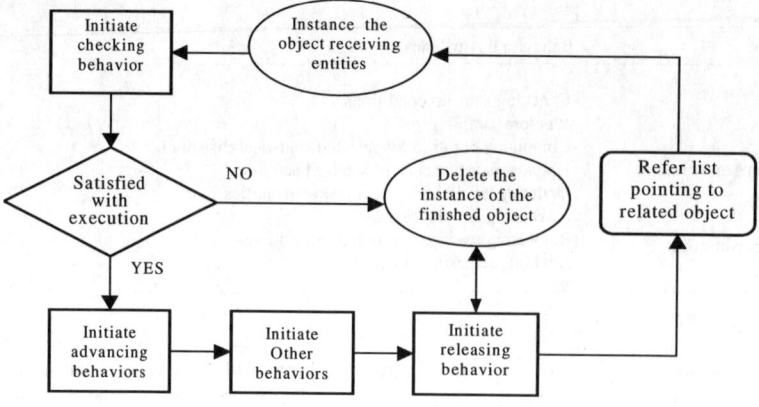

Figure 2: Activity Object-Oriented Simulation Algorithm

Through the visual-modeling environment, some complex construction operations can be modeled in the simple activity elements without the requirements of simulation programming. These modeling features available are summarized as following:

1. Stochastic duration of activities and stochastic arrival times
2. Probabilistic release of resources for multiple routes
3. Probabilistic inputs of the "affecting logic" to following activities
4. Probabilistic outputs of the "being affected logic" from multiple preceding activities
5. Conditional receiving resources by activities
6. Uneven matching of shared resources for an activity

Achieving these features for modeling complex operations require inputting relative definitions or parameters through the activity object-associated dialog box, as well as initiating relative behaviors as illustrated in Figure 2.

The visual modeling environment and the modeling features of the AOOVCSS will offer users a lot of advantages, such as ease of input manipulation, minimization of errors, and least amounts of learning time required. These are helpful not only to non-specialist simulation users, but also to the experienced simulation modelers.

5.2 *Example*

Figure 3 shows a graphical activity-based model for an earthmoving operation, which was built according to above methods, including the activity

object-oriented analysis method that views the earthmoving operation as being composed of eight activity objects (see Figure 1), as well as the graphical modeling through visual modeling environment.

The distribution pattern and parameters for stochastic duration of activity and stochastic arriving time are inputted through the sub-dialog box activated by the "Duration" button of the activity object-associated dialog box (see Figure 4). Probabilities in releasing of a resource for multiple routes require inputting for each following activity. After "Dump", the truck may require repairing or directly return, so the probabilities for repairing and returning should be specified. If one process entity affects multiple following activities (e.g. "Dump" affects "Spread" or "Crush"), the affecting logical relationship may be "OR", or "AND". The affecting probability to each following activity should be specified for "OR" logical affecting relationship. If one activity is affected by multiple preceding activities (e.g. "Spread" affected by "Dump" or "Crush"), the logical affected relationship may also be "OR", or "AND". The probabilities of being affected by each preceding activity should be specified for "OR" logical affected relationship. If a resource is conditionally received by an activity, the receiving condition may be specified in the "Receive" editing box. The "Release" editing box is used to specify attribute changes for activity execution, which may be the conditional attribute required by following activities (e.g. "Load" only receives empty trucks while "Haul" receives full trucks). The "Load" activity requires loaders and trucks . Suppose one unit truck is filled by four operation cycles of loader (not shown in Figure 3 and Figure 4), then the situation is called uneven

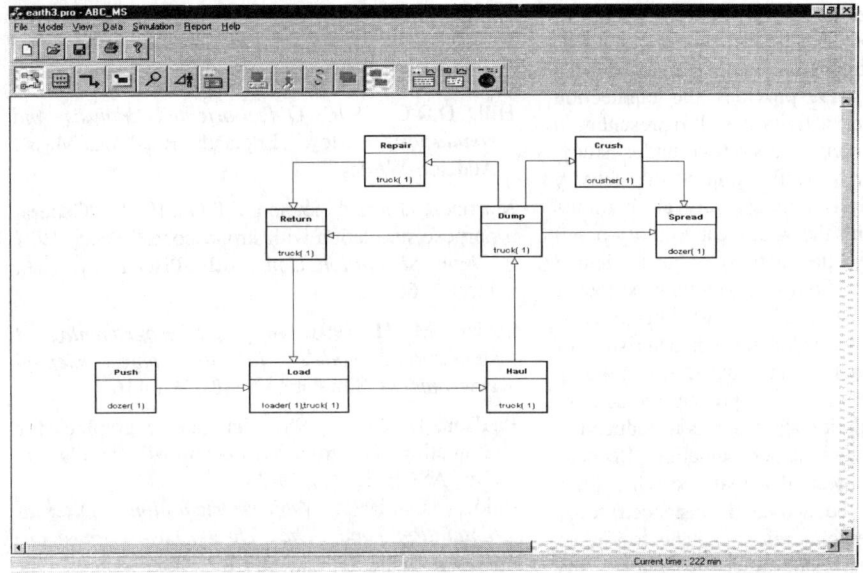

Figure 3. Graphical modeling interface of AOOVCSS.

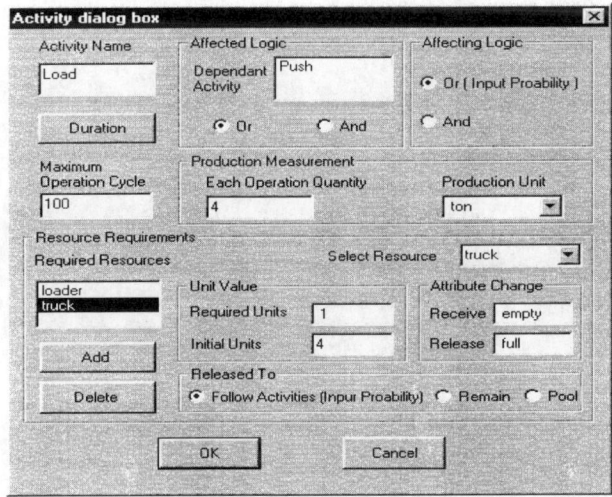

Figure 4. Dialog box for attributes associated with activity object.

matching allocation of resources. For this uneven matching allocation, the ¼ and 1 should be respectively specified in the "Required units" editing box for the truck resource and loader resource.

6 CONCLUSIONS

In order to develop the graphical activity-based modeling and simulation system with the object-oriented approach, the object-oriented analysis is

397

used from analyzing period to designing or implementing period. The object-oriented analysis views the real problems as being composed of a series of objects, so it provides the connection between the abstract activity-based representation and the real construction operation under study. Regarding the activity as the object, the activity object-oriented analysis remains very close to the modeling process in the graphical activity-based network. Therefore, the activity object-oriented analysis not only benefits the developing period of system, but also assists users in modeling process. On the other hand, based on the characteristics of activity-based model and dynamic binding mechanism of object-oriented approach, the activity object-oriented simulation algorithm is introduced to carry out the simulation experimentation. Because only those activity objects that may receive entities being released will be activated (instanced) only when the entities are being released, so as to initiate checking, advancing and other behaviors, the activity object-oriented simulation algorithm will increase the simulation experimentation speed compared with AS or three-stage simulation algorithm. Based on the activity object-oriented simulation strategy described above, the activity object-oriented visual construction simulation system has been developed (AOOVCSS). The visual modeling environment available in AOOVCSS provides assistance in modeling through conducting users to decompose the construction operations into activity objects that can be easily identified and represented through the activity object-associated dialog box. Accordingly, the AOOVCSS is able to simplify the modeling process, including modeling complex operation without requirement of programming, so that get the simulation system quick and easy to use.

REFERENCES

Chell A. Roberts, Yasser M. Dessouky. 1998. *An Overview of Object-Oriented Simulation.* SIMULATION 70:6, 359-368.

Germano Kienbaum, Ray J. Paul, 1994. *H_ACDNET: An object-oriented graphic user interface for simulation modeling of manufacturing system.* Simulation 2:141-157.

Graham, I.1994. *Object-Oriented Methods.* Wokingham, UK: Addison-Wesley.

Halpin, D.W. 1977."CYCLONE-A method for modeling job site processes." *J. Constr. Div.,* ASCE, (103(3), 489-499.

Halpin, D.W., and Riggs, L. S. 1992. *Planning and Analysis of Construction Operations.* John Wiley & Sons, Inc., New York, N.W.

Hill, D.R.C. 1996. *Object-oriented analysis and simulation.* Harlow, England; Reading, Mass.: Addison-Wesley.

Martinez, J., and Ioannou, P.G. 1994. "General purpose simulation with stroboscope." *Proc., 1994 Winter Simulation Conf.,* IEE, Piscataway, N.J., 1159-1166.

Ozden, M. H. 1991. *Graphical programming of simulation models in an object-oriented environment.* Simulation 56 (2): 104-116.

Paulson, B. C., Jr. 1987. "Interactive graphics for simulating construction operation." *J. Constr. Div.,* ASCE, 104(1), 69-76.

Pidd, M. 1995. *Object-Orientation, Discrete Simulation and The Three-Phase Approach.* Journal of the Operational Research Society. 46, 362-374.

Shi, J. 1999. "Activity-Based Construction (ABC) modeling and simulation method." *J. Constr. Engrg. & Mgmt,* ASCE, 125(9), 354-360.

Zeigler, B.P. 1990. *Object-Oriented Simulation with Hierarchical ,Modular Models.* Academic Press, Inc.

Information technology support for dimensional control in building construction

R. Wu & G. Maas
Eindhoven University of Technology, Netherlands

F. Tolman
Delft University of Technology, Netherlands

ABSTRACT: To ensure the predefined dimensional quality, a plan of dimensional control must be designed, on the basis of building drawings and specifications delivered by architects, before the building is constructed. The dimensional control plan must provide site personnel with information on, among others, setting out and assembling building components, which can often be done by means of Total Stations. The essence of designing a dimensional control plan is to find out which points should be used as positioning points, which points should be set out in advance or controlled afterwards, and why. To support designing such a plan, we have developed a tool that fits into the current trend of using standardized electronic product models as the main carrier of construction information and is capable of advising the user on choosing the appropriate points for dimensional control aspects.

1 INTRODUCTION

Dimensional control in the building industry can be defined as the operational techniques and activities that are necessary, for the assurance of the defined dimensional quality of a building (Hoof, 1986). In order to achieve precise and efficient dimensional control, a plan of dimensional control, that includes a tolerance plan, an assembling plan, a setting out plan and a dimension monitoring plan, must be designed on the basis of building drawings and specifications, before the building is constructed. In our previous papers (Wu et al 1997, 1998), we have emphasized the role of dimensional control in building design and construction and given an introduction to the building construction process of direct importance to ensure dimensional quality. We have also described the setting out process and positioning process including the use of Total Station and other often used equipment. We have also introduced the concept of main control points, references, setting out points, positioning points, product measure, setting out measure and positioning measure. We have pointed out that the essence of designing a dimensional control plan is to find out which points should be used as positioning points, which points should be set out in advance or controlled afterwards, and why. These questions are in

fact largely related to the aspect of dimensional quality. Therefore, we must be able to predict the dimensional deviation limits which show the dimensional quality.

We have worked out the theoretical principle of predicting deviation limits for a structural element. To predict the dimensional deviation limits, we view the construction process as a stochastic process and under certain circumstances a normally distributed process. When we predict the deviation limit of each object, it is essential that we consider the deviation as a result of chain processes. Especially, the processes of prefabricating production, setting out and positioning should be taken into account. These three processes cause respectively the product deviation, the setting out deviation and the positioning deviation. They should be added on top of one another. The final "place deviation" which can be defined as the deviation related with the place of a point of the building part, is equal to the sum of these separate deviations. See Figure 1.

In our previous papers (Wu et al 1998, 2000), we have mentioned two methods of predicting the dimensional deviation limits of each point of an object and explained the simulation method including calculation principle focusing on setting out deviation and positioning deviation. In this paper, we focus on the simulation of product deviations.

Information Technology (IT) has been widely applied in the building industry, ranging from Product Data Technology (PDT), CAD systems, internet technology to virtual reality and so on. In the field of PDT, STEP is still ongoing and IAI-IFC (International Alliance for Interoperability - Industry Foundation Class) is emerging. Another development is the Unified Modeling Language (UML) which provides system architects working on

object analysis and design with one consistent language for specifying, visualizing, constructing, and documenting the artifacts of software systems, as well as for business modeling.

In the following, we will explain the principle of simulating product deviations. We will also show our UML model representing the building structural element and predicting principle, and the implementation of the model as well.

2 SIMULATING PRODUCT DEVIATIONS

For the simulation of concrete elements, we use a so called randomizer that is able to choose a number randomly. Random (10) means that it takes a number randomly between 0 and 10. The chance distribution is then constant, that is to say, the chance for each number between 0 and 10 to be taken is equal. There are two alternatives to make a computer to choose a number by which the chance distribution is not constant but is according to, for example, a Gauss-distribution. We can use a table which consists of numbers that are representative for a Gauss-distribution (The number 5 comes more often than the number 1 and 9). We can also use a filter which processes mathematically the numbers that are chosen by the randomizer so that a Gauss-distribution still exists.

The simulation of concrete elements will be made by determining the various sorts of dimension deviations separately and then adding them together. Floor slabs, beams, walls and columns are all characterized by their rectangular form and consist of 6 surfaces, 3 symmetry axis's and further 8 conceptual corner points and 12 edges. We say "conceptual" because in fact corner points and edges don't exist. They are always in more or less measure rounded off.

To orient, we choose a coordinate system in which the X, Y and Z direction is in accordance with the width, the thickness and height direction of the element respectively. The origin is the center gravity point of the element.

Agreed is that each surface is defined by 9 (grid)points. In fact, the number of points doesn't matter. The principle keeps the same for either more or less points. It matters however for the accuracy, that is to say, the more points, the more accurate. The middle (grid)point is coincident with the cutting point of the surface with an axis of the coordinate system. The other points are equally divided so that the distance to the edge (r) of the element in each case is same. The place of these points on the surface doesn't in fact matter either. We will show that at the end the dimension deviation of these points is only perpendicular to the surface.

The element is now defined by 6*9 = 54 points. For the simulation, the element is divided in 3 separate segments. Each segment consists of 2 either opposite or parallel surfaces, combined with each other by one of the axis's of the coordinate system. For the clarity, a column is used as an example, but it can also be a wall, beam or floor element.

We consider now randomly 2 parallel surfaces of the element, connected by one of the axis's of the coordinate system, the so called connection axis. We

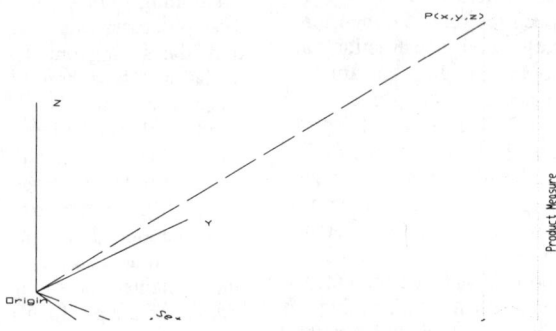

Figure 1 Basic elements causing place deviation

assume that the deviation in the direction of the connection axis of the 18 points on these 2 parallel surfaces is zero and bring the various sorts of deviations in the direction of the connection axis in calculation for these points.

For a three dimensional object, we can distinguish the various sorts of deviations also in three dimensions. The following deviations are valid for a segment of the element:

1 Dimensional: The distance between the opposite surfaces, measured along the connection axis, characterized the dimensions (length, width, height or thickness) of the element. A deviation on this distance moves itself along the line of the connection axis.

2 Dimensional: The direction of the surface in regard to the connection axis, is determined by 2 randomly unparalleled lines, that lie in the surface. The cutting point of the surface and the connection axis can be viewed as a spherical hinge point around which the surface can rotate. A deviation on the direction of the surface characterized the (un)squareness of an individual surface in regard to the connection axis and the (un)parallelness of the surfaces with respect to each other.

3 Dimensional: The form of the individual surface. Due to the deviation in the flatness of the surface, it is a 3 dimensional object. The surface is in fact not a surface. The unflatness is caused by deviations that can be characterized as curving, unstraightness and surface roughness.

The above mentioned deviations that are defined as the characteristics of the individual element, will

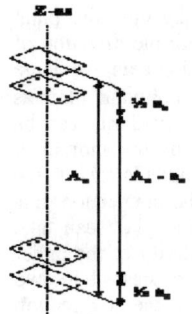

Figure 3 Unperpendicularity

be simulated in the computer model by determining them individually and then adding on top of one another. The segment that consists of the undersurface, the Z axis and the topsurface is graphically illustrated as an example. The principles that apply to this segment are also valid for the other two segments.

By dimension of the element, we mean actually the length, height, thickness or width of the element. In the simulation model, we define the dimension of the element as the distance perpendicular to and between 2 parallel surfaces at which the cutting points of these 2 surfaces with the connection axis serving as measuring points. The distance between the undersurface and the top surface of the element is thus defined by the length of the Z axis between the cutting points of the undersurface and above surface with this axis. In the simulation model, the deviation on this distance is determined by a random number and can be considered as a constant deviation that is valid for both surfaces. The chance distribution of this pick is in accordance with a predefined distribution. For the nominal distance (A), it is defined that the maximal allowable deviation a_{max} mm (98% certainty) may be carried. Whenever the computer randomly chooses a value (a) within this area, then the top surface and the undersurface gets a deviation of 0.5a respectively seen from the origin. The points in the surface all get therefore the same positive or negative deviation of 0.5a in the direction of the connection axis. See Figure 2.

We mean by the direction of the surface in fact the direction of the surface in regard to the connection axis. That is in the case of the undersurface and the top surface the Z axis. A deviation in the direction of the surface can be described by the unperpendicularity of the individual surface and the parallelness of both surfaces in regard to the connection axis.

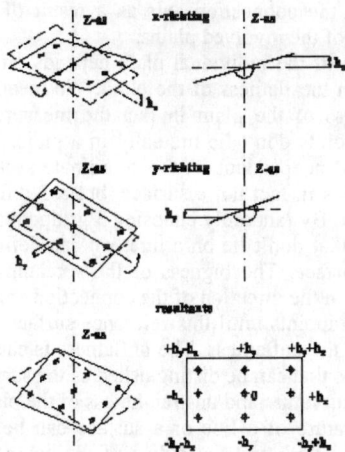

Figure 2   1D Deviation

401

If a plain independent from the opposite plain gets a deviation in connection with the direction of the plain in regard to the connection axis, we name this the unperpendicularity of the individual plain as shown in Figure 3. The unperpendicularity can be defined as a variable deviation of the individual plain. We simulate the unperpendicularity by the cutting point of the plain with the connection axis seen as spherical hinge point. The plain can take each direction. The unperpendicularity of the under plain and the top plain can be expressed in two numbers: the unperpendicularity in the X direction and Y direction. In the simulation model, the unperpendicularity of the individual plain is defined by two random numbers. The chance distribution of this pick meets a predefined chance distribution. The nominal unperpendicularity of a rectangular element is 0. However, since it is not practical, the maximal allowable deviation for the unperpendicularity of the under plain and top plain in the X and Y direction is defined respectively as h_{xmax} and h_{ymax} mm/m (98% credibility). Whenever the computer chooses randomly two values (h_x and h_y) within these limits, then the plain gets an unperpendicularity of these two values. For two opposite plains, four independent numbers are chosen. The points in a plain all get an unequal positive or negative deviation in the direction of the connection axis as a result of the unperpendicularity of the involved plain. In the following figures, we show how the various deviations in the Z direction for the points in the under plain or top plain interrelates.

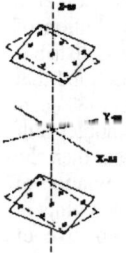

Figure 4 Unparallelness

The parallelness of two opposite plains is analog to the unperpendicularity of two opposite plains. This is concerned with a deviation in the direction of both plains in regard to the connection axis. The parallelness can be considered as a constant deviation of the opposite plains. We simulate the parallelness by the cutting points of the opposite plains with the connection axis seen as spherical hinge points.

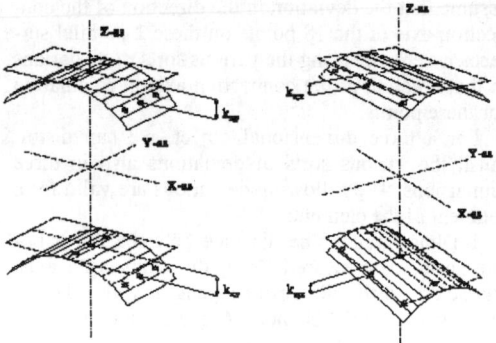

Figure 5 Common curvature

See Figure 4. The plains can thus take each direction. The parallelness of the under plain and top plain can be expressed in two numbers: the parallelness in the X and Y direction respectively. In the simulation model, the parallelness of these opposite plains are defined by two random numbers.

The chance distribution of this pick meets the predefined distribution. The maximal allowable deviation for the parallelness of the under plain and top plain in the X and Y direction is defined respectively as e_{xmax} and e_{ymax} mm/m (98% credibility). Whenever the computer chooses randomly two values (e_x and e_y) within these limits, then the plain gets a parallelness of these two values. For two opposite plains, two independent numbers are chosen that are valid for both plains. The points in these plains all get an unequal positive or negative deviation in the direction of the connection axis as a result of the parallelness of the involved plains.

The form of the individual plain depends on the deviations in the flatness of the plain. We mean by the unflatness of the plain in fact the measure in which the points don't lie mutually in a plain. The unflatness is independent on the coordinate system. The surface is in fact not a surface, but a 3 dimensional object. By randomly choosing 3 gridpoints on the surface that don't lie on a line, we can define a reference surface. The bigness of the spreading of the distance in the direction of the connection axis of the other gridpoints until this reference surface is a measure for the unflatness. The unflatness is caused by deviations that can be distinguished as the surface roughness, curvature and unstraightness of the plain.

The curvature of a line or a surface can be defined by its length and the radius of the belonging circle or sphere.

Regarding the curvature of a segment of the element, we consider the common curvature of both surfaces, the curvature of the individual surface and the curvature of a gridline of the individual surface.

The common curvature of two opposite surfaces can be considered as a constant deviation that is valid for both surfaces. See Figure 5. In this simulation model, we define the curvature of the opposite surfaces of a segment in two directions.

For the under surface and top surface, the curvature is expressed in these two directions in two numbers: the curvature on the X direction and Y direction respectively. In the simulation model, the curvature of these opposite surfaces is defined by two random numbers. The chance distribution of this pick meets the predefined distribution. For the curvature of the under surface and top surface, the maximal allowable deviation in the X and Y direction may have $k_{z;x,max}$ mm/m and $k_{z;y,max}$ mm/m (98% credibility) respectively. Whenever the computer chooses randomly two values ($k_{z;x}$ and $k_{z;y}$) within these limits, then these surfaces get a curvature with a result of these two values. For two opposite surfaces, two independent numbers are chosen that are valid for both surfaces. All the gridpoints in the opposite surfaces, except the cutting point with the connection axis, get therefore an unequal positive or negative deviation in the direction of the connection axis as a result of the common curvature of the opposite surfaces.

The curvature of the individual surface can be considered as a variable deviation of both surfaces. All the gridpoints in a surface, except the cutting point with the connection axis, get therefore an unequal positive or negative deviation in the direction of the connection axis as a result of the curvature of the individual surface.

The curvature of a gridline of the individual surface can be considered as a variable deviation of all gridlines. All gridpoints in a surface, except the cutting point with the connection axis, get therefore an unequal positive or negative deviation in the direction of the connection axis as a result of the curvature of the gridlines in the individual surface.

The warp of a surface can be defined by the measure in which 4 points don't lie in a plain. The plain is in more or less degree tortured. The warp can be measured by an imagined plain drawn by three random corner points. The distance between the fourth corner point with the corner point of the imaginary plain is a measure of the warp. By dividing the warp over 4 corner points, we will find no warp within the original plain. This plain creates the starting point for the simulation of the warp in this model. For the warp of the plain of a segment, we consider the common warp of both plains and the individual warp.

For the form of the surface, there are more characteristics such as straightness and warp to be considered. Also there can be more directions for the curvature of the surface. However, apparently the parameters given in the above have a satisfied simulation of the practice.

3 MODEL AND ITS IMPLEMENTATION

We have created the UML model representing the building elements and predicting principle. In Figure 6, the object classes are connected with five types of relations: unidirectional relations (with cardinalities, single roles and open arrows), bi-directional relations (with cardinalities and roles), generalization (with closed arrows), aggregation (with diamonds) and dependency (with dashed lines and arrows). Each class may contain attributes and operations. The reason for using UML is that we can generate Visual Basic code and use the class model as our primary representation.

A floorplan describes the relevant physical objects on one particular floor in a World Coordinate System (WCS). A floorplan has one main control point which is defined in WCS. A floorplan consists of a set of Typical Structural Assemblies. A Typical Structural Assembly (TSA) consists of several Positioning Objects. A Positioning Object (a physical object that has to be positioned) is either a structural element, or formwork. Each Positioning Object refers to an Object Coordinate System (OCS) that can be transformed to WCS, and consists of an Origin, and has Positioning points. A Positioning Object is represented by a rectangular box with height, width and length. A Positioning Object has an ID. Positioning Points refer to Reference Points. Reference Points refer to Setting out Points referring to Main Control Point. Positioning Points have Product Tolerance, Setting out Tolerance and Positioning Tolerance. Positioning Points have Product Deviation, Setting out Deviation and Positioning Deviation. Product Deviation can be 1D, 2D and 3D. 1D Deviation includes the deviation in Length, Width and Height. 2D Deviation includes Unparallelness and Unperpendicularity. 3D Deviation includes the deviation of Curvature and Warp. Positioning Points can be selected or not. Selected positioning points have relationship causing Rotation and Translation of the Origin which in turn affects the Deviation of Positioning Points.

With the development of CAD systems and product data technology, AutoCAD Architectural Desktop introduces the IFC utility and have some intelligent objects such as walls, columns and openings. It can thus read and write IFC file partially. Among other software packages, Allplan and Speedicon have even more intelligent objects. AutoCAD also introduces the concept of exposing AutoCAD objects through an ActiveX interface and programming those objects using the Visual Basic for Applications programming environment. and chosen AutoCAD Architecture Desktop, Access and Excel as the platform using VBA as the programming language to implement the model.

For the implementation of the product model, considering the compatibility with AEC/IFC objects, we use AutoCAD objects and AutoCAD ADT (Architectural Desktop) objects for graphical objects, such as wall, window, opening, and grid. We can also use 3dbox to represent other rectangular objects. Additional non-graphical attributes can be stored in Access database. Attributes can also be extracted to Excel and statistical analysis can be performed. All these can be done by programming in VBA which is found as a component in AutoCAD

and many Microsoft Office applications including Access and Excel. The software prototype displays the typical structural situation, generates a collection of points, predicts the deviation limits by putting together all kinds of product deviations, setting out deviation and positioning deviation as well, and also gives the possibility distribution of deviation limits. Coordinates related information can be transferred into a total station, or a fieldbook if necessary. In this way, the system advises the user on choosing the appropriate points for dimensional control aspects. Figure 7 and 8 are respectively the screen capture of the designed system and the partial running result of the system.

4 CONCLUSION AND RECOMMENDATION

We think the product modeling approach is a good candidate for structuring the dimensional aspects of building component data. It allows us to represent and store procedural and mathematical calculation knowledge. Also, the principle of predicting deviation limits is valid for any rectangular building component and it can be extended when we

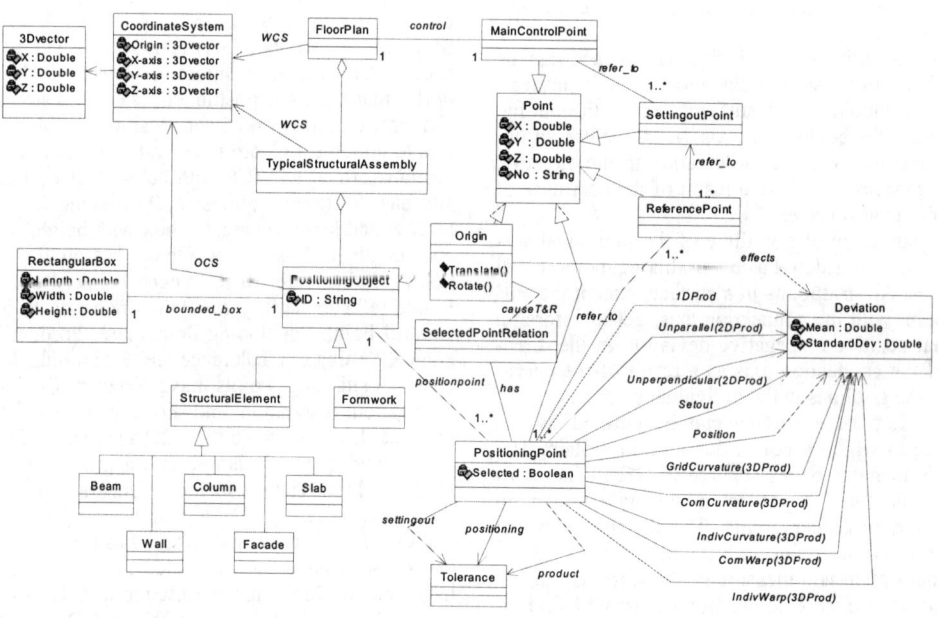

Figure 6 Simplified UML model

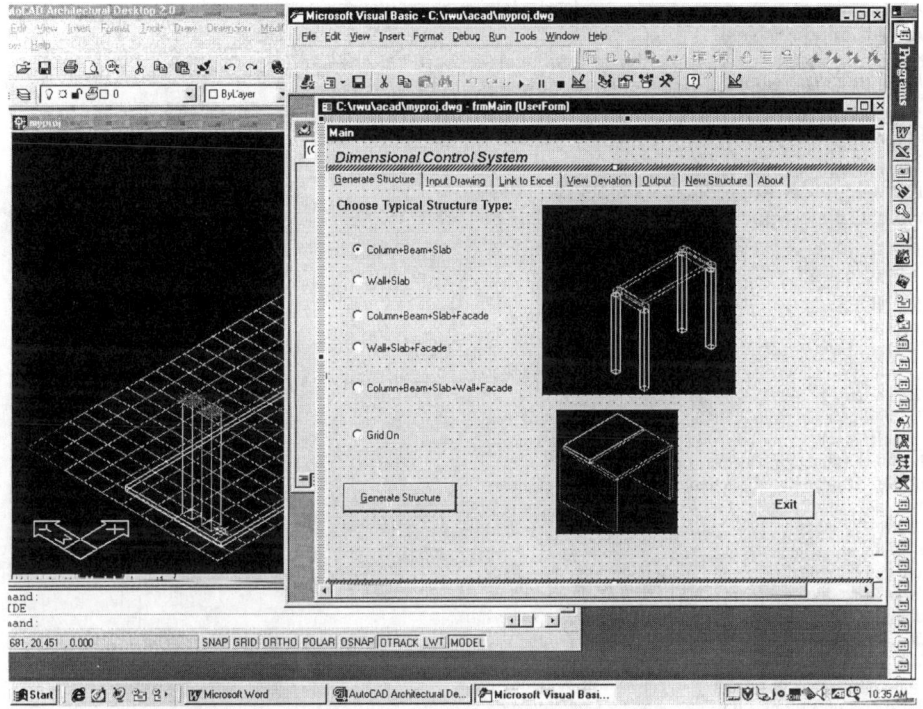

Figure 7 Screen capture of the designed system

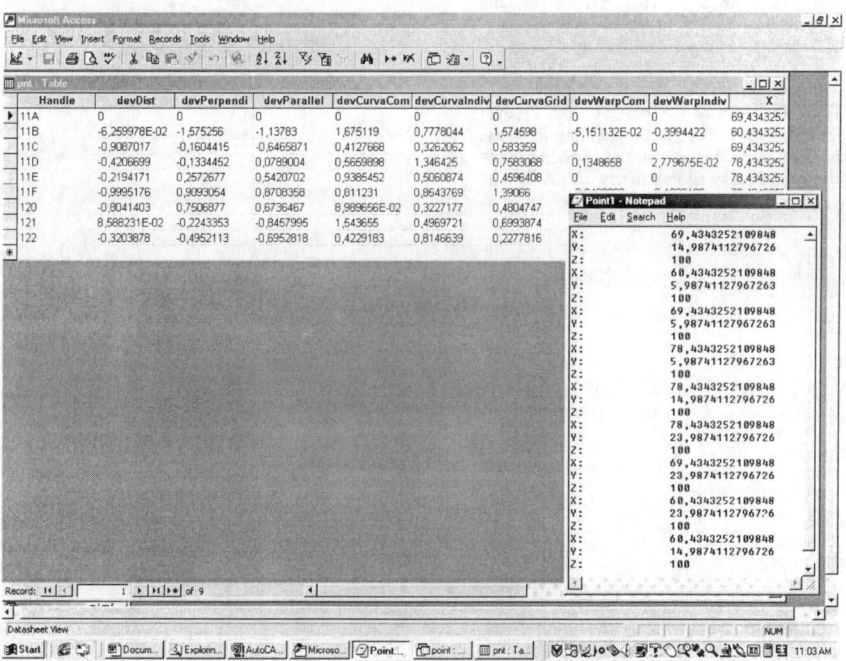

Figure 8 Partial running result of the system

405

consider the joint of two or more elements and when an element is in other shape.

The limiting factor seems to be the general availability of IFC based product models. Though this is true - it will take some years before product modeling and the application of IFCs is generally accepted in the construction industry - it is not a severe limitation. The solution is to produce a simple instantiation tool that supports the translation of AutoCAD to IFC data.

REFERENCES

Autodesk 1999. AutoCAD 2000 User's Guide, 00120-090000-5011, February 5, 1999.

Reins, J. 1999. Maatcontrole in de bouw: De ontwikkeling van een simulatiemodel voor het voorspellen van kritieke maten in betonnen draagconstructies (in Dutch). *M.Sc thesis.* Eindhoven University of Technology, Eindhoven, Netherlands.

Boggs W. & M. Boggs 1999. Mastering UML with rational rose. Sybex Inc., Alameda, U.S.A.

van Hoof, P. 1986. Maatbeheersing in de bouw, een ontwikkeling van uitzetmethoden (in Dutch), *Ph.D. Dissertation,* Eindhoven University of Technology, Eindhoven, Netherlands.

Wu, R., P. van Hoof, G. Maas & F. Tolman 1997. Towards a computer aided system for dimensional control on building sites. *Proceedings of the First International Conference on Construction Industry Development: Building the Future Together:* 272 – 279, Singapore.

Wu, R., P. van Hoof, G. Maas & F. Tolman 1998. Product modeling for dimensional control in the building industry. *Proceedings of First International Conference on New Information Technologies for Decision Making in Civil Engineering:* 389 – 400, Montreal, Canada.

Wu, R., P. van Hoof, G. Maas & F. Tolman 2000. Product modeling for dimensional control in the building industry. *Accepted by the journal of Computer-Aided Civil and Infrastructure Engineering,* Blackwell Publishers, U.S.A.

Authors Index